4th International Conference Recent Trends in Structural Materials (COMAT 2016)

IOP Conference Series: Materials Science and Engineering
Volume 179

Pilsen, Czech Republic
9 – 11 November 2016

Editors:

Jan Dzugan
Zbysek Novy

ISBN: 978-1-5108-3717-1
ISSN: 1757-8981

Printed from e-media with permission by:

Curran Associates, Inc.
57 Morehouse Lane
Red Hook, NY 12571

Some format issues inherent in the e-media version may also appear in this print version.

Printed by Curran Associates, Inc. (2017)

For permission requests, please contact the Institute of Physics
at the address below.

Institute of Physics
Dirac House, Temple Back
Bristol BS1 6BE UK

Phone: 44 1 17 929 7481
Fax: 44 1 17 920 0979

techtracking@iop.org

Additional copies of this publication are available from:

Curran Associates, Inc.
57 Morehouse Lane
Red Hook, NY 12571 USA
Phone: 845-758-0400
Fax: 845-758-2633
Email: curran@proceedings.com
Web: www.proceedings.com

Table of contents

Volume 179

4th International Conference Recent Trends in Structural Materials
9–11 November 2016, Pilsen, Czech Republic

Accepted papers received: 14 February 2017
Published online: 1 March 2017

Preface

011001
OPEN ACCESS
4th International Conference Recent Trends in Structural Materials

011002
OPEN ACCESS
Peer review statement

Papers

012001
OPEN ACCESS
Features of the physico-mechanical behavior of UFG low-alloyed bronze Cu-1Cr-0.08Zr produced by severe plastic deformation

D A Aksenov, R Asfandiyarov, G I Raab and G B Isyandavletova.....1

012002
OPEN ACCESS
Investigations of influence of vibration smoothing conditions of geometrical structure on machined surfaces

D Bańkowski and S Spadło.....6

012003
OPEN ACCESS
Increase of chromium yield by slag reduction during production of chromium steels

J Bažan, L Socha, V Kurka, P Jonšta and M Sušovský.....12

012004
OPEN ACCESS
Finite element analysis regarding the forming behaviour of symmetric hybrid structures consisting of two sheet metal outer layers and a thermoplastic core

B-A Behrens, A Bouguecha, C Bonk and H Schulze.....19

012005
OPEN ACCESS
Numerical and experimental investigations on an extrusion process for a newly developed ultra-high-carbon lightweight steel for the automotive industry

B-A Behrens, A Bouguecha, C Bonk, D Yarcu and M Kazhai.....26

012006
OPEN ACCESS
Numerical analysis of a deep drawing process with additional force transmission for an extension of the process limits

B-A Behrens, C Bonk, N Grbic and M Vucetic.....33

012007
OPEN ACCESS
Simulation model of Al-Ti dissimilar laser welding-brazing and its experimental verification

M Behúlová, E Babalová and M Nagy.....39

012008
OPEN ACCESS
Numerical simulation of multi-rifled tube drawing – finding proper feedstock dimensions and tool geometry

P Bella, P Buček, M Ridzoň, M Mojžiš and L' Parilák.....46

012009
OPEN ACCESS
New heat treatment process for advanced high-strength steels

D Bublíková, Š Jeníček, I Vorel and B Mašek.....54

012010
OPEN ACCESS
Magnetic properties of 42CrMo4 steel

T Bulin, E Svabenska, M Hapla, P Roupcova, C Ondrusek and O Schneeweiss.....62

012011
OPEN ACCESS
Automated microstructural analysis of titanium alloys using digital image processing

A Campbell, P Murray, E Yakushina, S Marshall and W Ion.....68

012012
OPEN ACCESS
Properties and microstructure of twin-roll cast Al-Mg alloy containing Sc and Zr

M Cieslar, J Bajer, M Zimina, M. Šlapáková and O Grydin.....74

012013
OPEN ACCESS
Heat treatment of nitrided layer formed on X37CrMoV5-1 hot working tool steel

A Ciski, P Wach, J Tacikowski, T Babul and P Šuchmann.....81

012014
OPEN ACCESS
Elements for the modeling of the thermal process in heating furnaces for steel forming

D Constantinescu and A B Carlan.....89

012015
OPEN ACCESS
Experimental comparison between upsetting characteristics of porous components prepared by Fe-based sintering technology

W Depczyński, T Miłek and Ł Nowakowski.....95

012016
OPEN ACCESS
Bainite austenitization in 51CrV4 spring steel: accelerated cementite spheroidisation

J Dlouhy, D Hauserova and P Motycka.....102

012017
OPEN ACCESS
Improvement of wear resistance for C45 steel using plasma nitriding, nitrocarburizing and nitriding/ manganese phosphating duplex treatment

T V Doan, D Kusmič, M Pospíchal and D Dobrocký.....108

012018
OPEN ACCESS
Structure and mechanical characterization of Mg-Nd-Zn alloys prepared by different processes

D Dvorský, J Kubásek, D Vojtěch, I Voňavková, M Veselý and M Čavojský.....116

012019
OPEN ACCESS
Mechanical properties determination of AM components

J Dzugan, M Sibr, P Konopík, R Procházka and M. Rund.....122

012020
OPEN ACCESS
Mini-tensile specimen application for sheets characterization

J Džugan, M Rund, A Prantl and P Konopík.....128

012021
OPEN ACCESS
A study of carbide dissolution in bearing steels using computational thermodynamics and kinetics

E Erişir, O G Bilir and A E Gezmişoğlu.....135

012022
OPEN ACCESS
Study of martensitic-ferritic dual phase steels produced by hot stamping

E Erişir and O G Bilir.....141

012023
OPEN ACCESS
Sheet necking prediction in forming limit diagrams with the anisotropy influence incorporation

P Farahnak, A Prantl, J Dzugan, P Konopik and R Prochazka.....147

012024
OPEN ACCESS
Prediction of thinning of the sheet metal in the program AutoForm and its experimental verification

M Fedorko, M Urbánek and M Rund.....153

012025
OPEN ACCESS
3D printed porous stainless steel for potential use in medicine

M Fousová, J Kubásek, D Vojtěch, J Fojt and J Čapek.....160

012026
OPEN ACCESS
Strain-amplitude dependent cyclic hardening of 08Ch18N10T austenitic stainless steel

J Fumfera, J Džugan, J Kuželka, R Procházka and M Španiel.....166

012027
OPEN ACCESS
Sintering and heat treatment of $Al_{15}Ti_5Co_{35}Ni_{25}Fe_{20}$ high-entropy alloy

K Górecki, J Zýka, J Malek, J Horvát, J Čapek and P Bała.....172

012028
OPEN ACCESS
Mechanical properties and structure of AZ61 magnesium alloy processed by equal channel angular pressing

O Hilšer, S Rusz, T Tański, P Snopiński, J Džugan and M Kraus.....179

012029
OPEN ACCESS
The effect of the material surface oxidation on laser light absorption

M Hruška, J Tesař, M Vostřák and E Smazalová.....186

012030
OPEN ACCESS
Testing of T91 steel in heavy liquid metals

M Chocholoušek, Z Fulín, J Janoušek and Z Špirit.....192

012031
OPEN ACCESS
Reinforcement of the bulk crowns excavator technology by dispersion hardening of titanium carbide to increase wear resistance

I Chumanov, A Anikeev and V Chumanov.....197

012032
OPEN ACCESS
Creep test with use of miniaturized specimens

E Chvostová and J Džugan.....203

012033
OPEN ACCESS
Using of material-technological modelling for designing production of closed die forgings

K Ibrahim, I Vorel, Š Jeníček, J Káňa, D Aišman and V Kotěšovec.....209

012034
OPEN ACCESS
Fatigue performance of blade steel T552 in a corrosive environment

J Janoušek, S Hřeben, Z Špirit, J Strejcius and J Kasl.....215

012035
OPEN ACCESS
Effect of silicon on stability of austenite during isothermal annealing of low-alloy steel with medium carbon content in the transition region between pearlitic and bainitic transformation

Š Jeníček, I Vorel, J Káňa, K Ibrahim and V Kotěšovec.....221

012036
OPEN ACCESS
Semi-solid processing of high-chromium tool steel to obtain microstructures without carbide network

H Jirková, D Aišman, K Rubešová, K Opatová and B Mašek.....227

012037
OPEN ACCESS
High-strength state of ultrafine-grained martensitic steel produced by high pressure torsion

M V Karavaeva, M A Nikitina, A V Ganeev and R K Islamgaliev.....233

012038
OPEN ACCESS
Characterization of TiAl based alloys with various content of carbon

A Klimová, J Lapin and T Pelachová.....242

012039
OPEN ACCESS
Correlation between standard Charpy and sub-size Charpy test results of selected steels in upper shelf region

P Konopík, J Džugan, T Bucki, S Rzepa, M Rund and R Procházka.....248

012040
OPEN ACCESS
Comparison of irradiated 15Kh2MFA material mechanical properties using conventional testing methods and innovative approach of small punch testing (SPT) and automated ball indentation (ABIT)

R Kopriva, P Petelova, I Eliasova, M Kytka and M Culek.....254

012041
OPEN ACCESS
Impact of quenching temperature and isothermal holding time during austempering on bainite content in high-silicon steel

V Kotěšovec, I Vorel, Š Jeníček, J Káňa and K Ibrahim.....260

012042
OPEN ACCESS
Influence of the input parameters on the efficiency of plaster sanding with alundum abrasive discs

D Krajcarz, S Spadło and P Młynarczyk.....265

012043
OPEN ACCESS
Aluminium alloys with transition metals prepared by powder metallurgy

V Kucera, F Prusa and D Vojtech.....270

012044
OPEN ACCESS
High ductility of bainite-based microstructure of middle carbon steel 42SiMn

L Kučerová, M Bystrianský and Š Jeníček.....276

012045
OPEN ACCESS
Modeling plasticity of materials with nanostructure

N R Kudinova.....282

012046
OPEN ACCESS
Tribological and corrosion properties of plasma nitrided and nitrocarburized 42CrMo4 steel

D Kusmic and D Van Thanh.....289

012047
OPEN ACCESS
Evaluation of the effectiveness of the use of horizontal and vertical rolls in the "Rolling-pressing" process on the basis of the stress-strain state studying

S Lezhnev, A Naizabekov, E Panin, T Koinov, I Mazur and A Arbuz.....297

012048
OPEN ACCESS
Hydrogen effect on compression mechanical properties of TiNb alloys at elevated temperatures

M Losertová, O Štefek, I Schindler, L Szkandera and P Kudela.....303

012049
OPEN ACCESS
Influence of rotary swaging and subsequent age hardening on properties of EN AW 6082 aluminium alloy

L Maleček, J Palán, J Nacházel and J Dlouhý.....309

012050
OPEN ACCESS
Processing of bulk Al7075 alloy by spark plasma sintering

P Málek, O Molnárová, J Cinert, F Lukáč and T Chráska.....315

012051
OPEN ACCESS
FEM model and experimental measurement of clinched joint

P Malý, F Lopot and J Sojka.....321

012052
OPEN ACCESS
A method for assessing the wear of sanding abrasive discs differing in the type of grains (SiC, Al2O3(95A))

P Młynarczyk, S Spadło and D Krajcarz.....327

012053
OPEN ACCESS
Effects of the pouring temperature on the formation of the bonding zone between AZ91 and AlSi17 in the compound casting process

R Mola, T Bucki and A Dziadoń.....332

012054
OPEN ACCESS
Experimental and computational evaluation of rolling bearing steel durability

L Nohál and M Vaculka.....338

012055
OPEN ACCESS
The effect of annealing on mechanical and structural properties of UFG titanium grade 2

J Palán, T Kubina and P Motyčka.....342

012056
OPEN ACCESS
Implementation of American weld connection standards into finite element computations

M Peč, P Vosynek, F Šebek and T Návrat.....349

012057
OPEN ACCESS
Fe-Mn(Al, Si) TWIP steel – strengthening characteristics and weldability

P Podany, M Koukolikova, T Kubina, R Prochazka and A Franc.....356

012058
OPEN ACCESS
Prediction of fracture in the shearing process using DEFORM and MARC software

I Poláková, M Urbánek, T Kubina and J Džugan.....363

012059
OPEN ACCESS
Evaluation of in-situ deformation experiments of TRIP steel

J Procházka, L Kučerová and M Bystrianský.....369

012060
OPEN ACCESS
Strain controlled cyclic tests on miniaturized specimens

R Procházka and J Džugan.....376

012061
OPEN ACCESS
Development of geometry of forming tools for extrusion of strip sheet by SPD process

S Rusz, M Salajka, J Džugan, O Hilšer, J Bořuta, M Pastrňák and J Švec.....382

012062
OPEN ACCESS
Influence of relative rolling reduction and thickness layers bimetallic plate at the non-uniformity of the strain after rolling process

D Rydz, G Stradomski and H Dyja.....393

012063
OPEN ACCESS
Influence of specimen dimensions on ductile-to-brittle transition temperature in Charpy impact test

S Rzepa, T Bucki, P Konopík, J Džugan, M Rund and R Procházka.....400

012064
OPEN ACCESS
Effect of mechanical treatment on intergranular corrosion of 6064 alloy bars

P Sláma and J Nacházel.....406

012065
OPEN ACCESS
Mechanical testing of anisotropy in ODS steel tubes

Z Špirit, M Chocholoušek and M Šíma.....413

012066
OPEN ACCESS
Computer simulation of combination extrusion of ENAW1050A aluminum

P Thomas.....417

012067
OPEN ACCESS
Distribution of porosity and macrosegregation in slab steel ingot

M Tkadleckova, P Jonsta, Z Carbol, M Susovsky, K Michalek, L Socha, J Sviželová and J Zwyrtek.....423

012068
OPEN ACCESS
Effect of silicon content on microstructure of low-alloy Q&P-Processed steels

I Vorel, Š Jeníček, J Káňa, K Ibrahim, V Kotěšovec and B Mašek.....429

012069
OPEN ACCESS
Anticorrosion nitrided layers on unalloyed and alloyed steels

P Wach, J Michalski, K Burdyński and A Ciski.....435

012070
OPEN ACCESS
The influence of heat treatment on properties of cold rolled alloyed steel and nickel superalloys sheets used in aircraft industry

K Zaba, I Dul and S Puchlerska.....444

Preface

COMAT history is relatively new, we started in 2010. COMAT deals with recent trends in structural materials that cover production technologies (casting, forging, rolling…), heat and thermo-mechanical treatments, computer modelling and physical simulations and materials characterisation. The conference was launched by the research institute COMTES FHT a.s. that is very active in the fields covered by COMAT. The idea was to create a base for knowledge sharing between industry and research and academic experts. It is located in the heart of the Pilsen region that is a traditional heavy industry centre in the Czech Republic with rich history spanning over 150 years.

COMAT 2016 was held at Angelo Hotel in Plzen on November 9th – 11th. The conference was organized by company TANGER s.r.o. and COMTES FHT a.s. COMAT 2016 was the fourth of the biannually held conferences. The subtitle of the conference this year was the 70th anniversary of Prof. Jozef Zrnik, who has been working in the field of material science for many years. It was great pleasure to welcome some of Prof. Zrnik's students who came for the conference and made the atmosphere of the anniversary even more pleasant. We were really pleased to welcome more than 140 people from 14 countries attending the conference (Algeria, Czech Republic, Denmark, Germany, Kazakhstan, Poland, Portugal, Russian Federation, Romania, Slovakia, Thailand, Turkey, Spain, United Kingdom). The participants presented almost 60 lectures and 48 posters. The conference was divided into seven sessions and a poster session. It the current proceedings, 70 papers are selected. We hope the papers presented here will be interesting for readers and a good base for inspiration for future developments. We are looking forward to seeing more participants in 2018.

Yours sincerely

Jan DŽUGAN and Zbyšek NOVÝ

January 2017

IOP Conf. Series: Materials Science and Engineering **179** (2017) 011001 doi:10.1088/1757-899X/179/1/011001

Organisators:

Partner

Sponsors:

ŠKODA JS a.s.

IOP Conf. Series: Materials Science and Engineering **179** (2017) 011002 doi:10.1088/1757-899X/179/1/011002

Peer review statement

All papers published in this volume of *IOP Conference Series: Materials Science and Engineering* have been peer reviewed through processes administered by the proceedings Editors. Reviews were conducted by expert referees to the professional and scientific standards expected of a proceedings journal published by IOP Publishing.

Features of the physico-mechanical behavior of UFG low-alloyed bronze Cu-1Cr-0.08Zr produced by severe plastic deformation

D A Aksenov[1,2], R Asfandiyarov[1,2], G I Raab[1] and G B Isyandavletova[1]

[1]Ufa State Aviation Technical University, 12 K. Marx str., Ufa 450008 Russia

[2]Institute of Molecule and Crystal Physics, Ufa Research Center RAS, 71 pr. Oktyabrya, Ufa 450054 Russia

E-mail: a.r.n@list.ru

Abstract. The authors analyze the effect of the accumulated strain during severe plastic deformation as well as the temperature and aging time on the physico-mechanical properties of low-alloyed chromium-zirconium bronze Cu-1Cr-0.08Zr. The following SPD processing techniques were employed in the work: high-pressure torsion (HPT) and equal-channel angular pressing (ECAP). As a result of the investigation there have been revealed the most reasonable regimes of thermomechanical treatment, leading to the maximum increment in mechanical properties, while preserving a high level of electrical conductivity.

1 Introduction

Low-alloyed bronzes with high electrical and thermal conductivity are the most important group of Cu-based conductor alloys. The total content of alloying elements in alloys of this type is within a range of 0.1 to 3...5% [1, 2, 3]. These alloys possess a combination of such properties as high electrical conductivity and high-temperature strength, which cannot be attained in materials based on other metals [1]. The level of properties is influenced by both the total content of alloying elements in the alloy and the method of its thermomechanical treatment. Traditionally, properties in such alloys are formed by a three-stage treatment, including high-temperature heating and quenching (to fix the supersaturated solid solution with the alloying elements of the solid solution), deformation processing at room temperature (rolling, drawing) and aging (precipitation of fine-dispersed phases from the solid solution).

The decomposition kinetics of the supersaturated solid solution depends on the type and regimes of preliminary treatments [1, 2, 3]. In recent time, severe plastic deformation (SPD) has become greatly developed. This type of processing drastically transforms a material's structure to the ultrafine-grained one or even to the nanostructured state [4, 5]. A number of papers have demonstrated the positive effect of such structural states on the decomposition kinetics of the supersaturated solid solution and the formation of enhanced physico-mechanical properties [3, 6]. In this connection, in the present investigation SPD with different accumulated strains during the processing was used at the stage of cold deformation. This approach enables revealing the efficiency of SPD processing and, consequently, the effect of the accumulated strain level and the regime of subsequent aging on the formation of physico-mechanical properties. The following SPD processing techniques were employed in the study: high-pressure torsion (HPT) and equal-channel angular pressing (ECAP).

2 Material and experimental procedure

As the material for the investigation, the low-alloyed Cu-1Cr-0.08Zr (wt%) was used. This alloy is used in production of electrodes for resistance spot welding and seam welding, railway contact wires, as well as other parts which must combine high electrical conductivity and high–temperature strength [1, 7, 8]. The samples used in the study were in the shape of rods with a diameter of 10 mm, length of 60, and in the shape of disks with a diameter of 10 mm, thickness of 6 mm. In the course of the investigation, the samples were subjected to thermomechanical treatment comprising three stages.

At the first stage, high-temperature treatment was performed – holding the samples for 1 hour at a temperature of 1050° C, with subsequent water quenching.

At the second stage, the samples were subjected to two different types of SPD processing:

- equal-channel angular pressing (ECAP) – with the channels' intersection angle 90°, route Bc, 1 and 4 passes;
- high pressure torsion (HPT), at the pressure of 6 GPa, 1 and 5 revolutions.

At the third stage, for the sake of comparison, search for reasonable regimes of aging was conducted for the samples with a coarse-grained (CG) and ultrafine-grained (UFG) structure at different temperatures and different holding times. The range of aging temperatures of 400° C, 450° C, 500° C was selected on the basis of previous studies, which demonstrated that temperatures below such a range provide a very insignificant increment in mechanical properties, as compared to the initial ones [1, 3, 6, 9]. The aging time was 10, 20, 30 min, and subsequently 1 hour – 10 hours, with an interval of 1 hour. This interval of time was selected in order to show the decomposition kinetics of the solid solution.

The Vickers microhardness tests (under a load of 0.1 kg, holding time - 10 s) were performed using the EMCO-Test DuraJet 10 microhardness tester. Electrical conductivity was measured using the VE27-NTs eddy-current conductivity meter (with an accuracy of measurement 0.5 MS/m).

3 Results and discussion

As a result of the performed studies, graphs and diagrams have been produced, which demonstrate the tendency in variation of microhardness and electrical conductivity depending on different processing parameters (time, temperature) and the accumulated strain level during ECAP and HPT processing (figure 1, figure 2). Analysis of the obtained data has revealed that SPD processing leads to a noticeable enhancement in strength, and this process runs especially actively when HPT processing is employed. This is connected with the fact that during HPT the hydrostatic pressure increased up to 6 GPa is applied, which provides accumulation of high strain values without the samples' damage as well as formation of an ultra-disperse structure with high homogeneity with an average size of structural elements 220±25 nm. For instance, after one revolution (e=30) and five revolutions (e=270) hardness reaches 1840 MPa, in contrast to the initial hardness of 860 MPa, and the increment in hardness is 980 MPa. A similar situation, showing an increase in hardness, is observed after ECAP processing, but the increment in hardness is much lower (figure 1). Subsequent heat treatment (aging) leads to a further growth in hardness, and this process runs most actively after HPT processing under both e=30 (1 revolution) and e=270 (5 revolutions). It has been established that after 10 minutes of annealing at 500° C there is observed an intensive hardness growth, then hardness declines, and for the sample with e=30 this decline in hardness is more active and in 10 hours hardness reaches the initial material's level. The hardness values attained after 10 minutes of aging are superior and are 1.4 times higher than the hardness values after the conventional treatment. Such an active growth of microhardness is the regular one as the HPT process leads to the refinement of initial grains and to the increase of disclocation density. At the stage of thermal treatment the process of dissolution of the solid solution with precipitation of nanostructured particles in the ultrafine grained matrix takes place, which leads to strengthening. High non-equilibrium character of grains of such structures and high temperature most probably contributes to the accelerated dissolution of the solid solution and active coagulation of these particles and, accordingly to the sudden increase, and then decrease of hardness, which has been experimentally proved. For other aging temperatures, the hardness values attained after HPT processing are lower, but still rather high, for example after 5 HPT revolutions followed by aging of the alloy at 450° C for 2 hours, the alloy's microhardness is 2650 MPa.

After ECAP+HT, the best values of hardness (мах. 2320 MPa) are exhibited by the samples produced by 4 processing cycles and at the aging temperature of 450° C (figure 1). The distinctive

feature of this processing regime is the high thermal stability of the hardness value, as the decrease in hardness during 10 hours of holding time at the aging temperature was only 60 MPa. Strains at ECAP are considerably lower as compared to the HPT strains. The scheme of deformation is different in case of ECAP as well, which leads to formation of a considerably elongated structure towards the shear direction. The average cross plane size of structural elements in this case makes up 250±25 nm, but structural elements up to 1 μm are also observed. That is why the hardness increment due to the structure refinement is lower than after HPT. Strengthening due to the dissolution of the solid solution at a temperature of 450° C occurs slower, the overaging is not observed even after 10 hours of holding time, the contribution of such dissolution to strengthening is larger.
After one cycle of ECAP followed by aging at 450° C, the hardness values do not exceed 2000 MPa and exhibit thermal stability for only up to 5 hours of holding time (figure 1).

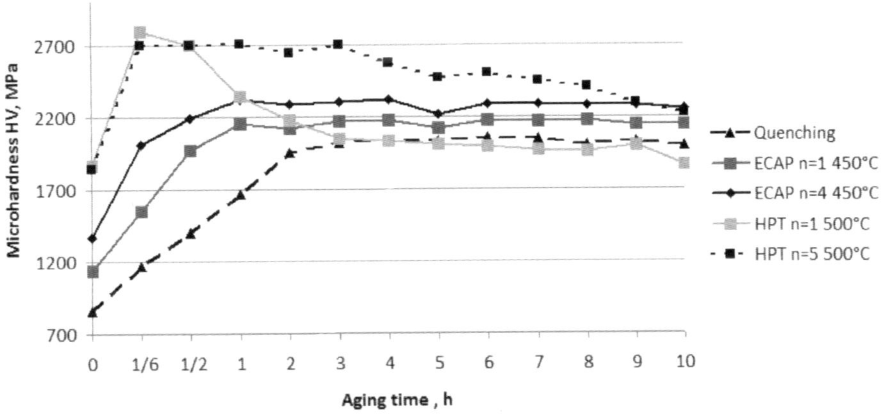

Figure 1. Variation in microhardness depending on the time of aging.

Figure 2 shows the most characteristic graphs of variation in electrical conductivity depending on the time of aging. It has been established that after 3 hours of treatment stabilization of this parameter is observed, while after 10 hours of treatment the electrical conductivity value is close for all processing regimes, but slightly higher as compared with the conventional treatment. A similar situation is observed for the aging temperature of 450° C, but the electrical conductivity value stabilizes after 5-6 hours of holding time.

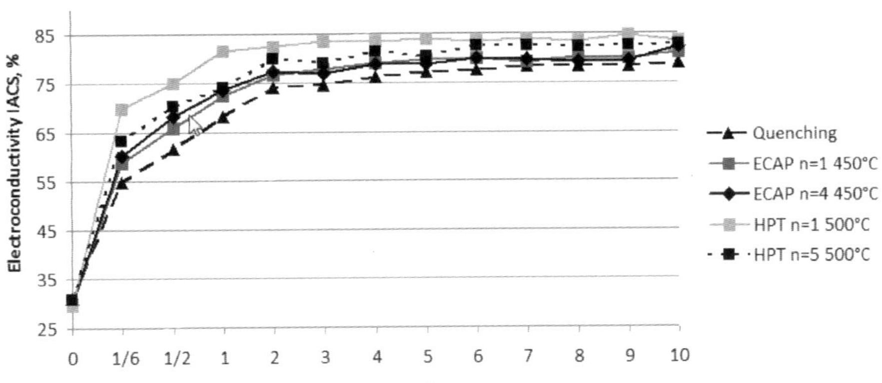

Figure 2. Variation in electrical conductivity depending of the time of aging.

The performed studies have revealed several most reasobale regimes demonstrating a combination of strength and electrical conductivity that is noticeably better than the traditional one. The most reasonable regime of ECAP processing is 4 passes with subsequent aging at a temperature of 500° C with a holding time of 1 hour. In this case, it is possible to attain a combination of microhardness of 2320 MPa and electrical conductivity of 80% IACS. The reasonable regime of HPT processing, including 5 HPT revolutions with subsequent aging of the alloy at 450° C for 2 hours, leads to an increase in hardness to 2650 MPa and an increase in electrical conductivity to 80% IACS. These regimes, in comparison with the conventional ones, are shown in figure 2.

When comparing the characteristics (figure 3) of the strain-hardenable Cu-1Cr-0.08Zr alloy, it may be noted that the microhardness after deformation processing by ECAP and HPT is higher than the microhardness after the commercial (industrial) treatment. Besides, the increment in microhardness after aging of the alloy is much higher in case of processing by SPD techniques. The material's electrical conductivity after ECAP and HPT processing reaches values of about 30% IACS. The solid solution decomposition at the stage of aging leads to the restoration of electrical conductivity to 80% IACS.

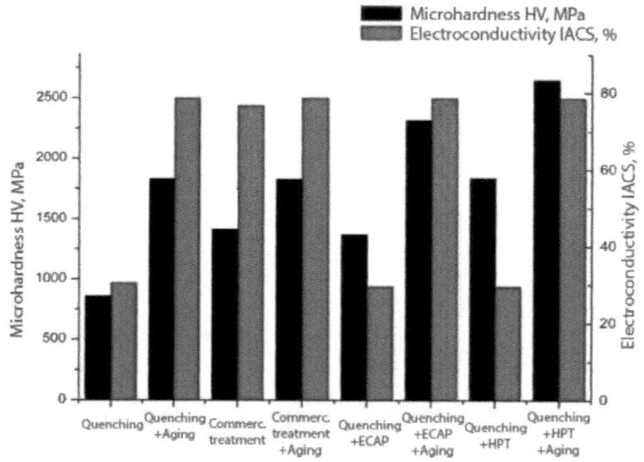

Figure 3. Variation in microhardness and electrical conductivity, depending on the type of processing.

In the course of the investigation, the contributions of the deformation component (work hardening) and of the thermally-activated component (fine-dispersed precipitation of second phases from the solid solution) into strength were analyzed. The results of this analysis are presented in figure 4.

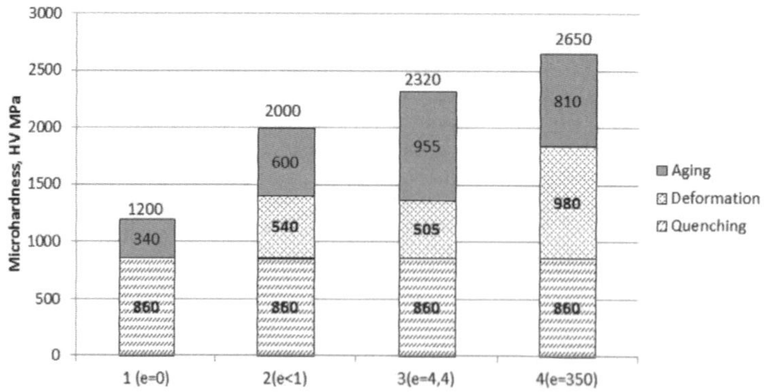

Figure 4. Diagram showing the contributions into strength after SPD processing and aging, where: 1 – quenching + 450° C 1 h (e=0); 2 – quenching + commercial treatment (CT) (e<1) + 450° C

1 h ; 3 – quenching + ECAP 4 passes (e=4.4) 500° C 30 min; 4 – quenching + HPT 5 revolutions (e=350) 450° C 2 h.

Note should be made that the maximum increment in strength due to the thermal activation (aging) is observed after ECAP processing (955 MPa), while the maximum increment in strength due to the deformation action is observed after HPT processing (980 MPa).

4 Conclusions

1. The study has revealed that the use of SPD processing leads to a noticeable enhancement of physico-mechanical properties of the Cu-1Cr-0.08Zr alloy. The regimes have been experimentally established that enables producing the best combination of strength and electrical conductivity in the Cu-1Cr-0.08Zr alloy. For ECAP processing, such a regime is deformation processing through 4 passes and aging at 500° C for 1 hour (strength 2320 MPa and electrical conductivity 80% IACS). For HPT processing, such a regime is 5 revolutions with subsequent aging at 450° C for 2 hours (strength 2650 MPa and electrical conductivity 80% IACS) .
2. The best thermal stability under the aging temperature of 450° C is ensured by the SPD processing regime including 4 cycles of ECAP processing at room temperature via route Bc and with the channels' intersection angle of 90 degrees.
3. The largest contribution into the increment of strength due to the thermal activation (aging) is observed after ECAP processing (955 MPa), while the maximum increment of strength due to the deformation action (work hardening) is observed after HPT processing (980 MPa).
4. An increase in accumulated strain during SPD contributes to attaining higher strength values after both ECAP processing and HPT processing.

Acknowledgements

The authors acknowledge the financial support by the Ministry of Education and Science of Russian Federation under Grant agreement No. 14.586.21.0025 (unique identification number RFMEFI58616X0025).

References

[1] Osintsev O E 2004 Copper and Copper Alloys. Domestic and Foreign Grades: Reference Book (Moscow: Mashinostroenie) p 336 (in Russian)
[2] Batra I S, Dey G K and Kulkarni U D 2001 Microstructure and properties of a Cu–Cr–Zr alloy *Journal of Nuclear Materials* **299** pp 91–100
[3] Faizov I A, Raab G I, Faizova S N, Aksenov D A, Zaripov M G, Gunderov D V, Golubev O V 2016 Dissolution of the second particles in Cu-Cr-Zr alloy upon the equal channel angular pressing *Tambov University Reports. Series: Natural and Technical Sciences* **21** pp 1387–91 (in Russian)
[4] Valiev R Z, Zhilyaev A P and Langdon T G 2014 Bulk Nanostructured Materials: Fundamentals and Applications (Hoboken, New Jersey: John Wiley & Sons, Inc.) p 456
[5] Utyashev F Z and Raab G I 2006 The model of structure refinement in metals at large deformations and factors effecting grain sizes *Rev. Adv. Mater. Sci.* **11** pp 137–151
[6] Vinogradov A, Patlan V, Suzuki Y, Kitagawa K and Kopylov V 2005 Structure and properties of ultra-fine grain Cu-Cr-Zr alloy produced by equal-channel angular pressing *Acta Mater.* **50** pp 1639–51
[7] Nikolaev A K 2001 Low-alloyed copper alloys. Features of the compositions and production technology *Tsvetnye Metally* **5** pp 84–88 (in Russian)
[8] Nikolaev A K 1978 Alloys for Contact Welding Electrodes (Moscow: Metallurgiya) p 96 (in Russian)
[9] Wang Z, Zhong Y, Rao X, Wang C, Wang J, Zhang Z, Ren W and Ren Z 2012 Electrical and mechanical properties of Cu–Cr–Zr alloy aged under imposed direct continuous current *Trans. Nonferrous Met. Soc. China* **22** pp 1106–1111

Investigations of influence of vibration smoothing conditions of geometrical structure on machined surfaces

D Bańkowski, S Spadło

Kielce University of Technology, Division of Materials Science and Ammunitions
Aleja Tysiąclecia Państwa Polskiego 7, 25-314 Kielce , Poland, EU

E-mail: dbankowski@tu.kielce.pl

Abstract. The paper presents influence of the type of abrasive media on the operating parameters of the surface of components made of composite ceramic-glass, coated with a layer of antiferromagnetic. The research presents the possibilities offered by the use of resin bonded media with different intensities abrasive, understood as different content of abrasive grains. There were used Rollwasch smoothing media PB series – resin plastic media. Media were cone-shaped, with different abrasive properties (10%, 50% and 85%). The process was carried out by using chemical compounds - liquid supportive ME series L100 A22 / NF. Attention has also been given to the relation between the properties of abrasive media and surfaces that can be obtained after polishing with porcelain media and deburring with resin bonded media. As the output surface used the disc of hard drive. Then, to analysis of the possibility various kinds of abrasive media in vibratory finishing the analysis of surface texture with an optical profilometer Talysurf CCI Lite - Taylor Hobson were done. As a result of the effects polishing and deburring using vibratory finishing were compared.

1 Introduction

In a rapidly growing industry, increasing requirements for the final product the manufactures still seek how to reduce manufacturing costs and improve the aesthetics of the finished items. Often the last step in the production of the products is sometimes machining, EDM, laser cutting, water-jet abrasive, etc [1, 2]. In the case of objects which require greater dimensional accuracy and a low surface roughness polishing processes are needed. In the case of mass production, finishing work using the loose abrasive media becomes increasingly important [3]. Undoubtedly, one of the biggest advantages of this type of surface treatment is the possibility of finishing process of small objects and complex shapes, which poses a challenge to traditional methods of polishing.

2 Vibratory finishing process

The term vibratory finishing is usually understood as a machining process, which is based on the chemical-mechanical interaction of charge in the form of loose media in the presence of liquid on the support surface of the workpiece in order to achieve the desired characteristics of the geometrical structure of the surface [4, 5]. Treatment with loose abrasives media is one of the varieties of finishing machine parts. The essence of the process is micromachining uneven surface by the action of the abrasive grains [6]. The mechanism of the process is caused by the vibration of the container, wherein there are media and shaped workpieces [7]. Spatial vibratory movements cause movement of the input media and workpieces. As a result, of the vibration then moves the elements feedstock, which is combined with the occurrence of mechanical wear causes the machined surface [8, 9]. A relatively simple technological devices used in the treatment of vibratory finishing proposed a wide range of media machining cause large potential applications. One of the most important factors influencing the effects of the treatment are media processing. They exist in different geometrical forms having different dimensions. The bonding material abrasive grains are a ceramic binder or resin. In addition, the offer are polishing fittings, usually made of porcelain or polishing-enhancing made of metal [10].

Figure 1. The vibrating machine Rollwasch SMR-D25.

The factor improving the efficiency of processing are liquid aids in the form of aqueous solutions of chemical compounds. Thanks to the wide range of available media in the form of media abrasive can manipulate the degree of intensity of wear. The use of media with high intensities abrasive (understood as a large percentage of abrasive grains) increases the intensity of the treatment of the surface with shallow appreciation. The use of media with small abrasive intensities can slow the stock removal is small. One of the advantages of smoothing the container is the ability to work with small objects as well as complex shapes [4, 11] involved in the process finishing considerable expenditures of manual work. Vibratory finishing machining is sometimes used on an industrial scale as a finishing treatment to remove burr from the edges smoothing the surface. It also enables the removal of oxide layers, traces of heat treatment, smoothing the surface down to a brightening [12, 13]. In many cases, the final result of the treatment is to be improved reflectivity. In the experiments the authors show that the process should be carried out in two stages pre-smoothing using the media abrasive and polishing with porcelain or steel media.

3 Object of study

As a reference surface the authors chosen the surface of disc from hard drive disc. Hard drive formerly was made of duralumin. This idea was not too beneficial for physical reasons: vibrations are created at the edges of plates and in addition coefficient expansion duralumin is relatively high. Currently, these discs are made of a ceramic-glass composite, which is coated with a layer of antiferromagnetic [AFC]. AFC consists of two layers of cobalt separated by three atomic layer of ruthenium. The material from which the rings are made of is a special aluminum alloy characterized by minimal dimensional changes with temperature. The same aluminum discs are not be carriers of information; they must be coated with a special layer having ferromagnetic properties. This layer must also comply with the relevant (very high) conditions of mechanical resistance. It is mostly cobalt or ceramic iron so ferrites. The application process of this layer is one of the most critical and costly production steps hard disk. These layers are generally very thin[14].
The authors proposed for this purpose in the loose abrasive machining in the vibrating containers - vibratory finishing. Uses the device's Rollwasch SMR D25 with a capacity of 25 liter (dm^3). The frequency of vibration of the tumbler established at 2500 Hz. For each time of the tumbler were poured into about 15 kg of machining liquid in the form of abrasive media and in addition approx. 200 ml liquid FE-L120-B32 / R aimed at lightening and the wetting of the surface and improving the reflectivity of workpieces. The first part of the experiment was to check how the media of porcelain series EB 0410V are able to 'destroy' the surface and what results we get on the surface. The duration

of treatment was set to three hours. From preliminary studies showed that longer processing times do not cause significant changes on the surface.

Figure 2. Porcelain media EB 0410V for polishing.

The next stage of the research was to examine and compare the extent to the resin bonded media of various intensities abrasive (understood as the percentage of abrasive grains) act on the surface. There were used a plastic media with greatest potential abrasive PB 14 KT (85%), next PB 14KR (ca.50%) and then PB 14KB (ca. 15%). The duration of the vibratory smoothing was 180 minutes

Figure 3. Resin bonded media PB 14 KT.

4 Tests results

Results of the studies concern measurements of surface geometrical parameters using optical microscope Tylor Talysurf CCI Lite (Taylor Hobson) and obtained results are presented in Table 1. Also morphology of obtained surfaces after each stage of machining in 3D configuration are depicted (Figures 4-8). It must define the parameters geometrical product specifications contained in Table 1. According to ISO 25178 Sa parameter is defined as the arithmetic average surface height [15]. Parameter Sz is the maximum height of the surface, and parameter Sv is the maximum cavity surface. Parameter highest peak area Sp is the difference between Sz and Sv. Sq parameter is the root mean square of surface roughness [16]

Table 1. The values of the basic parameters of the geometrical surface structure obtained for the different media machining.

	Surface output	EB 0410V porcelain media	PB 14KB plastic media	PB 14KR plastic media	PB 14 KT plastic media
Sa, μm	0,0039	0,1011	0,1587	0,1599	0,1969
Sq, μm	0,049	0,1642	0,2031	0,2250	0,2577
Sz, μm	0,0370	1,9448	2,2787	3,1943	3,5000
Sv, μm	0,0183	1,5343	0,9997	2,0680	1,6463
Sp, μm	0,0187	0,4105	1,2790	1,1263	1,857

Results of the studies presents comparison surface of the hard disc drive (initial stage) with the surface after polishing by porcelain medium, and deburring using plastic medium with different degree of abrasiveness.

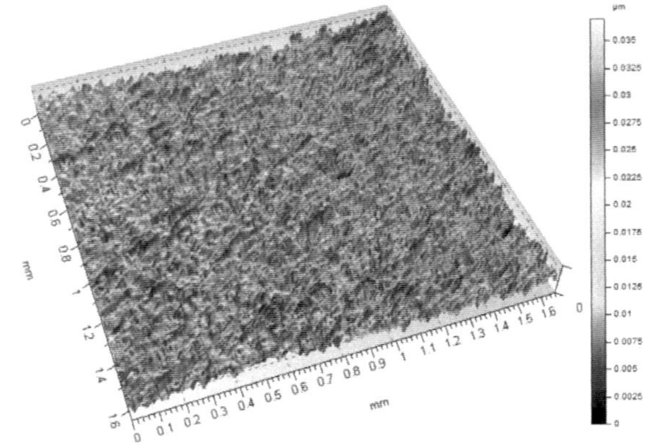

Figure 4. 3D optical surfaces of the hard drive disc, Sa =0,004 μm.

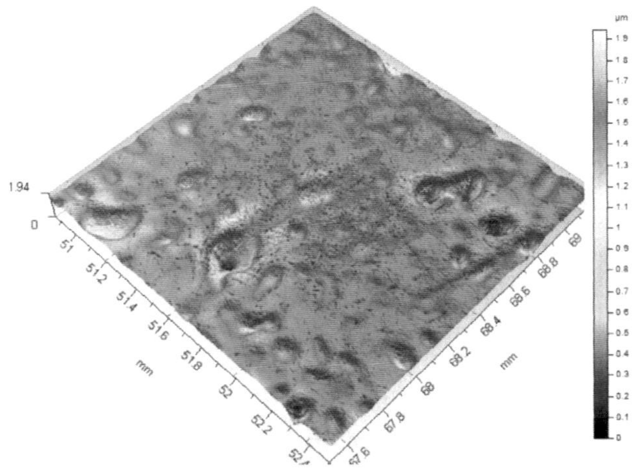

Figure 5. 3D optical surfaces of the hard drive disc after polishing (EB 0410V porcelain media) on 180 minutes, Sa =0,10μm.

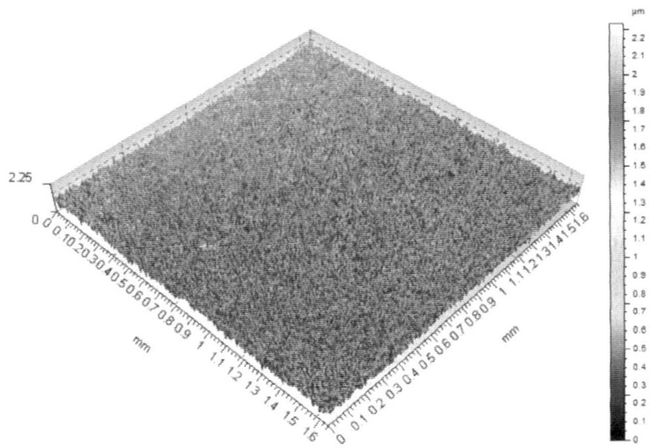

Figure 6. 3D optical surfaces of the disc after debburing with PB14KB on 180 minutes, Sa =0,15μm.

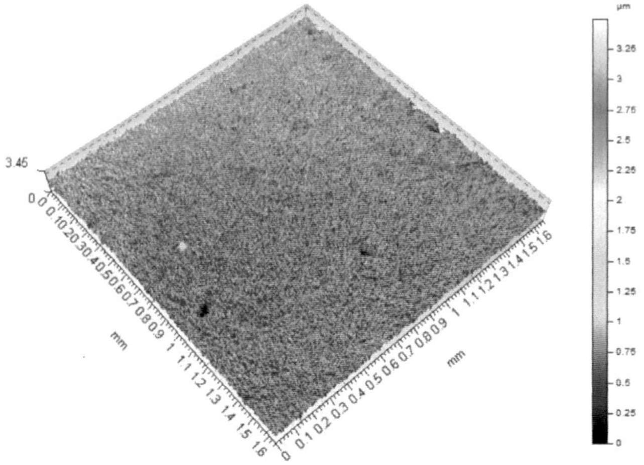

Figure 7. 3D optical surfaces of the disc after debburing with PB14KR on 180 minutes, Sa =0,16μm.

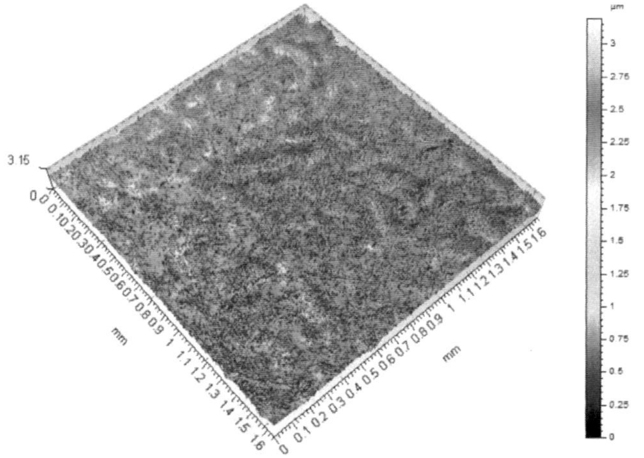

Figure 8. 3D optical surfaces of the disc after debburing with PB14KT on 180 minutes, Sa =0,20μm.

5 General remarks

The surface layer of hard disk is very smooth and reflective. Observations with using the optical microscope showed an isotropic structure, which results from the interaction of electromagnetic head on rotating at high speed rotary disc plate (of up to 15,000 rev/min). The hard disc surfaces are polished to a high gloss, which is confirmed by measurements of surface texture for which the parameter Sa is only 0.004μm. Vibrators finishing treatment is applied, wherein the workpieces and the media processing move freely relative to each other in the working tumbler and leads to a surface anisotropy. The use of media porcelain dedicated to polishing processes of vibratory finishing cause draw the surface. On the other hand, the use of media plastic dedicated to smoothing and debarring, edge rounding interacts on a test surface in more intensively. A significant impact on the final result is the aggressiveness media plastic used, understood as the percentage of abrasive grains in the volume of elementary media. The aim of the study was to examine the possible effects of surface roughness performing vibratory machining with the use of media, porcelain and plastic about the various abrasive possibilities. As the output surface a low surface roughness was used to limit the time of the experiment. Rougher surfaces require longer vibratory finishing treatment.

6 Conclusion

Higher percentage content of abrasive understood as the possibility of abrasive media used in the treatment of loose produces the surface with higher surface roughness.

Vibratory finishing with porcelain media allows for the arithmetic average surface roughness Sa approx. 0.10 µm.

Using media aggressive PB14KT (85% of abrasive grains) leads to obtaining the arithmetic average surface roughness Sa approx. 0.20 µm.

The longer treatment times do not have a significant change in the parameters of geometric structure of the surface.

Each time, before the process the appropriate choice of media machining, chemical compounds - media support is important to determine the proper ratio of media to workpieces, the oscillation frequency of the tank and the duration of treatment.

References

[1] Filipowski R., Marcinak M., (2000), Processing techniques and mechanical erosion, Oficyna Wydawnicza Politechniki Warszawskiej, Warszawa 2000 (in Polish)

[2] Spadło S., Młynarczyk P., Łakomiec K., (2016), Influence of the of electrical discharge alloying methods on the surface quality of carbon steel, The International Journal of Advanced Manufacturing Technology, DOI 10.1007/s00170-016-9168-1

[3] Spadło S., Młynarczyk P., Bańkowski D., (2014), Analysis of the Effect of Processing Vibro-abrasive Finishing on the Geometric Structure Surface Scales Ammunition and Sharp Edges, Journal of Achievements in Materials and Manufacturing Engineering, Tom: 66/1, pp: 39-44

[4] Spadło S., Bańkowski D., (2015), Influence Of The Smoothing Conditions In Vibro-Abrasive Finishing And Deburring Process for Geometric Structure of the Surface Machine Parts Made Of Aluminum Alloys EN AW-2017A; pp: 1062-1068, Proceedings of 24th International Conference on Metallurgy and Materials.

[5] Brinksmeier E, A. Giwerzew,(2005) "Hard gear finishing viewed as a process of abrasive wear". Wear 258: pp. 62-69

[6] Starosta R.(2008) Surface treatment, Wydawnictwo Akademii Morskiej w Gdyni, Gdynia (in Polish)

[7] Oryński F., Synajewski R., Bechciński G., (2013) A physical model of a vibratory grinding surfaces in the transverse direction. Mechanik, 1, pp. 30-34, (in Polish)

[8] Kacalak W., Tandecka K., (2012), Effect of superfinishing methods kinematic features on the machined surface. J. Machine Engineering 4, pp. 35-48

[9] Suzuki J., Miyake S., Watanabe J.: (1977) Fundamental Properties of Chemical-Mechanical Polishing with Powder Collision. Bull. Japan Soc. of Prec. Eng. Vol. 11, No. 4,

[10] http://www.rollwasch.it/en Vibro Dry Experience PL, Rollwasch Italiana S.p.a (02.06.2016)

[11] Harasymowicz J., Wanatuch E.:(1994) Machining lapping. Wydawnictwa Politechniki Krakowskiej, Kraków (in Polish)

[12] Rao, Suren B., (2009), "Repair of Aircraft Transmission Gears via Isotropic Superfinishing," May 2009, Gear Technology

[13] Xiao L., B.-G Rosen, Naser Amini, P.H. Nilsson. (2004)"A study on the effect of surface topography on rough friction in roller contact". Wear 254: pp. 1162-1169

[14] http://forum.pcformat.pl/Z-jakiego-materialu-jest-wykonana-tarcza-dysku-twardego-t (30.07.2016)

[15] Adamczak S., Miko E., Cus Franc; Strojniski Vestnik, (2009) A model of surface roughness constitution in the metal cutting process applying tools with defined stereometry; -Journal of Mechanical Engineering, No 55, pp. 45-54

[16] Janecki D., Stępień K., Adamczak S., (2010), Problems of measurement of barrel- and saddle-shaped elements using the radial method; Measurement, Vol. 43/5, pp. 659-663

IOP Conf. Series: Materials Science and Engineering **179** (2017) 012003 doi:10.1088/1757-899X/179/1/012003

Increase of chromium yield by slag reduction during production of chromium steels

J Bažan[1], L Socha[1], V Kurka[2], P Jonšta[3], M Sušovský[3]

[1] VŠB-Technical University of Ostrava, Faculty of Metallurgy and Materials Engineering, Department of Metallurgy and Foundry, 17. listopadu 15/2172, 708 33 Ostrava-Poruba, Czech Republic, EU

[2] Materiálový a Metalurgický Výzkum s.r.o., Pohraniční 693/31, 703 00, Ostrava-Vítkovice, Czech Republic, EU

[3] VÍTKOVICE HEAVY MACHINERY a.s., Ruská 2887/101, Vítkovice, 703 00 Ostrava, Czech Republic, EU

E-mail: jiri.bazan@vsb.cz

Abstract. The paper is focused on the evaluation of the course of Cr_2O_3 reduction from slag to alloyed steel under laboratory conditions. The experiments were aimed at the evaluation of increase in the chromium content in the melt together with the studying the behaviour of Cr_2O_3 and the mechanism of reduction by means of three reducing agents. Anthracite, ferrosilicon and mixtures of anthracite and ferrosilicon belong among the selected reducing agents. The experimental melts were focused on the proposal of a theoretical calculation of the consumption of selected reducing agents, study of reduction under laboratory conditions at application of alloyed steel with content of chromium of 12.16 wt. %, temperatures of 1600 °C and 1650 °C, together with the change of amount of reducing agents and reduction time. The results indicated in the paper constitute basic information on the possibilities of Cr_2O_3 reduction from slag; they will be used for verification of results in the pilot plant and operation experiments which will simulate operating conditions in the electric arc furnace.

1 Introduction

In general, the content of chromium in alloyed steels ranges from 8 to 30 wt. %. During melting of a steel charge containing chromium during steel production in an electric arc furnace, chromium is lost through oxidation, being transferred into slag. Primary losses already occur during the charge melting in the EAF; the main proportion of chromium, however, is lost from the melt during the oxidation period in which oxygen is blown to ensure essential refining reactions and decarbonisation of the melt [1, 2, 3, 4].

A high content of chromium in the slag affects its properties, which is apparent from the creation of high-viscose slag crust, which reduces the reactivity of the slag and prevents the formation of foamy slag during steel production in the EAF. Efficient reduction of chromium oxides from the slag during production of alloyed steels depends on the selection of a suitable reducing agent and the reduction technology applied during processing of steel in the EAF. The applied technology and suitable reducing agent should provide for a high degree of chromium oxide reduction from the slag. However, selection of the reducing agent depends on the equipment and technological capacities of the EAF operated. The reducing agents used include: silicon, carbon and calcium carbide. Under

operating conditions, several technologies have been developed and applied; for an efficient reduction of chromium losses in the melt, however, it is necessary to minimize losses as early as during the melting period, and particularly within the oxidation period [5, 6, 7, 8].
The paper states the results of laboratory experiments focused on the study of behaviour of chrome and its oxides, as well as the mechanism of its reduction by means of *anthracite, ferrosilicon* and *mixtures of anthracite and ferrosilicon*. The experimental melts were aimed at the proposal of the theoretical calculation of the consumption of selected reducing agents, the study of reduction under laboratory conditions together with a change in the amount of reducing agents and reduction time. The objective of the experiments was the verification of the basic possibilities of Cr_2O_3 reduction from slag with the use of various reducing agents.

2 Characteristics of the experiments

The laboratory experiments were implemented with the application of a charge representing alloyed steel containing 12.16 wt. % Cr and slag containing approx. 11.5 wt. % Cr_2O_3. These samples were taken after melting of the charge and short oxidation with oxygen from the EAF under operating conditions. The basic chemical composition of the alloyed steel and slag used is indicated in **table 1** and **table 2**.

Table 1. Production chemical composition of chromium-alloyed steel taken from the EAF.

Chemical composition of steel (wt. %)							
C	Mn	Si	P	S	**Cr**	Ni	Mo
0.02	0.62	0.31	0.023	0.003	**12.16**	4.1	0.42

Table 2. Chemical composition of slag containing Cr_2O_3 taken from the EAF.

Chemical composition of slag (wt. %)							
CaO	SiO_2	Al_2O_3	MnO	MgO	**Cr_2O_3**	FeO	V_2O_5
8.2	28.6	5.0	3.8	35.7	**11.5**	4.9	0.16

Three reducing agents: anthracite, ferrosilicon and a mixture of anthracite and ferrosilicon in the proportion 1:1 were chosen for the reduction of slags containing Cr_2O_3. Chemical composition of particular reducing agents is given in **table 3**.

Table 3. Chemical composition of reducing agent: *anthracite, ferrosilicon* and mixture of *anthracite and ferrosilicon (1:1)*.

Anthracite – C (wt. %)				
C	S	Humidity	Ash	Volatility
90.08	0.97	5.82	6.80	3.12
Ferrosilicon – FeSi (wt. %)				
Si	S	P	Mn	C
73.5	0.01	0.03	0.28	0.06

Before the laboratory melts, the consumption of the selected reducing agents was calculated theoretically. For the actual calculation, it is necessary to know the following parameters representing values acquired under operating conditions: *chemical composition of the chromium-alloyed steel, oxygen activity in the melt* $a_{[O]}$, *chemical composition of slag containing* Cr_2O_3, *metal and slag weight.*
The theoretical amount of reducing agents consumed was calculated from the given chemical composition of the slag as indicated in **table 1**. The actual calculations were performed on the

assumption of 100 % reduction of Cr_2O_3 oxides and remaining easily reducible FeO, MnO oxides from the slag. The calculated theoretical amount for particular reducing agents for 100 g of slag is given in **table 4**.

Table 4. Theoretical amount of reducing agents calculated according to the chemical composition of slag.

Reduction agent ($g \cdot 100g\ slag^{-1}$)		
Anthracite	Ferrosilicon	Anthracite + Ferrosilicon
4.65	6.70	2.32+3.35

For the laboratory experiments, an induction furnace was used, connected to a high-frequency generator GV 22. For the actual experimental laboratory melts, a block was assembled and located inside the inductor, as illustrated in **figure. 1**.

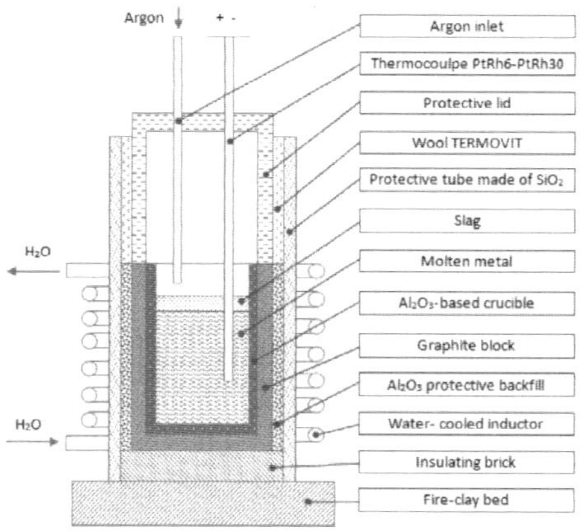

Figure 1. Schematic sectional view of the working block.

The working block consisted of a protective tube made of SiO_2 and a hollow graphite block for indirect heating of the charge. The gap between the protective tube and graphite block was filled with a protective filling of Al_2O_3 dust. The protective tube was sealed with fire-resistant TERMOVIT wool. The Al_2O_3-based working crucible was inserted in the hollow graphite block. The working space was closed with a magnesite lid with an inlet hole of argon inert gas. During melting, the inert gas was fed to the block from a pressure cylinder with a constant flow of 0.5 $l \cdot min^{-1}$. Besides the argon inlet, there was also a thermocouple of type B located here to continuously measure temperature.

In total, 16 relevant melts were executed under laboratory conditions. The laboratory experiments proceeded in the following way: firstly, a steel charge was added within the range of 200 to 300 g. After the steel charge had been melted and tempered at the temperature of 1600 °C or 1650 °C, a steel sample was taken. The obtained sample was used to determine the initial content of chromium in the steel. Subsequently, crushed reducing slag containing Cr_2O_3 was added to the steel weight in the amount of 10 wt. %. To this slag 10 wt. % of metallurgical spar CaF_2 was added to ensure fluidity of the reducing slag. 60 seconds after addition of the reducing slag the selected reducing agent, *anthracite* or *ferrosilicon or a mixture of anthracite and ferrosilicon*, was added. Additions of the reducing agents were executed in two variants:

- *variant I – corresponding to the 1×theoretical amount providing for 100 % reduction of Cr_2O_3 oxides from the slag;*
- *variant II – corresponding to the 2×theoretical amount providing for reduction of Cr_2O_3 oxides from the slag.*

The overall acting time of reducing agents was determined at 600 s. During the experiments, samples of steel were taken to determine the chromium content. The steel samples were taken in the middle of the experiment at 300 s and at its end at 600 s. Upon completion of the experiments, the individual steel samples were analysed to ascertain efficiency under laboratory conditions.

3 Results and discussion

Assessment of the Cr_2O_3 reduction from slag to alloyed steel under laboratory conditions was implemented in several stages. Firstly, evaluation of the course of increase in the chromium content in the steel was executed together with evaluation of the degree of reduction by means of three reducing agents, change in the amount of reducing agents and temperatures of 1600 °C and 1650 °C. The resulting course of the change in chromium content is indicated in **figure 2a** and **figure 2b**, the achieved degrees of reduction in **figure 3a** and **figure 3b** and complex evaluation of the degrees of reduction in **figure 4a** and **figure 4b**.

By comparing the results regarding the increase in chromium content at the application of three reducing agents from **figure 2** it was found that at *variant I* an increase in chromium content was in the range of 12.44 to 12.80 wt. %, which represents the degree of slag reduction according to the **figure 3** in the range of 31.80 % to 81.40 %. In the case of *variant II*, the increase of chromium content was in the range of 12.49 wt. % to 12.85 wt. %. It represents the degree of slag reduction in the range of 41.90 % to 87.70 %.

It is obvious from the results in the **figure 2a** to **figure 4b** that the highest increase in the chromium content of 0.69 wt. %, together with the achieved degree of reduction of 87.7 %, was found when the *mixture anthracite and ferrosilicon* was used in the proportion 1:1 together with the double the theoretical amount of reducing agent applied and at the experiment temperature of 1650 °C. In the contrary, the lowest increase in the chromium content of 0.28 wt. %, together with the achieved degree of reduction of 31.8 %, was found when *anthracite* was used, the theoretical amount of reducing agent applied and at the experiment temperature of 1600 °C.

Comparing the results from **figure 2a** to **figure 4b** with a view to the individual reducing agents, it can be stated that *the mixture of anthracite and ferrosilicon* seems to be more efficient. In the case of the *mixture of anthracite and ferrosilicon*, better results were achieved, with regard to both the chromium content increase in the range of 0.15 wt. % to 0.21 wt. % and the achieved degree of reduction being in the range of 19 % to 26 %. These results are valid at application of the double the theoretical amount of reducing agent and at the experiment temperature 1650 °C.

In addition to the reducing agents, the effect of an increase in the theoretical amount of reducing agents providing for reduction of Cr_2O_3 oxides from the slag was also tested. In this case, the application of *variant II* with the 2× theoretical amount of reducing agent resulted in an increase in the chromium content in steel together with an increase in the degree of reduction, as evident from **figure 2a** to **figure 4b**.

A temperature increase from 1600 °C to 1650 °C had a positive influence on the efficiency of Cr_2O_3 reduction from the slag. The increase in the theoretical amount of reducing agent and extension of the reduction time from 300 s to 600 s had a similar effect, as it is evident from **figure 2a** to **figure 4b**.

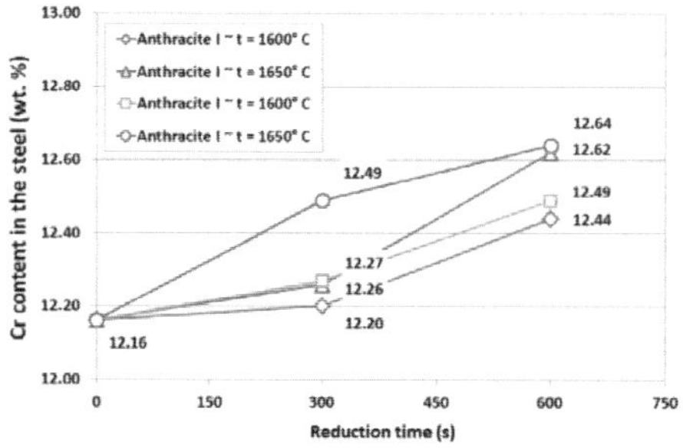

a) reducing agent anthracite and variants I and II

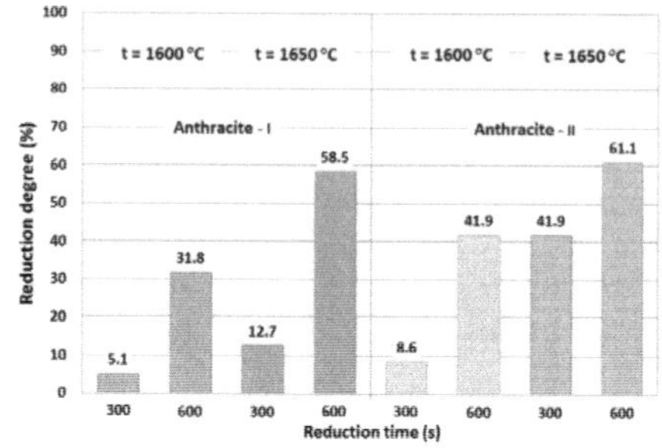

a) reducing agent anthracite and variants I and II

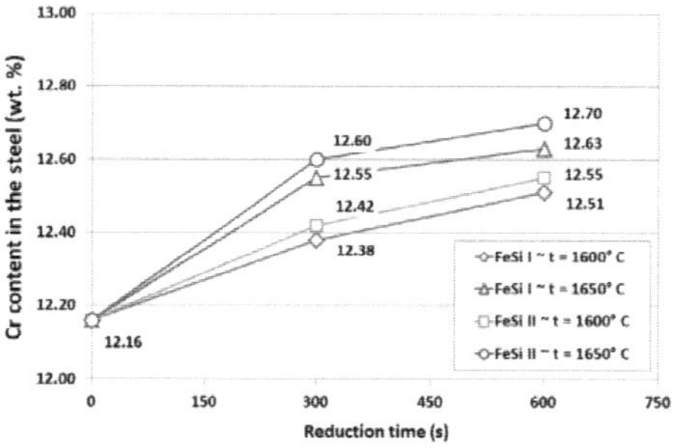

b) reducing agent ferrosilicon and variants I and II

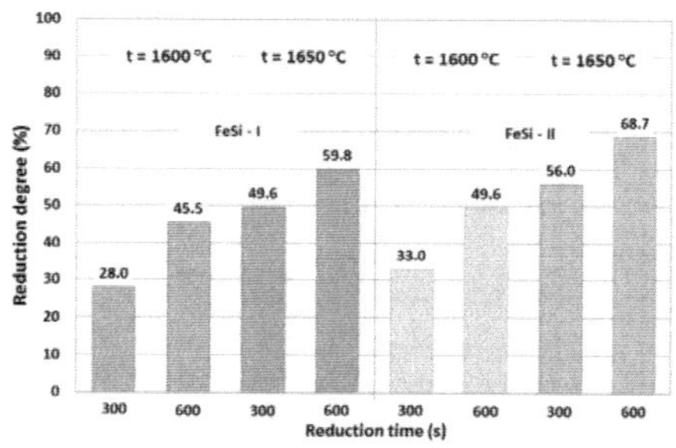

b) reducing agent ferrosilicon and variants I and II

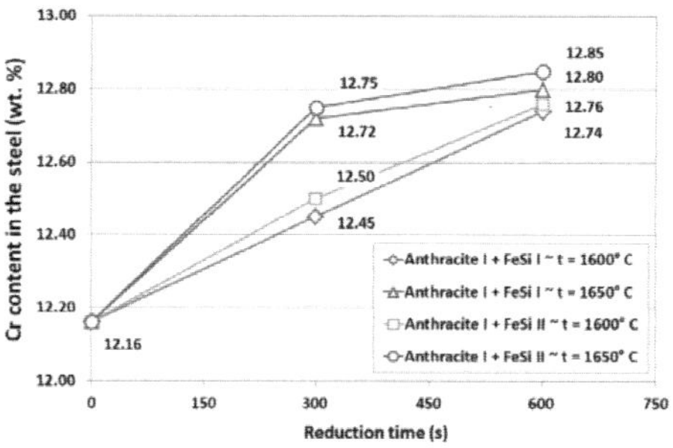

c) reducing agent anthracite + ferrosilicon and variants I and II

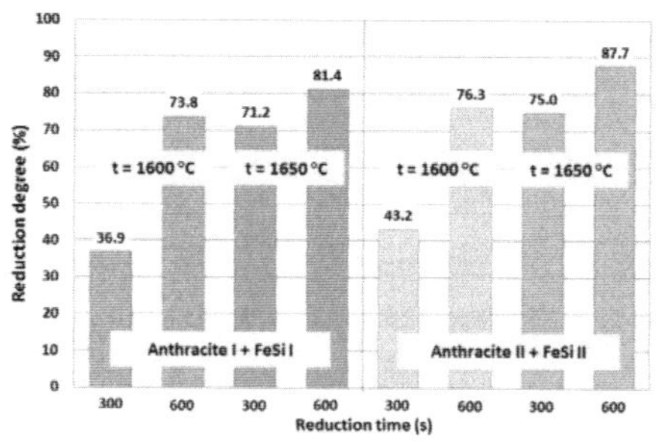

c) reducing agent anthracite + ferrosilicon and variants I and II

Figure 2. Changes in chromium content in the metal when using the reducing agent.

Figure 3. The degree of reduction of Cr_2O_3 oxides from the slag when using the reducing agent.

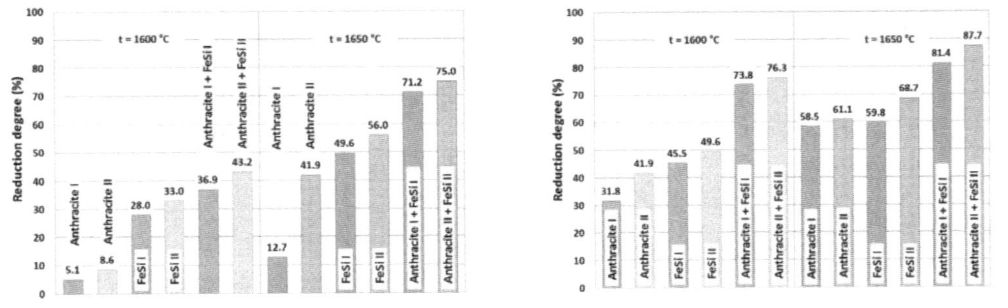

a) reduction time 300 s *b) reduction time 600 s*

Figure 4. Aggregate comparison of achieved degrees of reduction in Cr_2O_3 oxides when using the reducing agent.

4 Conclusions

Pursuant to the results achieved by means of the laboratory experiments, it is possible to state:

- *trend of reduction of Cr_2O_3 oxides to steel at different efficiency was verified at all three reducing agents;*
- *the mixture of anthracite and ferrosilicon in the proportion 1:1 seems to be more efficient reducing agent than separately used ferrosilicon and anthracite;*
- *in the case of steel with the Cr content 12.16 wt. % and Cr_2O_3 11.5 wt. % in the slag, the reducing agents can be ordered in the following sequence from the best: the mixture of anthracite and ferrosilicon, ferrosilicon and anthracite;*
- *with the usage of the mixture of ferrosilicon and anthracite in the proportion 1:1, together with the application of the double the theoretical amount of reducing agent and the temperature 1650 °C the efficiency of Cr_2O_3 reduction to 87.7 % was achieved;*
- *the increase of temperature from 1600 °C to 1650 °C had a positive influence on the efficiency of Cr_2O_3 reduction from slag;*
- *in addition to temperature increase, the increase in theoretical amount of reducing agent together with the extension of reduction time from 300 s to 600 s has also a positive influence on reduction efficiency;*
- *it can be also assumed that with the increase in Cr_2O_3 content in the slag a higher viscosity of slag connected with the crust creation can occur which decreases the reduction efficiency. This assumption will be verify with the subsequent series of experiments;*
- *in the subsequent series of experiments, the possibility of combination of various reducing agents at application of steel with various content of Cr and Cr_2O_3 in the slag under similar laboratory conditions will be verify which will be a part of following papers.*

The acquired laboratory findings will subsequently be used during the half-operational and operational experiments that will imitate the operating conditions in the EAF with the aim of designing and testing new technology for the reduction of chromium oxides from the slag using selected reducing agent.

Acknowledgements

This paper was created in the project the reg. number **TA04010036** and it was carried out in the financial support of TA ČR. The work was carried out in the support of projects of **Student Grant Competition** numbers **SP2016/89** and **SP2016/103**.

References

[1] Görnerup M, Lahiri A K 1998 Reduction of electric arc furnace slags in stainless steelmaking, Part 1, Observations *Ironmaking and Steelmaking* **25:4** 317-322

[2] Görnerup M, Lahiri A K 1998 Reduction of electric arc furnace slags in stainless steelmaking, Part 2, Mechanism of CrO_x reduction *Ironmaking and Steelmaking* **25:5** 382-386

[3] Arh B, Tehovnik F 2007 The oxidation and reduction of chromium during the elaboration of stainless steels in an electric arc furnace. *Materials and Technology* **41:5** 203-211

[4] Arh B, Tehovnik F. 2011 *Metalurgija* The oxidation and reduction of chromium of stainless steels in an electric arc furnace **50:3** 179-182

[5] Zhang Y, Fruehan R J 1995 Effect of the bubble size and chemical reaction on slag foaming *Metallurgical Transaction* **26B:4** 803-812

[6] Ito K 1989 Slag foaming in new iron and steelmaking processes *Metallurgical Transaction* **20B:6** 515-521

[7] Gutiérrez-Paredes J, Romero-Serrano A, Hernandez M A Chavez-Alcalá E, Zeifert B. 2005 Reduction of chromium oxide from slags *Revista de Metalurgija* **41:1** 443-446

[8] Juhart M, Peter M, Koch K, Lamut J, Royman A 2001 Foaming behaviour of slags from stainless steel production in the electric arc furnace *Stahl und Eisen* **121:9** 35-41

Finite element analysis regarding the forming behaviour of symmetric hybrid structures consisting of two sheet metal outer layers and a thermoplastic core

B-A Behrens[1], A Bouguecha[1], C Bonk[1], H Schulze[1,2]

[1] Institute of Forming Technology and Machines, Leibniz Universität Hannover, An der Universität 2, 30823 Garbsen

[2] E-mail: schulze@ifum.uni-hannover.de

Abstract. To face challenges like damping effects or weight reduction in the automotive sector, new hybrid material combinations are developed. One possibility is the combination of several symmetric material layers with varying material characteristics to achieve in the component production less weight and appropriate stiffness in comparison to components produced with sheet metal. This article deals with the characterization of deep drawing behaviour of layered sandwich structures. The behaviour of the several layers and the layer interaction have been taken into account for the technical design of a deep drawing process. A material layer characterization is performed. Instabilities as interlaminar failures, ruptures or wrinkling of the structure have been investigated as part of additional experimental characterization tests on the basis of various deep drawing process parameters. Finally, the experimental data is used as input for the numerical modelling and simulation of layered structures. The FE simulation includes the material behaviour of the layers and layer interactions with cohesive zone modelling. Based on the results an important contribution for prediction accuracy in the numerical simulation has been provided.

1 Introduction

The development of new materials is one approach to face the challenges in the lightweight sector. By the concept of symmetrical sandwich structures different materials with varying behaviour are combined. In this paper a symmetric sandwich structure with outer sheet metal layers with a thickness of s_0 = 0.25 mm and a polymer core (s_0 = 0.7) is used for the investigations. To achieve a precise description of the behaviour for the sandwich structure the sheet metal and the polymer core are characterized separately. In addition delamination tests for the sandwich structure are carried out.

2 Material characterization

The sheet metal layer in of 0.25 mm thickness was characterized with different test procedures. In order to obtain the flow curves and the anisotropic parameters for the material, quasi-static tensile tests with cut-out sample configurations in the three directions 0°, 45° and 90° according to EN ISO 6892-1 were carried out. For an accurate description of the material hardening behaviour, the characteristics of the sheet metal at high strains are required. For the FE analysis the true stress - true strain curve

IOP Conf. Series: Materials Science and Engineering **179** (2017) 012004 doi:10.1088/1757-899X/179/1/012004

determined in the uniaxial tensile test needs to be extrapolated beyond the point of the uniform elongation. Using an inadequate extrapolation process an inaccurate description of the material behaviour at high strain can occur. To ensure a realistic description of the flow behaviour at high strains the hydraulic bulge test according to DIN ISO 16808 is used and the deformation of the blank is continuously measured with an optical system [1], [2].

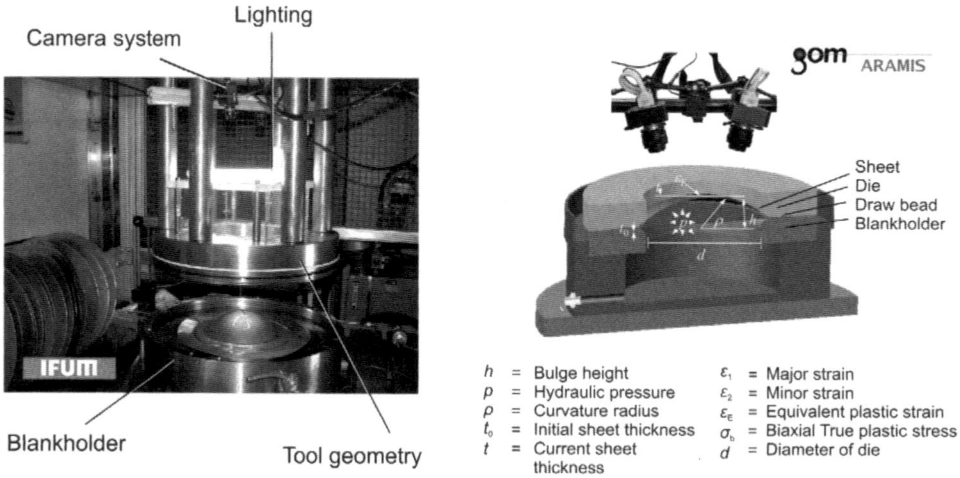

Figure 1. Experimental set-up (left) and functional sketch (right) for biaxial bulge test.

In figure 1 on the left side the experimental set-up and a schematic sketch regarding the realization of the hydraulic bulge set-up are shown. The tool consisting of die, blankholder, pressure chamber and pressure load cell is integrated in a hydraulic forming machine. The sheet is clamped between blankholder and die and the material flow is prevented by a draw bead. The displacement of chosen material points on the sheet surface are recorded with the optical measurement system Aramis by the company GOM mbH with a stochastic and continuous grid and the hydraulic pressure in the center of the pressure chamber is measured as well. The biaxial true stress - true plastic strain curve from the bulge test at room temperature was transformed to the uniaxial stress state with a mathematical approach based on the principle of the equivalence of plastic work [3].

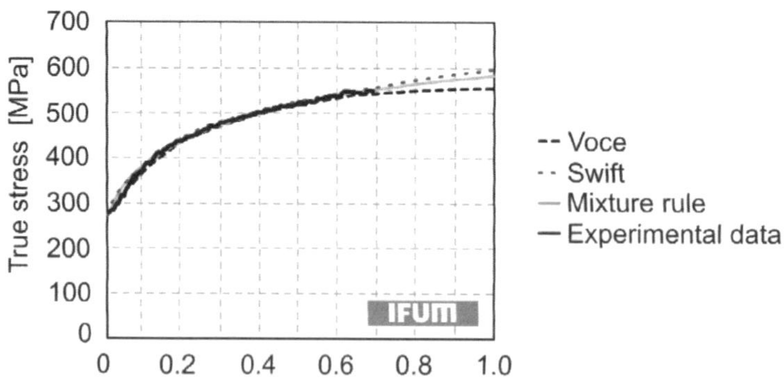

Figure 2. True stress - True plastic strain curve for sheet metal HC220 with thickness 0.25 mm, see also [4].

IOP Conf. Series: Materials Science and Engineering **179** (2017) 012004 doi:10.1088/1757-899X/179/1/012004

In figure 2 the true stress - true plastic strain behaviour for the sheet metal HC220 is shown with different extrapolation approaches, see also [4]. In addition to the material characterization of the sandwich structure investigations for the polymer layer are carried out. In this case the samples cutted from the sandwich structure are modified. Via die sinking both sheet metal layers are removed and the samples are used for tensile tests at room temperature with the dilatometer DIL 805A/D+T from TA Instruments, formerly Bähr GmbH. Different strain rates were taken into account.

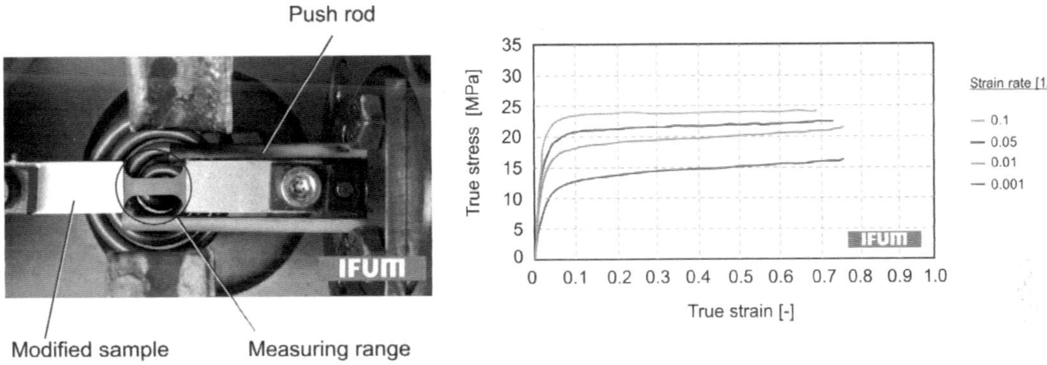

Figure 3. Modified tensile test sample (left) and True stress - True strain behaviour of the polymer layer with thickness 0.7 mm for different strain rates.

The experimental set-up and the modified sample concerning the measurement range for the tensile test with the dilatometer are shown in figure 3 on the left side. On the right side the true stress - true strain behaviour for different strain rates is illustrated. With an increasing strain rate the tensile strength of the inner layer increases. In comparison to the sheet metal the polymer layer shows low stress values, but high true strain values.

3 Investigation of interface failure

The development of new materials is one approach to face the challenges in the lightweight sector. With the concept of symmetrical sandwich structures different materials with varying behaviour are combined. Based on the layered structure, sandwich structures have different properties as compared to monolithic sheet metals. For this reason in addition to the behaviour of the single layer material the layer interaction has to be taken into account for the design of a deep drawing process. Experimental and numerical investigations for hybrid structures regarding the design of deep drawing processes have been examined in [4]. An important failure mechanism for composites and sandwich structures is delamination. Under combined loading conditions in a deep drawing process a crack occurs and propagates.

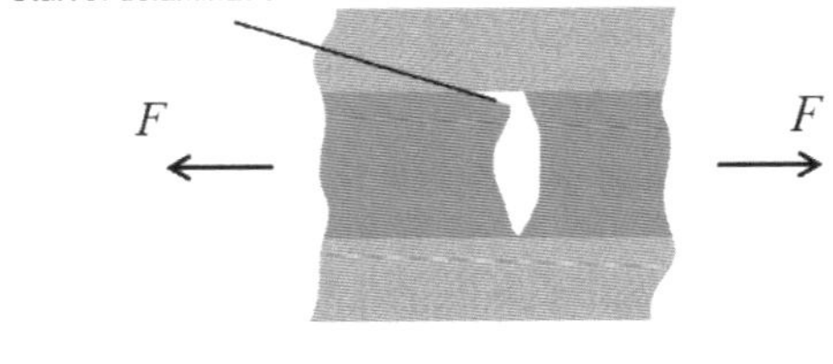

Figure 4. Delamination failure for sandwich structures, acc. to [5].

In figure 4 a sketch for a delamination process within a sandwich structure is shown. Delamination can start within the core, but the most critical case is the start of delamination at the interface. For composite materials delamination is a well examined failure mechanism and various test procedures can be found in literature. Subject of the article is the transfer of the existing methods for composites to investigations for sandwich structures. In this paper the climbing drum peel test [6] is used to characterize the interface behaviour between two components with different material properties.

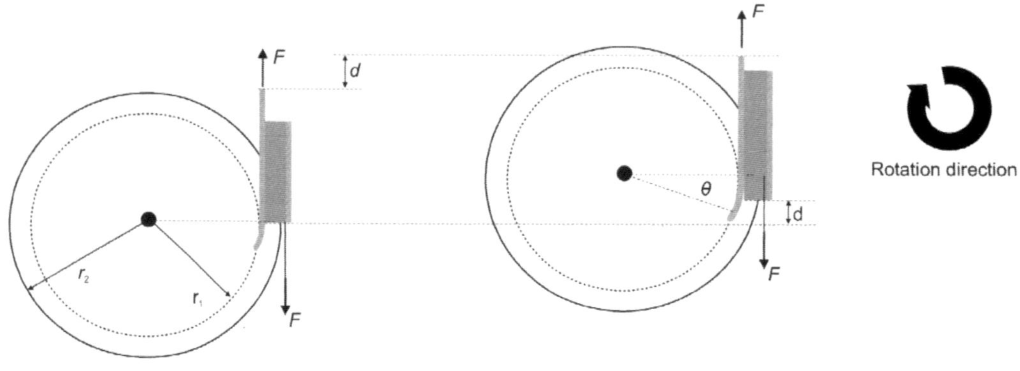

Figure 5. Climbing Drum Peel Test, acc. to [6].

In figure 5 the principle function of the climbing drum peel test according to [6] is shown. F is the applied pulling force, r_1 is the radius of the drum, r_2 is the radius of the flange, d is the crosshead displacement and θ is the occurring rotation angle of the drum. At the beginning a pre-crack is brought into the middle layer, the outer sheet metal layer gets clamped on the drum and the drum is integrated in a tensile machine. Afterwards the set-up undergoes a vertical loading and the drum component rotates clockwise. Based on the force-displacement data an analysis of the delamination behaviour can be carried out.

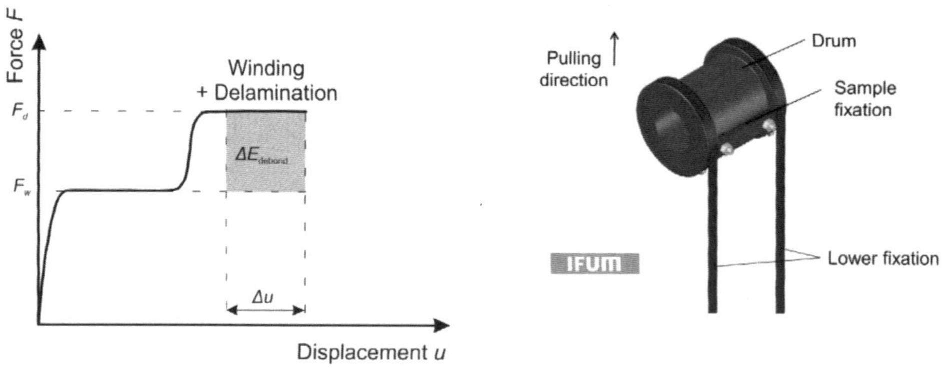

Figure 6. Force-displacmenet curve (left acc. to [7]) and experimetal set-up of climbing drum peel test (right).

The force-displacement curve for a climbing drum peel test is shown in figure 6. The first force increase until F_w is the force needed for the winding of the drum until the point where the pre-crack starts. The second increase up to force value F_d is the force for winding and delamination of the structure.

With the force-displacement data and the geometrical dimensions of the set-up it is possible to estimate the dissipated energy ΔE_{debond} for the delamination and with this a critical strain energy release rate G_C.

$$\Delta E_{debond} = (F_d - F_w)\Delta u \qquad G_C = \frac{\Delta E_{debond}}{\Delta A} = \frac{F_d - F_w}{w}\frac{r_2 - r_1}{r_2}$$

ΔA is the area which undergoes a rotation and w is the width of the sample. With the value G_C an arising delamination effect is described and this parameter can be used for the numerical modelling of sandwich structures under deep drawing loading conditions.

4 Numerical modelling and FE simulation

For the modelling of the sandwich structure the material properties determined by experimental material characterization procedures presented in section 2 are used. For the numerical modelling of the interface between the sheet metal and the polymer an established approach is the implementation of a traction-separation curve.

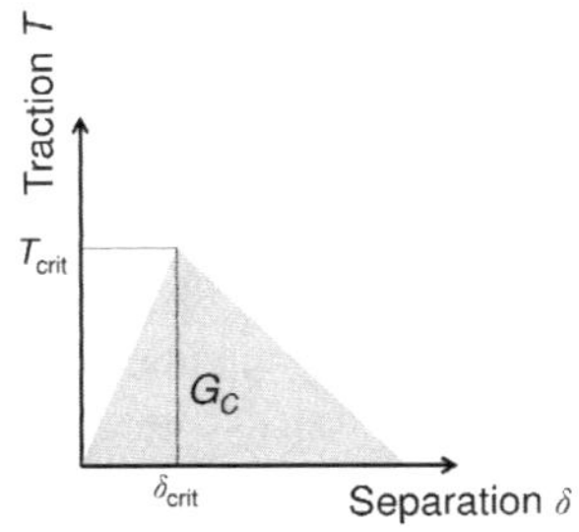

Figure 7. Traction-separation curve for numerical modelling of interface failure (acc. to [8]).

The traction-separation curve, e. g. in figure 7, is used to describe the failure mechanism for interfaces between two materials with different characteristics. The interface is modeled with cohesive elements and the behaviour of the elements is described with the traction-separation curve, which is represented by critical values for the traction T_{crit} and the separation δ_{crit} the arising of the delamination. The critical fracture energy G_C considered in section 3 defines the area of the triangle. For a detailed description of cohesive element modelling approach see [8], [9]. To generate a simple test case a tensile test with removed small part of the sheet metal outer layer is modeled to proof the modeling approach for the sandwich structure and the delamination effect. For the numerical modelling of sheet metal and polymer layer the results from the material characterization are implemented and the layers are modeled as solid elements. The interface is modeled with cohesive elements with a thickness of 0.01 mm.

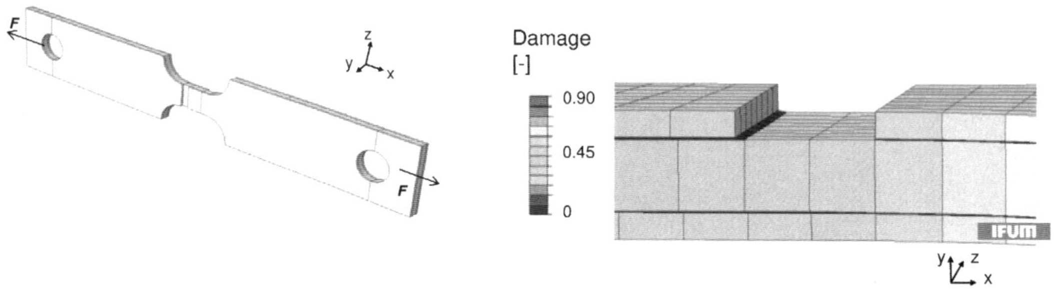

Figure 8. FE model for tensile test (left) and damage failure value (right).

On the left side of figure 8 the FE model for the tensile test is shown and on the right the damage value for the FE simulation are presented. The results of the FE simulation show that no significant delamination occurs and a coincidence with the experimental behaviour of the sandwich structure exists. This test case is chosen to proof the modeling approach with solid elements for the layers and cohesive elements for the interface.

5 Conclusion

The development of new materials is one approach to face the challenges in the lightweight sector. With the concept of symmetrical sandwich structures different materials with varying behaviour are combined. In this article the material characterization for the different layers, experimental investigations for the interface behaviour and a transfer to one modelling approach for FE simulation of sandwich structures is presented. The FE simulation tryout shows a satisfying prediction for the behaviour of sandwich structure and the interface. For more complex applications additional investigations are needed to describe the material and interface behaviour under different loadings, which can occur in a deep drawing process and influence the behaviour of the sandwich structures in different ways.

References

[1] Behrens, B.-A.; Rolfes, R.; Vucetic, M.; Peshekhodov, I.; Reinoso, J; Vogler, M.; Grbic, N.: Material characterization for FEA of the clinching process of short fiber reinforced thermoplastics with an aluminum sheet; Advanced Materials Research Vols. 966-967, 2014, pp. 557-568

[2] Behrens, B.-A.; Bouguecha, A.; Peshekodov, I.; Götze, T.; Huinink, T.; Vucetic, M.; Friebe, H.; Möller, T.: Numerical Validation of Analytical Biaxial True Stress - True Strain Curves from the Bulge Test; 8th International Conference and Workshop on Numerical Simulation of 3D Sheet Metal Forming Processes (Numisheet), 2011, Seoul, South Korea, pp. 107-114

[3] Sigvant, M.; Mattiasson, K.; Vegter, H.; Thilderkvist P.: A viscous pressure bulge test for the determination of a plastic hardening curve and equibiaxial material data; Int. J. Mater Form, 2009 pp. 235-242

[4] Behrens, B.-A.; Vucetic, M.; Neumann, A.; Osiecki, T.; Grbic, N. (2015): Experimental test and FEA of a sheet metal forming process of composite material and steel foil in sandwich design using LS-DYNA, 18th International ESAFORM Conference on Material Forming, 15.04.2015 – 17.04.2015, Graz, Key Engineering Materials Vols. 651-653 (2015), pp 439-445

[5] Schürmann, H., Konstruieren mit Faser-Kunststoff-Verbunden, Springer Verlag, 2007

[6] Nettles, A. T.; Gregory, E. D.; Jackson, J. R.: Using the Climbing Drum Peel (CDP) Test to Obtain a G_{IC} Value for Core/Face Sheet Bonds, J. of Composite Materials, Vol. 41, No. 24/2007, pp. 2863-2876

[7] Daghia, F., Cluzel, C.: The Climbing Drum Peel Test: An alternative to the Double Cantilever Beam for the Determination of fracture toughness of monolithic laminates; J. Composites Part A: Applied Science and Manufacturing, Vol. 78, 2015, pp. 70-83

[8] Hallet, S. R; Harper, P. W.: Modelling delamination with cohesive elements, Numerical Modelling of Failure in Advanced Composite Materials, Woodhead Publishing Series in Composite Science and Engineering, 2015, pp. 55-72

[9] van den Bosch, M. J.: Interfacial delamination in polymer coated metal sheet, Eindhoven, University of Technology, 2007

Numerical and experimental investigations on an extrusion process for a newly developed ultra-high-carbon lightweight steel for the automotive industry

B-A Behrens, A Bouguecha, C Bonk, D Yarcu, M Kazhai

Leibniz Universität Hannover, Institute of Forming Technology and Machines, An der Universität 2, 30823 Garbsen, Germany

kazhai@ifum.uni-hannover.de

Abstract. In this study the material flow of a newly developed ultra-high-carbon lightweight steel (uhc-steel) with a high amount of aluminum was investigated in an extrusion process. Cylinder compression tests were performed for material characterization and frictional behaviour was determined by using ring compression tests. Numerical simulations were carried to determine the optimal die geometry as well as to calculate the process loads and dominated stresses in the die occurring during the process. Based on the numerical results, an extrusion process was designed and implemented. Experiments showed that the uhc-steel can be formed by extrusion however it is associated with a high wear rate.

1 Introduction

The demands on nowadays processes are focused on resource efficiency, like in example reducing the process chain, as well as lightweight [1,2].

Especially in the automotive sector the lightweight aspect is in focus of actual researches, where in example components of the powertrain are tried to be reduced by weight. Lightweight can either be reached by geometrically optimizing the components (topology optimization) or by new materials. With regard to lightweight ultra-high-carbon steels are potential lightweight steels, which can be used for massive formed parts in the automotive to reduce the weight. The research on uhc-steels began in the early 70´s, where studies were carried out to develop new thermo-mechanical processes for the creation of ultra-fine ferrite graine-sizes [3,4,5,]. In these studies about uhc-steels, the superplasticity of the uhc-steels were observed. In the following years, many studies were carried out investigating the superplasticity and the suerplastic flow of uhc-steels. According to recent investigations the carbon content has to be increased to meet the requirements for the superplasticity [6,7]. Uhc-steels contain about 1 - 2.1 % carbon. Besides the superplasticity at a certain temperature range, the characteristic mechanical properties of uhc-steels were found by low ductility and fracture toughness at ambient temperature, which is unusual for common steels in an intermediate carbon content of 0.1-1 %. Further researches focused on alloying elements stabilizing the occurring phases. One of the most important alloying element is aluminum. On the one hand it stabilizes the existing ferrite phase, but in terms of lightweight and resource efficiency it also reduces its weight due to the lower density in comparison with common steel.

Although these uhc-steels show a good formability at high temperatures, they are difficult to process, which results in high die wear [8,9]. The reason is the formation of thick networks of iron carbides or κ-carbides. These κ-carbides are formed, by cooling down high carbon steels from high temperatures to intermediate temperatures [10]. The carbide networks are areas, at which cracks can be initiated, primarily. In the last years in many researches the formation of these carbide networks aiming to create a homogeneous and fine-sized grain to avoid this critical formation was investigated.

In this study the flow behaviour of a newly developed uhc-steel is investigated on an extrusion process. A material characterization to determine flow curves and the temperature dependent friction coefficients was carried out. The data was implemented into the commercial software system Forge (Transvalor) and numerical simulations were carried out designing an extrusion process for different die geometries. Experiments were carried out, so that the numerical model was validated. The results showed a good agreement comparing the numerical with the experimental results.

2 Numerical and experimental investigations

2.1 Material characterization

To determine the strain-, strain-rate- and temperature dependent flow curves of the uhc-steel, cylindrical compression test were carried out on a servo-hydraulic plastometer, see Figure 1.

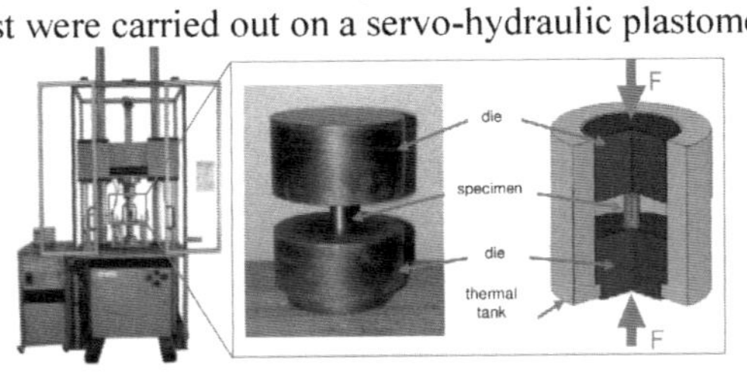

Figure 1. Servo-hydraulic plastometer (Instron).

Herefor specimens with Ø11x18 mm were used. The flow curves were determined at temperatures from 900 to 1200 °C in 100 degree steps. Strain rates of 1, 10 and 40s^{-1} were used, which cover a good range of strain rates occurring in most bulk metal forming processes. The compression tests were carried out up to a strain of 0.7, by which a homogeneous uniaxial forming can be assumed. To minimize the friction between the dies and the specimens, bornitrid was used as lubricant. The specimens were heated up in a furnace. To guarantee adiabatic conditions, the specimens were placed in a thermal tank.

In the figure 2 the exemplary flow curves of the uhc-steel are shown for the temperature range 900 – 1200 °C and the two strain-rates 1s^{-1} as well as 40s^{-1}. As can be observed, with increasing strain-rate, the flow stress level of the uhc-steel increases. At a temperature of 900 °C the flow stress increases from 280 MPa at a strain-rate of 1s^{-1} to 375 MPa at a strain-rate of 40s^{-1}. Furthermore, it can be observed that the flow curves at 900 °C, for both strain-rates, show a significant decrease over the strain. This is assumed to be linked with the high amount of aluminum in the uhc-steel.

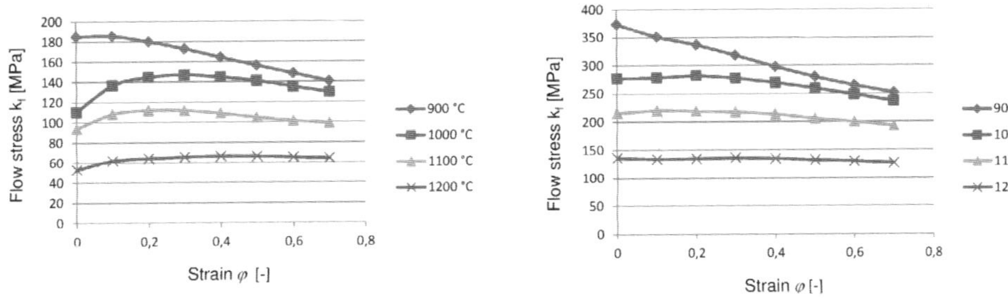

Figure 2. Uhc-steel flow curves at strain-rate 1s^{-1} (left) and 40s^{-1} (right).

To determine the friction behaviour of the uhc-steel, ring compression tests were performed using a screw press (Weingarten). Temperature dependent friction coefficients were determined. The billet temperature was varied between 800 and 1250 °C. The die temperature was varied between 150 and 250 °C, as die material the classic tool steel X40CrMoV5-1 (1.2344) was used. The dies were heated up with a heating jacket and the specimens were heated in a furnace. The tool setup for the ring compression tests is shown in figure 3, left. The specimen geometry, next to a forged part is shown in figure 3, left. As ratio for outer diameter to inner diameter to height, 6:3:2 was used. The geometry of the forged specimens were measured tactile and the friction factor was determined by avitzur method, calculating the friction coefficient with analytic approaches from the geometric dimensions after forging [11]. The results of the ring compression tests are shown in figure 3, right. As can be seen, the determined friction factors range between 0.26 and 0.44 depending on tool and specimen temperature. Furthermore, a significant influence of the die temperature can be observed at a die temperature of up to 200 °C. A dependency of the friction coefficient on rising specimen temperatures can be recognized at die temperatures above 225 °C.

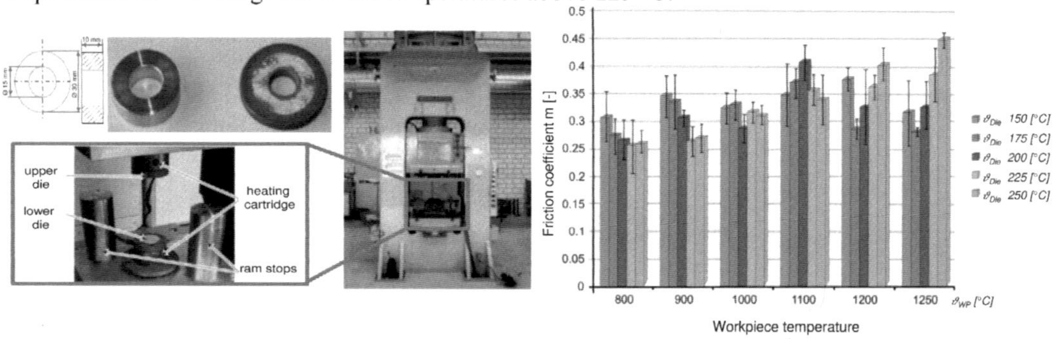

Figure 3. Tool setup for ring compression tests and specimen dimensions (left) with temperature dependent friction factors of the uhc-steel (right).

The data of the material characterization, which contains flow curves as well as temperature dependent friction coefficients were edited and implemented in the software system Forge (Transvalor) for carrying out numerical simulations.

2.2 *FE-aided design of the extrusion process and experiments*

Before carrying out the experimental investigations on the extrusion process, the tool system was numerically designed by carrying out simulations using the software system Forge (Transvalor).

To find the optimal die geometry geometric parameters, which were the tapering radius as well as the tapering, were varied and the stress state in the dies were determined. The tool system consists of an upper and a lower die, which are available in different geometries. The lower dies were investigated for two different taperings to reach local strains up to $\varphi = 1.5$ in the final part. In figure 4 the FE-model of the extrusion process next to forged part for the two different taperings is shown. The forging temperature was set to 1250 °C.

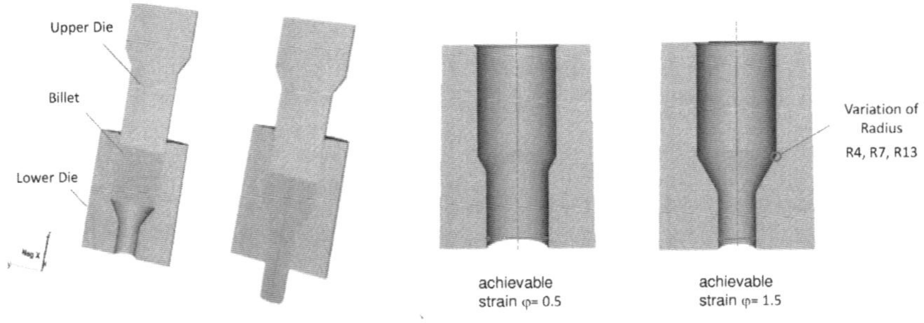

Figure 4. FE-model of the extrusion process

For the die stress analysis, the uncoupled method was used. In a first step a material flow simulation with rigid dies and an elasto-plastic material behavior is carried out. In a following step the loads from the material flow simulation are transferred and the real stress analysis is carried out using elastic modelled dies. Thus, the stress and strains within the dies can be determined and geometric changes of the dies can be performed if necessary. Regarding the die stress analysis results, only the exemplary results of the die with with the achievable strain of φ=1.5 in the part, which is more critical in terms of stresses, are shown in the following figure 5. To determine the optimal stress state for the dies, three different radii were investigated for the tapering of the lower tool, mentioned 13, 7 and 4 mm. To assess the loads on the dies, the von Mises stress as well as the first principal stress is taken into account. While the von Mises stress is an indicator for plastic deformations, the first principal stress shows the tension and compression stress and is an indicator for crack initiation. In figure 5 the results of the die stress analysis are shown for the lower die and the respective radii. As can be seen in figure 5 (left), the radii 4 mm as well as 13 mm show a high stress level up to 1000 MPa in the tapering area of the die. The die with the tapering radius of 7 mm show a comparatively lower stress level of about 900 MPa, which can be observed only in the upper tapering area. The comparison with the first principal stress, which is shown in figure 5 (right) shows qualitatively same results. The dies with 4 mm and 13 mm radii show stress levels of 950 MPa and 700 MPa. For the die with the radius of 7 mm a maximum stress of 520 MPa can be observed. In summary, the die with the radius of 7 mm showed the comparatively lowest stresses in the extrusion process, whereat all the stresses are under the tensile strength of the material. For further experimental investigations, the die with the 7 mm tapering radius was chosen.

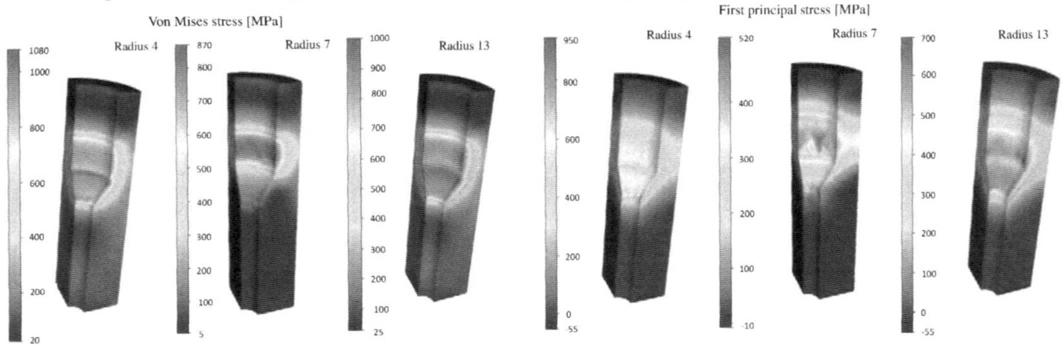

Figure 5. Von Mises stress (left) and first principal stress of the lower die for 3 different radii.

The experimental investigations were carried out on a screw press (Lasco), which is shown in figure 6. As lubricant the oil based lubricant Berulit 722 (Bechem) was used.

Figure 6. Screw press (Lasco).

Selected results and a comparison of the forged part with numerical simulations are shown in figure 7. In figure 7 (left) the numerically determined temperature distribution and the strain distribution are shown next to forged parts with the achievable strain of φ=0.5. The strain distribution in the center of the shaft is at the value of 0.5. At the outer areas of the shaft the plastic strain increase up to 1.6 due to the friction between the die and the billet. A comparison of the geometrical dimensions between the numerical results and the forged parts showed a good agreement. In figure 7 (right) the temperature distribution as well as the strain distribution of the part are shown next to forged parts with strain of φ=1.5. The strain in the shaft of the part is about 1.5 and increase up to 2.75 in the outer areas of the shaft due to friction.

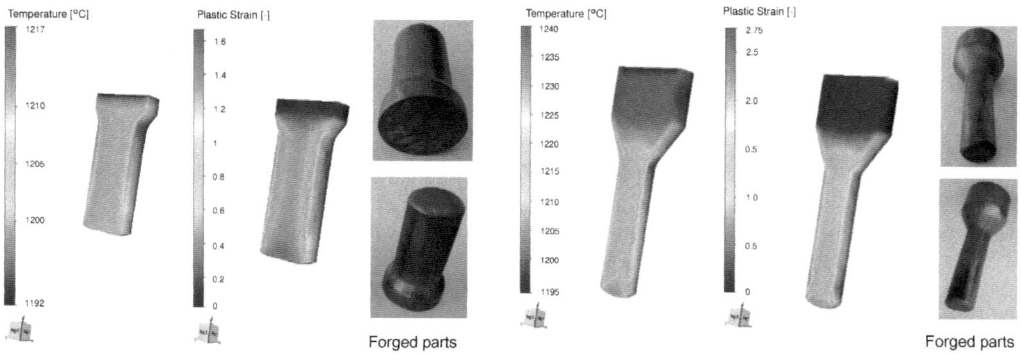

Figure 7. Temperature distribution, plastic strain and forged parts with the strains of φ=0.5 (left) and φ=1.5 (right).

The forged parts, shown in figure 7 indicate a high wear on the dies, which can be seen on the surface of the parts at their respective shafts. The numerical abrasive die wear determination, which is calculated from the sliding velocity and the normal stress applied on the surface, showed that a high die wear is expected. In figure 8 the numerical determined die wear is shown next to the dies after forging. An optical, qualitative evaluation of the die wear shows a good agreement of simulations with experiments. As can be seen, the dies show intense grooves at the tapering areas. This die wear behaviour is typical and one of the disadvantages of the uhc-steels. Here different lubricants and surface treatments of the dies should be investigated.

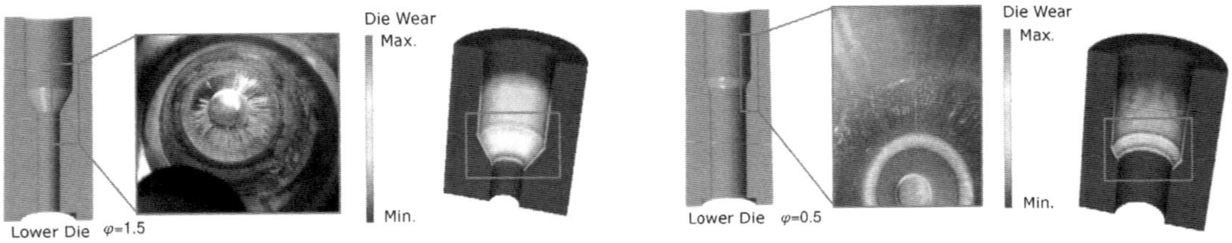

Figure 8. Die wear of the die with achievable strain of φ=1.5 (left) and the die with achievable strain of φ=0.5 (right).

The results show that the newly developed uhc-steel is suitable for processing. The properties of the forged parts are to be investigated in further investigations like tension test and metallographic investigations.

3 Summary and outlook

Uhc-steels are steels with a high amount of up to 2.1 % carbon.

In the hot forging temperature range uhc-steels show a high superplasticity, which was first observed and developed in the early 70´s. The alloying of elements like aluminum into the uhc-steel can reduce the density and thereby the weight of the material. In terms of lightweight bulk metal formed parts produced with the uhc-steel are in the focus of many investigations. Within this study a newly developed uhc-steel, which will be used for a process chain to produce piston pins, was investigated. A material characterization was carried out to determine the temperature-, strain- and strain-rate dependent flow curves via cylindrical compression tests. Furthermore, the temperature dependent friction behaviour was investigated using ring compression tests. An extrusion process with different taperings was numerically designed and experimental tests were carried out. The results show that the material can be processed at hot forging temperatures. The forged parts will be investigated for their mechanical properties with the help of tensile test and microsections. The knowledge gained, will be transferred to a process route for the production of piston pins in a backwards extrusion process. However, the high die wear has to be considered processing the investigated uhc-steel.

Acknowledgements

This research and development project is funded by the German Federal Ministry of Education and Research (BMBF) within the Framework Concept "Research for Tomorrow's Production" (funding number: 02PN2050 ff.) and managed by the Project Management Agency Karlsruhe (PTKA).

References

[1] Behrens, B.-A., Huskic, A., Kazhai, M.: Process-integrated heat treatment of hot forged components, 60 Excellent Inventions in Metal Forming, 1 January 2015, Pages 421-425
[2] Denkena, B., Park, H.-S., Behrens, B.-A., (...), Lüken, I., Klassen, A.: Development of a concept to optimize the energy efficiency in forging process chains, International Journal of Precision Engineering and Manufacturing
[3] Lesuer, D.R.; Syn, C.K.; Goldberg, A.; Wadsworth, J.; Sherby, O.D.: The Case of Ultrahigh-Carbon Steels as Structural Materials, J. Metals (JOM), Vol 45, 1993, S. 40
[4] Oyama, T.; Sherby, O.D; et al.: Application of the divorced eutectoid transformation to the development of fine-grained, spheroidized structures in ultrahigh carbon steels, Scripta Metallurgica 18 (8), 1984, S. 799-804
[5] Taleff, E. M.; Nagao, M.; et al.: High-strain-rate superplasticity in ultrahigh-carbon steel containing 10 wt.% Al (UHCS-10Al). Scripta Materialia 34 (12), 1996, S. 1919-1923

[6] Sherby, O.D.; Kum, D.W; et al.: UHCS Containing Aluminum, US Patent #4.769.214, 06.09.1988

[7] Sherby, O. D.; Oyama, T.; Wadsworth, J.: Divorced Eutectoid Transformation Process and Product of UHCS, US Patent #4.448.618,15.05.1984

[8] Lin, W., Fan, Y., Zhang, z., zhu, J., Liu Y.: Microstructure and Mechanical Properties of An Ultra- high Carbon Steel, Key Engineering Materials Vols 324-325 (2006) pp 907-910

[9] Pol, F.: Massivumformung dichtereduzierter UHC-Stähle unter nicht superplastischen Bedingungen (Massive forming close of UHC steel under not super+plastic conditions), Dissertation, RWTH Aachen, 2010

[10] Kim, W.-J., Ruano, O. A., Wolfenstine, J., Frommeyer, G.: Superplastic behavior of a kappa carbide material (Fe3AlCx), Journal of Material Research, 1997

[11] Avitzur, B.: Forging of hollow disks, Israel, J. of. Tech., 1964, S. 295 – 304

Numerical analysis of a deep drawing process with additional force transmission for an extension of the process limits

B-A Behrens[1], C Bonk[1], N Grbic[1,2], M Vucetic[1]

[1]Institute of Forming Technology and Machines, Leibniz Universität Hannover, An der Universität 2, 30823 Garbsen

[2]E-mail: grbic@ifum.uni-hannover.de

Abstract. By sheet metal forming processes the forming limits and part characteristics are defined through the process specific loads. In deep drawing processes the maximum deep draw ratios as well as the springback behaviour of the metal parts are depending on the stress distribution in the part material during the forming process. While exceeding the load limits, a failure in the material occurs, which can be avoided by additional force transmission activated in the deep drawing process before the forming limit of material is achieved. This contribution deals with numerical investigation of process effect caused by additional force transmission regarding the extension of the process limits. Here, the steel material HCT 600X+Z (1.0941) in thickness $s_0 = 1.0$ mm is analyzed numerically using the anisotropic model Hill48. This model is validated by the means of cup test by Swift. Both, the FEA of conventional and forming process with additional force transmission are carried out. The numerical results are compared with reference geometry of rectangle cup.

Keywords: deep drawing, process limit, deep drawing with additional force transmission, deep drawing with counter punch, FEA

1 Introduction and motivation

Increasing globalization, resulting in a rapid change of the industrial market and the society, strictly expected regulation of CO_2 emissions and the increasing scarcity of resources are the main drivers of the automotive industry. A great potential for decrease CO_2 emissions is to reduce the vehicles mass in the car body. Therefore, the continuous innovation of existing production processes are required, with the aim to develop processes with a high level of productivity, minimum number of production stages, maximum utilization of the material, low revision need, aspired energy balance and high shape or dimensional accuracy. By production of car body parts made from sheet metals, the deep drawing and bending are crucial. In order to expend the process limits by sheet metal forming, the development of innovative processes is required. Here, the finite element method is an indispensable tool.

2 State of the art

By sheet metal forming processes the force, applied in sheet metal, causes the modification of its geometry without removing any material. The stresses distributed over the metal part are beyond the yield strength of material inducing the plastic deformation in sheet metal without material failure.

As extensively used sheet metal forming process the deep drawing is applied in the production of different metal parts like pots and pans for cooking, containers, sinks, automobile parts, such as panels and gas tanks.

IOP Conf. Series: Materials Science and Engineering **179** (2017) 012006 doi:10.1088/1757-899X/179/1/012006

2.1 Conventional deep drawing

Conventional deep drawing is a sheet metal forming process which requires a blank, blank holder, punch, and draw ring (die) (see Figure 1). Here, the blank as a piece of sheet metal is radially drawn or stretched into the desired shape by the means of specific force transmission induced from the mechanical punch load. The important area of the punch is the radius, where the forming load is transferred through the drawn part wall into the deformation zone (flange). During the conventional deep drawing process the blank is clamped by the blank holder above the die. The external shape of the part is defined through the die cavity. The cross section of the drawn part can pass different shapes, with straight, tapered, or even curved walls. Here, the cylindrical or rectangular parts are most common.

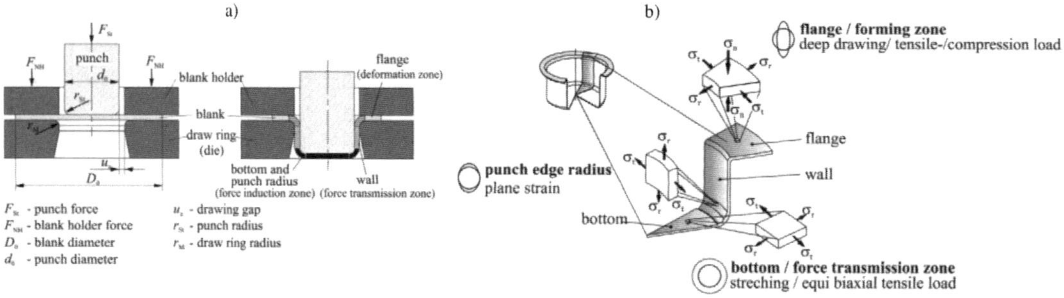

Figure 1. Conventional deep drawing a) positioning and forming, b) stress distribution.

Due to tensile forces effecting part wall, blank thinning is occurring. Based on the local stresses an uneven wall thinning appears. At the punch radius where the part wall loses contact with the punch, the part wall thickness shows lowest value. Here, the part failure can occur. Thus, the maximum stress, which can be transferred to the deformation zone (flange) is determined [1].

2.2 Deep drawing with additional force transmission

For extension of forming limits deep drawing processes are applied like hydro-mechanical deep drawing, hydro-form process or hydraulic deep drawing process [2]. The main commonality of these processes is the usage of active media for additional force transmission, by which the initial forming zone with the highest local stresses is reloaded. Thus, the additional forming load is transferred into the crucial deformation zone (flange). Hereby, a further material drawing can occur.

The extension of forming limits can be also achieved by the means of additional mechanical force transmission. Here, between different process approaches can be distinguished like additional material feeding [3], usage of differentiated force transmission through the variable surface pressure in the blank holder or counter-punch, as well as by an expansion punch.

By additional material feeding the crucial tensile stresses can be decreased when pulling the material in the flange area by additional mechanical load. Thus the punch force transmission is reloaded. Other approaches deal with specific adjustment of friction. Due to the reduction of the friction at draw ring radius a further material drawing from the flange area is realized. By processes with defined material flow a cup is first drawn and subsequently the material is pushed over the wall by means of a mechanical load, while the drawing punch transforms the metal part.

The usage of differentiated force transmission through the variable surface pressure in the blank holder allows the reduction of the friction between blank holder and die. Thus, the material flow is supported by deep drawing process. Here, by means of stiffness-optimized blank holder systems the local surface pressure in the flange area can be varied and suited for better process stability.

By differentiated force transmission using the expansion punch the force transmission takes place at whole lateral surface of the part wall. Here, the distribution of the forming load at part wall occurs by using a specific punch with grub-screw grooves, which are machined for force application into the inner part wall [2, 3].

2.3 New tool system developed for deep drawing with additional mechanical force transmission

The new tool system designed for deep drawing process with additional mechanical force transmission is developed at Institute of Forming Technology and Machines (Figure 2a). At punch radius and the transfer zone to the wall area, where the highest stretching of material occurs, the forming load is transmitted into the flange zone (forming zone) by the means of external pressure applicated on the part bottom (see Figure 2b).

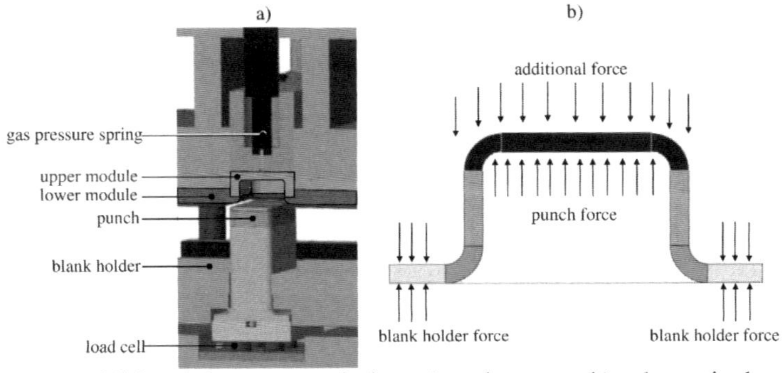

Figure 2. Additional force transmission, a) tool system, b) schematic description.

The punch is located on the base plate and the load cells are placed between. The blank holder device is mounted on an intermediate plate with guiding elements, which is located on the top of press quills. Thus, the deep drawing takes place in return. The closed die is designed as two-part module, the upper and lower die module. The lower die module, or the drawing ring, represents the rectangle cup. The upper die module represents the geometrical shape of the drawing part bottom. Here, besides the force in the part bottom also the force in radius between bottom and wall of the drawn part is transferred by the upper die module which works like a counter punch. This is initiated when the drawn part runs into the upper die, whereby a vertical additional force in the drawing process from bottom into the flange zone is transferred by means of a gas pressure spring. The gas spring is fixed to the counter punch.

3 Numerical Analysis

By FEA of the effect caused by the additional force transmission, the processes with and without the activation of the gas pressure spring were taken into account. For numerical investigation the rectangle cup part made of steel material HCT 600X+Z100 in thickness of s_0 = 1.0 mm is chosen.

3.1 Material modelling

For modelling of material behavior flow criterion by Hill [4] is used in Abaqus. Based on six material-specific coefficients calculated from anisotropic coefficients (r-values) the shape of the yield locus is generated. Here, quasi-static uniaxial tensile tests were performed and analyzed according to DIN EN ISO 10002. The specimens aligned at 0°, 45° and 90° to the rolling direction were tested at room temperature. Thus, the r-values r_0, r_{45} and r_{90} are determined.

a)

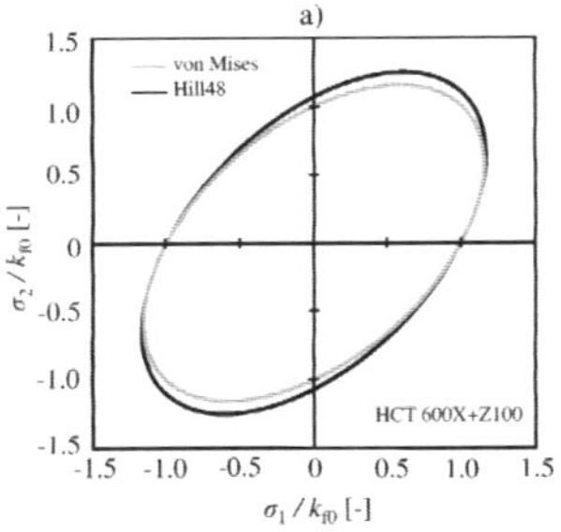

b)

$$F = \frac{1}{2}\left(\frac{1}{{R_{22}}^2} + \frac{1}{{R_{33}}^2} - \frac{1}{{R_{11}}^2}\right)$$

$$G = \frac{1}{2}\left(\frac{1}{{R_{33}}^2} + \frac{1}{{R_{11}}^2} - \frac{1}{{R_{22}}^2} +\right)$$

$$H = \frac{1}{2}\left(\frac{1}{{R_{22}}^2} + \frac{1}{{R_{33}}^2} - \frac{1}{{R_{11}}^2}\right)$$

$$N = \frac{3}{2{R_{12}}^2}; L = \frac{3}{2{R_{23}}^2}; M = \frac{3}{2{R_{13}}^2}$$

Figure 3. Material modelling by Hill48 for HCT600 a) yield locus b) Hill-coefficients.

In Figure 3 the yield locus according to Hill48 and von Mises for material HCT 600 is presented. For further investigations the yield locus modeled by Hill48 is used. For parametrization of the Hill criterion in Abaqus, a calculation of Abaqus-suited Hill coefficients is carried out (see Table 1).

Table 1. Parametrization of Hill material model for HCT 600

HCT600	r_0	r_{45}	r_{90}	G	H	F	N	L	M	R_{11}	R_{22}	R_{33}	R_{12}	R_{13}	R_{23}
[-]	0.932	1.042	1.197	0.5457	0.454	0.387	1.662	3/2	3/2	1.0	1.067	1.049	1.034	1.0	1.0

Beyond the point of uniform elongation the true stress - true plastic strain curve determined in the uniaxial tensile test needs to be extrapolated for the FEA. By material modelling with the extrapolation process an inaccurate description of the material behavior at high strain can occur. Thus, the steel material HCT 600 was analyzed in a hydraulic bulge test [5].

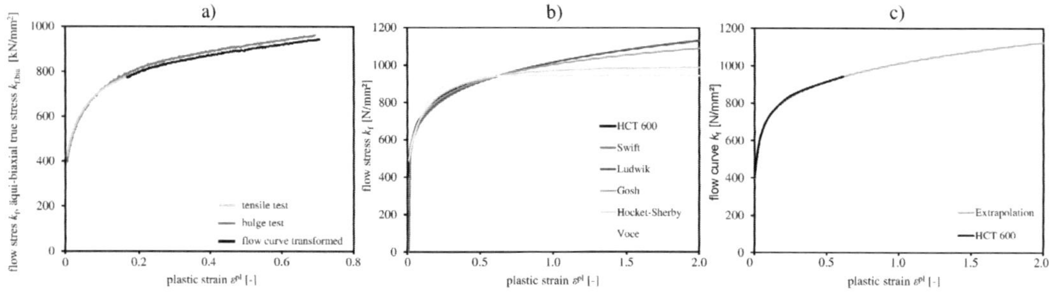

Figure 4. HCT 600 s_0=1.0 mm a) experimental results, b) extrapolation, c) best-fit extrapolation.

The biaxial true stress - true plastic strain curve from the bulge test is transformed to the uniaxial stress state with a mathematical approach based on the principle of the equivalence of plastic work (Figure 4a) [6]. Thus, more experimental data can be taken into account for the following extrapolation process (Figure 4b). Here, the deviation from the actual flow behaviour at high strains made by the extrapolation is minimized.

The extrapolation functions presented in Figure 4b) can not correctly reproduce the flow behavior of the experimentally determined flow curve for the HCT 600. In order to achieve an accurate description of the hardening behavior at high plastic strains a combined approach based on Swift und Hockett-Sherby is used (see Figure 4c) [7]:

$$k_f = \alpha_s \cdot \left(A_S \cdot \left(B_S + \varphi_{pl}\right)^{ns}\right) + (1 - \alpha_s) \cdot \left(A_{HS} - (A_{HS} - B_{HS}) e^{-C_{HS}\varphi_{pl}^{n_{HS}}}\right) \tag{1}$$

The extrapolation coefficients for the combined extrapolation functions by Swift and Hockett-Sherby are summarized in Table 2.

Table 2. Coefficients related to Swift / Hockett-Sherby combined approach for HCT 600

Swift / Hocket-Sherby	α_s	A_s	B_s	n_s	B_{HS}	A_{HS}	C_{HS}	n_{HS}
[-]	0.42	1800	0.01218	0.2	147.3	438.9	21.59	0.9397

3.2 Model validation

Using the cup test by Swift the material modelling approach Hill48 is validated quantitatively. The test is carried out on universal Roell+Korthaus machine, while the force and displacement are measured. Afterwards, the forming process based on Swift is analyzed numerically with the software Abaqus CAE 6.13-1 (Figure 5a). Here, the values force and displacement are calculated and compared with experimental results. The reaction forces of both, the test and the FEA, are showing a typical evaluation for a deep drawing process. Here, the punch forces determined by the experiments show a good agreement with the numerical results (Figure 5b). For further verification of the Hill model the geometry of the cup is measured optically with ATOS system (GOM mbH). Here, the experimentally and numerically determined CAD geometries of a round cup are compared digitally with software Geomagic® QualifyTM (Geomagic, USA) (see Figure 5c).

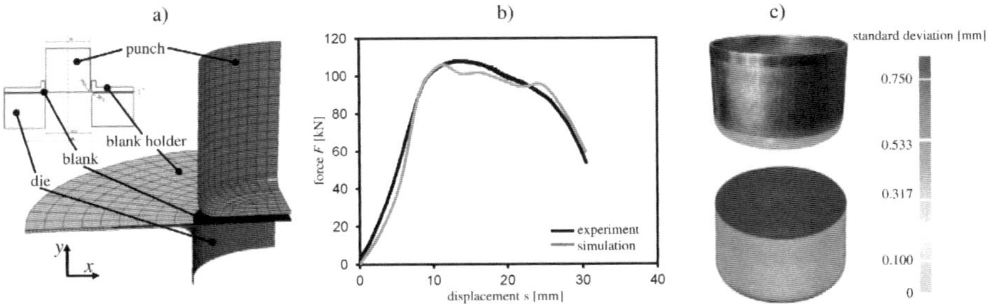

Figure 5. Quantitative model verification, a) FE model cup test by Swift b) comparison force-displacement curve, c) geometric comparison.

Two geometries, CAD from experiment and CAD from FEA, are aligned with each other and the best fit analysis of dimensional matching between the experimental and numerical results is carried out. For this analysis, the optically measured geometry is chosen as a reference. The dimensional deviation analysis shows a standard deviation of 0.089 mm regarding the whole dimensional deviation. The presented results show a sufficient agreement between the FE results and the real part. Hence, the material model by Hill48 will be used for the further numerical process analysis.

3.3 *FEA of deep drawing process with additional force transmission*

In order to investigate the effect caused by additional force transmission the FEA of deep drawing process, presented in section 2.3, is carried out. Hereby, both processes, the conventional process and process with additional force transmission are investigated numerically (see Figure 6a,b).

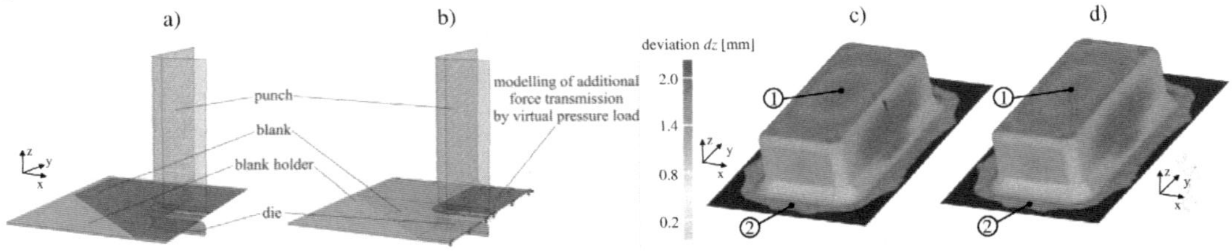

Figure 6. HCT 600, a) FE model conventional process, b) FE model deep drawing with additional force transmission, c) comparison of FEA results (conventional deep drawing) with reference CAD-geometry, d) comparison of FEA results (process with additional force transmission) with reference CAD-geometry

For 3D FE modelling, consisting of punch, die, blank holder and blank, a quarter model is applied. Here, the Hill48 model with the isotropic hardening rule is used. Based on calculation of effective E-modulus using the exponential approach by Yoshida [8] the springback behaviour is calculated with Abaqus/Standard implicit solver. The interaction between blank and active surface of the tool system is modelled with Coulomb's friction model. For FEA of deep drawing with additional force transmission the 3D FE model of conventional process (Figure 6a) is extended with a virtual pressure component (see Figure 6b). The numerical results are compared with reference CAD-geometry of the rectangle cup part (see Figure 6c,d). The positions 1 and 2 of the cup parts are considered for geometric 2D-comparison (Figure 7).

The corresponding positions of reference CAD-geometry and CAD from FEA are matched with each other and the best fit analysis of dimensional 2D-matching is carried out using the software Geomagic® QualifyTM. Regarding the bottom area (position 1) the dimensional deviation analysis shows a positive deviation of dz = 1.186 mm for conventional deep drawing and dz = 0.113 mm by process with additional force transmission, which means a geometrical improvement of the rectangle cup by 6.16% regarding the part bottom area.

Furthermore, a negative deviations of dz = -0.345 mm by conventional deep drawing and of dz = -0.294 mm by process with additional force transmission, in the flange area (position 2) are determined. This corresponds into a reduction of the flange springback by 14.78 %.

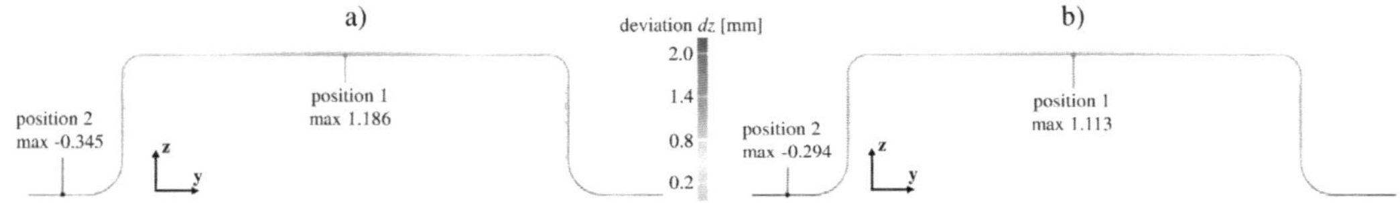

Figure 7. Geometric 2D-comparison, a) conventional process / reference CAD-geometry, b) process with additional force transmission / reference CAD-geometry,

Due to the activation of the additional force transmission, in the bottom area of the rectangle cup part, the effect of the reduction of the springback in the flange area is verified numerically. Based on numerical results an improved shaping of the bottom of the rectangle cup part can be expected by the means of additional force transmission.

4 Summary and Outlook

In order to expend the process limits by deep drawing, the development of innovative processes using the FEA is required. In earlier process design phase the tool system is developed, which allows the activation of additional force transmission in the bottom area of the drawn rectangle cup part with a counter punch. The deep drawing process with additional force transmission is investigated numerically. Here, the validated material model Hill48, based on anisotropic coefficient for describing the anisotropic material behaviour, is used. By the means of FEA a positive effect of additional force transmission, resulting in reduction of the springback in the flange area and improvement of shaping in the bottom of the rectangle cup part, can be determined numerically.

In next research step experimental investigations on the new tool system are planned. For model verification the comparison between numerical and experimental results will be carried out.

Acknowledgement

This paper is based on investigations of the project (BE169/139-1): "*Deep drawing with additional force transmission*", which is kindly supported by the German Research Foundation (DFG). The authors thank the DFG for project foundation.

References

[1] Doege E and Behrens B-A 2010 Handbuch der Umformtechnik, Springer ISBN 978-3-642-04248-5
[2] Siebel E 1953 WERKSTOFFMECHANIK, VDI-Sonderheft Werkstoffe I, VDI-Verlag Düsseldorf
[3] Otto M 2003 Erweiterung der Umformgrenzen beim Tiefziehen und Kragenziehen durch Nachschieben von Werkstoff, Dissertation, Universität Magdeburg
[4] Hill R 1948 A theory of the yielding and plastic flow of anisotropic metals, Proceedings of the Royal Society of London, Series A., Vol. **193**, 281-197.
[5] Hallfeldt T, Hotz W, Leppin, C, Keller S, Friebe H, Till E T, Müller R, Vučetić M and Vegter H 2014 Sheet Bulge Testing, Book Chapter in Comprehensive Materials Processing Volume **1** May 2014, Pages 85-93, ISBN: 978-008096533-8, Elsevier Ltd.
[6] Sigvant M, Mattiasson K, Vegter H and Thilderkvist P 2009 A viscous pressure bulge test for the determination of a plastic hardening curve and equibiaxial material data, original research, Springer, DOI 10.1007/s12289-009-0407-y
[7] Behrens B-A, Bouguecha A, Vucetic M and Peshekhodov I 2012 Characterisation of the quasi-static flow and fracture behaviour of dual-phase steel sheets in a wide range of plane stress states, in Archives of Civil and Mechanical Engineering 12 (**4**), pp. 397-406
[8] F. Yoshida, T. Uemori, "A model of large-strain cyclic plasticity describing the Bauschinger effect and workhardening stagnation", International Journal of Plasticity, PII S0749-6419(01)00050-X, 2001

IOP Conf. Series: Materials Science and Engineering **179** (2017) 012007 doi:10.1088/1757-899X/179/1/012007

Simulation model of Al-Ti dissimilar laser welding-brazing and its experimental verification

M Behúlová[1,1], E Babalová[1,2] and M Nagy[1,3]

Slovak University of Technology in Bratislava, Faculty of Materials Science and Technology in Trnava, Paulínska 16, 917 24 Trnava, Slovak Republic

E-mail: [1]maria.behulova@stuba.sk, [2]eva.babalova@stuba.sk, [3]mate.nagy@stuba.sk

Abstract. Formation of dissimilar weld joints of light metals and alloys including Al-Ti joints is interesting mainly due to demands on the weight reduction and corrosion resistance of components and structures in automotive, aircraft, aeronautic and other industries. Joining of Al-Ti alloys represents quite difficult problem. Generally, the fusion welding of these materials can lead to the development of different metastable phases and formation of brittle intermetallic compounds. The paper deals with numerical simulation of the laser welding-brazing process of titanium Grade 2 and EN AW 5083 aluminum alloy sheets using the 5087 aluminum filler wire. Simulation model for welding-brazing of testing samples with the dimensions of 50 × 100 × 2 mm was developed in order to perform numerical experiments applying variable welding parameters and to design proper combination of these parameters for formation of sound Al-Ti welded-brazed joints. Thermal properties of welded materials in the dependence on temperature were computed using JMatPro software. The conical model of the heat source was exploited for description of the heat input to the weld due to the moving laser beam source. The sample cooling by convection and radiation to the surrounding air and shielding argon gas was taken into account. Developed simulation model was verified by comparison of obtained results of numerical simulation with the temperatures measured during real experiments of laser welding-brazing by the TruDisk 4002 disk laser.

1 Introduction

Demands on the weight reduction and effort to achieve constantly better properties of components and complex large structures support the development in the field of joining of dissimilar materials. During last decades, intensive research has been targeted at the formation of joints from Al-Ti alloys [1-3]. The joining of these materials is interesting for automotive, aircraft, aeronautic, shipbuilding and other industries. However, fusion welding of Al-Ti alloys is quite difficult not only due to the large differences in melting temperatures and thermal properties of these materials but mainly owing to formation of brittle intermetallic compounds during weld metal solidification. Many methods have been investigated to join Al-Ti alloys, including arc and laser fusion welding [4-10], brazing [11-13], friction stir welding [14-16], explosion welding [17-19] or diffusion bonding [20-21] with the aim to suppress or minimize the growth of the brittle intermetallic phases.

In this paper, the laser welding-brazing process is applied to join sheets from titanium Grade 2 and EN AW 5083 aluminum alloy using the 5087 aluminum filler wire. The simulation model for numerical analysis of the dissimilar laser welding-brazing of Ti-Al alloys is developed and verified using temperature measurement during experimental joint formation.

2 Problem description

Laser beam welding-brazing process is intended to exploit for joining of two sheets with the thicknesses of 2 mm from commercially pure titanium Grade 2 and the EN AW 5083-H111 aluminum alloy. The filler wire with the diameter of 1.2 mm from the 5087 aluminum alloy was chosen as a filler material. The chemical composition of welded and filler materials is given in the Table 1 and 2, respectively. Argon with the flow rate of 18 l.min^{-1} is supposed to be used as shielding gas to protect the weld bead and weld root during laser beam welding.

Table 1. Chemical composition of the titanium Grade 2.

Element	Fe	C	O	H	N	Ti
wt. %	0.3	0.1	0.25	0.015	0.03	balance

Table 2. Chemical composition of the EN AW 5083-H111 aluminum alloy and filler material.

Element	Mg	Si	Fe	Mn	Cu	Cr	Zn	Zr	Ti	Be	Al
Al alloy	4.7	0.4	0.31	0.26	0.19	0.13	-	-	-	-	bal.
Filler	4.5 – 5.2	≤ 0.25	≤ 0.4	0.7 – 0.1	≤ 0.05	0.05 – 0.25	≤ 0.25	0.1 – 0.2	≤ 0.15	≤ 3.10^{-4}	bal.

The following welding parameters were chosen for initial trials of laser welding-brazing process:

- the laser power from 1.8 kW to 2.0 kW,
- the welding speed from 25 mm.s^{-1} to 30 mm.s^{-1} and
- the laser offset of 300 μm from the weld centerline towards the aluminum sheet.

The influence of these parameters on the quality of dissimilar joints was studied numerically applying computer simulation of welding-brazing process in programme code ANSYS and also experimentally. The welding trials were performed with continuous wave TruDisk 4002 disk laser with the maximum power of 2.0 kW, the wavelength of 1.03 μm and the beam quality (BPP) of 8 mm.mrad. Laser beam was transported to the BEO D70 focusing optics via laser light cable with the core diameter of 400 μm. The disk laser beam was focused at +2 mm above the surface of welded samples. During experiments, the temperatures were measured by two thermocouples of the K-type placed on the top side of titanium plate and bottom side of aluminum plate, respectively, in the distance of 5 mm from the weld centerline. Results of experimental temperature measurements were used for the verification of results of numerical simulations.

3 Simulation model for the analysis of temperature fields during laser welding-brazing

Transient temperature fields developed during laser welding-brazing process can be described by the heat diffusion partial differential equation in the form [22]

$$\rho c_p \frac{\partial T}{\partial t} = \left[\frac{\partial T}{\partial x}\left(\lambda_x \frac{\partial T}{\partial x} \right) + \frac{\partial T}{\partial y}\left(\lambda_y \frac{\partial T}{\partial y} \right) + \frac{\partial T}{\partial z}\left(\lambda_z \frac{\partial T}{\partial z} \right) \right] + q_v . \tag{1}$$

where $T(x, y, z, t)$ is the temperature [°C, K], t is the time [s], ρ is the density [kg.m^{-3}], c_p is the specific heat [J.kg^{-1}.K^{-1}], λ_x, λ_y, λ_z are the thermal conductivities [W.m^{-1}.K^{-1}] in the x, y, z directions of Cartesian coordinate system and q_v is the volumetric density of internal heat sources [W.m^{-3}], i. e. the heat generated in the unit volume of material per unit time. To solve the Fourier-Kirchhoff's partial differential equation (1), it is necessary to define geometrical, physical, initial and boundary conditions, i. e. geometrical shapes and dimensions of welded components, thermo-physical properties of welded materials, intensity of internal heat sources, initial temperature distribution at the beginning of the welding process and also the conditions and thermal effects at the interface of welded structures and environment.

Geometrical model consists of two plates with the dimensions of 50 mm × 100 mm × 2.0 mm. The geometry of a weld joint including the filler material (figure 1a) was proposed on the base of previous similar experiments [23]. The 3D finite element mesh (figure 1b) was generated using the element type of SOLID 70 in the ANSYS software [24].The mesh density was higher in the region of the weld pool and the heat affected zone where the largest temperature gradients are supposed. The finest mesh density perpendicular to the weld line was 0.005 mm. The length of elements in welding direction was constant with the value of 0.2 mm.

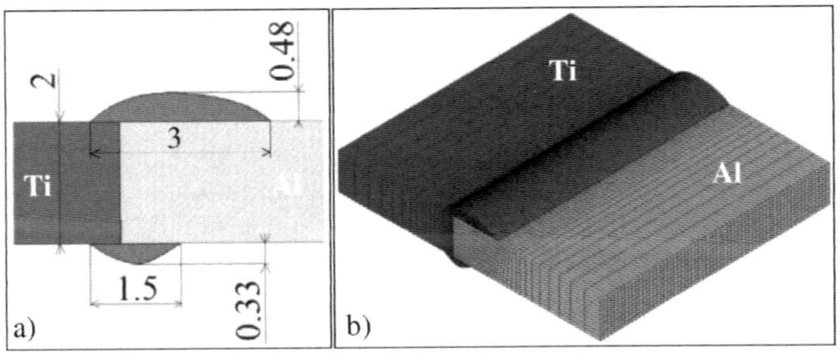

Figure 1. Geometry of the weld joint (a) and a detail from the generated FE mesh (b).

Thermo-physical properties of the Ti Grade 2, Al base material and filler wire in the dependence on temperature (figure 2) were computed using the JMatPro software [25]. Titanium Grade 2 represents commercially pure titanium (99.2 wt. %) with the solidus temperature of T_S = 1665 °C and the liquidus temperature of T_L = 1713 °C. The equilibrium solidus and liquidus temperatures for the AW 5083 alloy were computed to be 575 °C and 635 °C, respectively. The transformation temperatures for filler material are similar with the values of T_S= 573 °C and T_L= 635 °C. The enthalpy of fusion was modeled applying the method of modified specific heat [24].

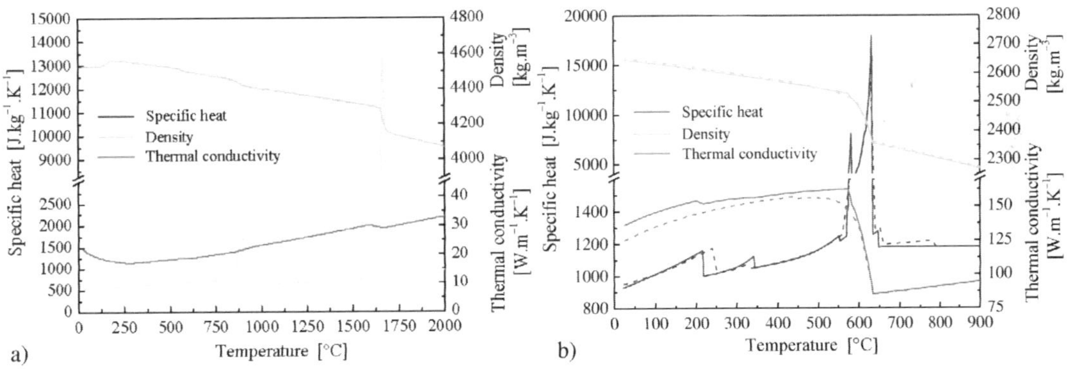

Figure 2. Thermal properties of the a) Ti Grade 2 and b) AW 5083-H111 alloy and filler material. in the dependence on the temperature

The initial condition can be written in the form

$$T(x, y, z, t = 0) = T_0(x, y, z) \tag{2}$$

where T_0 is the initial temperature of welded components. The initial temperature of plates was supposed to be 20 °C.
The boundary condition of the 3rd kind [22] was exploited for the definition of welded plates cooling by convection and radiation to the argon shielding gas and surrounding air

$$-\lambda \frac{\partial T}{\partial \mathbf{n}} = h(T_w - T_f) \tag{3}$$

where T_w [°C, K] is the sample surface temperature, T_f [°C, K] is the surrounding fluid temperature and h [W.m^{-2}.K^{-1}] represents the combine heat transfer coefficient given by the sum of convection heat transfer coefficient h_C and radiation heat transfer coefficient h_R

$$h = h_C + h_R = h_C + \frac{\varepsilon \sigma_0 \left(T_w^4 - T_f^4\right)}{T_w - T_f} \tag{4}$$

where ε [-] is the emissivity and σ_0 is the Stefan-Boltzmann's constant. The values of the combine heat transfer coefficient for the top and bottom surfaces of welded plates were computed using classical criterial equations [22] and entered to the ANSYS software in the dependence on the surface temperature.

The heat input in fusion welding can be modeled by different methods including moving point or line heat sources, surface or volumetric heat sources. For presented numerical simulation of laser welding-brazing of Ti-Al plates, the 3D conical volumetric heat source model (figure 3) was applied. The volumetric density of internal heat source q_v [W.m^{-3}] is given by the function [26]:

$$q_v(x, y, z) = \frac{9\eta Q_0 e^3}{\pi (e^3 - 1)} \frac{1}{(z_e - z_i)(r_e^2 + r_e r_i + r_i^2)} \exp \frac{3(x^2 + y^2)}{r_0^2(z)} \tag{5}$$

where Q_0 [W] is the maximum heat source intensity, η is the efficiency, r_e [m] and r_i [m] are the surface radii in planes, $z = z_e$ and $z = z_i$, respectively and x, y, z are the coordinates of the instant position of the heat source.

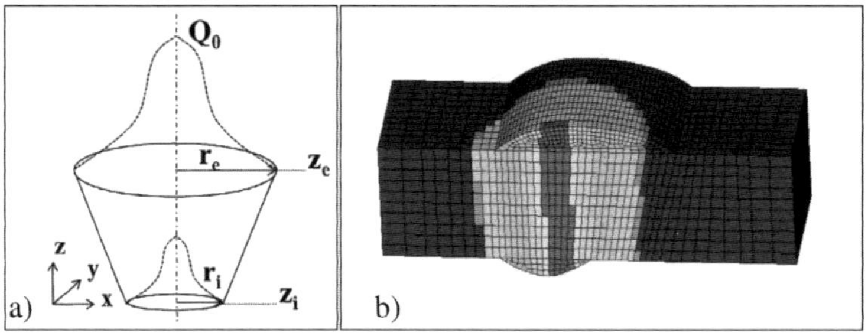

Figure 3. 3D conical heat source [27] and the scheme of its application to the FE model.

4 Results of numerical simulation and their comparison with experiments

Numerical simulations of laser beam welding-brazing of plates from dissimilar Ti-Al alloys were performed using the ANSYS software with the laser power of 1.8 kW and the welding speed of 25 mm.s^{-1} in the experiment 1 and the laser power of 2.0 kW and the welding speed of 30 mm.s^{-1} in the experiment 2. In both cases the laser offset was set to 300 μm from the weld centerline towards the aluminum sheet. Adjusting of parameters defining the conical heat source model led to setting the values of r_e and r_i to 1.3 mm and 1.1 mm, respectively. According to the previous experience, the efficiency of laser heat source was supposed to be 63 %.

For illustration, the details of computed temperature fields in the time of 0.6 seconds are shown in figure 4. The top views depict the asymmetric temperature distribution in the in the Al and Ti plates developed during the laser welding-brazing process. The asymmetry of temperature fields results mainly from the differences in thermal properties of welded materials. Some influence can be assigned also to the laser beam offset towards the Al plate. For both considered welding speeds, the laser power seems to be too high as the maximum temperatures exceed the liquidus temperature of both joint materials. The width of molten zone is comparable for both experiments as it can be seen from the temperature fields in the cross-sectional area in figure 4.

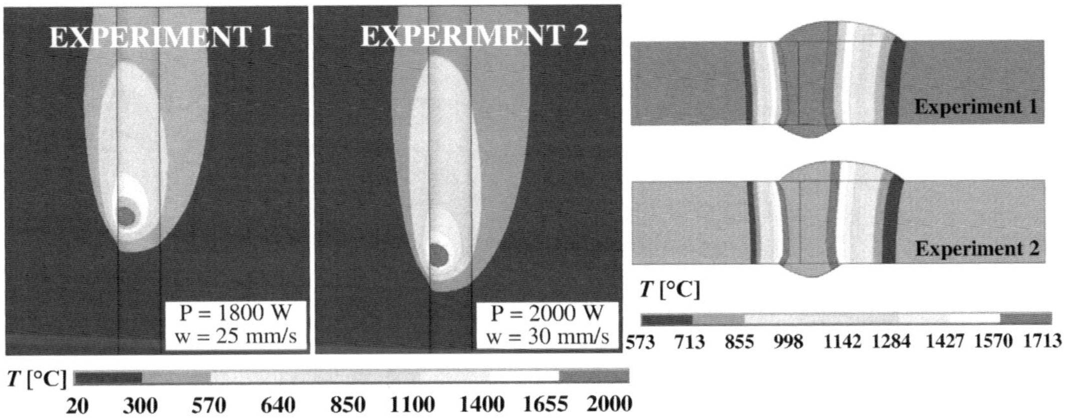

Figure 4. Details from the temperature fields in the time of 0.6 seconds – top view and the weld cross-sections perpendicular to the weld line.

Figure 5 illustrating the macrostructures of welded-brazed joints prepared in experiment 1 and 2 confirms the melting-down both joint materials. It is necessary to note that the weld joint during successful welding-brazing of Ti Grade 2 and AW 5083-H111 aluminum alloy should be formed by molten filler material and Al-alloy and molten filler and solid titanium. By this way it is possible to avoid the development of brittle intermetallic compounds during rapid solidification of molten and mixed Ti and Al alloys. In this reason, the decrease in laser power is recommended for following computations and also experimental joining if the welding speed remains in the interval from 25 mm.s^{-1} to 30 mm.s^{-1}. Another possibility is to increase the welding speed. This alternative can lead to the positive consequences concerning reduction of molten and heat affected zones. Moreover, the laser beam offset from the weld centerline towards the aluminum sheet which was 300 μm can be increased to avoid melting of the Ti Grade 2 sheet.

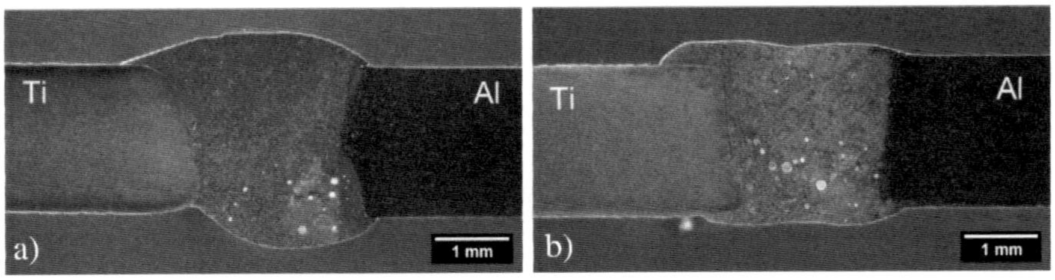

Figure 5. Macrostructures of welded-brazed joints prepared with different welding parameters in a) experiment 1 and b) experiment 2.

In figure 6, the time dependences of computed and measured temperatures for the experiment 1 and experiment 2 are plotted. Comparison of computed and experimental results indicates a good agreement between the numerically and experimentally obtained temperatures on both, Al and Ti sides. However, the computed temperatures are slightly lower than the measured ones. The temperature differences result mainly from the accuracy of the thermocouples location. The next cause for higher measured temperatures is the higher initial and surrounding temperature during the experiments realized in hot summer-time.

In spite of good compliance of computed and experimentally measured temperatures, application of additional thermocouples placed at the distances closer to the weld centerline is advised for the following experiments in order to record the more accurate temperature profile during laser welding-brazing process.

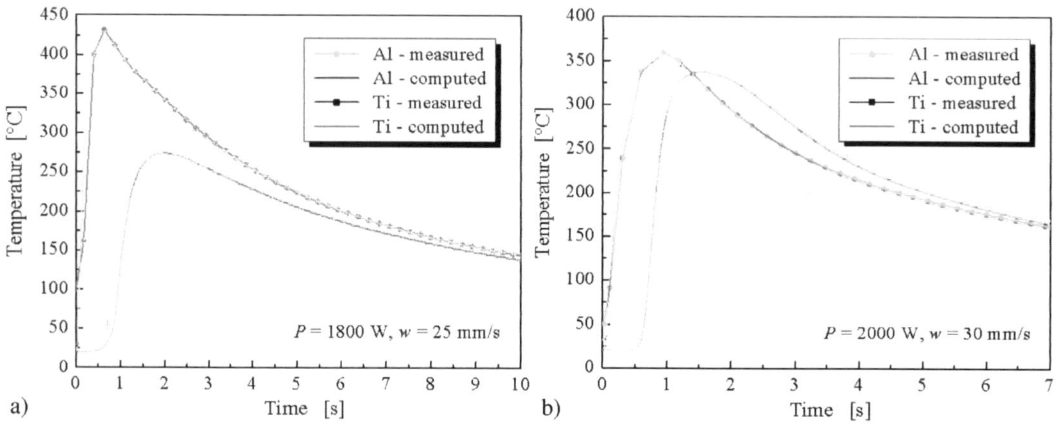

Figure 6. Comparison of computed and measured temperatures during the welding-brazing Ti-Al alloys a) experiment 1 and b) experiment 2.

5 Conclusions

The simulation model for the numerical analysis of the laser welding-brazing of sheets from titanium Grade 2 and EN AW 5083 aluminum alloy using the 5087 aluminum filler wire was developed and verified by the temperature measurements during the laser welding-brazing of experimental samples by the TruDisk 4002 disk laser using the values of laser power from 1800 W to 2000 W and welding speeds of 25 $mm.s^{-1}$ and 30 $mm.s^{-1}$, respectively. For both compared cases, a good agreement between computed and measured temperatures was achieved.

In the next work, the described simulation model will be used for the investigation of the influence of welding parameters on the temperature fields developed during the laser welding-brazing of dissimilar Ti Grade 2 and EN AW 5083 aluminum alloy plates in order to design optimal parameters for formation of sound dissimilar weld joints.

Acknowledgement

The research has been supported by the VEGA SR within the project No. 1/1010/16 and the Program to support excellent young research teams (PRO-MAT-ZVAR) of the STU Bratislava.

References

[1] Martinsen K, Hu S J and Carlson, B E 2015 Joining of dissimilar materials. *CIRP Ann-Manuf Techn* **642** 679-699. ISSN 0007-8506

[2] Kalaiselvan k and Elango A 2014 Laser beam welding of Ti/Al dissimilar thin sheets - a literature review *JAMME* **671** 39-44. ISSN 1307-6892

[3] Sánchez Amaya J M, Amaya-Vázquez M R and Botana F J 2013 Laser welding of light metal alloys: aluminium and titanium alloys. In: S. Katayama (Ed.) Handbook of Laser Welding Technologies, Woodhead Publishing Series in Electronic and Optical Materials (2013) 215. ISBN 13: 9780857092649

[4] Rao A. S., Reddy G M and Prasad K S 2013 Microstructure and tensile properties of dissimilar metal gas tungsten arc welding of aluminium to titanium alloy. *Mater Sci Tech* **27** (1) 65-70. ISSN 0267-0836

[5] Li U K, Li Y, Wei S and Wang J 2014 Interfacial Microstructural Characterization of Ti/Al Joints by Gas Tungsten Arc Welding *Mater Manuf Process* **29** (8) 969-974. ISSN 1532-2475

[6] Wei S, Li Y, Wang J and Liu K 2014 Influence of Welding Heat Input on Microstructure of Ti/Al Joint During Pulsed Gas Metal Arc Welding *Mater Manuf Process* **29** (8) 945-960. ISSN 1042-6914

[7] Tomashchuk I et al 2014 Direct keyhole laser welding of aluminum alloy AA5754 to titanium alloy Ti6Al4V *J Mater Process Tech* **217** 96-104. ISSN 0924-0136

[8] Majumdar B, Galun R, Weisheit A and Mordike B L 1997 Formation of a crack-free joint between Ti alloy and Al alloy by using a high-power CO2 laser *J Mater Sci* **32** 6191-6200. ISSN 0022-2461

[9] Lee S J, Takahashi M, Kawahito Y and Katayama S 2015 Microstructural Evolution and Characteristics of Weld Fusion Zone in High Speed Dissimilar Welding of Ti and Al *Int J Precis Eng Man* **16** (10), 2121-2127. ISSN 2234-7593

[10] Chen S H, Li L Q and Chen Y B 2013 Interfacial reaction mode and its influence on tensile strength in laser joining Al alloy to Ti alloy *Mater Sci Tech* **26** (2) 230-235. ISSN 1743-2847

[11] Chen S, Li L, Chen Y and Huang J 2011 Joining mechanism of Ti/Al dissimilar alloys during laser welding-brazing process *J Alloys Comp* **509** (3) 891-898. ISSN: 0925-8388

[12] Chen Y, Chen S and Li L 2009 Effects of heat input on microstructure and mechanical property of Al/Ti joints by rectangular spot laser welding-brazing method *Int J Adv Manuf Technol* **44** 265-272. ISSN 0268-3768

[13] Möller F et al 2011 Combined Laser Beam Welding and Brazing Process for Aluminium Titanium Hybrid Structures *Phys Procedia* **12** 215-223. ISSN 1875-3892

[14] Dressler U, Biallas G, Mercado U A 2009 Friction stir welding of titanium alloy TiAl6V4 to aluminium alloy AA2024-T3 *Mater Sci Eng* **526** (1-2) 113-117. ISSN 0921-5093

[15] Chen Y, Ni Q, Ke L 2012 Interface characteristic of friction stir welding lap joints of Ti/Al dissimilar alloys *T Nonferr Metal Soc* **22** (2) 299-304. ISSN 1003-6326

[16] Song Z, Nakata K, Wu A, Liao J, Zhou L 2014 Influence of probe offset distance on interfacial microstructure and mechanical properties of friction stir butt welded joint of Ti6Al4V and A6061 dissimilar alloys *Mater Design* **57** 269-278. ISSN 0261-3069

[17] Fronczek D et al 2016 Microstructure Changes and Phase Growth Occurring at the Interface of the Al/Ti Explosively Welded and Annealed Joints *J Mater Eng Perform* **25** 3211. ISSN 1059-9495

[18] Gurevich L M, Trykov Y P, Kiselev O S 2013 Formation of structural and mechanical inhomogeneities in explosion welding of aluminium to titanium *Weld Res Int* **28** (2) 128-132. ISSN 1754-2138

[19] Xia H, Wang S, Ben H 2015 Microstructure and mechanical properties of Ti/Al explosive cladding *Mater Design* **56** 1014-1019. ISSN 0261-3069

[20] Rajakumar S, Balasuramanian V 2016 Diffusion bonding of titanium and AA 7075 aluminum alloy dissimilar joints *Int J Adv Manuf Tech* **86** 1095. ISSN 1433-3015

[21] Ren J W, Li Y, Feng T 2002 Microstructure characteristics in the interface zone of Ti/Al diffusion bonding *Mater Lett* **56** 647-652. ISSN 0167-577X

[22] Incropera P F and De Witt D P Fundamentals of heat and mass transfer. J. Wiley and Sons, New York, 1996. ISBN: 0471304603

[23] Sahul M, Sahul Ma, M. Vyskoč M and Pašák M 2016 Disk laser weld brazing of AW5083 aluminium alloy with titanium Grade 2 *J Mater Eng Perform* (in print)

[24] ANSYS Mechanical APDL Theory Reference. ANSYS, Inc. Release 15.0. SAS IP, Inc., 2013.

[25] JMatPro Help, release 6.1, Sente Software Ltd., 2012.

[26] Wu C S, Wang G, Zhang Y M 2006 A new heat source model for keyhole plasma arc welding in FEM analysis of the temperature profile *Weld J* **85** 284-291. ISSN 0043-2296

[27] Chukkan J R, Vasudevan M, Muthukumaran S, Kumar R R, Chandrasekhar N 2015 Simulation of laser butt welding of AISI 316L stainless steel sheet using various heat sources and experimental validation *J Mater Process Tech* **219** 48-59. ISSN 0924-0136

Numerical simulation of multi-rifled tube drawing – finding proper feedstock dimensions and tool geometry

P Bella[1], **P Buček**[1], **M Ridzoň**[1], **M Mojžiš**[1,2], **Ľ Parilák**[1,3]

[1]ŽP VVC s.r.o., Kolkáreň 35, 976 81, Podbrezová;
[2]Ústav výrobných technológií, Materiálovotechnologická fakulta STU (Trnava), Bottova 25, 917 24, Trnava;
[3]Katedra výrobných technológií, Fakulta výrobných technológií (Prešov), Technická univerzita v Košiciach

bella@zelpo.sk

Abstract. Production of multi-rifled seamless steel tubes is quite a new technology in Železiarne Podbrezová. Therefore, a lot of technological questions emerges (process technology, input feedstock dimensions, material flow during drawing, etc.) Pilot experiments to fine tune the process cost a lot of time and energy. For this, numerical simulation would be an alternative solution for achieving optimal parameters in production technology. This would reduce the number of experiments needed, lowering the overall costs of development. However, to claim the numerical results to be relevant it is necessary to verify them against the actual plant trials. Searching for optimal input feedstock dimension for drawing of multi-rifled tube with dimensions Ø28.6 mm × 6.3 mm is what makes the main topic of this paper. As a secondary task, effective position of the plug – die couple has been solved via numerical simulation. Comparing the calculated results with actual numbers from plant trials a good agreement was observed.

1 Introduction

Multi-rifled tubes, i.e. tubes with multiple inner grooving are used in heat exchangers and boilers. Over the last few years the effect of multi-rifled boiler tubes has been highly valued mainly in power generation at super critical pressures in coal-fired power plants. Rifled tubes are used in heat exchangers and boilers to provide highly efficient heat transfer.

In high pressure boilers, air bubbles are formed in the liquid close to the inner surface of the tubes. This is caused by the heat transfer from the tube walls and tends to create a steam film between the tube wall and the heated liquid. This steam film inhibits heat transfer between the tube and the liquid. If this exceeds a certain limit, the temperature of the tubes will increase rapidly, leading to the damage of the tubes. In order to prevent steam film formation on the tube walls, the mass flow rate of the liquid should be increased or the fluid should swirl to maintain a good contact between the liquid and the tube wall [1]. Rifled tubes in a conventional boiler provide a larger margin of safety for natural circulation and assisted circulation along with many economic advantages such as: reduction of weight for downcomers, headers, drum and connecting pipe, reduction of pumping power, and reduction of framework weight [2, 3]. The presence of internal rifling induces centripetal forces in the helical mass flow, thereby separating the water from the steam fraction and forcing the water towards the tube wall. As a result of this phenomenon, following advantages appear: improved heat transfer rate even at higher steam quality levels, very good heat transfer even at low mass flux levels, reduction in mean metal temperature of the tube wall, ability to increase heat transfer rate by optimizing the rifling geometry.

The multi-rifled precision seamless steel tubes are produced in Železiarne Podbrezová by means of cold drawing, utilizing multiple drawing operations (sequences) with intermediate heat treatment. The final drawing operation produces a typical grooving pattern on inner surface of the tube by means of a multi-rifled plug. During production it is very important to optimize the following geometrical parameters: rib height, rib width, and lead angle. The rifling itself has a key effect on flow regime inside the tube.

During production runs it is very important to establish and maintain four most crucial factors:

a) dimensions of input tubular feedstock,
b) drawing velocity,
c) geometry of the multi-rifled plug,
d) position (i.e. axial offset) of the plug with respect to the die.

After the ribbing, the tubes are heat treated to meet the specified metallurgical requirements. Before cutting to final lengths, the tubes are 100% checked by eddy current method and /or ultrasonic test, depending on customer's requirements. Resolving all these factors is time consuming and therefore expensive, therefore some preliminary decisions can be made with the help of numerical simulation [4, 5].

This mail goal of this article was to develop a suitable numerical model (**Figure 1**) of multi-rifled tube drawing with output dimensions of Ø28.6 mm × 6.3 mm. As a secondary task, the optimum position of the multi rifled plug with respect to the drawing die has been investigated. Both tasks have been elaborated using commercial FEM software package.

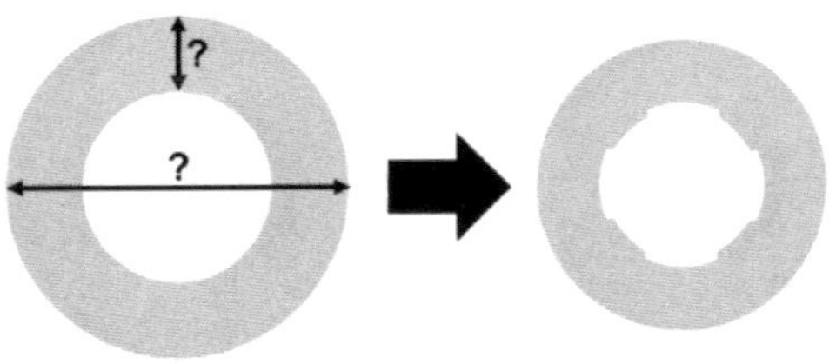

Figure 1. Input feedstock → multi-rifled tube.

2 Material, simulation and experiment

The drawing process was modelled in DEFORM-3D using a finite element model. The objects considered for modelling were: input feedstock, drawing die, multi-rifled plug, and the drawing carriage [6, 7]. The input feedstock dimensions were limited by the geometry of the preliminary forming tools. Five types of input feedstock were selected as a candidates for numerical simulation (**Table 1**).

Table 1. Models of input feedstock for numerical simulations.

Feedstock model number	[Outer Diameter] x [Wall Thickness]
1	34 x 7,0
2	35 x 7,5
3	36 x 7,5
4	36 x 8,0
5	36 x 8,5

Simulation parameters and settings were exactly the same for all models considered. The input feedstock was defined as a plastic object, which is suitable for modelling of cold forming processes under certain conditions. A piece of tube with ¼ symmetry was 70 mm long and described by 30 000 linear tetrahedrons. Smaller elements were created on the inner tube surface where high intensity of plastic deformation is takes place. After the drawing simulation the number of elements was increased 10-folds. This was necessary to ensure the strict dimensional requirements on inner grooving of the tube (ribs, grooves, rounded edges, etc.) This approach is very effective and less time consuming, as it requires fewer elements to describe less important regions, placing finer elements to the more important regions according to the meshing weights [8, 9]. The forming tools (die, plug, carriage) were considered rigid and wear-free objects. Initial model setup is shown in **Figure 2**. During drawing, the multi-rifled plug revolves around its own axis due to the helical shape of the grooves, determining the flow of the material. The Coulomb friction coefficient between the die and the tube and between the tube and the plug was estimated to be 0.08. The FEM formulation was of Lagrangian incremental type [10]. The drawing carriage has its velocity of drawing (v = 30 mm/s). The drawing carriage and the tube were connected through non-separable (so called "sticking") condition. The position of the plug with respect to the die was so that the plug face flushes the end of the bearing length.

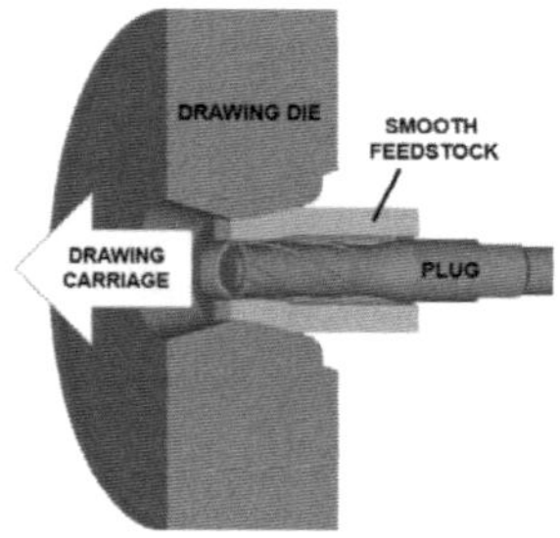

Figure 2. Initial model setup for numerical simulation of multi-rifled tube drawing.

3 Results and discussion

3.1 *Models of input feedstock*

Final dimensions of multi-rifled tubes from numerical simulation were measured in cross and longitudinal sections, as specified in **Figure 3**. The results are in **Table 2**.

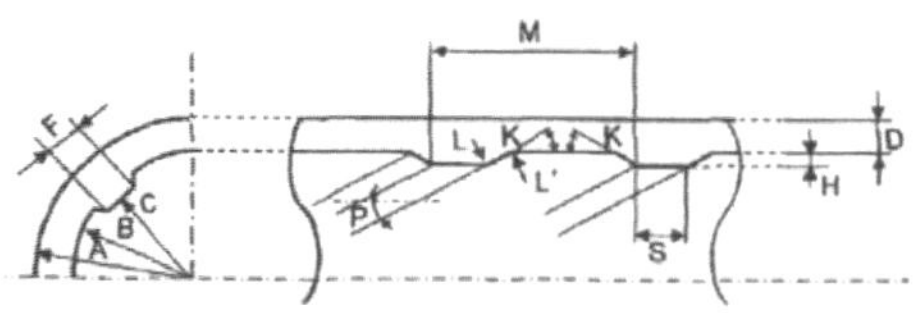

Figure 3. Key dimensions of multi-rifled tube – cross and longitudinal section.

Table 2. Models of input feedstock for numerical simulations.

Label	Description	Units	Nominal value	Tolerance	(34x7.0)	(35x7.5)	(36x7.5)	(36x8.0)	(36x8.5)
A	Outer diameter	mm	28.60	± 0.15	28.44	28.35	28.43	28.50	28.41
B	Inner diameter	mm	16.06	± 0.15	16.04	16.02	16.01	16.02	16.03

D	Wall thickness	mm	6.27	± 0.57	6.23	6.17	6.22	6.17	6.18
F	Rib width (cross)	mm	4.80	± 0.6	4.93	4.98	4.90	4.51	4.41
S	Rib width (longitudinal)	mm	8.30	± 1.04	7.80	9.22	9.05	8.98	9.16
H	Rib height	mm	0.70	± 0.15	0.30	0.52	0.56	0.71	0.71
K	Rib side angle	°	54	-10 / +15	11-15	45-60	55	50 - 56	56-68
L	Rib radius	mm	0.30	min 0.13 max 0.60	3.00	0.43	0.41	0.37	0.58
L´	Rib radius	mm	0.30	min 0.13 max 0.60	2.00	0.65	0.43	0.27	0.42
M	Rib pitch	mm	21.85	± 3.18	21.9	22.17	22.08	22.57	23.43

Model No. 1 (34×7.0):
The grooving profile showed the worst filling up. The rib height was just 0.3 mm and other dimensions (K, L, L´) were well off the tolerance, too. Larger diameter and thicker wall of the feedstock are necessary to achieve desired dimensions.

Model No. 2 (35×7.5):
A slightly better material flow was observed, but we still haven't reached all the dimensions specified. The rib height, rib radius and other dimensions were out of tolerance (red font values in **Table 2**).

Model No. 3 (36×7.5):
Similar results as in model No. 2. The rib height was at the lower tolerance limit. The rest of the dimensions were acceptable.

Model No. 4 (36×8.0):
All dimensions are within the tolerances.

Model No. 5 (36×8.5):
The material flow was similar to the model No. 4, but the rib angle and rib radius were just below the tolerance limit. Other dimensions were out of tolerance.

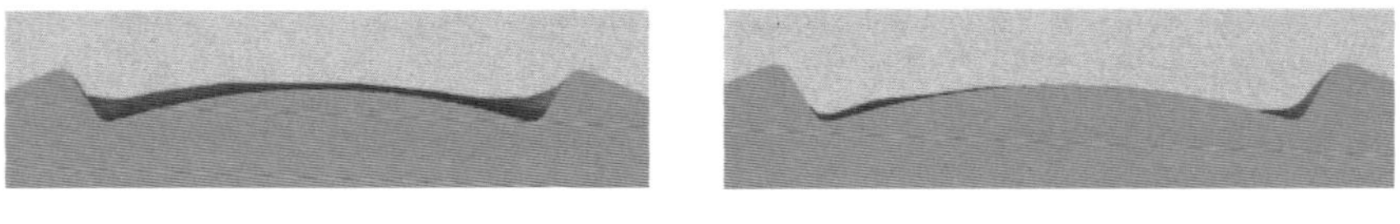

Figure 4. Cross section of multi-rifled tube for model No. 1 (left) and model No. 4 (right).

3.2 *Position of the plug with respect to the die*

The position of the plug has a profound effect on the material flow because of the change in contact area between the plug and the tube. We have evaluated 3 different plug positions (**Figure 5**) to obtain the best material flow. Simulation settings were the same as in previous cases and the input feedstock used came from model No. 4 (36×8.0).

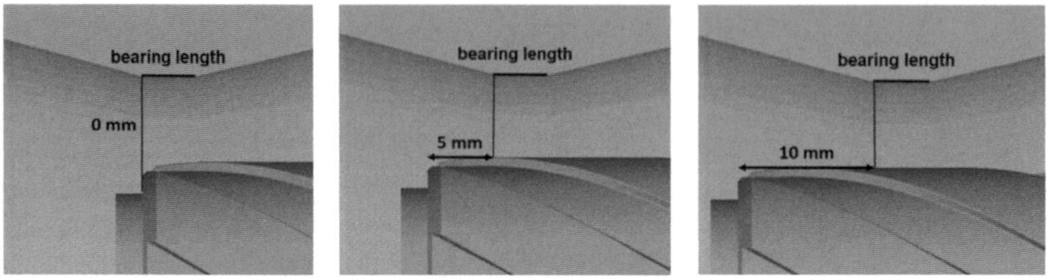

Figure 5. Position of the plug with respect to the die.

0mm plug-die axial offset:
The contact length was 11 mm (**Figure 6**, left). This seemed to be the ideal position for all the dimensions were within tolerances (see **Table 3**). The material flow (and thus deformation) was rather uniform (see **Figure 7**, centre).

5mm plug-die axial offset:
The contact length was 16.5 mm (**Figure 6**, centre). Dimensions were outside the tolerances.

10mm plug-die axial offset:
The contact length was 22 mm (**Figure 6**, right). The material flow is depicted in **Figure 7** (top). This particular configuration position is again undesirable for there was the "rib radius" violation: the material flow (and thus deformation) at the inner tube surface was quite heterogeneous. Higher deformation was observed in location L´ whereas insufficient deformation (thus insufficient groove filling) was observed in location L.

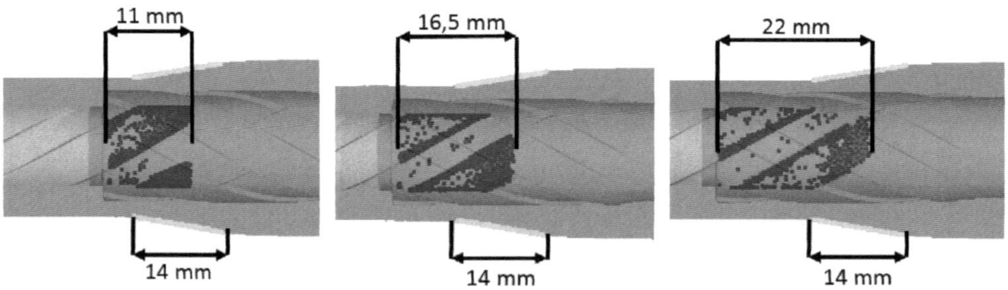

Figure 6. Calculated contact area between the tube and the forming tools for various plug-die axial offsets.

Table 3. Comparison of calculated tube dimensions according to various plug-die axial offsets

Label	Description	Units	Nominal value	Tolerance	*0 mm*	*5 mm*	*10 mm*
K	Rib side angle	°	54	-10 / +15	50 - 56	42 - 50	31 - 45
L	Rib radius	mm	0,3	min 0,13 max 0,60	0,37	0,62	1,1
L´	Rib radius	mm	0,3	min 0,13 max 0,60	0,27	0,45	0,43

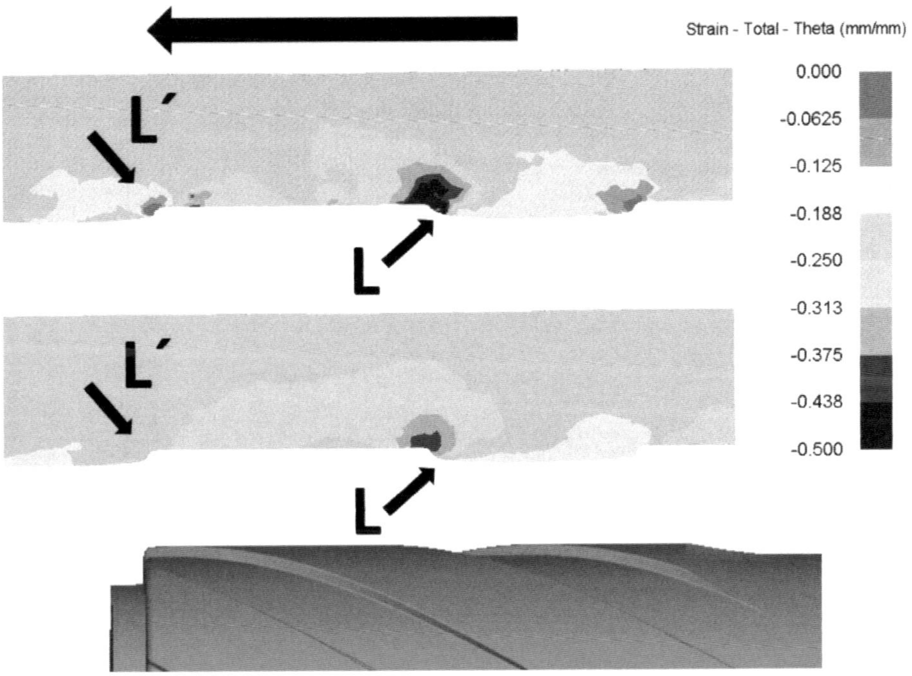

Figure 7. Longitudinal section of a multi-rifled tube: 10mm plug-die offset (top), 0mm plug-die offset (centre).

3.3 *Model validation*

Based on simulation results the obvious winner of the contest is the model No. 4. The plug-die axial offset was set to 0mm. To validate our numerical results, experimental tube drawing took place in Železiarne Podbrezová tube drawing plant.

As an input feedstock, Grade-T12 steel tube with dimensions of Ø36 × 8 mm was used. The drawing velocity was 1.85 m/min. The experiment was conducted at room temperature. The tube dimensions after drawing on a multi-rifled plug were measured with 3D optical scanner GOM ATOS II and the data obtained were processed in order to obtain reverse-engineered CAD model of the tube [11, 12]. Dimensions of the multi-rifled tube, both experimental and simulated, were measured in the cross and longitudinal section of the tube, as specified in **Figure 3**. Experimental and simulated values are compared in **Table 4**.

Table 4. *Comparison of experimental and simulation results*

	Description	Units	Nom. value	Tolerance	VALIDATION		Difference [mm]	Difference [%]
					3D scanning	Num. simulation		
A	Outer diameter	mm	28,6	±0,15	28,52	28,50	-0,02	-0,07
B	Inner diameter	mm	16,06	±0,15	16,00	16,02	0,02	0,13

D	Wall thickness	mm	6,27	±0,57	6,3	6,17	-0,13	-2,06
F	Rib width (cross)	mm	4,8	±0,6	4,34	4,51	-0,17	-3,92
S	Rib width (longitudinal)	mm	8,3	± 1,04	9,38	8,98	-0,4	4,26
H	Rib height	mm	0,7	± 0,15	0,71	0,71	0	0,00
K	Rib side angle	°	54	-10 / +15	67	50 – 56	- 6	8,96
L	Rib radius	mm	0,3	min 0,13 max 0,60	0,33	0,37	- 0,06	18,18
L′	Rib radius	mm	0,3	min 0,13 max 0,60	0,34	0,27	- 0,13	17,65
M	Rib pitch	mm	21,85	± 3,18	22,26	22,57	+ 0,31	1,39
P	Lead angle	°	30		25,20	26,30	+ 1,1	4,37

4 Conclusion

Solving technological optimization problems by means of numerical simulation proves to be a vital approach towards further innovations in tube production.

In this paper, numerical simulation approach was used to estimate the most suitable input feedstock dimensions along with optimum plug-die position for production of multi-rifled tubes with given dimensions. Furthermore, calculated dimensions from numerical simulation were compared with the actual tube dimensions from experimental drawing trials. The results agreement was acceptable, further promoting the idea of numerical simulation as the vital part of production technology design process in Železiarne Podbrezová.

References

[1] Mohankumar T C and Thomachan N 2013 CFD Studies on Multi Lead Rifled [MLR] Boiler Tubes, *Journal of Engineering Research and Applications*, Vol. 3, No. 5, pp. 24-26.

[2] Ajay N I 2012 CFD Analysis Of Super Heater In View Of Boiler tube Leakage, *International Journal of Engineering and Innovative Technology (IJEIT)*, Vol. 1, No. 3, pp. 29-31.

[3] Gangadhara D, Kumar K and Naveen K 2016 Numerical Analysis of Heat Transfer Enhancement in a Boiler Tube, *The International Journal of Innovative Research in Science, Engineering and Technology,* Vol. 5, No. 3, pp. 3417-3425.

[4] Meindersa T, Burchitza A, Bontea A and Lingbeeka R A 2008 Numerical product design, Springback prediction, compensation and optimization. *International Journal of Machine Tools and Manufacture*. Vol. 48, Issue 5, pp. 499–514.

[5] Tamizharasan T and Senthil K S 2002 Optimization of cutting insert geometry using DEFORM - 3D, Numerical simulation and experimental validation. *Original scientific paper*. ISSN 1726-4529. Vol. 11 No. 2, pp. 65-76.

[6] Bella P et al 2011 Numerical Simulation of Extrusion Elbows. *Acta Metallurgica Slovaca*, Vol. 17, No. 3, pp. 200-206.

[7] Kočiško R et al 2015 The influence of ECAP geometry on the effective strain distribution. *Advanced Materials Research,* Vol. 1127, pp. 135-141.

[8] Neves F O, Button S T, Caminaga C and Gentile F C 2005 Numerical and Experimental Analysis of Tube Drawing With Fixed Plug, *J. of the braz. soc. of mech. sci. & eng,* Vol. 27, No. 4, pp. 426 – 431.

[9] Rajput B S and Victor R 2014 Study of Different Parameters in Seamless Cold Drawing Process Using Finite Element Simulation, *international journal of engineering sciences & research* technology. Vol. 3, ISSN: 2277-9655, pp. 437–446.

[10] Swiatoniowski A, Szostak J and Gomulka K 2014 Analysis forming the diameter reduction of the tube to the cylinder hydraulic shock absorber, *Journal of KONES Powertrain and Transport*, Vol. 21, No. 1, pp. 281 – 286.

[11] Morovic L and Milde J 2015 CAD Model Created from Polygon Mesh. *Applied Mechanics and Materials*, ISSN: 1662-7482, Vol. 808, pp. 233-238.

[12] Ridzoň M et al 2012 The Effect of Technological Parameters Influencing the Properties of Seamless Cold-Drawn Tubes, *1st ed. (Scientific monographs) Köthen: Hochschule Anhalt*. ISBN 978-3-86011-048-5.

IOP Conf. Series: Materials Science and Engineering **179** (2017) 012009 doi:10.1088/1757-899X/179/1/012009

New heat treatment process for advanced high-strength steels

D Bublíková, Š Jeníček, I Vorel, B Mašek

University of West Bohemia, RTI- Regional Technological Institute, Univerzitní 22, CZ – 306 14 Pilsen, Czech Republic

E-mail: dagmar.bublikova@seznam.cz

Abstract

Today's advanced steels are required to possess high strength and ductility. It can be achieved by choosing an appropriate steel chemistry which has a substantial effect on the properties obtained by heat treatment. Mechanical properties influenced the presence of retained austenite in the final structure. Steels of this group typically require complicated heat treatment which places great demands on the equipment used. The present paper introduces new procedures aimed at simplifying the heat treatment of high-strength steels with the use of material-technological modelling.

Four experimental steels were made and cast, whose main alloying additions were manganese, silicon, chromium, molybdenum and nickel. The steels were treated using the Q-P process with subsequent interrupted quenching. The resulting structure was a mixture of martensite and retained austenite. Strength levels of more than 2000 MPa combined with 10-15 % elongation were obtained. These properties thus offer potential for the manufacture of intricate closed-die forgings with a reduced weight. Intercritical annealing was obtained structure not only on the basis of martensite, but also with certain proportion of bainitic ferrite and retained austenite.
Keywords: Q&P process, retained austenite, AHSS, intercritical annealing

1 Introduction

Today, there is a strong demand for simplified heat treatment of high-strength low-alloy steels. Modern heat treatment routes include processes that lead to TRIP-type microstructure, and Q&P processing. The concept of TRIP steels relies on a mixture of bainite, ferrite and retained austenite [Fig. 1] which is produced by intercritical annealing followed by controlled cooling and isothermal holding at the bainitic transformation temperature [1]. This route can provide ultimate strengths up to 1500 MPa and elongation levels of about 25 %. The other method, the quenching and partitioning (Q&P) process, allows strengths of more than 2000 MPa to be achieved, together with an elongation level of about 10 % [2]. Neither of them can be developed under real-world conditions. The reason is that both of them involve complex processes, in which a number of parameters must be optimized. In such case, material-technological modelling carried out in a thermomechanical simulator becomes a useful tool [Fig. 3]. With this method, small amounts of material can processed under laboratory conditions approaching the real-world process.

Stabilisation of retained austenite is an important factor in both of these heat treatment routes [Fig. 2]. Q&P processing can lead to high fractions of retained austenite. It is characterized by rapid cooling from austenite region to a temperature between the M_s and M_f temperatures, during which martensite forms, while a portion of austenite remains untransformed. During subsequent isothermal holding, the retained austenite becomes stabilised thanks to carbon which migrates from the super-saturated martensite to austenite. According to current knowledge, this austenite, which forms during the Q&P process, exists primarily in the form of thin films between martensite laths or plates [3]. It differs in this respect from the granular austenite that exists in steels which were heat treated to obtain the TRIP microstructure [4]. In high-strength steels, final microstructure and mechanical properties upon heat treatment are dictated mainly by their alloying concept. In order to stabilise retained austenite, it is important to increase the nickel content. To shift the M_s and M_f temperatures toward lower values, high-strength steels are alloyed with molybdenum and manganese. In addition, the levels of chromium and silicon are important for achieving sufficient mechanical properties.

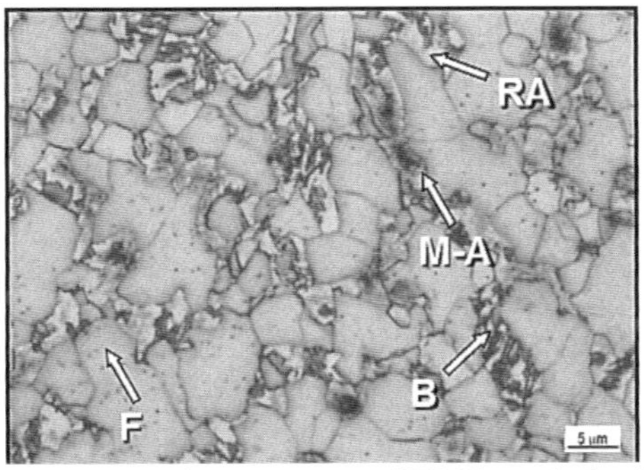

Figure 1. Microstructure of TRIP steel obtained by intercritical annealing.

Figure 2. Prior austenite grain identified in a martensitic matrix after Q&P processing.

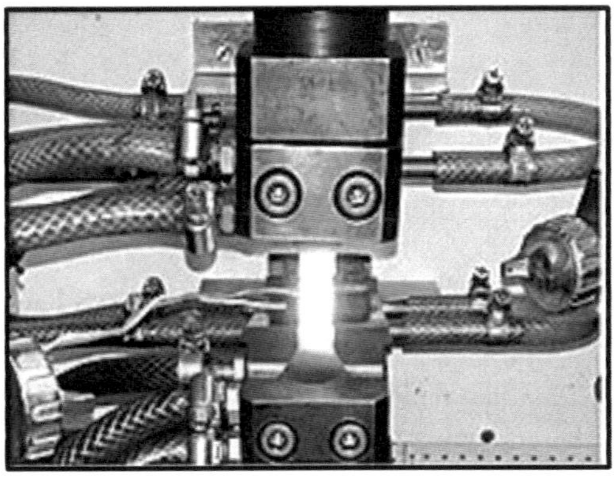

Figure 3. Physical simulation in a thermomechanical simulator.

2 Experimental

For this experiment, four new experimental steels were created whose special chemistries lower the M_s and M_f temperatures (Tab. 1). The aim was to carry out Q&P processing and heat treatment for TRIP microstructure on these steels to ascertain whether these chemistries can lead to different final microstructures, and to acceptable amounts of stabilised retained austenite.

In all these steels, the M_s and M_f temperatures were depressed predominantly through additions of manganese, silicon, and chromium. Silicon was chosen in order to prevent carbide formation and thus to provide adequate super-saturation of martensite with carbon. Molybdenum was employed to reduce the M_s and M_f temperatures and to shift the start of ferritic and pearlitic transformations towards lower cooling rates. Nickel was added in small amounts to stabilise austenite during cooling, to enhance hardenability, and to provide solid solution strengthening. The carbon content was the same in all steels: between 0.42 and 0.43 %.

These chemistries were developed with the aid of the JMatPro software which had been used for calculating the approximate transformation temperatures. In the AHSS-1 steel, the manganese level was 2.5 %, and the silicon level was 2.03%. The calculated M_s and M_f temperatures were 218 °C and 88 °C, respectively. In order to find whether molybdenum affects mechanical properties and transformation temperatures, its content in the AHSS-2 steel was set to 0.16%. However, this molybdenum content has not altered the M_s and M_f temperatures in any substantial way. The M_s and M_f temperatures were 214°C and 83°C, respectively. In AHSS-3, nickel level was 0.56 % to provide the desired hardenability and to depress martensitic transformation temperatures. The M_s and M_f temperatures were 209 °C and 78 °C, respectively. In AHSS-4, the nickel and molybdenum levels were identical to the previous two steels. This led to the lowest transformation temperature M_s of 204 °C, and the M_f of 73 °C.

The largest difference between the M_f temperatures found by calculation was a mere 15 °C. Nevertheless, it can still affect the treatment and the choice of quenchants.

Table 1. Chemical compositions of experimental steels AHSS-1–4 [wt. %.]

	C	Mn	Si	P	S	Cu	Cr	Ni	Al	Mo	Nb	M_s	M_f
AHSS-1	0.43	2.5	2.03	0.005	0.003	0.07	1.33	0.07	0.008	0.03	0.03	218	88
AHSS-2	0.428	2.48	2.03	0.005	0.003	0.07	1.46	0.08	0.004	0.16	0.03	214	83
AHSS-3	0.419	2.45	2.09	0.005	0.002	0.06	1.34	0.56	0.005	0.04	0.03	209	78
AHSS-4	0.426	2.46	1.99	0.005	0.002	0.06	1.33	0.56	0.005	0.15	0.03	204	73

2.1 Q&P Process

Several heat treatment routes were designed (Tab. 2). The appropriate austenitizing temperature (T_A), quenching temperature (QT), and partitioning temperature (PT) were chosen. For this treatment method, the quenching temperature is normally between the M_s and M_f. The partitioning temperature, at which carbon migrates from the super-saturated martensite to austenite, is of importance as well.

Table 2. Heat treatment parameters and resulting mechanical properties.

Route number	**T_A[°C]/t_A [s]**	**Cooling rate [°C/s]**	**QT [°C]**	**PT [°C] /t_{PT} [s]**	**HV [1]**	**R_m (UTS) [MPa]**	**A_{5mm} [%]**	**RA [%]**
1	850/100	1	100	150/600	573	2401	7	9
2	850/100	1	150	200/600	679	2308	10	12
3	850/100	16	150	200/600	693	2354	7	9

2 Intercritical Annealing

The other heat treatment method which was tested on the two experimental steels was intercritical annealing for bainite (Tab. 3). Using the JMatPro software, three routes were designed with various austenitizing temperatures (T_A), rates of cooling to the holding temperature for bainitic transformation (T_B), the rate of cooling to a temperature above M_s, and the rate of cooling to room temperature. The heating temperature was between the critical temperatures A_1 and A_3. Holding at the bainitic transformation temperature was important because at that stage austenite becomes stabilised and the strength of the material increases. The resulting fractions of phases are also influenced by the holding time. The rate of cooling to room temperature affects the final amount of retained austenite.

Table 3. Heat treatment parameters and amounts of retained austenite.

Route number	T_A[°C]/t_A [s]	Cooling rate to T_B [°C/s]	T_B[°C]/t_B [s]	Cooling rate to 250 °C [°C/s]	Cooling rate to RT [°C/s]	HV [1]	RA [%]
01	850/100	14	400/240	0.1	14	475	10
02	850/100	14	400/60	0.1	14	688	9
03	850/100	14	400/60	1	14	682	7

After both heat treatment routes, the amounts of retained austenite in selected specimens were measured using X-ray diffraction (Tab. 2., Tab. 3.).

3 Discussion of Results

In all steels, Q&P processing led to predominantly martensitic microstructures with some amounts of bainite and retained austenite. The highest hardness was 693 HV10 (Figs. 1–4). Retained austenite was probably located along the boundaries between martensite needles. The largest amount of retained austenite in the martensitic matrix, almost 12 %, was found in the AHSS-3 steel which had a higher nickel level (Tab. 2). Using transmission electron microscope, the morphology of this retained austenite was studied. In selected diffraction spots under dark field illumination, retained austenite was identified along the boundaries of prior austenite grains, subgrains, and martensite laths (Figs. 5 and 6).

IOP Conf. Series: Materials Science and Engineering **179** (2017) 012009 doi:10.1088/1757-899X/179/1/012009

Intercritical annealing for bainite in route 01 resulted in a tempered bainitic microstructure with martensite islands along prior austenite grains and a hardness of 475 HV10 (Figs. 7 and 8). Route 02 had shorter holding time at the bainitic transformation temperature than route 1. This led to martensitic-bainitic microstructure with some islands of bainitic ferrite (Figs. 9 and 10). The amount of martensite was thus larger and hardness was higher: 688 HV10. Cooling rate was increased in route 03 (Tab. 3), which led to martensitic-bainitic microstructure. As in previous cases, the largest fraction of retained austenite was found in the AHSS-3 steel (which had higher nickel level). Route 01 led to 10 % retained austenite. The specimen produced by route 01 was examined by transmission electron microscopy as well. Using dark field illumination, retained austenite was identified mainly along the boundaries between martensite laths.

Figure 1. Route 2: AHSS-3, cooling rate: 1 °C/s.

Figure 3. Route 2: AHSS-2, cooling rate: 1 °C/s.

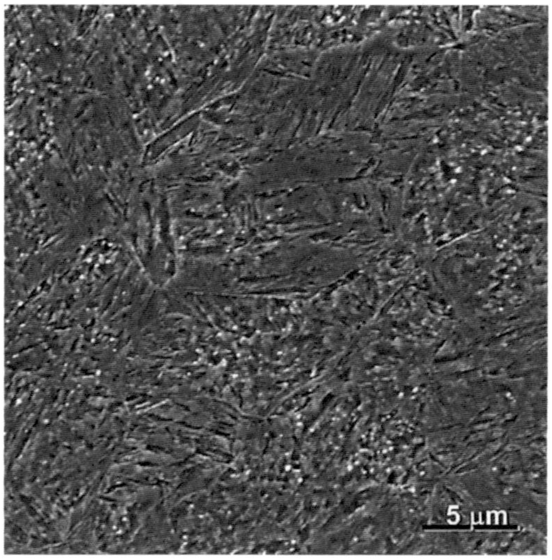

Figure 2. Route 2: AHSS-3, martensitic-bainitic structure, detail scanning electron micrograph.

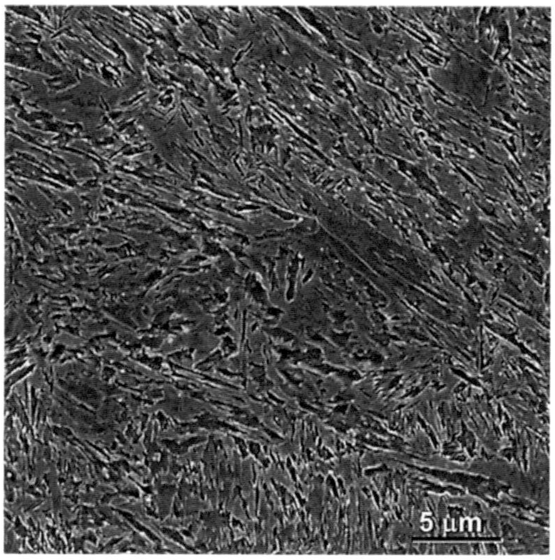

Figure 4. Route 2: AHSS-2, martensitic-bainitic structure, detail scanning electron micrograph.

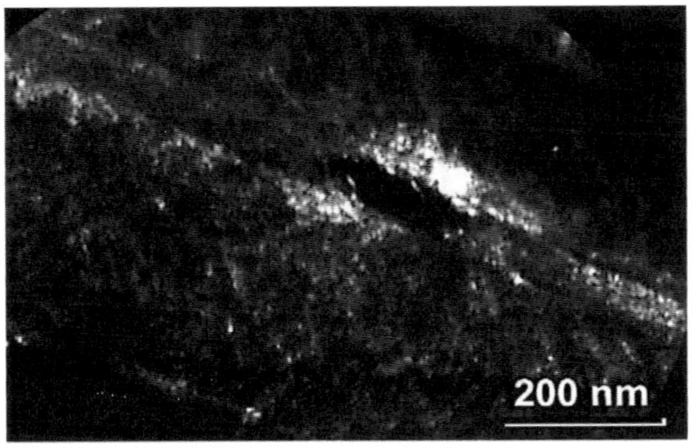

Figure 5. Route 2, AHSS-3, dark-field micrograph of retained austenite distribution.

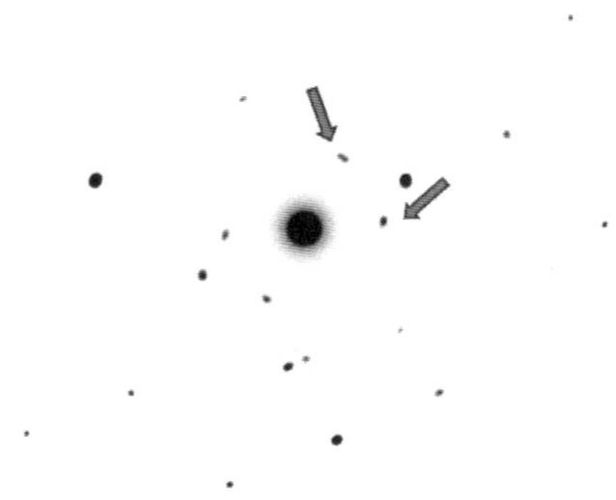

Figure 6. Route 2: AHSS-3, diffractogram of martensitic matrix with indicated austenite diffraction spots under dark field illumination. Zone axis of diffraction: z= [1 1 1].

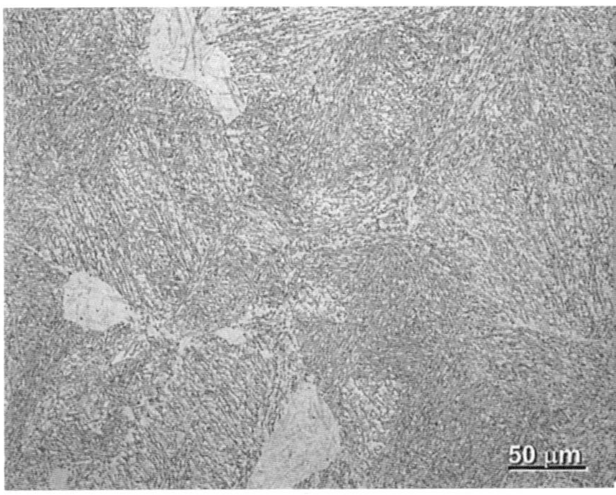

Figure 7. Route 01: rate of cooling to T_B: 14 °C/s, holding time: 240 s.

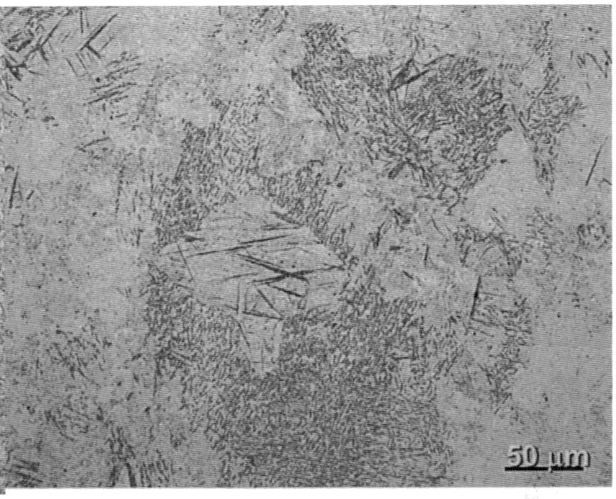

Figure 9. Route 02: rate of cooling to T_B: 14 °C/s, holding time: 60 s.

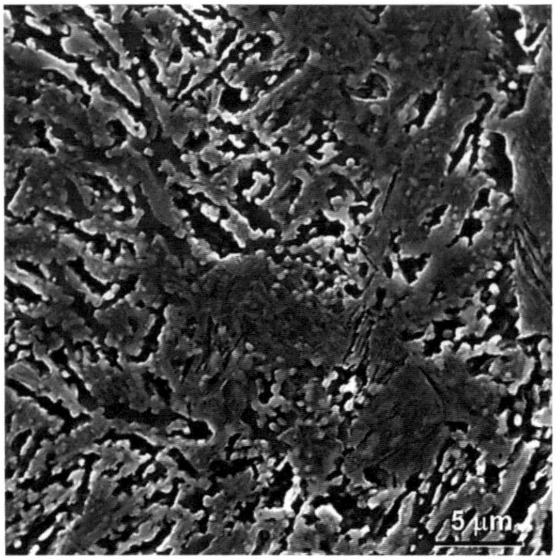

Figure. 8: Route 01, bainitic structure with martensite islands along prior austenite grain boundaries – detail scanning electron micrograph.

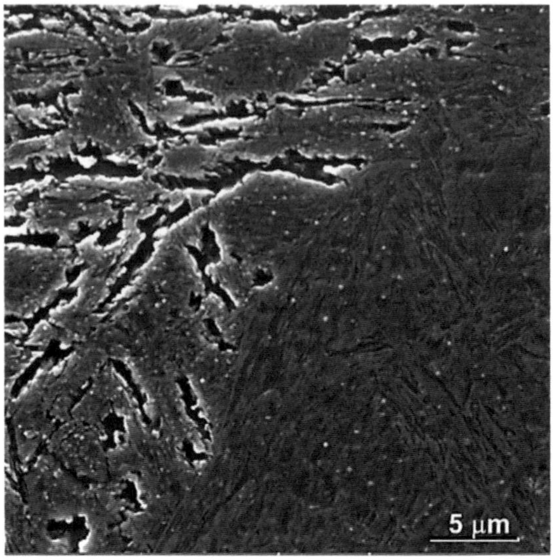

Figure. 10: Route 02: martensitic-bainitic structure, scattered areas of bainitic ferrite – scanning electron micrograph.

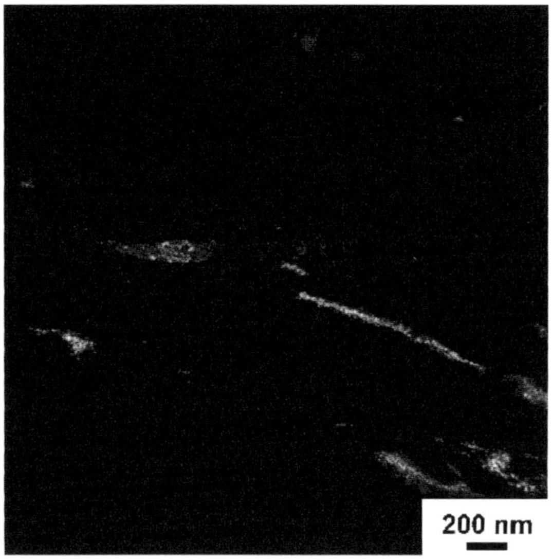

Figure 11: Route 01, AHSS-3, dark-field micrograph of retained austenite distribution.

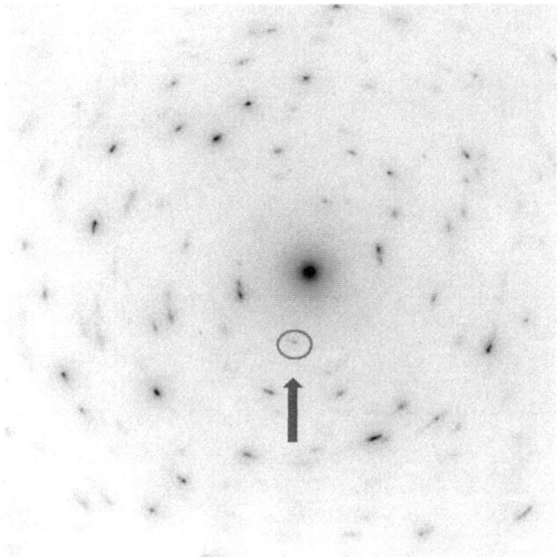

Figure 12: Route 01: AHSS-3, diffractogram of martensitic matrix with indicated austenite diffraction spots under dark field illumination and the indicated spot $(004)\gamma+(022)\alpha$.

4 Conclusion

Two heat treatment methods, Q&P process and intercritical annealing for bainite, were tested on newly-developed AHSS-type low-alloy steels containing manganese, silicon, chromium, molybdenum and nickel. This experiment with several different heat treatment routes has shown that the microstructure that forms in these high-strength steels is not exclusively martensite-based hardening microstructure. As expected, intercritical annealing produced bainitic ferrite. Mechanical properties and findings obtained by transmission electron microscopy suggest that Q&P process is a better alternative for these steels. In the material whose nickel content was increased to 0.56%, Q&P processing produced almost 12 % retained austenite in the martensitic matrix. Retained austenite was found on boundaries of prior austenite grains and subgrains and along the boundaries of martensite laths. In this material, intercritical annealing for bainite produced approximately 10 % of retained austenite. It was found mainly along the boundaries between martensite needles.

Acknowledgements
This paper includes results created within the projects LO1502 Development of Regional Technological Institute and SGS-2014-022 New Martensitic Structures - Process Parameters and Properties. The projects are subsidised by the Ministry of Education, Youth and Sports from specific resources of the state budget for research and development.

References

[1] Xiong, Z.P.; Kostryzhev A.G.; Saleh A.A.; Chen L.; Pereloma E.V. *Microstructures and mechanical properties of TRIP steel produced by strip casting simulated in the laboratory*, Materials Science & Engineering A, 2016, vol. 664, pp. 26–42.

[2] Mašek, B., Jirková, H., Hauserová, D., Kučerová, L., Klauberová D. *The Effect of Mn and Si on the Properties of Advanced High Strength Steels Processed by Quenching and Partitioning. Materials Science Forum*, 2010, vol. 654-656, pp. 94-97.

[3] De Moor, E., J. Gibbs, P. at al.: *Strategies for Third-Generation Advanced High-Strength Steel Development*, Iron & Steel Technology November, Volume 7, November 2010, PR-PM1110-6, Pages 1–7

[4] Qian, Z.; Lihe, Q.; Jun T.; Jiangying M.; Fucheng Z. *Inconsistent effects of mechanical stability of retained austenite on ductility and toughness of transformation-induced plasticity steels*, Materials Science & Engineering A, 2013, vol. 578, pp. 370–376.

[5] Jirková, H. et al.: *Influence of metastable retained austenite on macro and micromechanical properties of steel processed by the Q-P process*, Journal of Alloys and Compounds, available online, Journal of Alloys and Compounds, 615 (2014) S163–S168

Magnetic properties of 42CrMo4 steel

T Bulin[1], E Svabenska[2], M Hapla[2], P Roupcova[2], C Ondrusek[1] and O Schneeweiss[2,3]

[1] Brno University of Technology, Faculty of Electrical Engineering and Communication, Technická 3058/10, 616 00 Brno, Czech Republic

[2] Institute of Physics of Materials ASCR, Žižkova 22, 616 62 Brno, Czech Republic

[3] CEITEC IPM, Institute of Physics of Materials ASCR, Žižkova 22, 616 62 Brno, Czech Republic

E-mail: xbulin01@stud.feec.vutbr.cz

Abstract. Low alloyed high-grade chrome-molybdenum ferritic steel was investigated from the point of views of magnetic properties in dependence on heat and mechanical treatment. This steel can be used as components of magnetic circuits or some parts in electrical equipment. The basic information on structure and phase composition was obtained by optical and scanning electron microscopy, X-ray Powder Diffraction and Mössbauer Spectroscopy. The temperature stability of the material was proved by measurements of temperature dependences of magnetic moment. The magnetic parameters were obtained by measuring of magnetic hysteresis loops in dependence on saturation field and their frequencies. The results are discussed from the point of view of possible applications as a magnetic material in the very extremely environment, where high mechanical stresses and elevated temperatures can occur.

1 Introduction

The knowledge of magnetic properties of steel, which is intended for electrical machine, is very important because it can be applied as part of magnetic circuit and these parameters play important role. It is required the knowledge of parameters such as saturation magnetization, coercivity, magnetic losses, permeability and electrical resistivity in dependence on temperature and heat&mechanical treatment history. The content of datasheets is seldom for magnetic properties of construction low alloyed steels [1-3]. The aim of this study is to describe magnetic behavior of 42CrMo4 steel. Possible applications of this steel can be as parts of magnetic circuits or some parts in electrical equipment [4, 5].

2 Experimental

The samples of the 42CrMo4 steel were obtained from a rod and cut to smaller cylinders and toroids. The rod was heat treated and soft annealed. Chemical composition from producer was confirmed by Glow discharge optical emission spectrometry (GDOES) and the same elements were found by scanning electron microscope Tescan LYRA with EDX detector. The sample for microscopy was cut from cylinder (diameter 55 mm and thickness 4 mm) using spark erosion cutting. Surface of sample was ground on SiC paper to 2500 grid and polish by colloidal silica and it was etched first in Villea-bain and consequently in 2% Nital for better resolution. The chemical composition according to material description and identified by GDOES was following:

Table 1. Chemical composition in wt. %.

Element	Datasheet	Result GDOES
C	0.38 – 0.45	0.42
Si	max. 0.40	0.24
Mn	0.6 – 0.9	0.83
P	max. 0.025	0.010
S	max. 0.035	0.000
Cr	0.90 – 1.20	1.12
Mo	0.15 – 0.30	0.19

Samples for magnetic measurement were made from cylinder with diameter 60 mm by lathe and no additional adjustments were done. The aspect ratio of these samples is different to verify measured parameters on different sized toroid. The ideal ratio with homogenous field is 1.1 and border aspect ratio is 1.25. If the ratio is higher, the magnetic flux is not spread homogenous in entire cross section [6].

Table 2. Parameters of toroid samples

Dimensions of the sample	1	2	3	4	5	6	7	8	
Outer diameter	54.95	44.32	44.37	54.97	54.97	54.92	54.92	54.92	mm
Inner diameter	47.3	40.4	38.07	47.42	45.02	39.95	32.95	24.96	mm
Thickness	1	2	3.07	3.8	5	7.52	10.96	14.92	mm
Mass	4.6	3.94	9.44	17.67	30.12	64.94	129.54	218.74	g
Primary windings	109	97	95	104	110	92	126	97	threads
Secondary windings	20	40	20	40	20	40	40	20	threads
Ratio	1.16	1.10	1.17	1.16	1.22	1.37	1.67	2.20	

Structure and phase composition was investigated using optical microscopy (OM), scanning electron microscopy (SEM), X-ray powder diffraction (XRD) and ^{57}Fe Mössbauer spectroscopy (MS). The surfaces of the samples were examined by optical microscope Neophot 32 by Carl Zeiss Jena and scanning electron microscope Tescan LYRA 3 XMU FEG/SEMxFIB equipped with X-Max80 EDX detector for X-ray microanalysis and EBSD detector Nordlys by Oxford Instruments with Aztec control system. The X-ray powder patterns were collected on X'Pert diffractometer and CoKα radiation with qualitative analysis by HighScore® software and the JCPDS PDF-4 database. For a quantitative analysis HighScore plus® with Rietveld structural models based on the ICSD database was applied. Chemical analysis of the material was done by Glow discharge optical emission spectrometry using GD-Profiler 2 instrument with the Quantum™ XP software (Horiba Jobin Yvon, France) operating in radio frequency at 13.56 MHz. ^{57}Fe MS spectra were measured at room temperature in scattering geometry ~ 80 MGy ^{57}Co(Rh) radioactive source and calibrated against α-Fe as the standard. The values of the isomer shift are related to α-Fe at room temperature. The computer processing of the spectra yielded the values of the relative spectrum area I and values of the hyperfine parameters including isomer shift δ, quadrupole splitting Δ, and hyperfine induction B_{hf}. Changes in magnetic properties by high temperature were investigated by temperature dependence of magnetic moment. The temperature dependence of the magnetic moment was measured under vacuum (~ 10^{-2} Pa) using vibrating sample magnetometer in an external magnetic field of 5 mT and in a temperature range of 25 – 800 °C with a sweep of 4 °C·min^{-1} in vacuum. The dependence of magnetic moment on external field (hysteresis loop) was measured before and after annealing at room temperature in external fields ± 1 T. The magnetic parameters were acquired by equipment Remagraph – Remacomp C-710 (Magnet-Physik Dr. Steingroever GmbH). It is measuring the magnetic flux through the coil and it is based on the principle of Faraday law of electromagnetic induction. Saturation magnetization and hysteresis losses were measured on toroid shaped samples. Part equipment Remacomp was used for the measurement with frequency and Remagraph without frequency. The magnetic properties were measured in frequency range 0 – 2000 Hz at room temperature.

3 Results and discussion

Information on the structure and phase composition were obtained using OM, SEM (**Figure 1., 2.**) and XRD. The ferrite (matrix) structure parameters were unambiguously recognized from the XRD data (**Figure 33.**). The presence of carbide $Cr_{23}C_6$ and $(Fe, Mn)_3C$ in the steel was also confirmed but intensities of their diffraction peaks are close the detection limits. Content of these elements is about 3.6 wt.% $(Fe, Mn)_3$ and 1.5 wt.% $Cr_{23}C_6$.

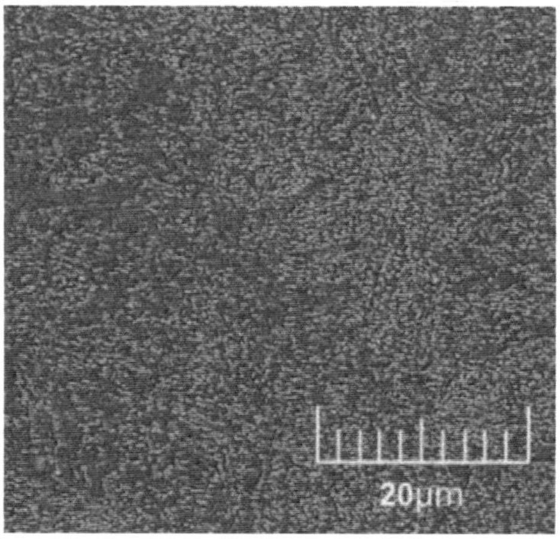

Figure 1. Scanning electron microscopy of the 42CrMo4 steel sample 2000 magnification.

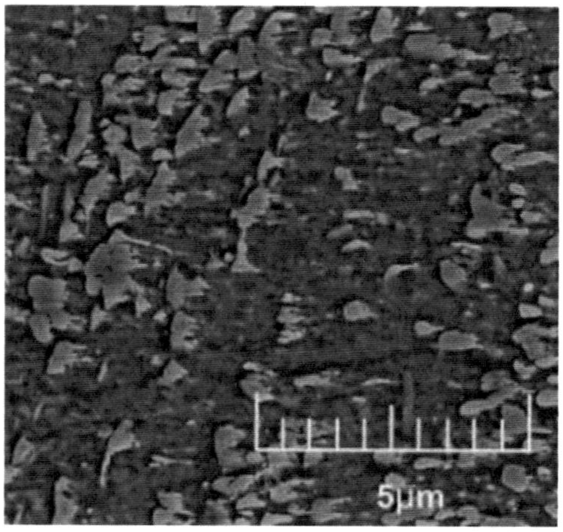

Figure 2. Scanning electron microscopy of the 42CrMo4 steel sample 10 000 magnification.

The Mössbauer spectrum of the steel sample is shown in **Figure 4.** The crosses denote experimental points and the line the fitted function [7]. Above the spectrum three main fitted components are drawn.

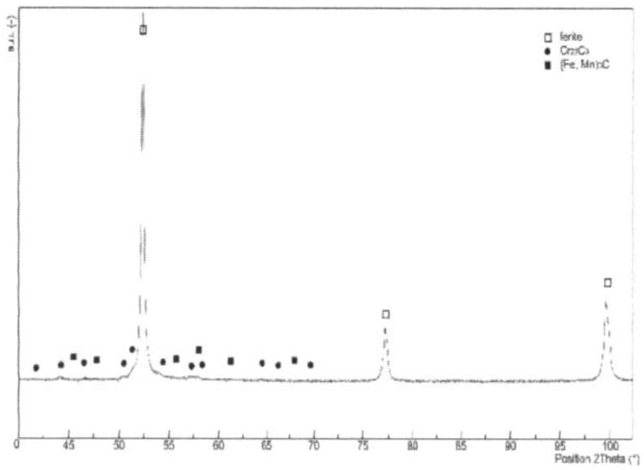

Figure 3. X-ray pattern of 42CrMo4 steel (□ferrite, ■ $(Fe, Mn)_3C$, and ♦ $Cr_{23}C_6$) obtain at room temperature.

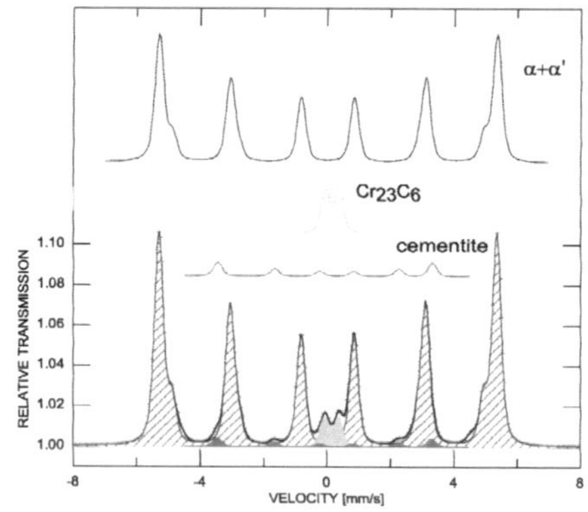

Figure 4. Room temperature temperature Mössbauer spectrum of the 42CrMo4 steel sample.

The experimental data were fitted by three sextet with mean hyperfine induction $\langle B_{hf} \rangle = 32.7 \pm 0.1$ T and relative content c = 0.94±0.01 which represent α phase. The sextet with hyperfine induction B_{hf} =

21.0±0.1 T, δ = 0.114±0.005 mm·s^{-1}, QS = -0.374±0.010 mm·s^{-1} and c = 0.02±0.01 represents cementite Fe_3C. The analyzed doublet (δ = 0.160±0.007 mm·s^{-1}, Δ = 0.427±0.010 mm·s^{-1} and c = 0.04±0.01) can be ascribed to $Cr_{23}C_6$ carbide [8, 9].
The temperature dependence of magnetic moment of the 42CrMo4 sample is drawn in **Figure 5.** From this dependence, we can observe magnetic transformations (Curie temperatures) at ~180 °C of cementite and at ~ 775 °C of ferrite matrix [10]. The hysteresis loops are given in **Figure 6.** They show that the samples remained stable after the vacuum annealing during the measurement of the above mentioned temperature dependence of the magnetic moment up to 800 °C in vacuum. From the saturation state, we can derive saturation magnetization which is 191.0 ± 0.4 emu·g^{-1}., i.e. 191 A·m^2·kg^{-1} or 2.40·10^{-3} Wb·m·kg^{-1}.

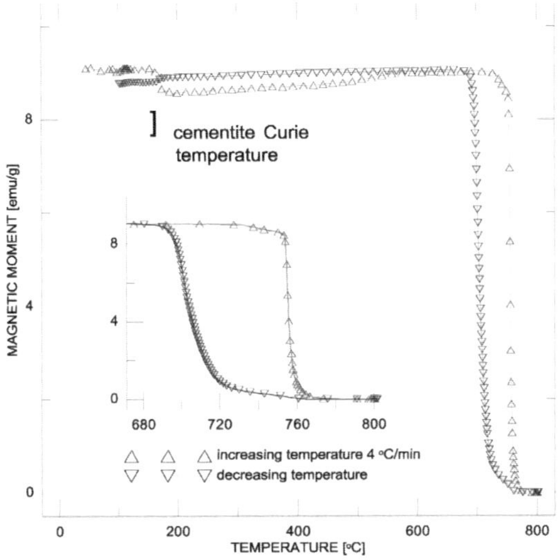

Figure 5. Temperature dependence of magnetic moment of the 42CrMo4 steel sample measured in vacuum. The detailed insert shows the temperature hysteresis of $\alpha \rightarrow \gamma \rightarrow \alpha$ magnetic phase transformation.

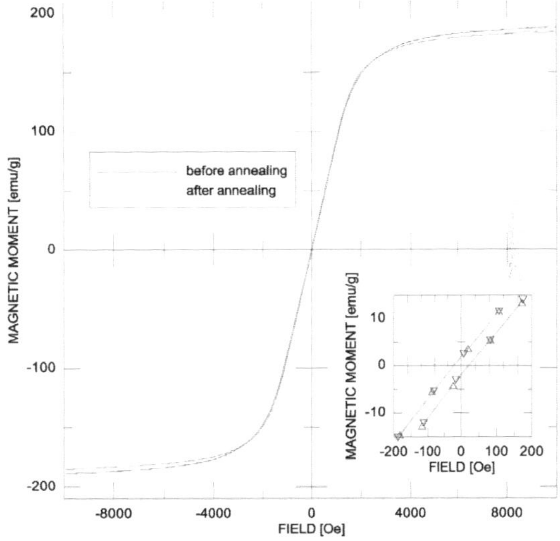

Figure 6. DC hysteresis loops of the 42CrMo4 steel sample before and after annealing.

The results of magnetic measurements on toroid shaped samples are shown in **Figures 7. – 11**. The basic comparisons are made among hysteresis curves measured without frequency because there are minimal eddy currents. In **Figure 7.** there is comparison among samples with the same ratio between outer and inner diameter but there are different thicknesses. Sample 1 is too thin and cross-section is not square. Hysteresis curve of sample 1 has the highest slope in contrast with sample 4 which has only one difference in the setting data, thickness. Sample 3 has smaller dimensions like sample 4 but the ratio is the same and slope of hysteresis curve is between sample 1 and 4.
Any significant differences on BH curves are not seen between the same outer diameter but slightly different ratio 1.10 and 1.17. The biggest differences may be expected between sample 4 and 7 but it is not. Sample 4 has the higher magnetic field strength. Distribution of magnetic flux through the cross-section is uneven but it has not significant influence on the shape of BH curve.

IOP Conf. Series: Materials Science and Engineering **179** (2017) 012010 doi:10.1088/1757-899X/179/1/012010

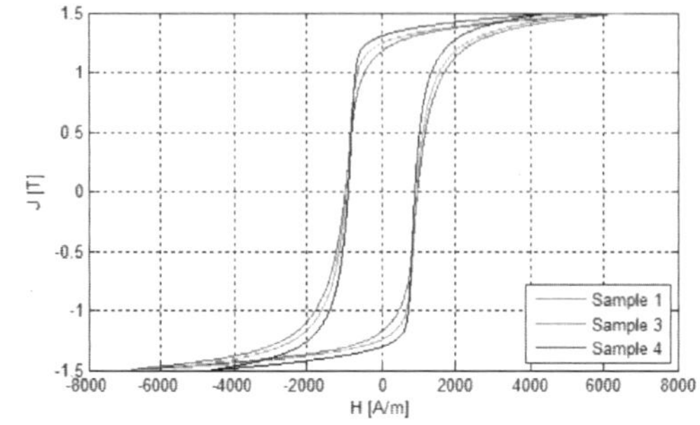

Figure 7 Comparison hysteresis curves at 0 Hz for sample 1, 3 and 4.

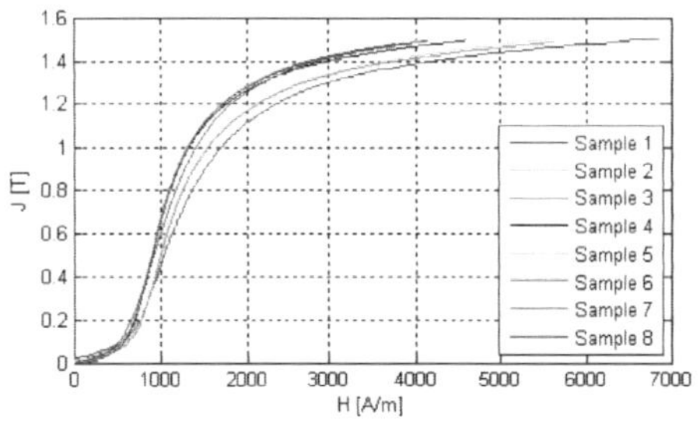

Figure 8. Initial magnetization at 0 Hz.

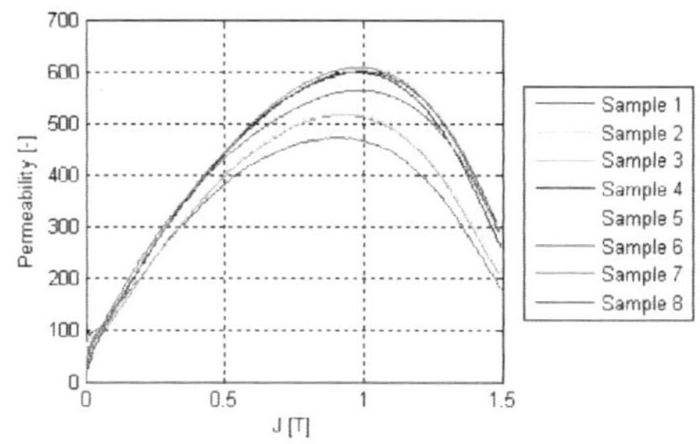

Figure 9. Permeability at 0 Hz.

Initial magnetization curves of all samples can be seen on **Figure 8**. The biggest differences are between samples with cross-section which is not square or with different outer diameter. Differences between samples can be seen better on **Figure 9.**, where is shown permeability of each sample. The worst permeability has sample 1 than sample 2 and 3 and rest of samples have similar permeability except for sample 8. This sample has the biggest cross-section and the worst ratio of diameters but permeability is worse than rest of the slightly smaller samples. Measurement errors can occur due to difficulties to put on enough primary winding on inner diameter which results in an uneven distribution of the magnetic flux through the cross-section. It was done measurement with frequencies besides quasistatic measurements and values of total magnetic losses were investigated. Results of these measurements can be seen on **Figures 10. and 11.** Samples with bigger volume have higher total magnetic losses and they are increasing exponentially as expected. It was not possible to measure Samples 7 and 8 on higher frequencies because of smaller number of windings which can be possible to put on these toroid samples and limitation of measuring device.

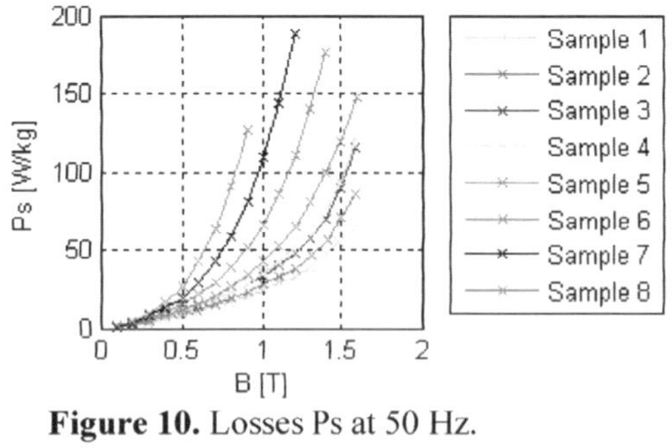

Figure 10. Losses Ps at 50 Hz.

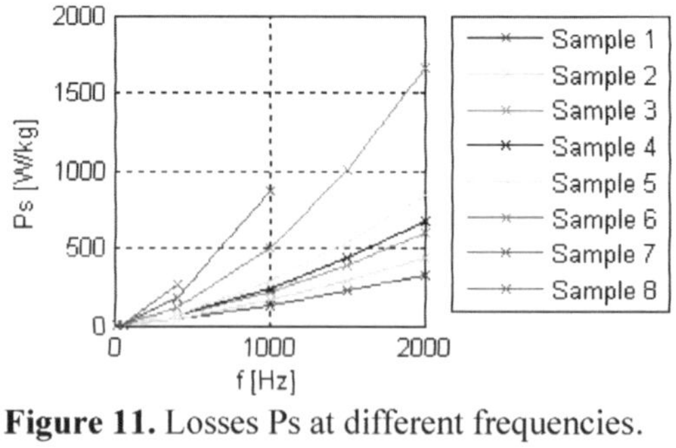

Figure 11. Losses Ps at different frequencies.

4 Conclusions

Obtained results show from XRD and Mössbauer spectroscopy that the samples consist of ferrite and (Fe, Mn)$_3$C and $Cr_{23}C_6$ carbides. The measurements of temperature dependences of magnetic moments confirm phase stability up to Curie temperature of ferrite (~775 °C). The saturation magnetization at room temperature is 191.0 ± 0.4 emu·g^{-1}., i.e. 191 A·m^2·kg^{-1} or $2.40 \cdot 10^{-3}$ Wb·m·kg^{-1} which corresponds with hyperfine induction $B_{hf} = 32.7 \pm 0.1$ T derived from the ferrite component in Mössbauer spectrum. Magnetic parameters were measured on toroid samples which were made from solid material of 42CrMo4 steel. It can differ in addiction on dimensions of these samples. Bigger samples give better permeability but it is difficult to measure them in higher frequencies. Too small samples have disadvantage of difficult demagnetization before quasistatic measurements. The AC measurements show that the total magnetic losses are dependent on frequency and excitation exponentially. For example, the power losses were 34.86 W·kg^{-1} in sample 4 at 50 Hz. The knowledge of the size of total magnetic losses is important at higher frequencies than 50 Hz in magnetic circuits. The temperature of the steel parts is increasing because of these losses and therefore it is appropriate to calculate their size. DC permeability reaches the maximum ~ 611 at magnetic field strength ~ 1279 A·m^{-1}. The obtained data on phase composition and magnetic behavior indicates, that the 42CrMo4 steel has good potential for applications in magnetic circuits where good mechanical properties together with soft magnetic properties are requested.

Acknowledgements

This research work has been carried out in the Centre for Research and Utilization of Renewable Energy (CVVOZE). Authors gratefully acknowledge financial support from the Ministry of Education, Youth and Sports of the Czech Republic under NPU I programme (project No. LO1210). They also acknowledge partner collaboration in frame of project of Technology Agency of the Czech Republic No. TE02000232.

References

[1] Tanner B K 1989 Nondestructive Testing and evaluation 5 pp 9–15
[2] Pyrhönen J, Jokinen T and Hrabovcova V 2008 Design of rotating electrical machines John Wiley & Sons ltd. pp 183–218
[3] Wilson J W, Karimian N, Liu J, Yin W, Davis C L, Peyton A J and Mafn J 2014 *Magn. Mater* 360 pp 52–58
[4] Lähteenmäki J 2002 Design and Voltage Supply of High-Speed Induction Machines *Doctoral thesis* (Helsinki: Helsinki University of Technology)
[5] Barta J, Uzhegov N, Ondrusek C and Pyrhönen J 2016 High-Speed Electrical Machine Topology Selection for the 6 kW, 120 000 rpm Helium Turbo-Circulator *International Review of Electrical Engineering (IREE)* vol 11 no. 1
[6] Fiorillo F 2010 Measurements of magnetic materials *Metrologia* vol 47 no. 2 pp S114–S142
[7] Zak T and Jiraskova Y 2006 Surface and Interface Analysis vol 38 no. 4 pp 710–14
[8] Schaaf P, Krämer A, Wiesen S and Gonser U 1994 Acta metal. mater. vol 42 pp 3077–81
[9] Gonser U, Ed. 1975 Mössbauer Spectroscopy vol 1 (Berlin: Springer)
[10] CRC Handbook of Chemistry and Physics 1999 vol 80 (New York: CRC Press) pp 12–119

IOP Conf. Series: Materials Science and Engineering **179** (2017) 012011 doi:10.1088/1757-899X/179/1/012011

Automated microstructural analysis of titanium alloys using digital image processing

A Campbell, P Murray, E Yakushina, S Marshall, W Ion

Advanced Forming Research Centre, University of Strathclyde, Glasgow, UK

Email: andrew.j.campbell@strath.ac.uk

Abstract. Titanium is a material that exhibits many desirable properties including a very high strength to weight ratio and corrosive resistance. However, the specific properties of any components depend upon the microstructure of the material, which varies by the manufacturing process. This means it is often necessary to analyse the microstructure when designing new processes or performing quality assurance on manufactured parts. For Ti6Al4V, grain size analysis is typically performed manually by expert material scientists as the complicated microstructure of the material means that, to the authors knowledge, no existing software reliably identifies the grain boundaries. This manual process is time consuming and offers low repeatability due to human error and subjectivity. In this paper, we propose a new, automated method to segment microstructural images of a Ti6Al4V alloy into its constituent grains and produce measurements. The results of applying this technique are evaluated by comparing the measurements obtained by different analysis methods. By using measurements from a complete manual segmentation as a benchmark we explore the reliability of the current manual estimations of grain size and contrast this with improvements offered by our approach.

1 Introduction

Titanium is a widely used material that is useful in a variety of applications ranging from aerospace to medicine [1]. Its popularity is due to several valuable properties. Titanium is very light with a density of approximately 60% that of iron yet it is still very strong. It also maintains its strength at high temperatures and, due to a high reactivity with oxygen, can form an oxide layer on its surface giving titanium a high corrosive resistance [2]. Titanium alloys are used to improve some of these properties even further. Ti6Al4V is particularly popular in the aerospace sector as it can replace both steel and aluminium parts, offering a weight saving against steel and both a strength and operating temperature improvement versus aluminium [3]. The aerospace sector represents around 50% of the world's titanium usage [3] and Ti6Al4V is the most commonly used alloy [2], making it valuable to research.

The properties of the Ti6Al4V alloy depend on its microstructure. This will vary based on the manufacturing processes used in production. An accurate and reliable analysis of the microstructure therefore benefits quality assurance of manufactured parts as well as research into new manufacturing techniques. Features such as the volume fraction of alpha grains, the size of alpha grains, the volume fraction of globular alpha phase and the width of Widmanstätten alpha laths are often measured as they are known to effect the mechanical properties of the alloys [4]. Much of this analysis is currently carried out by a manual process. For example, measurements of grain size require expert materials scientists with knowledge of the materials microstructure to measure axis lengths of hundreds of grains, a laborious and inefficient process. Existing software packages can provide some assistance by measuring the distance between mouse clicks, meaning axis measurements can be produced and stored simply by selecting points on the grain boundary. However, this process is still slow with users making subjective decisions about the location of grain boundaries. We propose using image processing to improve speed and repeatability.

In image processing software, measuring grain parameters, such as length and square area, are relatively and easy once the grain has been identified. The more difficult task is the process of labelling the image such that it is known to which grain, if any, each pixel belongs. This process of partioning an image into a set of labelled regions, where pixels share some measure of homogeneity, is commonly referred to as image segmentation [5]. Despite considerable research being conducted into generic image segmentation algorithms, the complexity of the task means a single technique cannot segment every image suitably for all applications [6]. Typically, combinations of different methods must be combined alongside expert knowledge of the application to produce accurate segmentation results.

Ti6Al4V alloys have complex microstructures, containing alpha and beta grains and often contain clusters of grains of the same phase separated only by unclear inter-grain boundaries. Commercial microstructural analysis software, such as Beuhler OmniMet or Clemex Vision PE, struggle to reliably identify these boundaries where grains of the same phase meet. This means their automated measurements are not reliable, particularly when the volume fraction of alpha phase is high.

Other research has proposed uses of image processing techniques to improve the analysis of titanium alloys. Tiley et al [7] proposed methods of measuring the Widmanstätten laths. This was incorporated into a list of procedures to semi-automate the analysis of α/β titanium alloys, proposed by Searles et al. [8,9]. In these procedures a Photoshop extension called Fovea Pro is used to obtain measurements of the microstructural features of a material. For example, thresholding techniques are used to determine the volume fraction of alpha phase, a technique the authors of this paper have observed effective use of within industry. However, methods to automatically measure the size of alpha grains have very limited success due to errors in the automatic segmentation methods. In an extension of this work published by P.C. Collins [4] automatic grain segmentation is abandoned and alpha grains are selected manually due to these errors.

Our research combines knowledge from the materials science domain with innovative image processing techniques to create a new algorithm that is suitable for segmenting Ti6Al4V microstructural images. This segmentation will allow individual measurements of alpha grains to be made, which will enable the automatic calculation of grain size and the volume fraction of globular alpha grains. For this study images of Ti6Al4V alloy were used and have been obtained by a Scanning Electron Microscope (SEM) Quanta FEG 250. The Watershed Algorithm [10] is then used for segmentation. This algorithm considers an image as a topographic surface where pixel values represent the height of the surface. The algorithm then floods the image from its shallowest regions with the locations these floods meet being marked as grain boundaries. This has been successfully applied to segment images in a range of applications from analysis of sea ice floes [11] to medical images [12]. However, the Watershed algorithm is prone to over-segmentation as flooding from small local minima, caused by noise in the image, results in that region being incorrectly detected as a grain [6]. Pre-defined regions in an image, known as markers, reduce over-segmentation by setting the start points for the flooding process. By combining the Watershed Algorithm with a range of pre and post-processing steps, and using our own method to compute markers for the Watershed, we improve the segmentation accuracy compared to existing techniques, when applied to this application. We also show how the algorithm could be implemented within a software tool that allows the user some intuitive control over results. Finally, the measurements produced by the software are compared with measurements produced by expert materials scientists to demonstrate that the software can achieve accurate results in less time.

2 Method

This section describes the method proposed to segment the grains within the microstructural images, examples of which are shown in Figure 1. The flow chart in Figure 2 shows the steps in this algorithm.

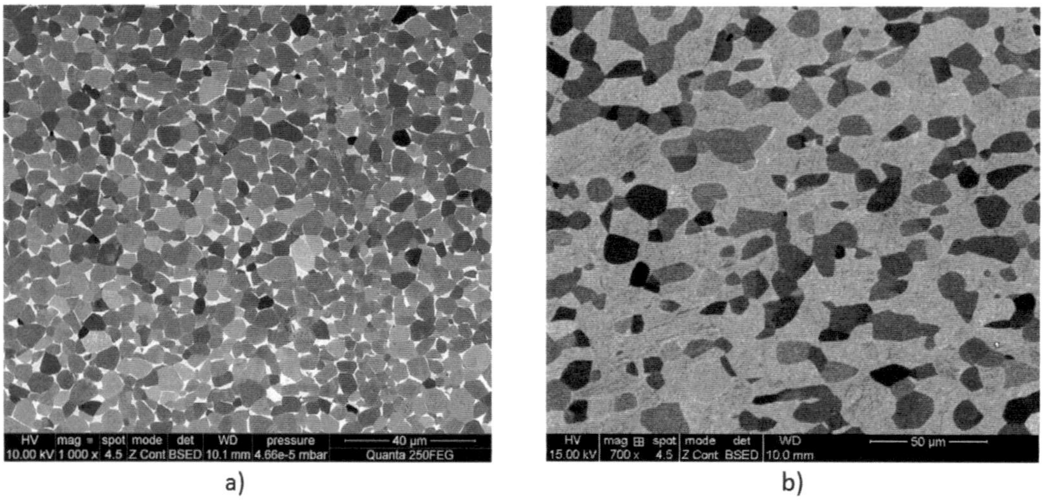

Figure 1. Examples of Ti6Al4V microstructure: a) magnification x1000; b) magnification x700 (SEM).

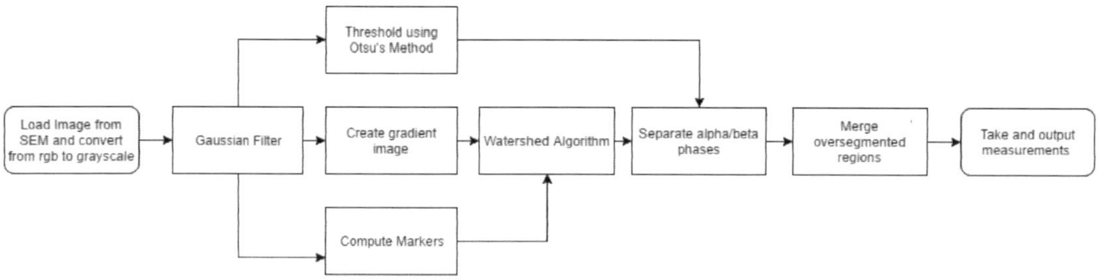

Figure 2. Flow chart of automated microstructural analysis process.

The Watershed Algorithm is used as the segmentation function in our system. As the Watershed Algorithm is known to often oversegment images we propose a combination of techniques to improve segmentation accuracy. For α/β microstructures we must also separate the phases before taking measurements. As the contrast of the phases differ, we can use thresholding [6] to estimate the locations of each phase. Otsu's method [13] is used to automatically compute a suitable threshold based on the properties of the image under study such that the threshold parameter is dynamic and can be easily adjusted for images of different brightness and contrast. However, noise in the image causes some pixels to be misidentified in this process. Therefore, to improve the reliability of phase separation, we assess each region from the Watershed segmentation based on the percentage of its constituent pixels that are classified as each phase. We label each region as belonging to either the alpha or beta phase to obtain a robust phase separation.

The Gaussian filter [15] reduces noise in the image by smoothing the intensity change between adjacent pixels. This helps minimise over-segmentation and increase the reliability of edge detection and thresholding. The effect of this filter is limited so as not to obscure grain boundaries.

Markers, as explained in the introduction, are also used to reduce over-segmentation. Other researchers use marker based watersheds to achieve good image segmentation results [14, 16, 17]. However, none of the methods tested in this study produced good results for the microstructures in our data. A new method of computing markers was therefore used. Our method uses edge detection

[5] to mark grain boundaries, based on a variation in intensity, and morphological operators [6] to expand these edges leaving small regions within each grain, that we can use as a marker. Typically, we set edge detection to only detect obvious grain boundaries. However, it also provides a user-friendly way to adjust the segmentation as the number of detected edges affects the number of markers produced. Therefore, we can set the edge detection threshold as a parameter called "sensitivity" that would allow the user to control the segmentation level in a simply and intuitively. A split and merge approach is also applied in our proposed algorithm. Over-segmentation errors are resolved by merging adjacent regions in our segmentation when they meet a certain criterion. In our case: grain size, the difference in the mean intensity between adjacent grains and the length of the boundary between regions could be used to determine if those regions belong to the same grain. The effectiveness of our segmentation approach can be demonstrated by comparing the segmentation we produced to those produced by other segmentation algorithms [6, 11, 12, 14] when applied to the same data. Figures 3 b) and c) show results using implementations of established Watershed methodologies [6, 14]. Figures 3 d) and e) use recent algorithms to improve the segmentation accuracy for images within their study [11, 12]. All of these algorithms achieve accurate segmentations in certain applications but Figure 3 shows they fail to accurately segment the images in our study. This illustrates the complexity of the segmentation task and why a new method was required.

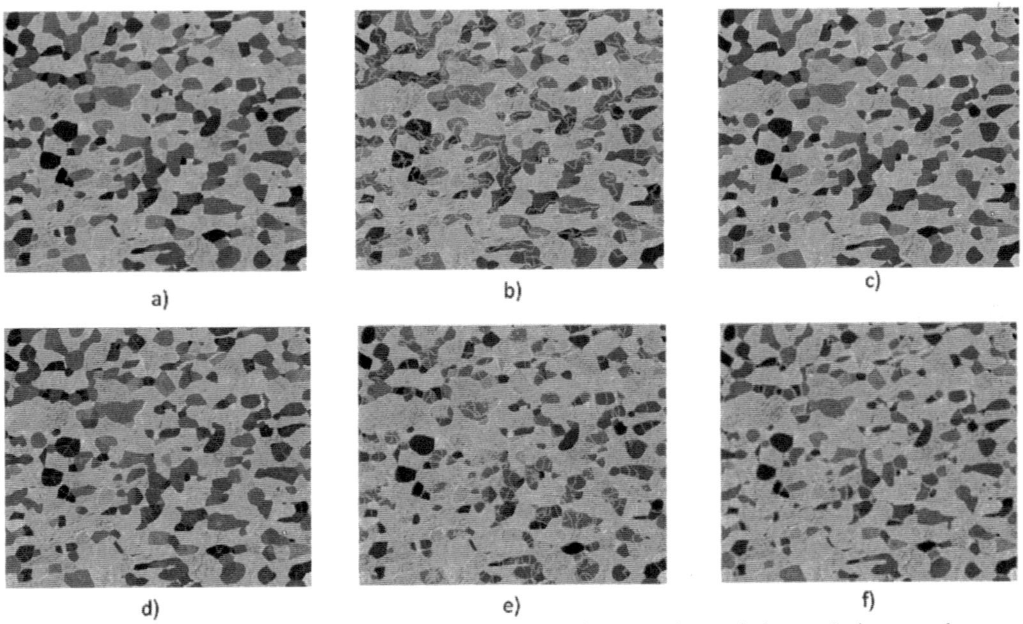

Figure 3. Comparison of segmentation of Ti6Al4V image using existing techniques and our method where a) original image, b) watershed using supressed minima [6], c) markers computed using reconstruction [14] d) watershed proposed by Al- Kofahi [12], e) algorithm proposed by Zhang [11], f) our algorithm.

3 Results and Discussion

The purpose of this research is to develop an accurate automated method of segmenting and measuring alpha grains in images of Ti6Al4V microstructures. When evaluating the proposed method, the potential for error in existing manual methods must be considered. We use the mean grain size for our comparison and measure the microstructure in Figure 1 a) using three methods. In one method a complete manual segmentation of the image was performed by an expert materials scientist. This allowed for every grain to be accurately measured giving what is as close as possible to the true grain measurements. 5 hours were taken to ensure this result was accurate, meaning this method would typically be impractical to use. Therefore, this result serves only as a benchmark with which to compare other measurements. The proposed automated method was then used to measure the image, as was a manual method for estimating of grain size, currently employed at the Advanced Forming Research Centre. In the manual method measurements are taken by loading the image into the ImageJ software package then measuring axes of the grain by clicking on the points where each axis would cross the grain boundary. This process was repeated until 300 grains were measured in order to obtain statistically valid results. For consistency each method would calculate the grain size as the average of the major and minor axis of the grain and report the average grain size for use in our comparison.

Table 1. Comparison of software and expert measurements.

	Manual Segmentation	Manual Estimation	Software
Average Grain Size (um)	4.28	4.75	4.16
Time	5 hours	1.25 hours	19 seconds

The results in Table 1 demonstrate the potential of our approach as they show the software achieved results closer to what is believed to be the true grain size than the existing manual estimation. A possible reason for this result is that the software measures every grain, while due to practical time constraints only a subset is measured during the manual estimation. In our experiment the software was measuring more than twice as many grains as were measured by the manual method. Another factor is that the boundaries of large grains were more obvious and easier to measure. It is possible that this would lead to a user being more likely to ignore small grains and unintentionally bias the result, although further research is required to confirm this hypothesis.

The results also show a significant time difference between the proposed automatic approach and manual methods. As this is a software method, our measurement is also repeatable. The proposed automatic method can therefore resolve key issues with existing manual processes provided it can be successfully applied to a variety of microstructural images. The volume fraction of alpha grains, the shape of grains and the presence of Widmanstätten alpha laths varies between different titanium microstructures which presents challenges to image segmentation software. Work is currently in progress to ensure our method can be applied to a variety of microstructural images.

Although mean alpha grain size was used in our tests, the automated approach also allows us to calculate a variety of other microstructural properties which are important for industrial and fundamental applications. The segmentation produced by our proposed algorithm allows all microstructural properties relating to alpha grain measurements to be taken. Knowing the axis lengths of each grain allows the distribution of grain size and aspect ratio to be calculated, as well as the volume fraction of globular grains. The total area of alpha phase could also be compared to the total area of beta phase to calculate the volume fraction of alpha grains. These calculations typically take no more than a few seconds once the segmentation is produced, depending on the number of grains measured. An independent software package that calculates such measurements has been produced to aid future efforts to validate our methods. The interface of this software is shown in Figure 4.

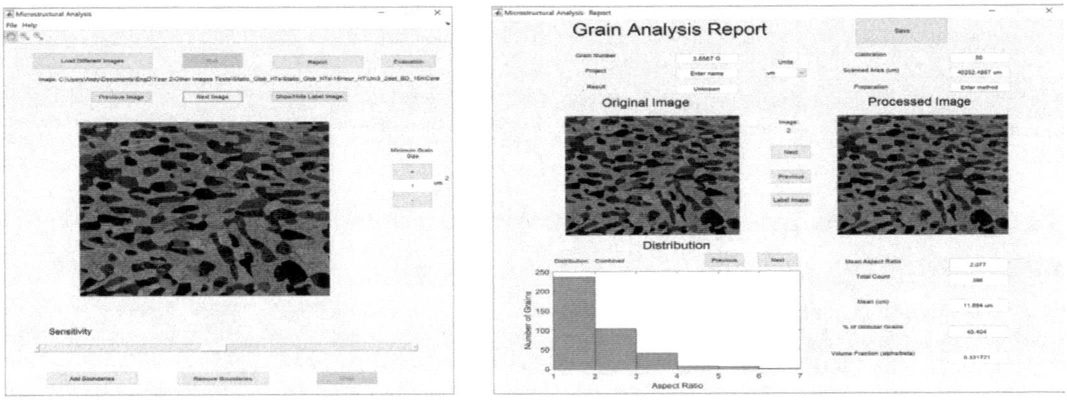

Figure 4. Example of Software Interface.

4 Conclusions

The results of this research demonstrates the potential of using image processing techniques for microstructural analysis of Ti6Al4V. Although current evaluation trials have been performed only on a limited dataset, results demonstrate the potential for highly accurate measurements to be produced. In our trial the software obtains a more accurate result than the existing manual process in far less time. We also demonstrated that the complexity of the Ti6Al4V microstructures cause errors in several other segmentation algorithms and that our method is capable of resolving these problems. Further work is in progress to validate the consistency of these results for a variety of microstructures exhibiting different properties.

References

[1] Welsch G, Boyer R and Collings EW 1993 *Materials properties handbook: titanium alloys.*
[2] Lütjering G and Williams JC 2003 *Titanium, volume 2*
[3] Polmear I 2006 *Light alloys from traditional alloys to nano-crystals*
[4] Collins PC, Welk B, Searles T, Tiley J, Russ JC and Fraser HL 2008 *Materials Science and Engineering* A 508 (2009) 174-182
[5] Gonzalez R.C, Woods R.E *Digital Image Processing* 1987
[6] Soille P 2013 *Morphological image analysis: principles and applications*
[7] Tiley J, Searles T, Lee E, Kar S, Banerjee R, Russ JC and Fraser HL *Materials Science and Engineering* pp191
[8] Searles T, Russ J and Fraser H 2005 *Stereology Procedures*
[9] Searles T, Tiley J, Tanner A, Williams R, Rollins B, Lee E, Kar S, Banerjee R and Fraser HL 2004 *Measurement Science and Technology* 16 (2005) 60-69
[10] Beucher S 1992 *Scanning Microscopy Supplement* pp 299
[11] Zhang Q and Skjetne R 2015 *IEEE Transactions on geoscience and remote sensing* pp2913
[12] Al-Kofahi Y, Lassoued W, Lee W and Roysam B 2010 *IEEE Transactions on biomedical engineering* pp841
[13] Otsu N 1975 *Automatica*, pp285
[14] Vincent L 1993 *IEEE Transactions on image processing* pp176
[15] Deng G and Cahill LW 1993 *Nuclear Science Symposium and Medical Imaging Conference* pp1615
[16] Martinez E, Jove X, Torre F and Santamaria E 1997 *16 Colloque sur le traitement du signal et des images*
[17] Krupka A, Riha A and Krizova L 2013 *Radioengineering Vol. 22* pp883

Properties and microstructure of twin-roll cast Al-Mg alloy containing Sc and Zr

M Cieslar[1], J Bajer[1], M Zimina[1], M. Šlapáková[1] and O Grydin[2]

[1]Faculty of Mathematics and Physics, Charles University, Ke Karlovu 5, 121 16 Prague 2, Czech Republic.

[2]Faculty of Mechanical Engineering, University Paderborn, Pohlweg 47-49, 33098 Paderborn, Germany.

E-mail: cieslar@met.mff.cuni.cz

Abstract. The influence of Sc and Zr on microhardness and microstructure of twin-roll cast Al-Mg alloy was studied. A significant effect of Sc and Zr on hardening of the material after annealing in a temperature interval between 250 °C and 400 °C was confirmed. Maximal hardness was obtained after annealing above 300 °C due to a precipitation of a fine dispersion of Sc and Zr containing precipitates. An exposure of the annealed material to higher temperatures suppresses the influence of these particles due to changes of the strengthening effect of precipitates. However, a partially positive role of remaining precipitates hindering recrystallization persists even up to 550 °C.

1 Introduction

Al-Mg-based alloys are widely used in aerospace industry due to their excellent weldability, reasonable corrosion resistance and possibility of superplastic forming. Additional alloying with Sc and Zr offer a great potential for developing of new lightweight structural materials with exceptional mechanical properties due to the presence of fine dispersion of coherent $Al_3(Sc,Zr)$ precipitates with the structure of the $L1_2$ type [1-7]. However, similarly to other high-strength Al-based alloys, sheets prepared from materials alloyed with Mg and Mn suffers from exfoliation corrosion [8-10]. Generally, exfoliation corrosion occurs when a highly directional microstructure typical for sheets and strips produced by rolling of initial ingots is present in the material. Therefore new manufacturing processes minimizing the formation of a typical pancake structure with flat and elongated grains are of a great interest. Twin-roll casting (TRC) is a typical example enabling directly cast strips with the thickness requested for semi-finished product. The most significant feature of this technology is a formation of rather equiaxed grains, in contrast to ones generated by a rolling procedure, which is an essential part of preparation of thinner strips from the originally direct chill (DC) cast ingots. Nevertheless, omission of steps as homogenization, scalping, rolling and intermediate annealing in several steps, which are the typical initial procedures of DC cast materials, leads to the formation of a structure, which exhibits markedly different features. Therefore also any final treatment of TRC material differs from ones established for DC cast materials. Nowadays, the processing of non-heat treatable Al-Mn and Al-Fe-Si-based alloys and high-strength Al-Mg-Si, Al-Zn-Mg-Cu, and Al-Mg-based alloys prepared by TRC is rather well established [11-13]. However, there are no information concerning the preparation and properties of TRC materials containing Zr and Sc. In the present contribution the microstructure of a model Al-Mg-Sc-Zr alloy prepared by TRC is described and compared with properties of TRC Al-Mg alloy with a similar level of Mg addition.

2 Experimental details

TRC alloys with a composition shown in Tab. 1 and thickness of 5 mm were studied. Twin-roll casting was realized with a vertical operation plane. No release agents were applied on the rolls surface before or during the process. Material was melted in a resistance furnace with mechanical stirring of the melt. Casting was done at 655 °C with 2.75 m/min (2.65 m/min for Al-Mg alloy) casting rate. More detailed description of applied experimental equipment is given in [14]. In order to investigate the changes of mechanical properties and microstructure at elevated temperatures samples were annealed in an air furnace and quenched after each annealing. Annealing was carried out using step-by-step isochronal heating scheme up to 550 °C with steps of 50 °C / 50 min or in an isothermal heating scheme at 300, 350 and 400 °C. Vickers microhardness measurements HV 0.1 with the load of 100 g at QNess 10A and residual electrical resistivity measurements (four point method in liquid nitrogen [15]) were done after each step. Light optical microscopy (LOM) (Olympus GX51) and in-situ transmission electron microscopy (TEM) (JEOL JEM 2000FX) observations with the same heating scheme as the one employed during isochronal annealing, were used for the material characterization. Specimens for electron microscopy were electropolished in 30 % HNO3 solution in methanol at -15 °C.

Table 1. Chemical composition of Al-Mg-Sc-Zr and Al-Mg alloys (in wt.%).)

Alloy	Al	Mg	Sc	Zr	Mn	Si	Cu	Fe	Zn
Al-Mg-Sc-Zr	Balance	3.24	0.19	0.14	0.16	0.11	0.024	0.21	<0.002
Al-Mg	Balance	3.67	<0.005	<0.001	0.18	0.11	0.025	0.24	<0.002

3 Results and discussion

The initial as-cast microstructure of both strips exhibits features typical for TRC aluminum and magnesium alloys with a strong influence of central segregation [16,17]. Finer grains of a size of about 50 to 100 microns appear in the central part of the strip and they are slightly elongated in the rolling direction, which is typical when an additional deformation of already solidified central part of the strip occurs in the gap between the rolls and grain coarsening is effectively inhibited by segregating primary phases rich in Fe, Si, Mg and Mn (figure 1).

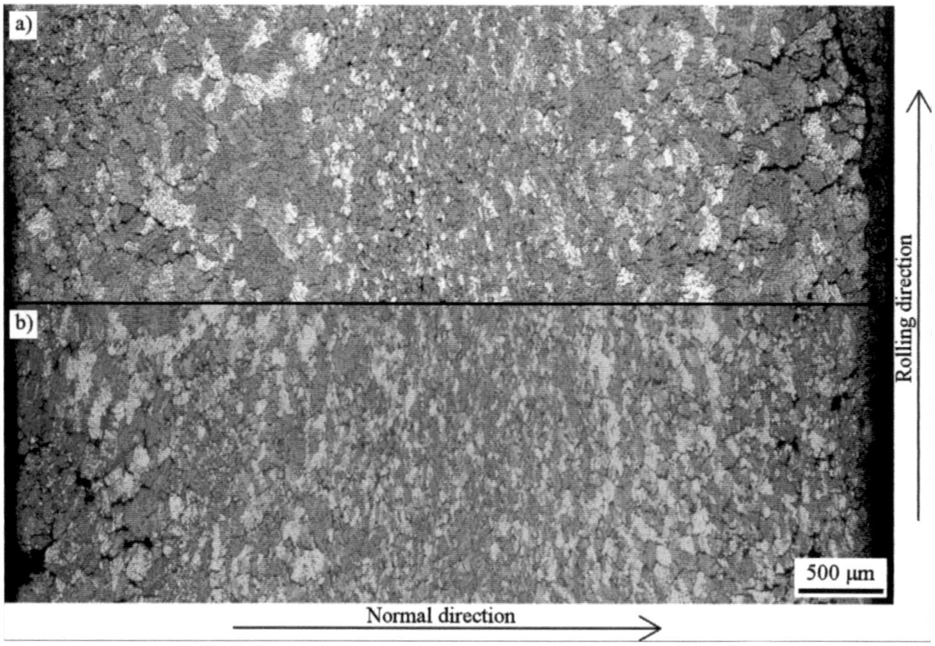

Figure 1. Microstructure of as-cast Al-Mg (a) and Al-Mg-Sc-Zr (b) alloys.

Coarser grains in the rest of the material (average grain size nearly 200 microns) and the presence of cracks reflect the inhomogeneity of material cooling during casting requesting further optimization of the process. This cooling starts first in the contact area between the melt and cold rolls and propagates toward the central part, which is always at higher temperature, accommodating thus any deformation imposed by rolls only in the central part of the strip. No significant difference in the grain size between Al-Mg and Al-Mg-Sc-Zr alloys is observed in initial states reflecting rather low influence of Sc and Zr on the solidification process. To a certain extent more pronounced centerline segregation observed in Al-Mg alloy (figure 2) is most probably an artifact of the presence of characteristic channels with segregated solutes as has been recently reported in TRC Al-Mn-Fe-Si alloy [18]. A positive role of slightly higher casting speed also could not be excluded.

The redistribution of solutes inside both materials was monitored by electrical resistivity measurements. The representation in the form of electrical resistivity spectra (negative temperature derivatives of resistivity evolutions) was chosen in order to highlight temperature ranges connected with the most significant redistribution of solutes (figure 3). A peak in positive values generally represents precipitation, coarsening of precipitates or their transformation connected generally with a solid solution depletion. Negative values are linked to particles or segregations dissolution and reversion of solutes into the solid solution. Only moderate changes are observed in both alloys below 200 °C. Above this temperature significant negative values appear in the Al-Mg spectrum, indicating a solid solution enrichment. Similar behavior in this temperature range was observed also in DC cast Al-Mg alloys with higher Mg concentration [19] suggesting that this enrichment occurs most probably due to the dissolution of particles in the Mg-rich areas segregated during casting in the central zone and at eutectic boundaries. The broad temperature interval moreover indicates that a long-range diffusion associated with the homogenization of the material and more uniform distribution of Mg occurs. On the contrary, two distinct peaks appear in the the Al-Mg-Sc-Zr alloy. The first one has its maximum situated near 280 °C, and the second one at 400 °C. Their position fits well with similar peaks observed by Vlach et al. [4,20] on DC cast Al-Sc-Zr and Al-Mn-Sc-Zr alloys and generally they are respectively attributed to Al_3Sc particles precipitation and Al_3Zr-shell formation in ternary Al-Sc-Zr alloy, while in alloys containing Mn the second peak is a convolution of the Al_3Zr-shell formation and precipitation of Mn-rich particles. Similar behavior is expected also in the TRC Al-Mg-Sc-Zr alloy. The homogenization of Mg distribution (as observed in the Al-Mg alloy) has most probably also the influence on the resistivity spectrum, however, is overlapped by massive resistivity changes caused by the depletion of the solid solution from Sc, Zr and Mn solutes.

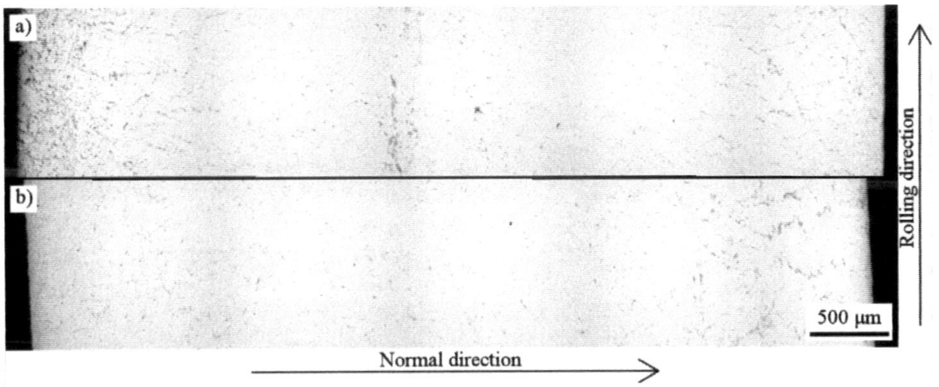

Figure 2. Distribution of primary phase particles in as-cast Al-Mg (a) and Al-Mg-Sc-Zr (b) alloys.

Figure 4 shows the evolution of microhardness in both alloys. Only sluggish decrease of microhardness in Al-Mg alloy reflects most probably only recrystallization and grain coarsening during annealing. On the contrary, the first peak in resistivity spectrum of Al-Mg-Sc-Zr alloy is directly connected with maximal hardening of the material, while microstructural changes occurring in the second one are responsible for a gradual degradation of microhardness.

TEM observations confirm the precipitation of two types of particles in the TRC Al-Mg-Sc-Zr alloy at higher temperatures (figure 5). A dense dispersion of $L1_2$ type precipitates appear in the material above 250 °C (most probably Al_3Sc), and their growth occurs (probably $Al_3(Sc, Zr)$ particles are formed) above 350 °C as well as the precipitation of Mn-containing particles.

The positive role of $Al_3(Sc, Zr)$ precipitates as an effective inhibitor of recrystallization is documented in figure 6. A significant grain growth is observed in the Sc and Zr-free alloy with the grain size reaching 1 mm after annealing at 550 °C. On the contrary remaining $Al_3(Sc, Zr)$ precipitates successfully hinder recrystallization in Al-Mg-Sc-Zr alloy even at highest annealing temperatures.

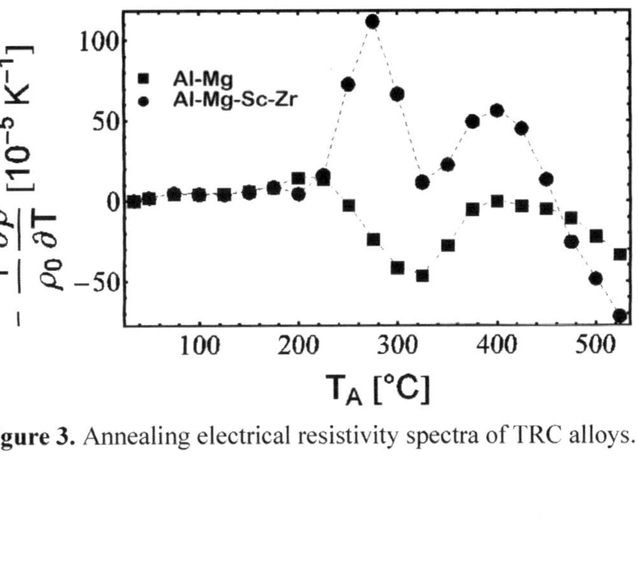

Figure 3. Annealing electrical resistivity spectra of TRC alloys.

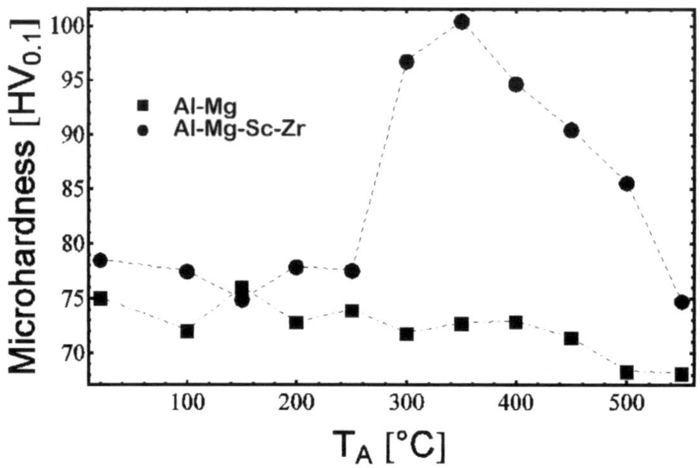

Figure 4. Evolution of microhardness of TRC alloys during annealing.

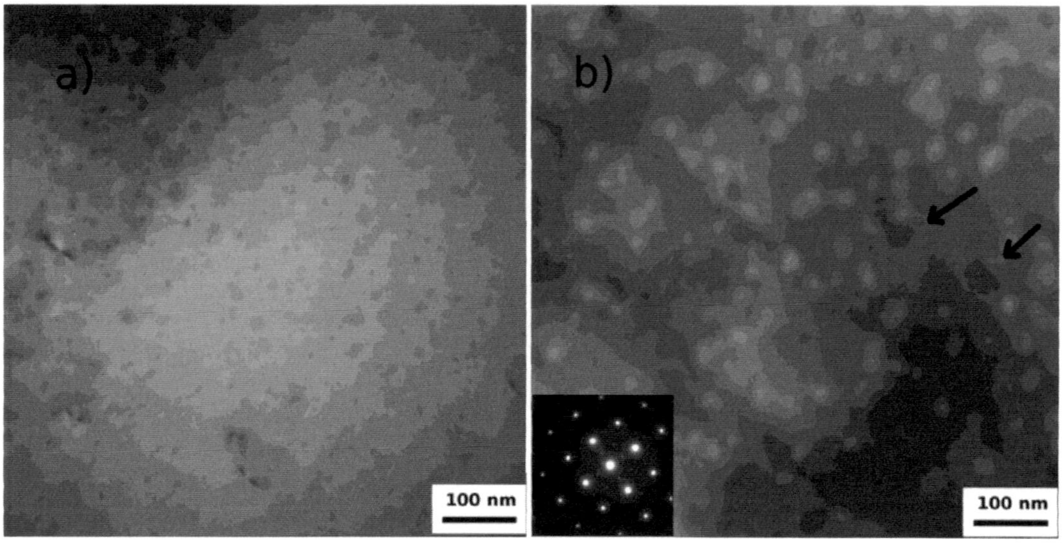

Figure 5. TEM micrographs of Al-Mg-Sc-Zr alloy after annealing up to 300 (bright field image close to $[100]_{Al}$ zone axis (a)) and 400 °C (dark field image in (100) $L1_2$ reflection close to $[001]_{Al}$ zone axis – see the inset (b)). Arrows indicate darker Mn-containing precipitates.

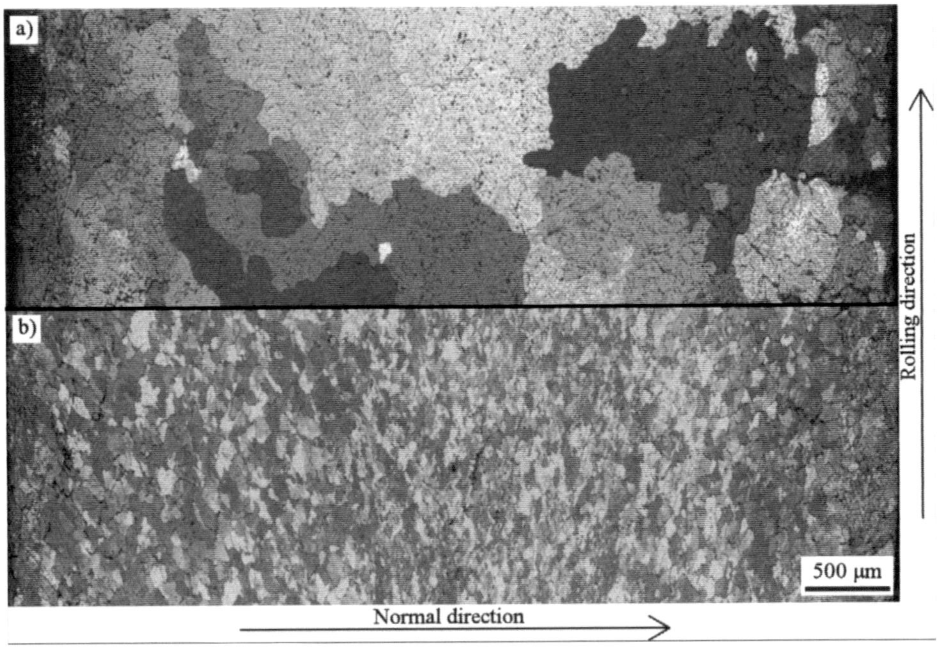

Figure 6. Microstructure of Al-Mg (a) and Al-Mg-Sc-Zr (b) alloys after isochronal annealing up to 550 °C.

4 Summary

Strips from Al-Mg-Sc-Zr and Al-Mg alloys were prepared by twin-roll casting method. The initial microstructure of both alloys exhibits centerline segregation and an inhomogeneous distribution of grain size with significantly smaller grains in the central zone of the strip. Annealing up to 550 °C results in a homogenization of Mg distribution within the material and formation of $Al_3(Sc,Zr)$ precipitates in Sc and Zr containing alloy. The formation of fine particles is responsible for a significant hardening of Al-Mg-Sc-Zr alloy, whereas their coarsening results in a significant drop of microhardness. Nevertheless, their positive role as inhibitors of recrystallization persist during the whole annealing process.

Acknowledgements
The financial support of the project of the Czech Science Foundation 16-16218S is gratefully acknowledged.

References

[1] Dewng Y, Yin Z, Zhao K, Duan J and He Z 2012 *J. Alloy. Compd.* **530** 71
[2] Fuller C B, Seidman D N and Dunand D C 2003 *Acta Materialia.* **51** 4803
[3] Lee S, Utsunomiya A, Akamatsu H, Neishi K, Furukawa M, Horita Z and Langdon T G 2002 *Acta Materialia* **50** 553
[4] Vlach M, Stulíková I, Smola B and Žaludová N 2010 *Mater. Charact.* **61** 1400.
[5] Knipling K E, Karnesky R A, Lee C P, Dunand D C and Seidman D N 2010 *Acta Mater.* **58** 5184
[6] Málek P, Turba K, Cieslar M, Drbohlav I and Kruml T 2007 *Mater. Sci. Eng. A* **462** 95

[7] Røyset J, Ryum N 2005 *Int. Mat. Rev.* **50** 19
[8] Liao M, Bellinger N C, Komorowski J P 2003 *Int. J. Fatigue* **25** 1059
[9] Peng Y, Li S, Deng Y, Zhou H, Xu G and Yin Z 2016 *Mater. Sci. Eng. A* **666** 61
[10] Oguocha I N A, Adigun O J and Yannacopoulos S 2008 *J. Mater. Sci.* **43** 4208
[11] Slámová M, Janeček M and Cieslar M 2007 *Mater. Sci. Eng.* **462** 375
[12] Birol Y and Karlík M 2006 *Scripta Materialia* **55** 625
[13] Xin S, Tan L and Guangming X 2015 *Rare Metal Mater. Eng.* **44** 581
[14] Grydin O, Ogins'kyy Y K, Danchenko V M and Bach F-W 2010 *Metallurgical and Mining Industry* **2** 348
[15] Hájek M, Veselý J and Cieslar M 2007 *Mater. Sci. Eng A* **462** 339
[16] Barekar N.S., Das S., Yang X., Huang Y, Omel El Fakir, Bhagurkar A.G., Zhou L. and Fan Z. 2016 *Mater. Sci. Eng. A* **650** 365
[17] Victoria-Hernandez J, Yi S, Bohlen J, Kurz G, Letzig D 2014 *J. Alloy. Compd.* 189
[18] Šlapáková M, Zimina M, Zaunschirm S., Kastner J, Bajer J and Cieslar M 2016 *Mater. Charact.* **118** 44
[19] Popović M and Romhanji E 2008 *Mater. Sci. Eng. A* **492** 460
[20] Vlach M, Stulíková I, Smola B, Žaludová N and Černá J 2010 *J. Alloy. Compd.***492** 143

Heat treatment of nitrided layer formed on X37CrMoV5-1 hot working tool steel

A Ciski[1], P Wach[1], J Tacikowski[1], T Babul[1] and P Šuchmann[2]

[1] Institute of Precision Mechanics, Duchnicka 3 St., 01-796 Warsaw, Poland
[2] COMTES FHT, Průmyslová 995, Dobřany, 334 41, Czech Republic

E-mail: aleksander.ciski@imp.edu.pl

Abstract. The paper presents the technology consisting of combination of the nitriding process with subsequent austenitizing at temperature above eutectoid temperature of the Fe-C system and further rapid cooling. Such treatment will cause formation of the martensite in the area of the primarily nitrided layer and the additional precipitation hardening by tempering of heat treated steel. The article shows that the heat treatment process of nitrided layer formed on X37CrMoV5-1 steel leads to strengthening of surface layer, the substrate and the core of nitrided part. Heat treatment of nitrided steel with the tempering in inert (nitrogen) or active (ammonia) atmosphere can increase the thickness of the layer formed by short-term nitriding process. After the nitriding process of X37CrMoV5-1 steel the nitrided layer had a thickness of about 160 μm, while a subsurface layer of iron nitrides had a thickness of 7 μm. After subsequent quenching and tempering processes, the nitrided layer undergoes additional diffusion and its thickness is increased to about 220 μm (inert atmosphere) or 280 μm (active atmosphere). If the tempering process is carried out in an inert atmosphere, the primarily formed layer of iron nitrides disappears. Tempering in an active atmosphere leads to forming of white layer with a thickness of 7 μm. Basic properties of nitrided layers formed in such way, like the hardness and the wear resistance, are presented.

1 Introduction

The overwhelming number of tool steels nitriding processes are long-duration processes carried out at low temperatures in the range of 500÷530°C. The reason for lower nitriding temperatures is tempering process carried-out before nitriding. This temperature of treatment is dependent on the precipitation processes occurring in these steels. For the hot work and high-speed steels tempering temperature ranges from 520 to 550°C. In order to increase the nitriding temperature, and thus producing a thicker nitrided layer in a shorter time, it would be convenient to develop nitriding process at the higher temperatures. In order to obtain good in-service properties not only of the nitrided layer, but also of the core of material, it becomes advantageous to use the heat treatment after the nitriding process (quenching and tempering) not only before the nitriding process.

Developed technology of hardening of nitrided layer, its substrate and the core of the nitrided part consists in the further soaking of workpiece at temperatures close to the austenitizing temperature and subsequent rapid cooling. Such treatment results in forming of a martensitic structure in the area of primary nitrided layer, which can be additionally hardened by aging. As a result of nitrogen migration occurring during austenitizing, the process performed in this way leads to forming in a nitrided subsurface zone of a (Fe,M)-C-N alloy with different structural features (chemical and phase composition) in comparison with the original nitrided layer. On the other hand, by choosing of an appropriate austenitizing temperature and proper cooling conditions there can also be achieved hardening of the nitrogen-free core of the material.

The already known solutions for hardening of the surface of the nitrided layer, namely, the nitrides layer substrate consist in the hardening of the diffusion layer zone. This can be achieved by rapid cooling after ferritic nitriding performed at the temperature close to eutectoid temperature of the Fe-N system (590°C), in order to achieve maximum supersaturation of a solution of Fe(N), followed by aging at temperatures of 80 ÷ 120°C, leading to strengthening of the Fe(N) solution as a result of formation of α" phase [1, 2]. This process gives visible strengthening effects in the case of carbon steels. In the case of alloyed steels such process is practically not effective because substantial impact on the strengthening of solution zone of the nitrided layer have special nitrides, mainly chromium nitrides, which are formed during the nitriding process [3, 4]. Another possible solution is the strengthening of the nitrided layer by austenitic nitriding combined with the rapid cooling [5-7]. The effect of the quenching is mainly strengthening of braunite or braunite/ferrite area by forming of nitrogen-martensite or nitrogen-carbon-martensite structure. This process can be used primarily for carbon steels, especially low carbon steels. In the case of alloy steel the process is less effective because the reduction of nitrides leads to decrease of nitrides layer hardness. Moreover, the hardness of strengthened core is reduced due to its further tempering [6]. The next solution leading to strengthening of the nitrided parts at their entire cross-section, ie. not only in the nitrided layer, but also in the core, is combination of the nitriding with subsequent austenitizing at temperature above eutectoid temperature of the Fe-C system and further rapid cooling. Such solution is known for bearing steels [8]. On the other hand, the data on the possibility of further precipitation hardening of iron and nitrogen alloys mostly come from laboratory studies [9-14].

Due to the properties of the nitrided layer and the substrate, it is favorable to carry-out the heat treatment after the nitriding process. Improved properties of layers can be obtained by dissolving of the subsurface layer of iron nitrides and subsequent nitrogen diffusion into the substrate as a result of austenitizing, further transformation of alloyed nitrogen austenite during the processes of deep cryogenic treatment (DCT) and the precipitation of fine alloyed carbonitrides in the internal nitriding zone during tempering. However, the heat treatment of alloy steel decreases hardness of the nitride layer and reduces dispersion of nitrides and carbonitrides, so it's advisable to use deep cryogenic treatment (after the quenching and before the tempering), which will cause grain refinement and obtaining of fine carbonitride precipitations of alloying elements. Sub-zero treatment of tools is widely used in the world, because the low temperatures technologies have a beneficial effect on the properties of heat-treated parts [15-18].

The paper presents the attempts to combine the heat treatment with DCT of nitrided layer formed on X37CrMoV5-1 hot work tool steel. Analyzes include basic properties such as microstructure, hardness, hardness at the cross section and wear resistance of steel. The aim of this work is to reduce overall treatment time and to form the nitrided layer with required thickness, characterized by the high properties of the surface of the material.

2 Experimental

Research was conducted using the X37CrMoV5-1 hot work tool steel with the chemical composition given in table 1 (declared composition according to PN-EN ISO 4957:2004).

Table 1. Chemical composition of the steel used in the research.

	Chemical elements content, mass %							
	C	Si	Mn	Cr	Mo	V	P	S
Declared composition	*0,33÷0,41*	*0,80÷1,20*	*0,25÷0,50*	*4,80÷5,50*	*1,10÷1,50*	*0,30÷0,50*	*max. 0,030*	*max. 0,020*
Actual composition	*0,35*	*1,29*	*0,33*	*4,91*	*1,31*	*0,29*	*0,03*	*0,008*

In order to study the effect of heat treatment on the nitrided layer the samples were processed in accordance with the following schemes (figure 1). Initially hardened and tempered samples samples were nitrided at 580°C for 4 hours. After the nitriding samples were quenched (austenitizing temperature of 1040°C) and tempered at 550°C. The optional DCT step (at -180°C for 8 hours) was implemented in order to determine if the additional transformation of alloyed nitrogen austenite will enhance the tribological properties of the nitrided steel. Tempering was performed in two various modes, with use of inert (nitrogen) or active (ammonia) atmosphere.

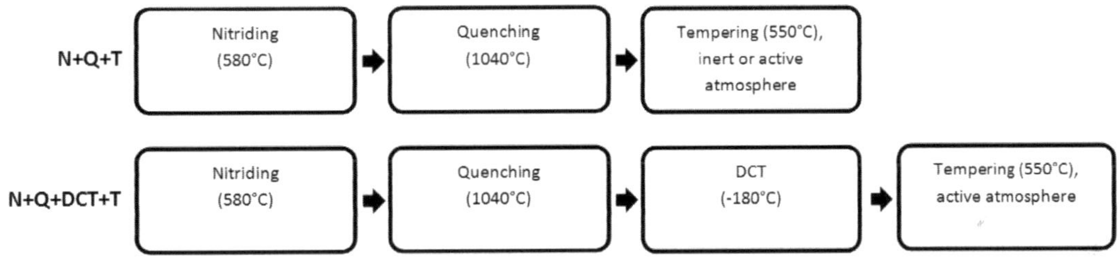

Figure 1. Block diagram of heat and thermo-chemical treatment routes.

Nitriding of steel samples was carried-out in NX609 Nitrex furnace. In this furnace it is possible to perform controlled gas nitriding in a fully automatic way. Computer control ensures repeatability of process parameters, which is an important element for the proper quality of the nitrided layers. Hardening of samples was carried out in a vacuum furnace with the high pressure gas quenching using nitrogen. The tempering of the samples was carried out in the same furnace in an atmosphere of nitrogen or in a nitriding NX609 furnace in an ammonia atmosphere.

Surface hardness measurements were made using Struers DuraScan-70 hardness tester under load of 0.5 kG. Hardness profiles were determined with aid of the same device.

Test of resistance to wear was carried out in a 3 rollers-cone system. The wear resistance of samples was compared with samples nitrided in the conventional manner (controlled gas nitriding at 520°C for 24 hours). Test was performed with a continuous lubrication of samples using Lux-10 oil, at a constant speed of counter-specimen rotating at a rate of 526 rpm and a constant unit pressure of 100 or 400 MPa. The depth of the wear, indicating the linear wear, was determined by measurements of diameters of the ellipses formed at the surface of each of the worn out rollers. Total friction time test was 100 minutes, wear marks were measured after every 10 minutes. After each measurement the load was increased in order to achieve constant unit pressure at the worn out surface. Conical counter-specimen was made of the X153CrMoV12 steel hardened to hardness of 60 HRC and ground to a roughness Ra = 0.32 micrometers.

3 Results and discussion

After the nitriding process of steel the obtained nitrided layer had a thickness of 160 µm with a subsurface layer of iron nitrides with a thickness of 7 µm. The surface hardness was 1235 HV0.5. The microstructure observed by LM is shown in Figure 1a.

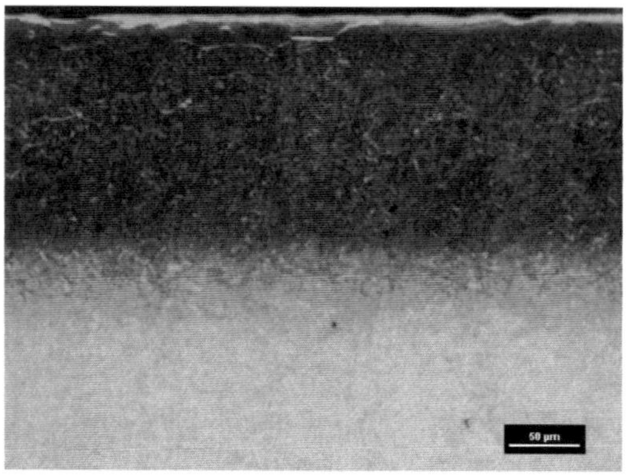

Figure 1. Microstructure of nitrided layer on X37CrMoV5-1 steel.

After subsequent quenching, deep cryogenic treatment and tempering processes (inert atmosphere) of nitrided X37CrMoV5-1 steel samples, the nitrided layer undergoes further diffusion and its thickness is increased to 225 µm (Figure 2a). There was no subsurface layer of iron nitrides and at the surface of samples was present the light area with a thickness of 60 microns from the surface. White area was soft and its hardness was 550 HV0.5. Whereas the dark layer observed under the white zone had a hardness in the range of 650 to 900 HV0.5 (Figure. 2a). After quenching, DCT and tempering in an active atmosphere of ammonia of samples, the correct surface hardness (1150 HV0.5) and high hardness of the core (640 HV0.5) was obtained, what is probably caused by the DCT process. The thickness of the subsurface layer of iron nitrides reached 7 µm, while the thickness of the nitrided layer was increased to 280 µm. It is impossible to assess the depth of diffusion zone based on the observed microstructure. Underneath the dark etched layer there is located the undisclosed by etching layer of matrix with high dispersion alloyed carbonitrides, influencing the hardness of diffusion zone. Microstructure of the nitrided layer after quenching ant tempering in an active atmosphere is shown in Figure 2b.

a) b)

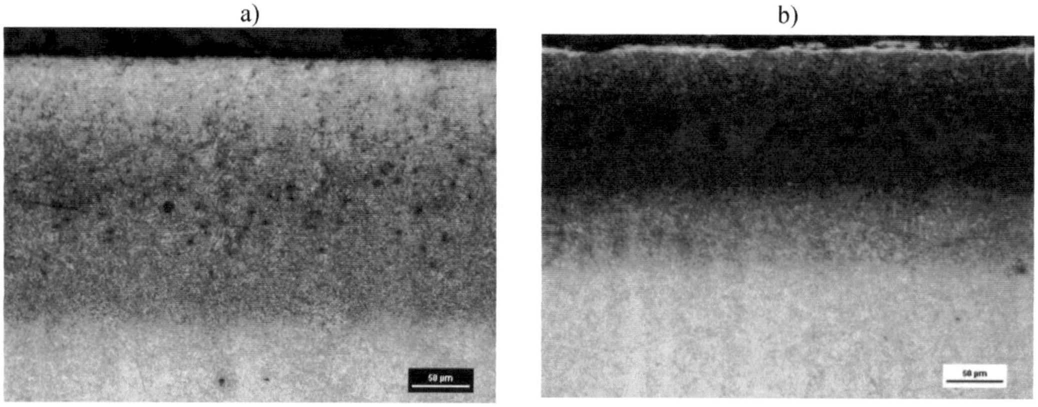

Figure 2. Microstructure of nitrided layers on X37CrMoV5-1 quenched, DCT and tempered in an inert (a) or active (b) atmosphere.

Figure 3 shows the distribution of hardness at a cross section of layers after nitriding or nitriding with subsequent heat treatment with tempering in an active atmosphere.

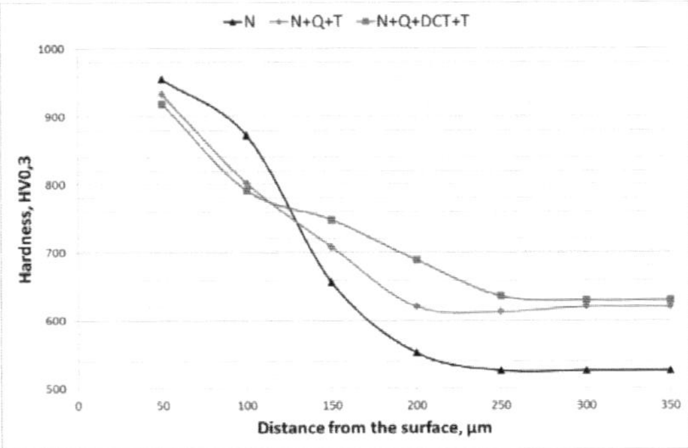

Figure 3. Comparison of hardness distribution of nitrided layer (N) with nitrided and heat treated layer (N+Q+T) or nitrided and heat treated layer with additional DCT (N+Q+DCT+T).

One can conclude that after the heat treatment of nitrided layer with properly selected parameters and use of active atmospheres during tempering (ammonia), it is possible to form in a short time a thick nitrided layer on the X37CrMoV5-1 steel (thickness up to 280 µm). The use of ammonia at a stage of tempering causes additional nitriding of the surface and forming of subsurface iron nitrides layer. Heat treatment cycle of nitrided layer, in comparison with steel simply nitrided, increases hardness of the core of material by about 100 HV (Figure 3). At the same time the use of additional operation of deep cryogenic treatment during the heat treatment cycle causes additional enhancement of the hardness level at the diffusion zone. Increased hardness is probably caused by the additional transformation of alloyed nitrogen austenite during the processes of deep cryogenic treatment and the precipitation of fine alloyed carbonitrides in the internal nitriding zone during tempering.

The results of wear resistance test in the form of diagrams of linear wear as a function of time are shown in Figures 4a (100 MPa) and 4b (400 MPa). All nitrided samples were characterized by the wear with a regular course, manifesting the occurrence of steady friction state in the entire course of wear, except running-in period.

At unit pressure of 100 MPa the lowest linear wear values in the first and last phase of test had the samples subjected to a conventional nitriding. The highest wear resistance of samples subjected to conventional nitriding is probably connected with presence of the hard compound layer at the surface of the material, influencing tribological properties of steel. Despite of the good wear properties of the surface of steel, such hard and therefore brittle layer is unfavorable from the point of view of fatigue strength of the tool made from the hot work tool steel.

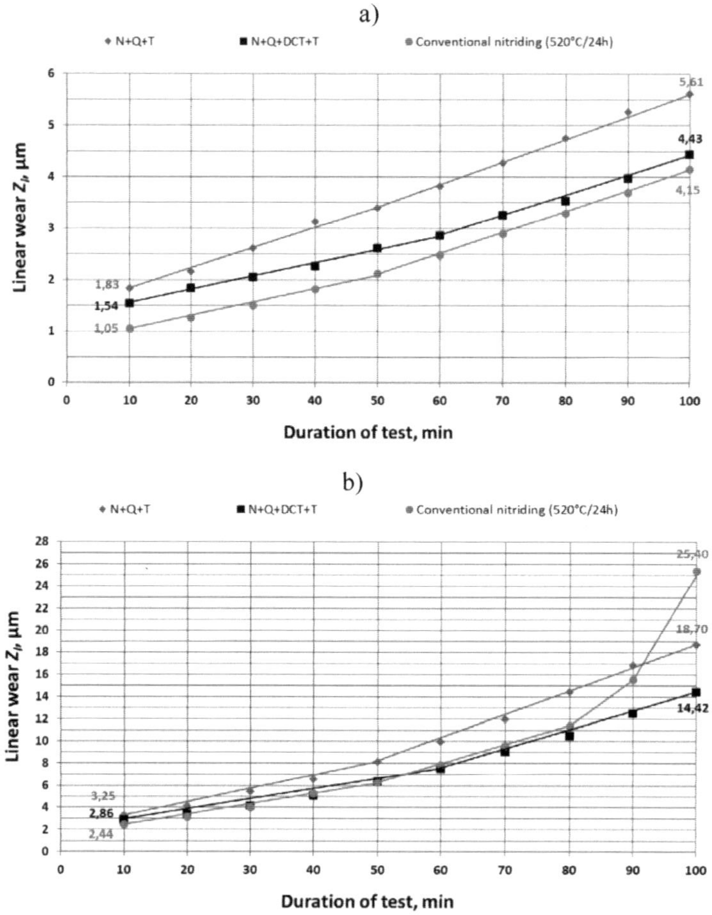

Figure 4. Wear courses at 100 MPa (a) and 400 MPa (b) of samples subjected to different modes of nitriding with or without heat treatment.

Wear test performed at the unit pressure of 400 MPa demonstrated the superior tribological properties of the heat treated nitrided layers (Figure 4b). For the conventional mode of treatment (controlled gas nitriding at 520°C for 24 h) the linear wear after 100 minutes of test was 25.40 µm, while after the heat treatment or heat treatment with DCT the linear wear was about two 26 % or 43 % lower (18.70 µm and 14.42 µm, respectively). The nitrided layer formed in a conventional manner manifested tendency to galling, with a visible phenomenon of accelerated wear at the last stages of the test (Figure 4b). The high tribological properties of the nitrided layer subjected to heat treatment are connected with proper structure of the compounds layer on the material's surface (Fig. 2b).

4 Conclusions

Heat treatment in an active atmosphere of nitrided X37CrMo5-1 hot work tool steel allows forming of thick nitrided layers by the short-term nitriding process. Developed technology allows to reduce the nitriding time necessary for forming the layers with a required thickness. After subsequent quenching and tempering processes the nitrided layer undergoes additional diffusion and its thickness is increased. If the tempering process is carried out in an inert atmosphere, the primarily formed layer of iron nitrides disappears. Tempering in an active atmosphere leads to forming of white layer with a thickness of 7 μm.

Shortening of the nitriding process enables reduction of the overall duration of heat an thermo-chemical treatment cycle, thus affecting the cost of production process of treated parts, as well lowers the emission of the environmentally harmful gases.

Heat treatment of nitrided layers increases the hardness in the core of the tool steel. Formed layers have high resistance to wear under high loading conditions and are resistant to galling.

Acknowledgements
This research was supported by Eureka Programme (Eureka E!9170 "CRYONITRIDE" Project).

References

[1] H. Krzymiński: Einfluss der Aushärtung durch Stickstoff in der Diffusionszone nitrierter unlagierter Stähle auf Festigeit, plastiche Verformbarkeit und das Verhalten beim umlaufender Biegung, Durferrit Hausmitteilungen, 1990, wyd. Deguss – Durferrit
[2] Ju. M. Lachtin, Ja. D. Kogan: Azotirovanie stali, wyd. Masinostrojenie, Moskva, 1976
[3] Ju. M. Lachtin, Ja. D. Kogan, H. J. Spies, S. Böhmer: Teoria I technologia azotirovania, Metallurgia, Moskva, 1991
[4] M. Fattah: Comparison of Ferritic and Austenitic Plasma Nitriding and Nitrocarburizing Behavior of AISI 4140 Low Alloy Steel, Materials & Design, no. 31(8), 2010, pp. 3915-3921
[5] Ju. M. Lachtin: Vysokotemperaturnoje azotirovanie, Metaloviedienie i Termicieska Obrabotka, 1991, no. 2, pp. 25-29
[6] Ju. M. Lachtin, G.N. Neustroev, N . I. Sologubova, L. P. Frolova: Nitricementacija pri 700°C s posledujuscej zakalkoj poverchnostnovo sloja, Metaloviedienie i Termicieska Obrabotka, 1987, no. 5, pp. 32-36
[7] S. Pakrasi, H. Jürgens, J. Betzold: Nitrocarburieren zwischen den eutoktoiden Temperaturen des Fe-N und Fe-C-Sytems, HTM, 1993, no. 5, pp. 215-219
[8] A.V. Gavrilova, P.P. Babenko, V.V. Dudarev: Strukturnyje prevvrascenia pri starenii azotistvo martensita splavov zezleza, legirovannych azotom I initridoobrazujuscimi elementami, Fiz. Met. I Metallov., 1989, t. 68, pp. 89-94
[9] E. J. Mittemeijer, A. Wierszyłowski: The isothermal and nonisothermal kinetics of tempering iron-carbon and nitrogen martensitesa and austenites, Zitung für Metalkunde, 1991, no. 6, pp. 419-429
[10] L. Cheng, A. Böttger and E.J. Mittemeijer: Tempering of Iron-Carbon-Nitrogen Martensites, Metallurgical Transactions. A, Physical Metallurgy and Materials Science, no. 23(4), 1992, pp. 1129-1145
[11] P.A. Beaven, E. P. Butler: Precipitation Nucleation on Disclocation in Fe-N, Acta Met., 1980, no. 1, pp. 1348-1359
[12] N.I. Kardonina, A.S. Yurovskikh, A. S. Kolpakov: Transformations in the Fe – N System, Metal Science and Heat Treatment, 52(9), 2011, pp. 457-467
[13] V.G. Gavriljuk, "Austenite and Martensite in Nitrogen-, Carbon- and Hydrogen-Containing Iron Alloys: Similarities and Differences, Materials Science & Engineering. A, Structural Materials: Properties, Microstructure and Processing, vol. 438-440, 2006, pp. 75-79
[14] Y. Wei, Z. Żurecki, R. D. Sisson Jr.: Termodynamiczny model przemian fazowych podczas międzykrytycznego azotowania austenitycznego, Inżynieria Powierzchni, 2013, no. 2, pp. 56-62
[15] M. Pellizzari, D. Caliskanoglu, A. Fernández, J. I. Barbero, B Pena, T. Uemit, R. Pizarro Sanz, R. Elvira Eguizabal, and L. A. Alava, Influence of different deep cryogenic treatment routes on the properties of high speed steel. HTM Journal of Heat Treatment and Materials: Vol. 67, no. 2, 2012, pp. 111-117.
[16] B. Podgornik, V. Leskovsek, J. Vizintin: Influence of deep cryogenic treatment on tribological properties of P/M high-speed steel, Journal Materials and Manufacturing Processes, 24 (2009),

pp. 734–738
[17] J. Jeleńkowski, A. Ciski, T. Babul: Effect of deep cryogenic treatment on substructure of HS6-5-2 high speed steel, "JAMME", Vol. 43, Issue 1, 2010, pp. 80-87.
[18] A. Ciski, J. Jeleńkowski, J. and T. Babul: Microstructural, thermal stability and wear resistance properties of X153CrMoV12 tool steel subjected to deep cryogenic treatment, Int. J. Microstructure and Materials Properties (2015), Vol. 10, No. 2, pp.140–148.

IOP Conf. Series: Materials Science and Engineering **179** (2017) 012014 doi:10.1088/1757-899X/179/1/012014

Elements for the modeling of the thermal process in heating furnaces for steel forming

D Constantinescu, A B Carlan

University Politehnica of Bucharest, Faculty of Science and Materials Engineering, Bucharest, Romania, EU

E-mail: dconstantinescu@ecomet.pub.ro; danco@ines.ro

Abstract. In the present paper, by "modelling of thermal process" will be understood the thermal techniques modelling, applied to the heating of steel billets in a large scale, in view of processing by forming. These technologies are correlated with the particularities of the thermal aggregates, having as main objective the reducing of energy consumptions and the optimizing of the aggregate design. When heating the steel billets in view of processing by forming, the duration and the quality of heating are influenced by the modality that the billets are receiving the thermal flow. The reception of the thermal flow depends on the heated surface exposed to the thermal radiation in compliance with their position on the hearth of the heating aggregate. The present paper intends to establish some parameters in view of optimizing the heating process. A basic point of the work is also the determination of some components of a mathematical model for the proposed heating technology. The authors have in view the complexity of the technical evolutions of the furnaces.

1 The objective of the article

The correlation between the thermal phenomena in a heating aggregate for the steel forming is very complex. We have to take this into consideration if we are expecting to obtain a good quality of heating and an energy and material reduction.

For example, the evaluation of the radiation heat exchange between the thermal isolation components may be calculated using the angular coefficient of radiation,φ, recommended by Heiligenstaedt [1], [2].

$$\varphi = \frac{1}{2\pi}\left(\frac{B}{\sqrt{1+B^2}}\cdot\arcsin\frac{L}{\sqrt{1+B^2+L^2}}+\frac{L}{\sqrt{1+L}}\cdot\arcsin\frac{B}{\sqrt{1+B^2+L^2}}\right) \quad (1)$$

where B and L refers to the ratio between the geometrical dimensions of the heated billets and the thermal aggregate (e.g. $B = h/b$, $L = l/b$) (figure 1)

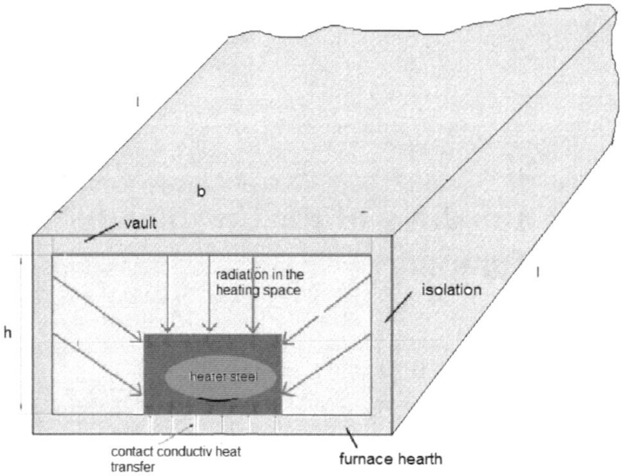

Figure 1. Heat transfer by radiation and contact conduction in the fourance with the dimanssion $b{\cdot}l{\cdot}h$.

2 Geometry and heat transfer

In order to analyse the heat exchange by radiation there were taken into consideration some cases for the heating furnaces. In figures 2 and 3 there are analysed the cases of the square and rectangular sections [2], [3]. In table 1 it is a presented a comparison between the two cases [3] and in table 2 it is analyzed the case of the billets with circular section.

Table 1. Comparison between the evaluation of the equivalent surface of heat transfer for the square and rectangular sections of the billets.

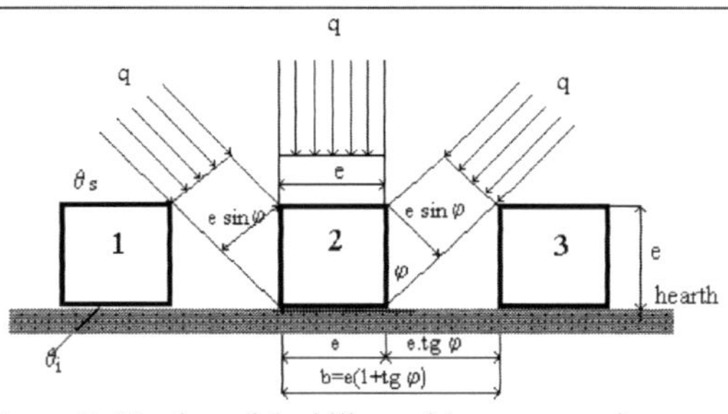 **Figure 2**. Heating of the billets with square section on the hearth of the furnace; q - thermal flow; φ - angle of radiation; θ_s – temperature of the upper surface; θ_i – temperature of the inferior surface of the billet; l – length of the billet; S – equivalent surface of heat exchange $S = e \cdot l + 2 \cdot e \cdot l \cdot sin\varphi = e \cdot l(1 + 2sin\varphi)$ (2) The difference of the temperature between the upper and inferior surface, $\Delta\theta$, will be: $\Delta\theta = \theta_s - \theta_i = \frac{q}{\lambda} \cdot (1 + 2 \sin \varphi)$ (3) λ – thermal conductivity of the steel If the billets are stuck, (as in case of the pusher-type furnace), heating by the upper face, then sin φ=0. If the billets are distanced one to each other, *(e<<e.tg φ)*, then *lim(sin* φ) *=1*. In this situation it was obtained a better uniformity of the temperature on the billet	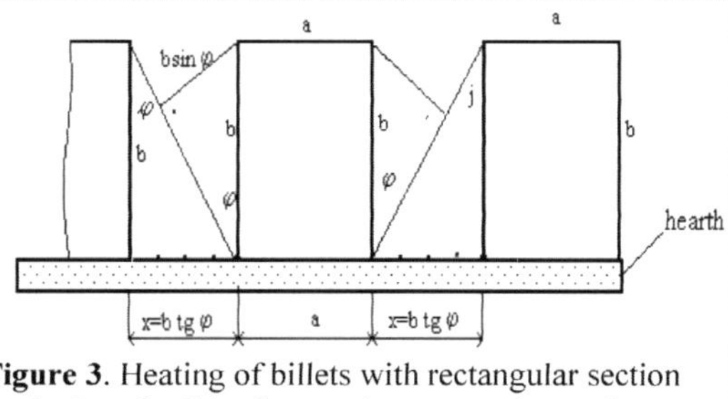 **Figure 3**. Heating of billets with rectangular section on the hearth of the furnace (same notation as for figure 2); $a/b = f$ $S = l \cdot b \cdot (j \cdot tg\varphi + 2 \sin \varphi)$ (8) j = 2 ÷0.4, depending on the distance between the billets Examples: *x=0.5·a* : *S=l·b·(2·*tg φ+*e·*sin φ) x=2.5·a : *S=l·b·(0,4·*tg φ+*2·*sinφ) Remarks: -the maximal values of equivalent surface in the conditions of *a = ct*, there are obtained at an incidence angle of the thermal radiation of 30^0; from these, the biggest value is obtained in the case of x=2.5a, for ratio of a/b = 0.25 -the smallest values of the coefficient of optimum

section than in case of the both faces heating in the pusher-type furnace. The heating mode equivalent to the case of the both faces heating in the pusher-type furnace is obtained for the case of the walking beam furnace, when $\varphi=60^0$.

The most favourable possibility, from the thermal point of view, in the case of the heating square billets, would be when $\varphi=60^o$. Having in view the requirement to provide a high degree of furnace hearth charging, as well as the same productivity, it must be taken into consideration the case when $\varphi=45^o$. This represents, on the basis of established data, practically the optimum supposed situation from the point of view of the heating uniformity, the heating time and furnace productivity.

To analyze easier the heating mode of the billets, it will be introduced the notion "*specific time of internal heating - STIH*",[3] representing the necessary time to heat a billet with effective thickness (which corresponds to the geometric thickness and is different from the *thermal thickness* which is reported to the Biot criteria; in the case of square section billets, $X=e$), reported to the thermal diffusivity, "a_0", suitable to the heating temperature:

$$t = \frac{X^2}{a_0} = \frac{X^2 \cdot c \cdot \rho}{\lambda} \tag{4}$$

For modelling, we propose to use the main relations, in the case of square section, presented below:

function of equivalent surface of heat exchange	$k_1 = 1 + 2sin\varphi$
function of heating duration	$k_2 = \frac{(1 + 4\sin^2\varphi)}{(1 + 2\sin\varphi)}$
function of the specific time of internal heating	$i = \frac{1}{(1 + 4\sin^2\varphi)}$
criteria of optimum distance between billets	$z = \frac{(1 + tg\,\varphi)}{(1 + 4\sin^2\varphi)}$
specific time of internal heating	$t = X^2 \cdot \frac{i}{a_0}$

The specific time of internal heating:

$$t = \frac{e^2 \cdot c \cdot \rho}{(1 + 4\sin^2\varphi)\cdot \lambda} \tag{5}$$

Example: STIH for reinforcing bars

X mm	80	100	120	140	200
t min	10.6	16.7	24.0	32.8	66.8

distance, z, are obtained for a distance between billets of x=1.5a........2a and for the values of the incidence angle of thermal radiation of $45......60^0$; in these conditions, the optimum value of the ratio between the section sides of billet must be 0.6.......1.2; at the distance of x>2.5a, the coefficient „z" may be considered constant.

- the specific time of heating is first of all influenced by the shape of the billet section: the minimum value is obtained for flat billets (f=5.5; i=0.05); for similar values of the ratio, the value of the STIH decreases by the increase of the distance x;
- for the interval considered optimum (x=1.5a.......2a), "i" has the value (0.3....0.4) for f=(0.6....0.5), ($\varphi=45^0$) and (0.18....0.22) for f=(1.1....0.9), ($\varphi=60^0$)
- for x>2a, the STIH for the same values of the incidence angle is not very modified.

STIH - represents the time necessary to heat a billet with effective thickness (which corresponds to the geometric thickness and is different from the *thermal thickness* which is reported to the Biot criteria; in this case, the section $X=b$), reported to the thermal diffusivity "a_0", suitable to the heating temperature.

For modelling, we are proposing to use the main relations, in the case of rectangular section:

function of equivalent surface of heat exchange	$k_1 = j \cdot tg\,\varphi + 2 \cdot \sin\varphi$
function of heating duration	$k_2 = \frac{j \cdot tg^2\varphi + 4\sin^2\varphi}{j \cdot tg\,\varphi + 2\sin\varphi} = \frac{1}{k_1 \cdot i}$
function of the specific time of internal heating	$i = \frac{1}{j \cdot tg^2\varphi + 4\sin^2\varphi}$
criteria of optimum distance between billets	$z = \frac{b \cdot (f + tg\,\varphi)}{j \cdot tg^2\varphi + 4\sin^2\varphi} = b \cdot (f + tg\,\varphi) \cdot i$
specific time of internal heating	$t = X^2 \cdot \frac{i}{a_0}$

The heating duration:

$$\tau = t \cdot \frac{\theta_f - \theta_i}{\Delta\theta} \cdot \frac{j \cdot tg^2\varphi + 4sin^2\varphi}{j \cdot tg\varphi + 2sin\varphi} \tag{9}$$

θ_f and θ_i: final and initial average temperatures of the heated material

The productivity "P" will be calculated in this case using the relation:

$$P = \frac{m \cdot L_C \cdot \Delta\theta}{t \cdot k_2 \cdot (\theta_f - \theta_i) \cdot b \cdot f} \tag{10}$$

Following [3], some data for the relations are:

x	φ	f	j	k_1	k_2
0.5a	26	1	1	1,88	0,94
	30	1.15	2	2,15	0,77
	45	2	2	3,41	1,17

The heating duration:

$$\tau = t \cdot \frac{\theta_f - \theta_i}{\Delta\theta} \cdot \frac{1 + 4\sin^2\varphi}{1 + 2\sin\varphi} \qquad (6)$$

The productivity "P" can be calculated using the relation:

$$P = \frac{m \cdot n \cdot \Delta\theta}{t \cdot k_2 \cdot (\theta_f - \theta_i)} = \frac{m \cdot L_c \cdot \Delta\theta}{t \cdot k_2 \cdot (\theta_f - \theta_i) \cdot e} \qquad (7)$$

where L_c is the length of the furnace

	60	3.5	2	5,19	1,75
	70	5.5	2	7,37	2,71
a	30	0.57	1	1,57	0,84
	45	1.00	1	2,41	1,25
	60	1.70	1	3,46	1,70
	70	2.75	1	4,62	2,40
1.5a	30	0.30	0.66	1,38	0,89
	45	0.60	0.66	2,07	1,68
	60	1.15	0.66	2,87	1,89
	70	1.80	0.66	3,69	2,46
2a	30	0.30	0.5	1,28	1,70
	45	0.50	0.5	1,91	1,74
	60	0.90	0.5	2,60	1,75
	70	1.40	0.5	3,25	2,21

Table 2. Heating of the billets with circular section (R-radius; l – length of the billet).

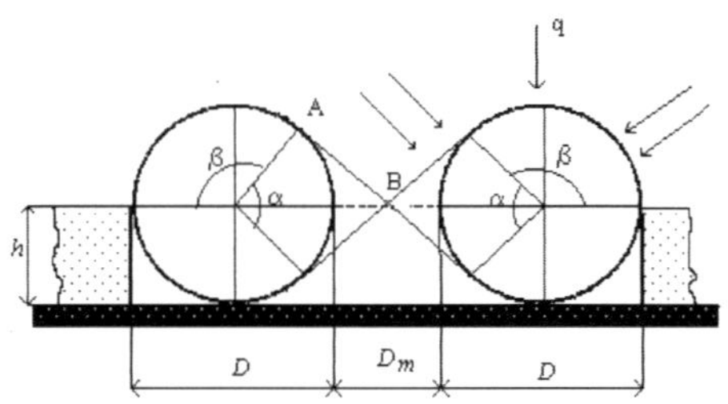

Figure 4. Heating of the billets with circular section in a roller heat furnace (a).

$\alpha = 2\arccos[D/(D_m + D)]$ (11)

$\beta = 180 - \arccos[D/(D_m + D)]$ (12)

$$S = R \cdot l \left[180 + \arccos\frac{D}{D_m + D}\right] \cdot \frac{\pi}{180} \qquad (13)$$

Case c) In case (a), if $h=D/4$, it is obtained :

$$\alpha = 2\arccos\frac{D}{D + D_m} \quad \text{and} \quad \beta = 210 - \arccos\frac{D}{D + D_m}$$

$$S = R \cdot l \cdot \left[210 + \arccos\frac{D}{D + D_m}\right] \cdot \frac{\pi}{180} \qquad (14)$$

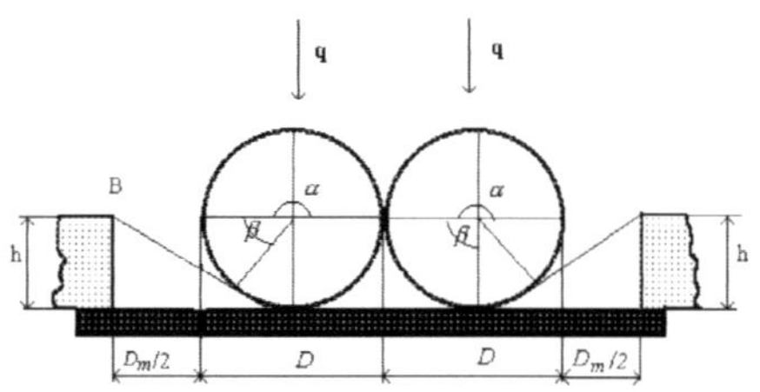

Figure 5. Heating of the billets with circular section on the hearth of the furnace (b).

$$\alpha = 180^o \qquad \beta = \arccos\frac{D}{D + D_m} \qquad (15)$$

$$S = R \cdot l \cdot \left[180 + \arccos\frac{D}{D + D_m}\right] \cdot \frac{\pi}{180}$$

(16)

Case d) In case when the billets do not support to the border and are tangent to the middle of the hearth , the equivalent surface will be given by :

$$S = R \cdot l \left[180 + arctg\frac{D}{2(D + D_m)} + \arccos\frac{2D}{\sqrt{4(D + D_m)^2 + D^2}}\right] \cdot \frac{\pi}{180} \qquad (17)$$

Case e) If over the two billets which support to the border ($h=0,5D$) it is placed the third one with the same diameter D, ($D \geq D_m$), the following relations for equivalent surfaces are obtained :

$$S = R \cdot l \cdot \left(180 - \arccos\frac{D + D_m}{2D}\right) \cdot \frac{\pi}{180} \qquad (18)$$

- for the ingot placed above :

$$S = R \cdot l \cdot \left(180 + 2\arccos\frac{D + D_m}{2D}\right) \cdot \frac{\pi}{180} \qquad (19)$$

The general form of relations for the equivalent surface of heat exchange will be :

$$\boxed{S = k_1 \cdot R \cdot l} \tag{20}$$

The geometric surface of billets is $S_0 = 6{,}28 \cdot R \cdot l$, thus S could be written :

$$\boxed{S = 0{,}159 \cdot k_1 \cdot S_0} \tag{21}$$

The heating time will be determined by relation:

$$\tau = \frac{\pi \cdot R \cdot \rho \cdot c \cdot (\theta_f - \theta_i)}{k_1 \cdot q} \tag{22}$$

where q is the thermal flow (kJ/m^{2}h); θ_f and θ_i: final and initial temperatures of the billet

In the table 3 are presented the values for the coefficient k_1 and for the equivalent surface of heat exchange.

Table 3. Values of the coefficient k_1 and the equivalent surface S.

Disposal mod of the billets	Ratio D_m/D	k_1	S
 case a) equ. (13)  case b) equ. (16)	D_m=4D	4,51	0,72S_0
	D_m=2D	4,37	0,69S_0
	D_m=1D	4,19	0,67S_0
	D_m=(2/3)D	4,07	0,65S_0
	D_m=0,5D	3,98	0,63S_0
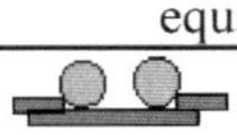 case c) equ. (14)	D_m=4D	5,05	0,80S_0
	D_m=2D	4,95	0,79S_0
	D_m=1D	4,82	0,77S_0
	D_m=(2/3)D	4,76	0,76S_0
	D_m=0,5D	4,71	0,75S_0
 case d) equ. (17)	D_m=4D	4,61	0,73S_0
	D_m=2D	4,54	0,72S_0
	D_m=1D	4,43	0,71S_0
	D_m=(2/3)D	4,36	0,69S_0
	D_m=0,5D	4,30	0,68S_0
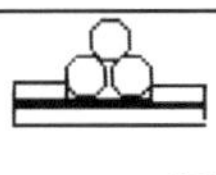 case e) equ.(18)	D_m=D	3,14	0,50S_0
	D_m=(2/3)D	2,56	0,41S_0
	D_m=0,5D	2,42	0,39S_0
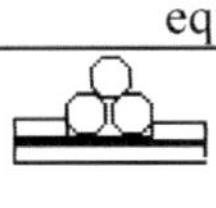 case e) equ.(19)	D_m=D	3,14	0,50S_0
	D_m=(2/3)D	4,31	0,69S_0
	D_m=0,5D	4,59	0,73S_0

4 Diagrams regarding the optimization of the heating

In order to select the best possibility for the billets disposal in a heating furnace we will use the criteria of the optimum distance, z (figure 6). Of course, we have to consider also the specific time of internal heating, function "i" (figure 7).

Applications for the optimization function in the case of square billets section are presented in figure 8.

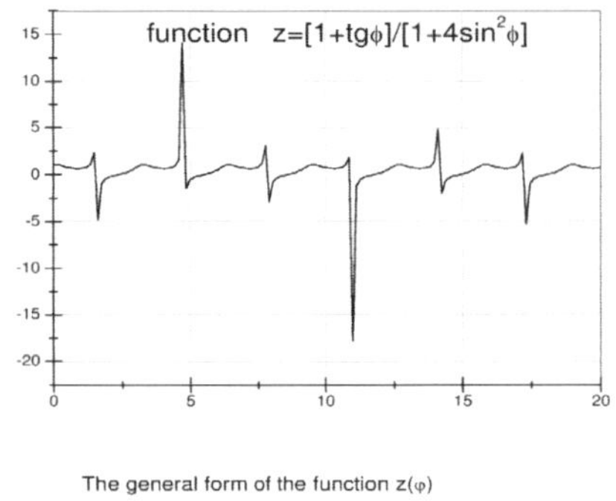

Figure 6. The graphic form of function "z".

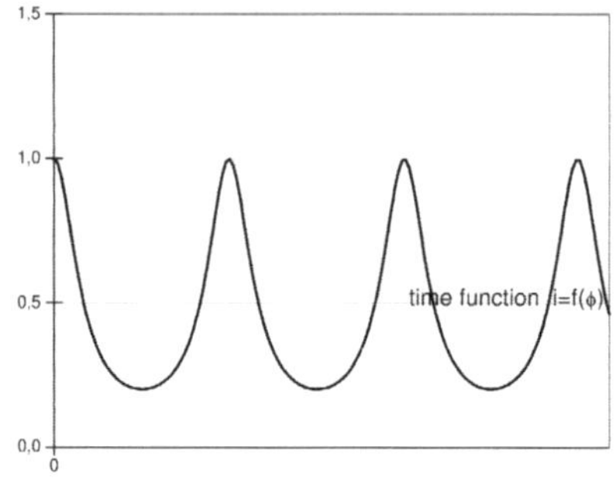

Figure 7. Function of the specific time of internal heating, i=f(φ).

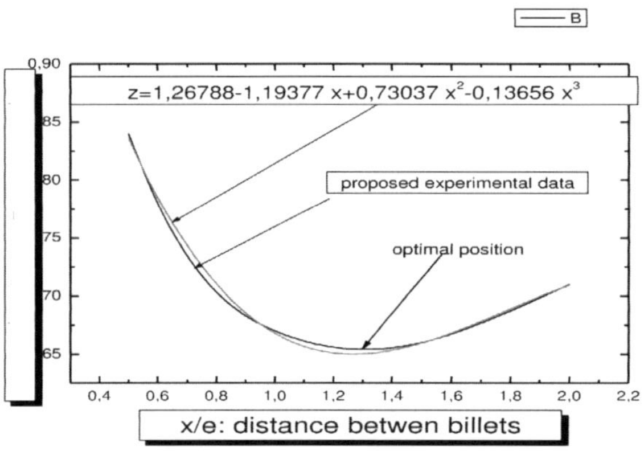

Figure 8. Optimal disposal of the square billets section in the furnace.

5 Conclusions

The coefficients regarding the disposal mode of the billets or ingots in a metallurgical heating furnace are basic to control the process of heat exchange between the flue gases, metallic material and the thermal isolation. The mathematical model can be established using the coefficient of the equivalent surface of heat exchange, the equivalent surface of heat exchange, coefficient of specific time of internal heating, criteria of optimum distance, calculation coefficient of the heating time. In case of square and rectangular section billets the optimum case (which correlates the charging mode, uniformity and time of heating with the output of the furnace) is given by the minimum value of the optimization function. In the case of circular sections, the equivalent surface of the heat exchange is established starting from the geometrical surface, using the special coefficient (k_1). The real heating time will be determined in case of rectangular sections billets by means of coefficient k_2 and in case of circular section billets by means of coefficient k1 and of equivalent surface of heat exchange.

References

[1] Heiligenstaedt W.: Thermique appliquee aux fours industriels, tom1, Dunot, Paris, 1971

[2] Constantinescu D, Carlan AB ; Particularities regarding the thermal radiation inside the heating furnaces for metal forming, METAL 2015, Brno, Editor: Tanger Ltd, Ostrava, CZ, pag.282-287 ; ISBN 978-80-87294-62-8];

[3] Constantinescu D, Aspects Regarding The Heat Transfer In Furnaces For Rolling Mills; WITpress, Heat Transfer VII, 2002, DOI: 10.2495/HT020271; ISBN 1-85312-9062; ISSN 1462-6063

Experimental comparison between upsetting characteristics of porous components prepared by Fe-based sintering technology

W Depczyński[1], T Miłek[1] and Ł Nowakowski[2]

[1] Kielce University of Technology, Faculty of Mechatronics and Machine Design, Department of Applied Computer Science and Armament Engineering, Al. Tysiąclecia Państwa Polskiego 7, 25-314 Kielce, Poland, UE
[2] Kielce University of Technology, Faculty of Mechatronics and Machine Design, Department of Mechanical Technology and Metrology, Al. Tysiąclecia Państwa Polskiego 7, 25-314 Kielce, Poland, UE

E-mail: wdep@tu.kielce.pl

Abstract. The investigation has been undertaken to develop and study the upsetting characteristics of iron-based sintered porous components by using elemental powders through powder metallurgical techniques. The paper presents experimental results of forming Fe-based sintered porous samples and upsetting it by two methods. The first method involves upsetting between two flat plates (anvils). The second method is based on the deformation of the material in closed-die. The study compares the graphs of variation of the force vs. displacement. The aim of the article is to compare the experimental characteristics of upsetting process of samples with different porosities. Metallic porous structure was obtained as a result of sintering processes Fe-based powders mixtures with addition of copper catalyst and Fe (III) oxide as a space holder. The sintering was proceeded at the temperature of 1130°C and 50 min. long, in dissociated ammonia protective gas. Some samples were sintered with addition of powdered activated carbon to increase its porosity properties and change chemical composition.

1 Introduction

Metallic foams are lightweight structural materials, however with enormous future potential for applications where lightweight combined with high stiffness and acceptable manufacturing costs are of prime interest. The performance of metallic foams, in particular those made of Fe, in various prototypes, such as foamed panels, sandwiches, complex 3-D - parts, casted and sintered foamed profiles has been discussed with respect to the expected and achieved goals. Scientific attempts to improve foam quality of any type of foam concentrate on the foam physics, i.e., bubble formation, foam nucleation, growth, stability, development or gas diffusion in the liquid state, where the foam structure is evolving [1-13]. Considering methods of production, metal foam structure characterises different level of homogeneity and comprises different characteristic features [1, 2]. Although they have very high potential, and wide fields of application, they are still not wide spread on the market. Some reviews about the applications of metallic foams and its several excellent properties of the material are available in the literature [14-19].

While structural applications take advantage of metal foams considering lightweight and specific mechanical properties, its functional applications were based on special functionality for instance a large open area in combination with very good thermal or electrical conductivity for heat dissipation or as electrode for batteries, respectively, and for overall application functionality [20-22].

Obviously applications of metal foams are strongly linked to the properties, and some of them are mainly related to those of the matrix metal itself, e.g., elasticity, temperature or corrosion resistance, etc., while others appear only in combination with the cellular structure, e.g., low density, large surface area or damping.

The foam structure is the characteristic feature of foam [23]. Mechanical properties depend mainly on the density, but are also based by the feature of the cellular structure i.e., cell connectivity, cell roundness, diameter distribution, etc.

Upsetting is defined as free forming, by which a workpiece segment is reduced dimension between usually plane, parallel platens (**Figure 1**). It also includes coining and heading. Closed-die forming, shape upsetting, or die heading involve tools which contain the intended shape wholly or in a part [24]. Based on elementary plasticity theory, upsetting forces can be determined approximately [24, 25, 26]. To do so, the distribution of the normal stress at the tool-workpiece interface is required. Substitution of the Tresca yield criterion [24]:

$$\sigma_x - \sigma_z = \sigma_p \tag{1}$$

With the help of the Tresca criterion the interface pressure is [24, 25]:

$$\sigma_z = -\sigma_p \exp\left[\frac{2\mu}{H}\left(\frac{d}{2} - x\right)\right] \tag{2}$$

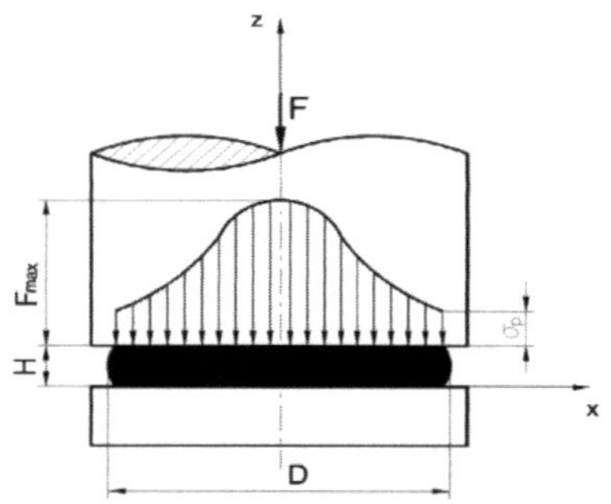

Figure 1. Normal stress distribution σ_z at the interfaces of axisymmetric upset cylinder (upsetting with friction, exponential distribution).

Figure 1 shows the interface pressure distributions provided by eq. 2.

The upsetting force is obtained by integration of the pressure over the contact area *A* [24]:

$$F_z = \int_A \sigma_z dA \tag{3}$$

The aim of the article is to compare the experimental characteristics upsetting for samples with two different materials and various porosities.

2 Methodology

2.1 Fe foam preparing

Metallic precursor composition was made of iron-based powders: Distaloy SE, ASC 100.29 (by Höganäs company), and Fe_2O_3 powder with different compositions. As a reaction catalyst, Cu was used and Fe_2O_3 was used as a foaming agent-space holder. The samples were affected only by external gravity reactions but weren't internally compacted. The mixture was sintered in a dissociated ammonia atmosphere at 1180°C for 45 minutes.

After sintering, samples situated in furnace, were moved to an area where it was cooled in a protective atmosphere. Cooling took place in with an average speed of about 25 degrees

Table 1. Chemical composition and average particle size of iron base powders.

Powder	Chemical Composition (Mas. %)					Average Particle Size (μm)
	C	Cu	Ni	Mo	Fe	
ASC 100.29	<0.01	-	-	-	Balanced	45–180
Distaloy SE		1.5	4	0.5	Balanced	45–180

per minute. To determine pore size and shape from the foam samples, the image analyser software was used. The properties of iron-based materials are shown in Table 1.

2.2 Machining

Placing an accurate sample matrix made of sintered foams required treatment which would give samples of repeating shape and dimensional accuracy. There's been conducted reconnaissance study, which concluded that among the three-studied treatment technologies, meaning: laser processing [28], EDM [29] and machining, only the milling process gave the desired precision net shape and machining stability and the process of shaping the small depth of cut [30]. The process of forming test samples made of sintered foam consisted in a finishing mill for milling the front surface of the outer cylindrical samples sized Ø15x10mm.

Figure 2. Samples before machining (a) and after machining (b).

Machining process was conducted without the use of coolant on the Hermle B300 milling center. A tool used to conduct the treatment was endmill, by Sandvik Coromant with a diameter of 10 mm. Trials cutting have been carried out with the following parameters: cutting speed v_c = 150 m / min, feed rate v_f = 300 mm / min, depth of cut a_p = 10 mm, infeed side a_e = 0.2 mm. The samples of Fe foam before and after machining are shown in **Figure 2.**

2.3 Upsetting

The experimental part of the research of upsetting was conducted at a special stand [27], which included the following:

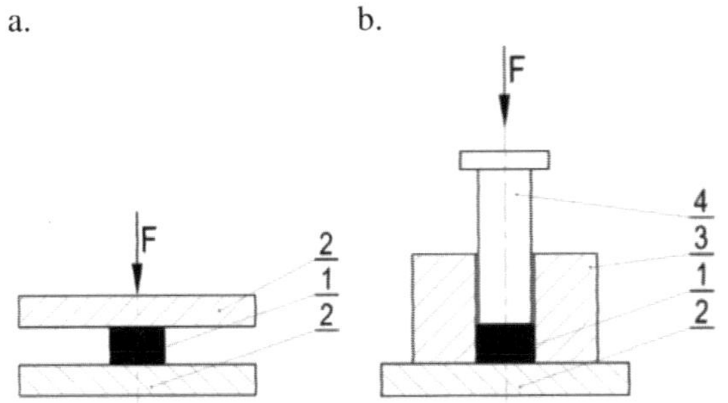

Figure 3. Diagrams of upsetting process (a) and closed-die forming (b), where: 1-axisymmetric cylinder, 2 – plate, 3 – die, 4 – punch, F – load.

- a tool for free upsetting and closed-die forming (**Figure 3**),
- LabTest 5.20SP1 testing machine (LABORTECH firm), 20kN force (machine calibrated by PN-EN ISO 7500-1:2005 and meets the metrological requirements for class 0.5),
- computer stand with Test&Motion software (LABORTECH) to measure forces and displacements.

The materials for experimental investigations were Distaloy SE and ASC 100.29 rod segments, whose diameter was D=15 mm and high h_0=10.3 mm, (which corresponded to the relative coefficient h_0/D=0.69).

3 Result and discussion

3.1 Porosity and microstructure

The microstructures of the porous metal foam were observed using an optical microscope from Nikon MA200. The view of samples obtained by means of optical microscopy is shown in **Figure 4**. The unetched samples were analysed. The obtained microstructures present open porosity and grain-to-grain connections. The measurement of porosity on the cross-section was obtained with NIS 4.20 imaging analysis system. The relative volume of particles was determined in accordance with the Cavalieri-Hacquert principle.

a. b.

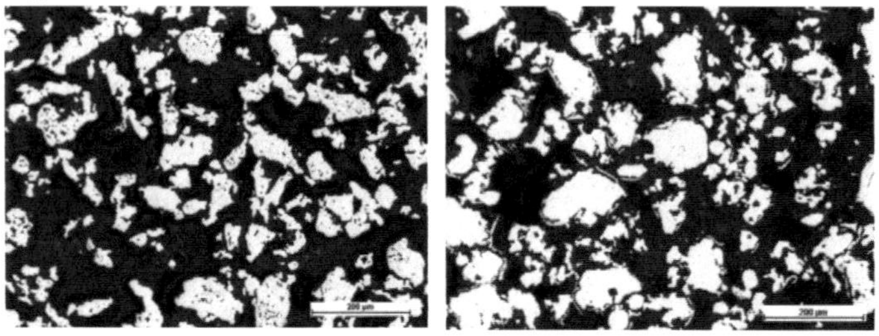

Figure 4. The microstructure of Fe foam samples made of Distaloy SE (a) and ASC 100.29 (b).

The research has shown that the porous metal foam with bulk density obtained from: ASC 100.29 resulted in a porosity of 62%, and obtained from Distaloy SE powder resulted in 75%. The shape of the sintered material particles and chemical composition affected the appearance of diffusion bridges creation. Earlier conducted studies consider metallic foams based on Fe [31, 32, 33] with the structure of considerably smaller pores. A similar distribution but for a much larger pore volume was obtained in using an organic foam precursor [34]. Nowadays, metallic porous materials have different structures and mechanical properties [35, 36]. To evaluate these properties, the test on samples of the obtained material should be performed. Most tests are conducted on samples of standard sizes. But when details of the porous materials have complex configurations of spatial elements, it's difficult to implement the standard tests [34] and experimental tests on the samples of the details are needed to obtain actual structural and mechanical properties of the porous material and its spatial configuration.

3.2 Upsetting

Upsetting characteristic of all materials demonstrated the typical behaviour of elastic-plastic foam under compression. Compression tests all the materials showed the typical behaviour of a flexible foam in a compressed state (Figure 5). The compression process based on highly porous sintered Fe includes three stages: stage linear elastic, plastic deformation stage, and the stage compaction [37, 38]. Force-displacement curves of the samples 1 and 2 made from DISTALOY SE and ASC 100.29 are shown in Figure 3 which differs from that described in the literature for [37,38]. At the beginning, the material has experienced a short-term elastic stage, the 5% compressive strain (which corresponded ratio $\Delta l/h_0$=0.05), where force increased linearly with increasing displacement Δl. Partially reversible deformation powders process takes place. Then the load slowly increase at higher displacement Δl in the next step of plastic deformation. As the load increases to 30% compressive strain, almost all of connections

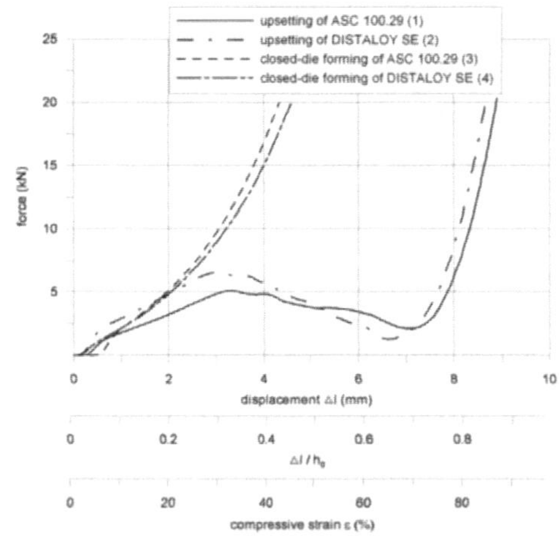

Figure 5. Force vs. displacement obtained for upsetting samples made from DISTALOY SE and ASC 100.29 at ratio h_0/D=0.69.

between the powder particles were plastically deformed. At this time, the pores become smaller and the shape of the pores changed irreversibly. Propagation strength within the material has been blocked due to the existence of pores in the sintering process, and therefore the load increased slowly with progressive displacement Δl. The maximum experimental values of loads obtained in upsetting of samples 1 and 2 made from Distaloy SE and ASC 100.29 at plastic deformation stage were analysed. The values were received at compressive strain ε=30%. The greatest force value was recorded in upsetting of sample 2 made from Distaloy SE (F=6529.36N). It was greater than the force for sample 1 made from ASC 100.29 (F=5097.98N) by approx. 30%. As a result of base analysis of changes in upsetting forces of samples 3 using ASC 100.29 in dies, it can be stated that together with an increase in relative ratio $\Delta l/h_0$ and compressive strain ε, the load increases exponentially for both examined materials, similarly with the Distaloy SE material.

4 Conclusions

Samples of porous sintered Fe with porosity up to 75 vol.% were successfully prepared using a powder metallurgical technique with Fe dioxide as a space-holder material. The influence of the initial iron powder size and iron oxide size on microstructural and mechanical characteristics was studied in details. The samples contained pores originating from bulk density and pores, which were formed by space holder (iron dioxide) decomposition. Increasing amounts of space holder in the initial mixture increased the total porosity. This clearly affects the course of loads during the upsetting and closed-die forming processes. In both cases, the diagrams characters of the curves of loads vs. strains are similar. The differences result from the use of a material with different mechanical properties and with various porosity.

References

[1] Ashby, M.F.; Evans, A.G.; Fleck, N.A.; Gibson, L.J.; Hutchinson, J.W.; Wadley, H.N.G. Metal Foams: A Design Guide; Buterworth-Heineman: Boston, MA, USA, 2000.

[2] Arwade, S.R.; Hajjar, J.F.; Schafer, B.W.; Moradi, M.; Smith, B.H.; Szyniszewski, S. Steel foam material processing, properties and potential structural applications. In Structural Materials and Mechanics, Proceedingsof the 2011 NSF Engineering Research and Innovation Conference, Atlanta, GA, USA, 4–7 January 2011;pp. 1–7

[1] Rack, A.; Helwig, H.M.; Bütow, A.; Rueda, A.; Matijašević-Lux, B.; Helfen, L.; Goebbels, J.; Banhart, J. Early pore formation in aluminium foams studied by synchrotron-based microtomography and 3-D image analysis. Acta Mater. 2009, 57, 4809–4821.

[3] Asavavisithchai, S.; Kennedy, A.R. In-situ oxide stabilization development of aluminum foams in powder metallurgical route. High Temp. Mater. Processes 2011, 30, 113–120.

[4] Garcia-Moreno, F.; Mukherjee, M.; Jiménez, C.; Rack, A.; Banhart, J. Metal foaming investigated by X-ray radioscopy. Metals 2011, 2, 10–21.

[5] Wang, Y.W.; Zuo, X.Q.; Lu, J.S.; Zhou, Y. Bubble nucleation of PM Al-9Si foam. Adv. Mater. Res. 2011, 183–185, 1682–1686.

[6] Asavavisithchai, S.; Kennedy, A.R. The role of oxidation during compaction on the expansion and stability of a foams made via a PM route. Adv. Eng. Mater. 2006, 8, 568–572.

[7] Liu, X.; Li, Y.; Chen, X.; Liu, Y.; Fan, X. Foam stability in gas injection foaming process. J. Mater. Sci. 2010, 45, 6481–6493.

[8] Vandewalle, N.; Caps, H.; Delon, G.; Saint-Jalmes, A.; Rio, E.; Saulnier, L.; Adler, M.; Biance, A.L.; Pitois, O.; Addad, S.C.; et al. Foam stability in microgravity. J. Phys. Conf. Ser. 2011, 327.

[9] Körner, C.; Arnold, M.; Singer, R.F. Metal foam stabilization by oxide network particles. Mater. Sci. Eng. A 2005, 396, 28–40.

[10] Mukherjee, M.; Garcia-Moreno, F.; Jiménez, C.; Banhart, J. Al and Zn foams blown by an intrinsic gas source. Adv. Eng. Mater. 2010, 12, 472–477.

[11] Mukherjee, M.; Garcia-Moreno, F.; Banhart, J. Solidification of metal foams. Acta Mater. 2010,

58, 6358–6370.
[12] Garcia-Moreno, F.; Solorzano, E.; Banhart, J. Kinetics of coalescence in liquid aluminium foams. Soft Matter 2011, 7, 9216–9223.
[13] Banhart, J. Manufacture, characterisation and application of cellular metals and metal foams. Prog. Mater. Sci. 2001, 46, 559–632.
[14] Davies, G.J.; Zhen, S. Metallic foams: Their production, properties and applications. J. Mater. Sci. 1983, 18, 1899–1911.
[15] Nakajima, H. Fabrication, properties and application of porous metals with directional pores. Prog. Mater. Sci. 2007, 52, 1091–1173.
[16] Kranzlin, N.; Niederberger, M. Controlled fabrication of porous metals from the nanometer to the macroscopic scale. Mater. Horiz. 2015, 2, 359–377.
[17] Qin, J.; Chen, Q.; Yang, C.; Huang, Y. Research process on property and application of metal porous materials. J. Alloys Compd. 2016, 654, 39–44.
[18] Lefebvre, L.P.; Banhart, J.; Dunand, D.C. Porous metals and metallic foams: Current status and recent developments. Adv. Eng. Mater. 2008, 10, 775–787.
[19] Lange K..: Handbook of metal forming. McGraw-Hill, Inc. USA, 1985.
[20] Piasecka M., Laser texturing, spark erosion and sanding of the surfaces and their practical applications in heat exchange devices, 2014, Advanced Material Research, vol. 874, pp. 95-100
[21] Piasecka M., Heat transfer mechanism, pressure drop and flow patterns during FC-72 flow boiling in horizontal and vertical minichannels with enhanced walls, 2013, International Journal of Heat and Mass Transfer, vol. 66, pp. 472-488
[22] Depczynski, W. Sintering of copper layers with a controlled porous structure. In Proceedings of the Metal 2014: 23rd International Conference on Metallurgy and Materials, Brno, Czech Republic, 21–23 May 2014; pp. 1219–1224.
[23] Depczynski, W.; Kazala, R.; Ludwinek, K.; Jedynak, K.: Modelling and Microstructural Characterization of Sintered Metallic Porous Materials, MATERIALS, Volume: 9 Issue: 7 Article Number: 567 DOI: 10.3390/ma9070567 Published: JUL 2016
[24] Lange K.:Handbook of metal forming. McGraw-Hill, Inc. USA, 1985.
[25] Kuczyński K., Erbel S.: Metal forming – laboratory. Wydawnictwa Politechniki Warszawskiej, Warszawa 1984 (in Polish)
[26] Miłek T.: Computer analysis of forming of round forged parts with mechanical press and hammer. Metal 2013, 22nd International Conference on Metallurgy and Materials, Conference Proceedings, Brno, Czech Republic, 15-17 May 2013; pp. 255-260
[27] Miłek T., Kowalik B., Kuliński B.: Evaluation of the possibility of performing cold backward extrusion of axisymmetric thin-walled aluminium die stampings with square section. Archives of Metallurgy and Materials 60, (4), 2015, pp. 3043-3049.
[28] Nowakowski Ł., Wijas M.,: The evaluation of the process of surface regeneration after laser cladding and face milling, Engineering Mechanics 2016 p. 430-433, 2016
[29] Nowakowski Ł., Miesikowska M., Blasiak M.,: Speech intelligibility in the position of CNC machine operator, Engineering Mechanics p. 422-425, 2016
[30] Nowakowski Ł., Miko E., Skrzyniarz M.,: The analysis of the zone for initiating the cutting process of X37CrMoV51 steel Engineering Mechanics 2016 p. 426-429, 2016
[31] Redenbach, C. Microstructure models for cellular materials. Com. Mat. Sci. 2009, 44, 1397–1407.
[32] Murakami, T.; Ohara, K.; Narushima, T.; Ouchi, C. Development of a new method of manufacturing iron foam using gases generated by reduction of iron oxide. Mater. Trans. 2007, 48, 2937–2944.
[33] Park, C.; Nutt, S.R. Effects of process parameter on steel foam synthesis. Mater. Sci. Eng. A 2001, 297, 62–68.
[34] Chang-Lin, L.; Hui, W.; Xiang-Yang, Z. Debinding of stainless steel foam precursor with 3-D

open-cell network structure. Trans. Nonferrous Metals Soc. China 2010, 20, 2340–2344.

[35] Bekoz, N.; Oktay, E. Mechanical properties of low alloy steel foams: Dependency on porosity and pore size. Mater. Sci. Eng. A 2013, 576, 82–90.

[36] Romero, C.; Noyola, J.C.; Santiago, U.; Valladares, R.M.; Valladares, A.; Valladares, A.A. A new approach to the computer modelling of amorphous nanoporous structures of semiconducting and metallic materials: A review. Materials 2010, 3, 467–502.

[37] Dewidar, M.M.; Khalil, K.A.; Lim, J.K. Processing and mechanical properties of porous 316L stainless steel for biomedical applications. Trans. Nonferrous Met. Soc. China 2007, 17, 468–473.

[38] Qiao, J.C.; Xi, Z.P.; Tang, H.P.; Wang, J.Y.; Zhu, J.L. Influence of porosity on quasi-static compressive properties of porous metal media fabricated by stainless steel fibers. Mater. Des. 2009, 30, 2737–2740.

IOP Conf. Series: Materials Science and Engineering **179** (2017) 012016 doi:10.1088/1757-899X/179/1/012016

Bainite austenitization in 51CrV4 spring steel: accelerated cementite spheroidisation

J Dlouhy, D Hauserova and P Motycka

COMTES FHT a.s., Prumyslova 995, 334 41, Dobrany, Czech Republic, EU

E-mail: jaromir.dlouhy@comtesfht.cz

Abstract. Spheroidisation is well studied process in steels with pearlitic structures. Bainite as initial microstructure for spheroidisation treatment is not so frequent. The article describes rapid carbide spheroidisation of bainitic structure of 51CrV4 spring steel. The experimental steel was treated in quenching dilatometer capable of rapid temperature changes. Isothermal austenitization was studied at different temperatures with respect to its kinetics and microstructural features. Special attention was paid to the morphology of growing austenite grains and carbide morphology during their dissolution in austenite. Rapid partial austenitization was repeatedly applied to achieve accelerated carbide spheroidisation. This phenomenon is known for pearlitic structures. The article shows that bainite can be spheroidised in similar way as pearlite. Accelerated carbide spheroidisation can be applied as replacement of conventional time-consuming spheroidisation annealing.

1 Introduction

Spring steels are one of the steel groups, where spheroidisation annealing is widely used. Examined steel belongs to the group of low alloyed steel with medium carbon content. Spheroidisation treatment is performed to decrease hardness, yield strength and increase ductility for technological reasons.

Conventional way to spheroidised carbides in steel is so-called “soft annealing”. This procedure employs tendency to minimize carbide/ferrite (or carbide/austenite) interface area [1]. The area minimization leads to formation of spherical carbides and their coarsening via mechanism of Ostwald ripening. This change of carbide shape and size brings steel structure closer to thermodynamic equilibrium. Sufficient temperature and time is necessary for successful spheroidisation which is usually defined as 95% of carbides in spheroidal form. However, morphological criteria are rarely set in engineering practice and the only criterion for spheroidisation annealing is the maximal permissible hardness of the treated material.

Soft annealing typically comprises of several hour soaking at spheroidisation temperature. This temperature is usually slightly below the A_1 temperature in case of hypo- and eutectoid steels and slightly higher than A_1 temperature in case of hypereutectoid steels. The main objective to choose the temperature is keeping the carbides in the matrix, whereas ferritic or austenitic, without dissolving them completely. The material is than cooled down slowly (usually in furnace) in order to retain globular carbides without austenite decomposition into lamellar pearlite. Whole soft annealing lasts several hours, even more than 24 hours in case of large batches (in order of 10 or 20 tons).

There is significantly faster way to spheroidise carbides in the structure. It was developed for spheroidisation of lamellar pearlite in low alloyed bearing steel 100CrMnSi6-4 [2, 3]. The principle of so called "Accelerated carbide Spheroidisation and Refinement" (ASR) is repeated austenitization and divorced pearlite transformation [4]. Three cycles are sufficient for complete spheroidisation of lamellar pearlite. Austenitization is in ASR process rapid, performed by i.e. induction heating in rate in order of tens degree per second. Ferritic matrix is austenitized and cementite lamellae only partially dissolved [5, 6]. The lamellae undergo fragmentation at the advancing ferrite/austenite front. Cementite remnants act as nuclei for cementite growth during subsequent cooling and austenite decomposition. Thus divorced pearlitic transformation occurs and no new cementite particles form during cooling.

ASR process produces spherical carbides in ferritic matrix in significantly smaller size (ca 3-times) than conventional soft annealing. This fact is reflected by the term "Refinement" in ASR title. Finer dispersion of carbides is highly beneficial for subsequent hardening [7]. Smaller carbides dissolve at higher rate than coarse ones. Lower quenching temperatures can be applied to gain the same hardness in comparison with soft annealed material. Finer dispersion means also shorter average distance of the particles. It causes more effective pinning of austenite grain boundaries during austenitization for quenching. Smaller austenite grain leads to finer martensite and higher hardness after hardening in comparison with soft annealed material.

Presented article describes application of ASR principle to the bainitic microstructure of spring steel 51CrV4. Alloying by chromium and vanadium slows down the kinetics of austenite decomposition. Thus bainitic structure can be often found in hot rolled semiproducts before spheroidisation treatment. The aim of experimental program was to investigate ability of bainite to be spheroidised via ASR process. The optimal parameters for ASR were set and dilatometric study of austenitization was performed.

2 Materials and methods

Experimental steel was grade 51CrV4, used widely for spring production. The chemical composition of the steel is in the table 1. The steel was delivered in form of hot-rolled bars 21 mm in diameter, air cooled after rolling, without any subsequent thermal treatment.

Table 1. Chemical composition of the experimental steel.

Element	C	Mn	Si	Cr	V	Cu	S	P	Fe
wt. %	0.51	0.96	0.27	1.07	0.14	0.02	0.017	0.015	bal.

Heat treatment was performed in quenching dilatometer Linseis L78 RITA. It enables dilatometric analysis of metallic cylindrical samples from 3 to 5 mm in diameter and typically 10 mm in length. Samples are heated by induction heating and cooled by gas flow. Maximal controlled cooling and heating rate is 200°C/s. Dilatometers chamber was evacuated and filled with argon. Specimens for heat treatment in the dilatometer were prepared in form of cylinders 4mm in diameter and 10 mm in length. These samples enables microstructural analyse, hardness measurement but also preparation of micro-tensile specimens [8, 9].

Microstructure of the samples was investigated on metallographic section prepared in longitudinal direction of the samples. Sections were prepared by mechanical grinding and polishing, microstructure was revealed by etching in 3% Nital solution. Sections were observed by optical microscope ZEISS Observer Z1m and scanning electron microscope (SEM) JEOL JSM 7400F. Hardness was measured by method HV10 on automatic hardness tester Struers Durascan 50.

ASR was performed as 3 cycles heating to temperatures from 750°C to 800°C. Every hold lasted 15 seconds. The regimes are shown in figure 1. Heating rate was 50°C·s^{-1}, cooling rate was 30 °C·s^{-1} to the temperature 730°C and 3 °C·s^{-1} to the 650°C or to the ambient temperature after the third heating cycle.

Isothermal austenitization was performed at temperatures from 750°C to 780°C. Samples with interrupted austenitization were prepared. The isothermal austenitization was stopped by quenching at moment with estimated amount of austenite in the structure reached 25 vol.%, based on the dilatometric record of austenitization. Metallography investigation of carbide morphology at ferrite/austenite interface was performed.

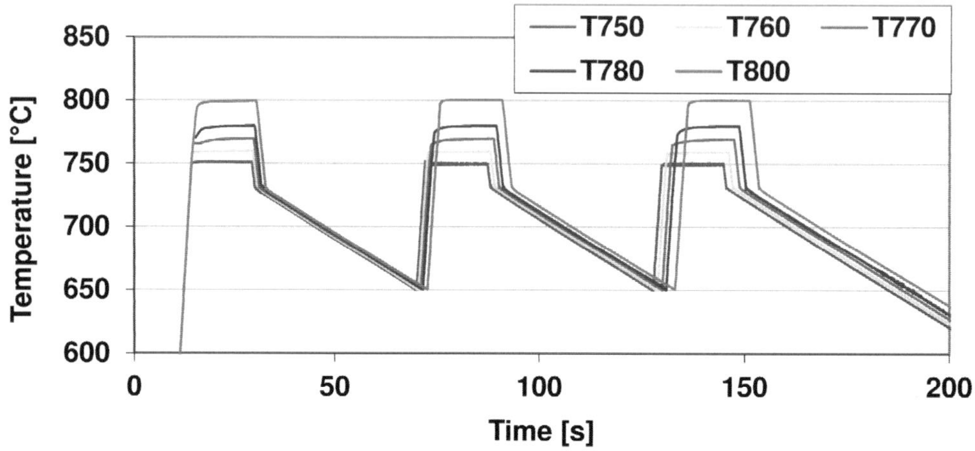

Figure 1. Regimes of ASR treatment performed in quenching dialtometer. Regimes labelled after the austenitization hold temperature.

3 Results and discussion

2.1 ASR via cyclic heating.

Initial microstructure of the bars was bainitic with hardness 345 HV10. Microstructure was homogeneous (figure 2), composed of ferritic matrix and carbides, emphasizing plate-like structure of the matrix (figure 3).

.

Figure 2. Microstructure of initial state observed in optical microscope.

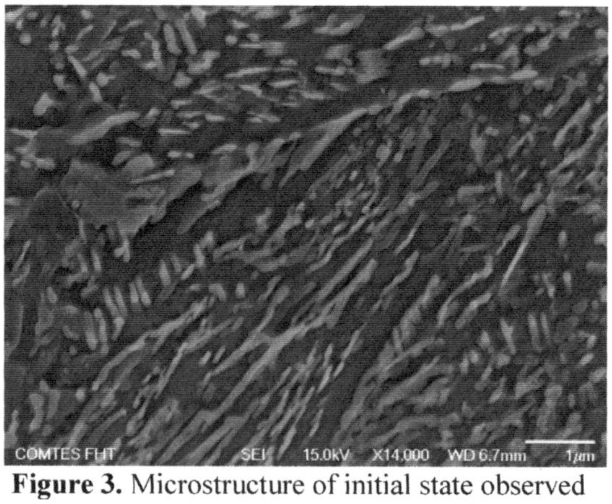

Figure 3. Microstructure of initial state observed in SEM in detail.

The microstructure after thermal cycles between 750°C and 650°C did not change in terms of carbide morphology. However, the hardness decreased rapidly from 345 HV10 to 287HV10 (table 2). This could be explained by recovery and recrystallization of the ferritic matrix.

ASR treatment was carried out successfully by thermal cycling to the 760°C. Carbides were almost completely spheroidised. Only traces of original elongated carbides were observed. Also formation

of new pearlitic lamellae during the last cooling occurred, as seen in the figure 4. This sample exhibited the lowest hardness due to the most complete carbide spheroidisation. Carbide globules size was from 100 to 300 nm.
Higher austenitization temperatures lead also to carbide spheroidisation. However carbide dissolution advanced too much in some areas, that new pearlite was formed there. This happened probably due to lack of undissolved carbides necessary for divorced pearlitic transformation. Thus, regular pearlitic transformation occurred with formation of cementite lamellae (figure 5). There were also visible martensite islands in the structure for samples T 780 and T800. Austenite in these areas was so stable that cooling rate 3 $C{\cdot}s^{-1}$ was higher than critical cooling rate for martensitic transformation. Austenite stability was increased by amount of dissolved carbon and alloying elements concentrated in carbides, mainly chromium. Hardness of samples rose with rising amount of new lamellar pearlite.

Table 2. Hardness and carbide morphology after cyclic heating. Regimes labelled after the austenitization hold temperature.

Regime	Initial state	T 750°C	T 760°C	T 770°C	T 780°C	T 800°C
HV10	345	287	254	265	301	350
carbides	bainitic	bainitic - intact	spheroidised, traces of new lamellae	spheroidised, ca. 5% of pearlite	spheroidised, ca. 30% of pearlite	pearlite, ca 20% spheroidise d carbides

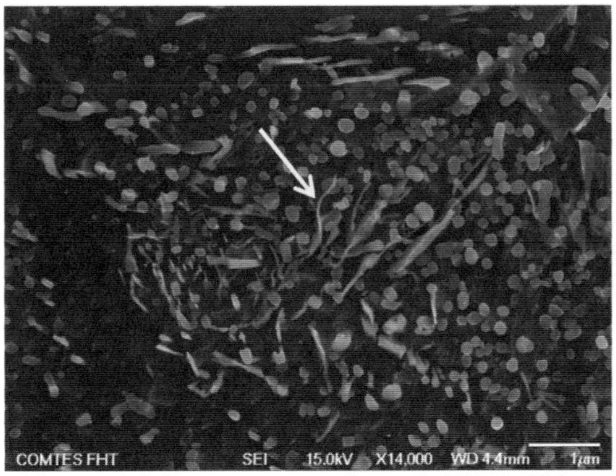

Figure 4. Microstructure of sample T760. Thin new pearlitic lamellae can be seen in the center (arrow).

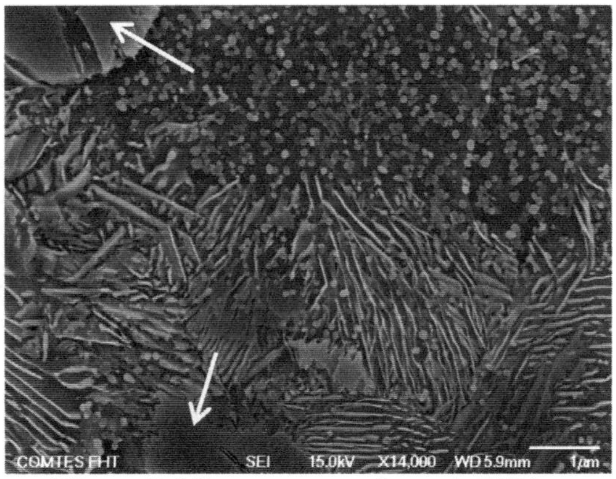

Figure 5. Microstructure of sample T780. Spheroidised carbides, new pearlite and traces of martensite (arrows).

Spheroidised carbide particles size decreased slightly with increasing austenitization temperature of ASR treatment. Globules diameter was in range of 50 to 200 nm in case of sample T800.

2.2 Dilatometric analyse of the austenitization
Isothermal austenitization was observed via sample dilatation. Figure 6 shows austenitization progress at temperatures from 750°C to 780°C. Curve for 750°C reveals reason why no carbide spheroidisation was observed after regime T 750. 15 second holds were too short to austenitize the sample. ASR based on reverse austenitization and divorced pearlitic transformation did not take place.

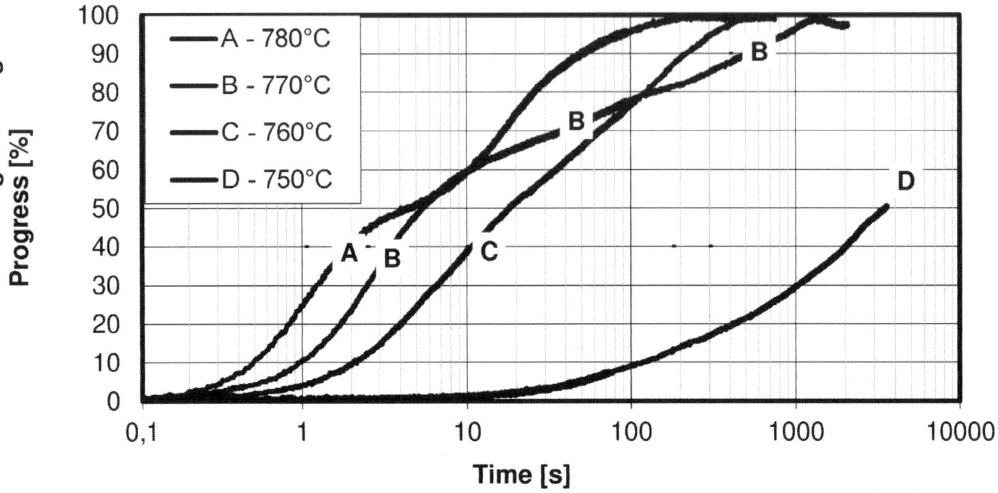

Figure 6. Length change of the material during isothermal holds at different austenitization temperatures. 100% at the y-axis represents 100% austenitization of the ferritic matrix.

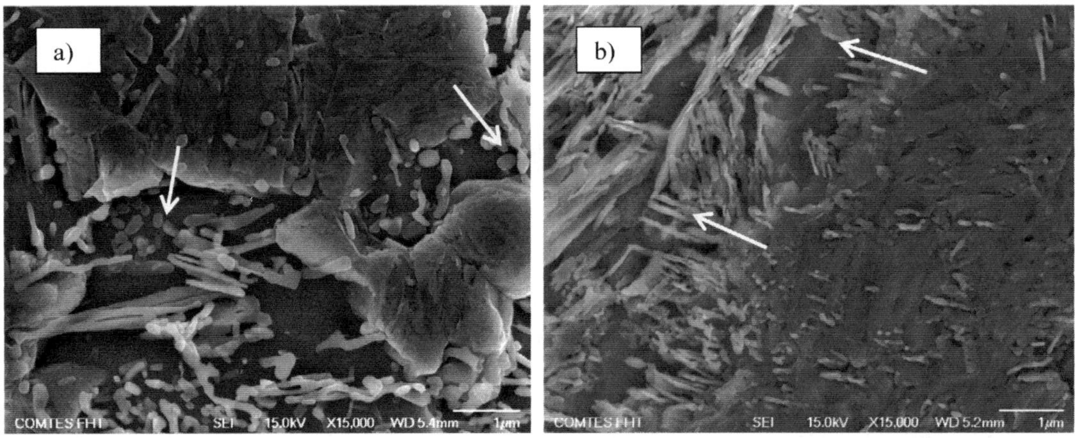

Figure 7. Austenite growing at temperature a) 750°C, b) 780°C. Arrows shows direction of austenite growth into bainite.

Samples with interrupted austenitization were prepared. Austenitization was stopped by rapid cooling of the sample at point, where ca 25vol.% of austenite was estimated to be present in the structure (according to curves from figure 6). Growing austenite was turned to martensite and the carbides "froze" in state they were at the beginning of the rapid cooling.

Figure 7 shows structure of samples with interrupted austenitization at temperatures 750°C and 780°C. Metallographic sections were etched by Nital. Martensite is etched at significantly slower rate than ferrite and carbides are not etched at all. Thus ferrite is etched deeply, revealing 3D shape of initial bainitic carbides. Martensite is etched only slightly showing surface relief reflecting individual martensite crystals.

There are clearly visible undissolved carbides in the growing austenite (turned to martensite). Carbides retained in austenite at temperature 780°C are smaller and seem to have higher spatial density. This corresponds with finer carbide globules observed after ASR performed at higher temperature.

4 Conclusion

Accelerated carbide Spheroidisation and Refinement (ASR) was successfully applied to the bainitic structure of 51CrV4 steel. Hardness dropped from initial 345 HV10 to 254 HV10. Vast majority of carbides was spheroidised. The spheroidisation regime comprised in three heating cycles to 760°C for 15 sec with intermediate cooling to 650°C.

Results showed that ASR process can be applied to the bainitic structure. The mechanism of carbide fragmentation and globules formation seems to be the same like in the case of pearlitic initial structure.

Spherical carbides with size from 100 to 300 nm were formed. Formation of finer carbides (from 50 nm to 200 nm) was observed after temperature cycling to higher temperatures; however lamellar pearlite also formed during cooling. Promotion of divorced pearlitic transformation can be achieved by prolonging the time of intermediate cooling. It is way to prevent formation of lamellar pearlite.

Acknowledgement

This paper was created by project Development of West-Bohemian Centre of Materials and Metallurgy No.: LO1412, financed by the MEYS of the Czech Republic.

References

[1] Bhadeshia H K D H 2012 Steels for bearings *prog. mater. sci.* vol **57** p 268-435
[2] Hauserova D, Dlouhy J and Novy Z 2014 Effect of heating rate on accelerated carbide spheroidisation (ASR) in 100CrMnSi6-4 bearing steel *Arch. Metall. Mater.* vol **59** p 1199-1203
[3] Dlouhy J, Hauserova D and Novy Z 2013 Accelerated carbide spheroidisation of chromium bearing steel by induction heating *Adv. Mat. Res.* vol **739** p 161-6
[4] Verhoeven J D and Gibson E D 1997 The divorced eutectoid transformation in steel *Metall. Mater. Trans.* A vol **29** p 1181-9
[5] Shtansky D V, Nakai K and Ohmori Y 1999 Pearlite to austenite transformation in an Fe-2 6Cr-1C alloy *Acta Mater.* vol **47** /9 p 2619-32
[6] Atkinson C, Akbay T and Reed R C 1995 Theory for reaustenitization from ferrite/cementite mixtures in Fe-C-X steels *Acta Metall. Mater.* vol **43**/5 p 2013-31
[7] Hauserova D, Dlouhy J and Novy Z 2015 Microstructure and properties of hardened 100CrMnSi6-4 bearing steel after accelerated carbide spheroidization and long-duration annealing Proc. Bearing Steel Technologies: 10th Volume Advances in Steel Technologies for Rolling Bearings p 389-409
[8] Džugan J, Procházka R, and Konopík P 2014 Micro-tensile test technique development and application to mechanical property determination," *Am. Soc. Test. Mater.* STP 1576 vol **6**, ed M A Sokolov and E Lucon p 12-29
[9] Konopik P , Dzugan J, Rund M 2014 Dynamic Tensile And Micro-Tensile Testing Using DIC Method, METAL 2012: 21st International Conference on Metallurgy and Materials (Ostrava: TANGER)

Improvement of wear resistance for C45 steel using plasma nitriding, nitrocarburizing and nitriding/ manganese phosphating duplex treatment

T V Doan, D Kusmič, M Pospíchal, D Dobrocký

Faculty of Military Technology, University of Defence in Brno, Kounicova 65, 662 10 Brno, Czech Republic

E-mail: thanhvan.doan@unob.cz, david.kusmic@unob.cz , ondrej.pilch@unob.cz, miroslav.pospichal@unob.cz, david.dobrocky@unob.cz

Abstract: This article focuses on effect of plasma nitriding, nitrocarburizing and nitriding/ manganese phosphating duplex treatments to wear resistance of C45 steel substrate. The wear test "ball on disc" was conducted to evaluate the coefficient of friction and wear rate using the BRUKER UMT-3 tribometer. The analysis results indicated that nitrocarburizing obtained the best wear resistance; the worst wear resistance was plasma nitriding. Manganese phosphating coating enabled to reduce the coefficient of friction enhanced wear resistance nitrided layer. The used surface treatments also improve non-equal wear of tempered surface over the sliding track.

1 Introduction

During the relative motion, the frictional ways present between two parts and it leads to wear and damage of those parts or one of them [1]. Although wear can be reduced by lubrication, in many cases we have to figure machine working in unlubricated condition (dry friction). In the history of surface treatments, hard chrome plating, a very effective method to enhance corrosion and wear resistance has been used widely in airspace, automotive and armament industry. However, during the chrome plating procedure, the used hexavalent chrome affects seriously to the human health and ambient environment. Amount of surface treatments is suggested to replace the hard chrome plating in industry.

Nitrocarburizing and plasma nitriding are among the great ways to increase corrosion resistance, wear resistance and service life of mechanical components. Mechanical and tribological properties of nitrided parts depend on process parameters, such as treatment duration, atmosphere, temperature, pressure, etc. and the composition of particular nitride phases has a direct influence to the wear resistance [2, 3]. During plasma nitriding process, two layers with distinct properties are formed: the outer compound layer consisting of ε-$Fe_{2-3}N$ and γ'-Fe_4N phase, and the thick inner diffusion layer. Phosphating is the most important surface treatment to improve corrosion resistance of material. In which, manganese phosphate coating exhibits high hardness, superior corrosion and wear resistance. Due to ability to absorb lubricants, manganese phosphating is mostly used for pistons and in weapon industry to ease the sliding of moving parts, which are running in lubricants by reducing the friction between metal parts.

C45 steel is a plain carbon steel for quenching and tempering, widely used in mechanical engineering and automotive components due to low cost and good combination of mechanical properties. In weapon manufacture, C45 steel is applied to product barrel of pistols or shotguns and other parts.

This article focused on investigating wear resistance of selected surface treatments for C45 steel: plasma nitriding, nitrocarburizing and duplex treatment of nitriding/manganese phosphate. The wear test "ball on disc'' during unlubricated sliding was conducted to investigate the coefficient of friction and wear resistance of substrate and surface treatments. The metallographic documentation was performed by optical microscopy. The surface hardness and microhardness test were carried out to determine formation of nitride layer and coating.

2 Methods and experiments

The samples were shaped as rotary disc with diameter of 70 mm, thickness of 6.6 mm and surface roughness Ra of 0.54 μm. The samples were quenched and subsequently tempered. Prior to nitriding process, samples were plasma cleaned and activated in plasma under gas mixture of $20H_2 : 2N_2$ (l/h) for 30 minutes. The plasma nitriding PN1 was treated under gas ratio of $3H_2 : 1N_2$ for 15 hours, nitriding process PN2 under ratio of $1H_2 : 3N_2$ (reverse ratio atmosphere) for 15 hours; the sample marked as DUPLEX1 represents duplex treatment of nitriding PN1 and manganese phosphating, likewise, sample DUPLEX2 is duplex treatment of PN2 and manganese phosphating. The nitrocarburizing Tenifer® was treated as two separate processes: nitrocarburizing in salt bath for 45 min under temperature 590 °C and followed oxidation process for 10 min at 430 °C.

The chemical composition of used steel was verified using the spectrometer LECO SA-2000 (on principle GDOES - Glow discharge optical spectroscopy). The surface hardness was measured by Vickers method with a load of 1 kg and 10 seconds dwell time. The structure observation and documentation was realized using the optical microscopy. Based on outstanding contrast between compound and diffusion layer, the thickness of compound layer could be determined. However, this method does not work in order to measure thickness of manganese phosphate coating and diffusion layer. Therefore, Vickers microhardness profile measurement was used to determine the thickness of diffusion layer with a load of 50 g and 10 seconds dwell time [4]. The thickness of manganese phosphate coating was investigated by the GDOES/QDP method (Glow discharge optical spectroscopy/ Quantitative Deep Profiling).

The wear test "ball on disc" was conducted using the tribometer BRUKER UMT-3 in accordance with ASTM G99-95a standard [5]. During the test, the normal load was kept in a value of 20 N. The used ball in diameter of 6.3 mm was made of carbide wolfram, which has the hardness value of 92 HRA. The test parameters were set as following: disc rotary velocity of 500 rpm corresponding to the circumferential velocity of 1.26 m/s, the cycle number of 13500 and humidity of 55%. The profilometer Taylor Hobson CLI 1000 with the software TALYSURF was used to evaluate depth and area of wear track profile.

The wear rate, which represents the wear resistance, is calculated according to the Archard equation [6]:

$$K = \frac{V}{W \cdot s} \quad (2)$$

where: K is the wear rate [$mm^3.N^{-1}.m^{-1}$],
W is the normal load [N],
s is the sliding distance [m],
V is the volume of removed material [mm^3] calculated from profile area of wear and rotary radius.

3 Results and discussions

3.1 *Chemical composition*

The results measured by GDOES method are in accordance with chemical composition of CSN 41 2050 standard (see table 1).

Table 1. Chemical composition of C45 steel (in wt%).

Element	**C%**	**Mn%**	**Si%**	**Cr%**	**Ni%**	**Cu%**	**P%**	**S%**
CSN 41 2050	0.42-0.50	0.50-0.80	0.17-0.37	max. 0.25	max. 0.30	max. 0.30	max. 0.040	max. 0.040
GDOES/BULK	0.45	0.635	0.233	0.168	0.12	0.22	0.0053	0.023

Parameters of GDOES/Bulk analysis: $U = 800$ V, $I = 30$ mA, $p(Ar) = 314$ Pa

3.2 Microstructure and microhardness of layers/ coating

The etched crosswise documented by optical microscopy are shown in figure 1 and figure 2. It can be seen the sorbitical structure of tempered substrate and the layer structure of nitrided and nitrocarburized samples. The nitrided layer is consisted of a very thin white colour so-called compound layer and diffusion layer, which has darker colour than substrate (see figure 1). In case of nitrocarburizing Tenifer ®, an oxide layer is formed outer of compound layer (see figure 2), while the DUPLEX1 and DUPLEX2 samples were covered by a manganese phosphating coating outer compound layer. The crystal structure of manganese phosphate coating is presented in figure 3.

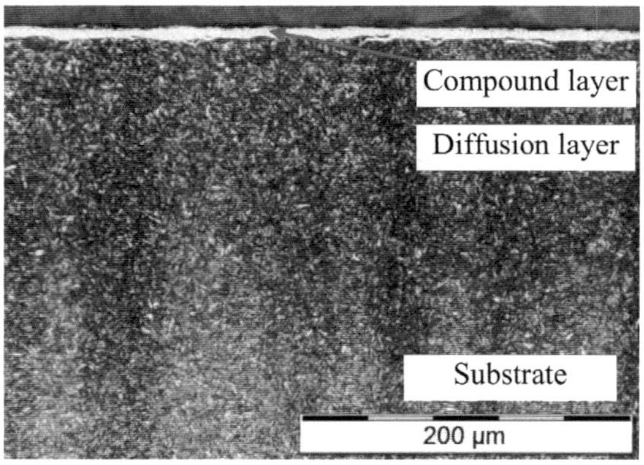

Figure 1. Microstructure of nitrided sample.

Figure 2. Microstructure of nitrocarburized sample.

The results of compound layer thickness and case depth (total layer thickness) are given in table 2. Nitrocarburizing sample reached the thickest compound layer, about three times greater than the nitrided ones. However, the case depth of nitrocarburized sample is markedly lower than plasma nitrided. The nitriding sample PN1 in classic nitriding atmosphere obtained thinner compound layer but greater case depth than the nitriding sample PN2 in reverse nitriding atmosphere. Moreover, in the same process parameters for the nitrided 42CrMo4 steel, nitriding PN1 also obtained greater case depth than nitriding PN2 [7].

Figure 3. The structure of manganese phosphating.

The surface hardness as average value of five measurements is summarized in table 2. Both the plasma nitriding and nitrocarburizing technologies increase surface hardness of substrate, in which the hardness of nitrocarburized sample is lower than nitrided ones and manganese phosphate coating reduces surface hardness of nitrided layers. Conversely, the microhardness of nitriding PN2 obtains greater surface hardness than PN1.

Table 2. Hardness and layer properties

Sample	Surface hardness [HV1]	Compound layer [μm]	Coating [μm]	Case depth [mm]
Tempered	303.86 ± 12.34	-	-	-
Nitrocarburized	444.62 ± 28.99	14.68	-	0.2562
PN1	461.34 ± 18.28	4.31	-	0.3938
DUPLEX1	357.86 ± 22.22	4.61	2.0	0.3497
PN2	523.06 ± 10.4	5.80	-	0.2815
DUPLEX2	389.92 ± 16.7	5.10	2.0	0.2980

The maximal obtained microhardness reached value of 510 HV0.05 (nitrocarburizing), the other samples have lower microhardness than 500 HV0.05, which is low for nitrided and nitrocarburized steel because C45 steel is a plain carbon steel, nitrogen just reacts with iron, not with alloying elements as alloyed steels. On the other hand, microhardness value reduces regard to the distance from surface to substrate. Substrate microhardness is in interval of 300-320 HV0.05.

3.3 *Tribological performance*

3.3.1 *The coefficient of friction*

The coefficient of friction of steel-carbide wolfram tribological system is plotted in the Figure 4. In general, the curve of friction coefficient gains the maximal value in a very short duration, i. e. running-in period, then the curve becomes steady. By a large number of cycles, the coefficient of friction may increase or oscillate hardly cause of the degraded contact surfaces.

Nitrocarburized samples exhibited the greatest coefficient of friction. About first 6000 sliding cycles, the friction coefficient of nitrocarburizing increased continually from 0.0001 to 0.81. Then the coefficient of friction became steady, but still the highest of all samples.

For the tempered sample and other surface treatments, the difference in friction coefficient is obvious in the running-in period. The tempered sample spent longer time (about 1200 cycles) than nitrided samples (about 450 cycles) to get the maximal value of μ (0.61). The duplex treated samples have lower coefficient of friction than tempered sample all the first 3100 cycles. For the nitrided samples, about first 450 cycles of sliding, the value of μ increases quickly up to 0.6 and then quickly decreases to value 0.4. Three samples nitrocarburizing, DUPLEX1 and DUPLEX2 do not exhibit so steep progress of coefficient of friction as nitrided and tempered samples. Manganese phosphate coating ensures the nitrided surface having lower values of friction coefficient at the beginning of wear test. After running-in period, the friction coefficient of all samples (except nitrocarburized sample) was in the narrow range of 0.5-0.6. The duplex treated samples exhibited lightly higher value of friction coefficient than nitrided samples.

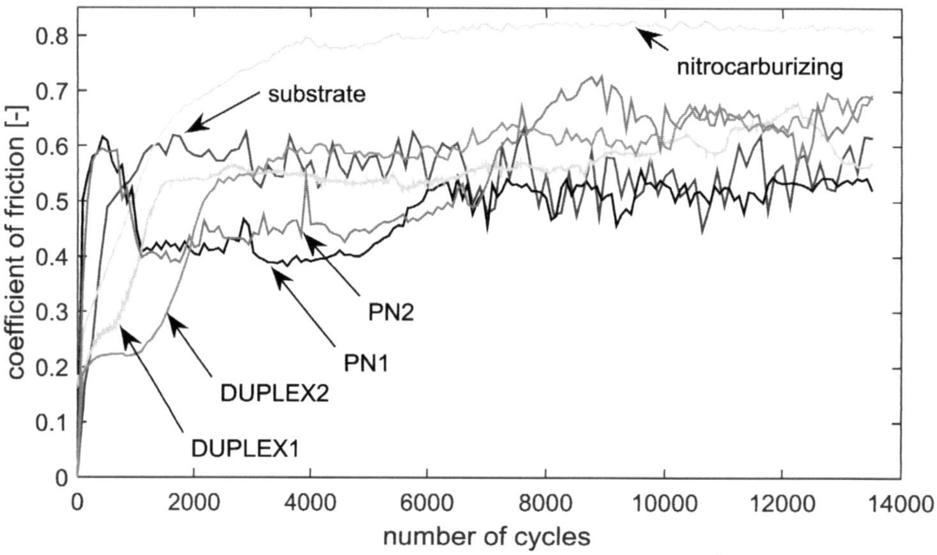

Figure 4. Progress of friction coefficient.

3.3.2 Wear

Wear resistance of the samples was evaluated by the shape and size of wear track. The wear track of tempered sample was significant and its width was changed along the track (see Figure 5) with uncertain period. Because during sliding, the particles were continually removed from material and transferred to other positions on over the trail of ball. By a number of cycles, the material particles were accumulated, the width of wear track increased. Parallel process with formation of particles was interactions between removed particles, ball and disc surfaces, e. g. plastic deformation and surface hardening. When the material particles were accumulated to a critical level, the ball could not transfer the particles forward, it overran the accumulated material. As consequence, the particles were detained and pressed down. Therefore, a wear track with non-uniform width was formed. In advanced, on surface of nitrided, nitrocarburized and duplex treated samples, more uniform wear track than tempered sample was created.

The width of wear track measured by microscopy, the wear depth and wear profile area measured by profilometer are given in table 3. The value of wear depth and wear profile area is sorted ascending.

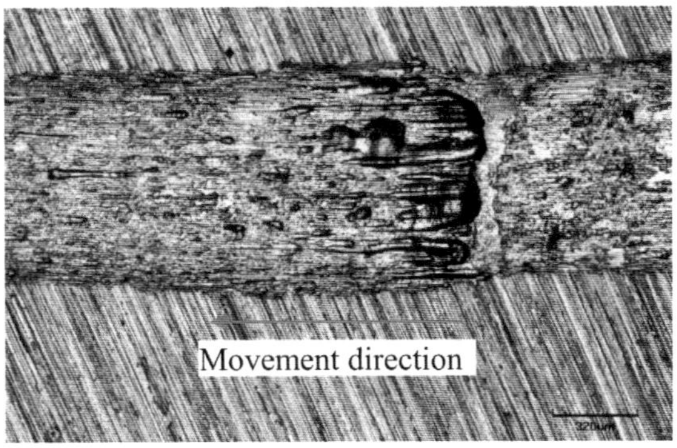

Figure 5. The wear track of tempered sample.

Table 3. The width and profile area of wear track

Sample	Width (μm)	Wear depth (μm)	Profile area (μm²)
Nitrocarburized	888 ± 26	3.25	1158.20
DUPLEX2	517 ± 14	4.04	1665.25
DUPLEX1	582 ± 43	4.68	1944.50
PN2	794 ± 35	5.21	2523.80
PN1	655 ± 16	6.45	3008.50
Tempered	814 ± 58	24.78	14188.50

The wear rate calculated according to equation (2) and the coefficient of friction (considered as an average value of stable section in the figure 4) is showed in figure 6. The sample is sorted by smaller to greater value of wear as following sequence: Nitrocarburizing, DUPLEX2, DUPLEX1, PN2, PN1 and tempered. Nitrocarburizing technology was the best treatment to enhance wear resistance of material (the wear rate value smaller than 12-times). Thanks to 2 μm thickness of manganese phosphate coating, duplex treatment reduce markedly coefficient of friction and wear rate (8-times). Plasma nitriding was the worst among used surface treatment but its wear rate was 5-times less than tempered sample.

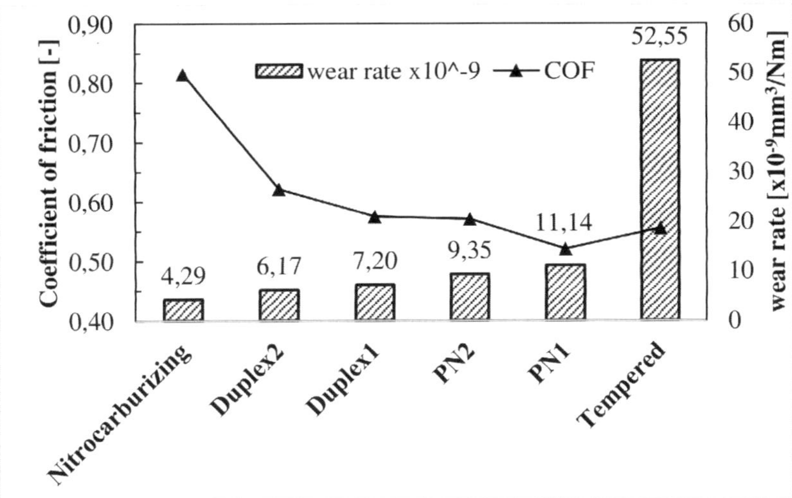

Figure 6. The coefficient of friction and wear rate.

Holemar et al. [9] pointed that, the ε-phase improves better friction properties than γ′ phase, and nitriding in enriched nitrogen atmosphere formed compound layer consisted of ε-phase. Presented results also confirmed that nitriding PN2 treated in reverse atmosphere $1H_2 : 3N_2$ (enriched nitrogen atmosphere) provides better wear resistance than plasma nitriding PN1 treated in classic atmosphere $3H_2 : 1N_2$. Therefore, the composition of nitriding atmosphere has an influence to coefficient of friction and wear resistance.

The coefficient of friction is considered very meaningful characteristics of tribological system. Generally, it is believed that the lower coefficient of friction leads to reduce of material degradation during sliding. However, in this article, the obtained results showed that, the wear rate (resistance) was not proportional to coefficient of friction: the tempered sample exhibited a relatively low friction but very extensive wear rate, in adverse, nitrocarburized sample reached the greatest coefficient of friction but the lowest wear rate (see figure 6), resp. the best wear resistance.

Figure 7 shows correlation of the wear rate to thickness and microhardness of compound layer. It is easy to determine that, the biggest bubble - the highest value of wear rate is tempered sample, on which is not formed any layer or coating, and its hardness is the least of all. In adverse, the smallest bubble – nitrocarburizing having the least wear rate is on the right top of graph (the greatest hardness and compound layer). The nitrided samples PN1 and PN2 have lightly greater wear rate than the duplex treated samples. In terms of experiment, the used load was relatively small (20N), the main mechanism was mild wear. It can be stated that, the thicker and harder is the layer, the better wear resistance of surface is reached. However, if the used load will higher, the main wear may change to severe mechanism, the hard and brittle layers can be peeled off and play a role as abrasive agent. The wear process may be accelerated.

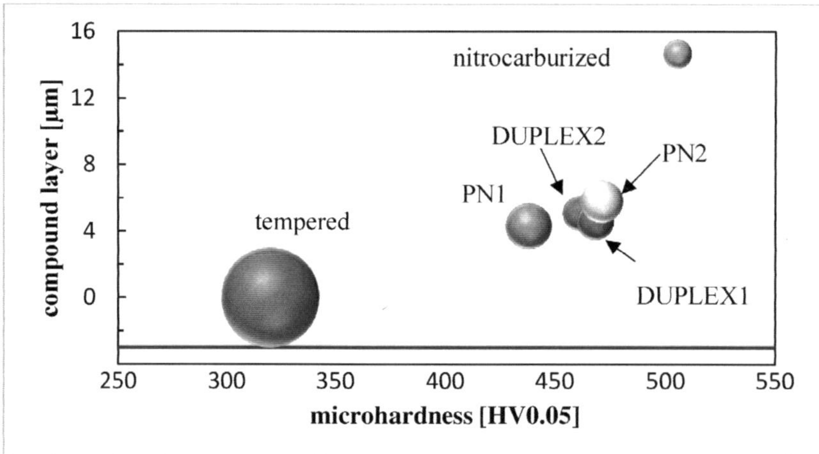

Figure 7. The wear rate in relation to microhardness and thickness of compound layer.

4 Conclusions

This article focused on evaluating the coefficient of friction and wear reisistance of C45 steel treated by plasma nitriding under different nitriding atmospheres, nitrocarburizing (Tenifer®) and nitriding/ manganese phosphating duplex treatment. Experimental results can be summarized in following conclusions:

During sliding of tempered sample, the removed particles were transferred, accumulated, pressed down and formed non-equal wear track.

Nitrocarburizing is the best technology in enhancing wear resistance of material (12-times better) though of high coefficient of friction. The duplex treatment nitriding/ manganese phosphating showed the best trend of friction coefficient and very high wear resistance. Plasma nitriding obtained the worst wear resistance but still 5-times better than tempered sample. Morever, the wear track of these treated surfaces created more uniform wear track than tempered sample, this characteristic may be useful in practice in improving vibration of machinery parts during operations.

The obtained results are an example to interpret that, the wear rate (resistance) may not be proportional to coefficient of friction: tempered sample exhibited a low friction but very high wear rate, on the contrary, nitrocarburized sample exhibited the greatest friction coefficient but the lowest wear rate.

The plasma nitriding and nitrocarburizing technologies formed on substrate nitrided layers with hardness about 100-200 HV higher than tempered surface. Two different gas atmospheres were used for plasma nitriding, in which the nitriding PN2 ($1H_2 : 3N_2$) obtains thicker compound layer, higher surface hardness than nitriding PN1 ($3H_2 : 1N_2$).

A relationship of wear rate to thickness of compound layer and microhardness is evaluated. The thicker compound layer with greater hardness could reduce the wear rate. The optimal thickness of compound layer is a subject remained for other studies.

Acknowledgements

The paper has been prepared thanks to the support of the project The Development of Technologies, Design of Firearms, Ammunition, Instrumentation, Engineering of Materials and Military Infrastructure "VÝZBROJ (DZRO K201)" and "Surface technology in applications special techniques SV16-216".

References

[1] Rabinowicz E 1995 Friction and wear of materials (Canada: John Wiley & Sons, Inc), ISBN 0-471-83084-4, p 124-129.

[2] Pokorny Z, Hrubý V and Studený Z 2016 Effect of nitrogen on surface morphology of layers *Metallic Materials*, vol. 54, no. 2, p 119–124.

[3] Pye D 2003 Practical nitriding and ferritic nitrocarburizing (ASM International Materials Park, Ohio, USA) ISBN 0-87170-791-8, p 32-37.

[4] DIN 50190-4: 1999, Hardness depth of heat-treated parts – Part 4: Determination of the diffusion hardening depth and diffusion depth.

[5] ASTM G99-95a, Standard Test Method for Wear Testing with a Pin-on-Disk Apparatus, ASTM International, West Conshohocken, PA, 2000, www.astm.org.

[6] W STACHOWIAK G (ed.) 2005 Wear - materials, mechanisms and practice(John Wiley & Sons, Ltd) ISBN 9780470017029, p 4.

[7] Doan T V, Kusmič D and Pospíchal M 2015 Surface Treatment Technologies for Wear Resistance Increasing of 42CrMo4 Steel. *Manufacturing Technology*, vol. 15, no. 3, ISSN 1213-2489, p 303-307

[8] Czichos, H., & Saito, T. (2006). *Springer handbook of materials measurement methods* (Vol. 978). L. Smith (Ed.). Berlin: Springer p 688.

[9] Holemar A, Hruby V 1989 Iontová nitridace v praxi (Praha: 1. vyd. Státní nakladatelství technické literatury), ISBN 80-03-00001-7, p 90-93.

Structure and mechanical characterization of Mg-Nd-Zn alloys prepared by different processes

D Dvorský[1], J Kubásek[1], D Vojtěch[1], I Voňavková[1], M Veselý[2] and M Čavojský[3]

[1]Faculty of chemical technology, Department of metals and corrosion engineering, University of chemistry and technology Prague, Technická 5, 166 28 Praha 6 – Dejvice, Czech Republic.

[2]Faculty of chemical technology, Department of Organic Technology, University of chemistry and technology Prague, Technická 5, 166 28 Praha 6 – Dejvice, Czech Republic.

[3]Institute of Materials and Machine Mechanics SAS, Slovak Academy of Sciences, Dúbravská cesta 9, 845 13 Bratislava, Slovak Republic

E-mail: dvorskyd@vscht.cz

Abstract. Magnesium alloys containing about 3 wt. % of Nd and 0.5 wt. % of Zn are considered as promising materials for application in transport and medical industry. Properly treated materials can reach ultimate tensile strength (UTS) higher than 300 MPa. Also the corrosion resistance of these alloys is superior to many other magnesium-based materials. Present work is focused on the preparation of Mg-3Nd-0.5Zn magnesium alloy by classical casting and subsequent thermal treatment. As-cast material was extruded at 400 °C, with extrusion ratio equal to 16 and velocity of 0.2 mm/s. The effect of thermal treatment and also strong plastic deformation during extrusion on final structure conditions and mechanical properties is specified. Present results confirm significant improvement of tensile yield strength (TYS) and UTS after extrusion process as a consequence of fine-grained structure combined with precipitation strengthening. Beside, texture strengthening in the direction parallel to the extrusion has been observed too.

1 Introduction

Thanks to low density and improved mechanical and corrosion properties specific magnesium-based alloys are of a great interest in aviation and automotive industry [1]. The main drawback is application of magnesium alloys at high temperatures due to their low yield strength [2]. However, the superior magnesium alloys containing rare earth elements and yttrium preserve their mechanical properties up to 300 °C [3]. Another utilization magnesium alloys find in medicine as materials for biodegradable implants. There is tendency to reduce or substitute alloying elements in order to reduce price and increase biocompatibility of this material [4]. In many works yttrium is replaced with cheaper zinc [1], [2], [5]. Zinc is biocompatible element essential for human body [6]. It can improve mechanical properties by solid solution hardening and age hardening. However amounts higher than 1 wt.% leads to another precipitation sequence, which is associated with hardness decrease [1], [2]. Neodymium is relatively abundant rare earth element with maximum solubility in Mg 3.6 wt.% at eutectic temperature [4]. It seems to be good alloying element because it significantly improves mechanical properties by solid solution strengthening and by precipitation hardening [7]. It also increases corrosion resistance by forming Nd_2O_3 oxide layer on the surface [8]. Pure magnesium is highly susceptible to galvanic corrosion especially with contained impurities like Fe, Ni, Cu, Co. Fe is insoluble in Mg-matrix and there is tolerance limit about 170 ppm of Fe in pure Mg. However, the influence of impurities works differently in Mg-RE alloys because rare earth elements trap impurities in intermetallic compounds, reducing the threat of galvanic corrosion [9].

Reduction in amount of alloying elements is beneficial for increasing biocompatibility of magnesium alloys, which are considered as materials for biodegradable implants. Those materials will degrade in a body after fulfilling their function without the necessity of subsequent removal. This will reduce the cost of healing process and increase comfort of the patient. Magnesium and its alloys have ideal mechanical properties for this purpose. The elastic modulus of magnesium is about 42 GPa, which is very close to the natural bone (3-20 GPa). Such values prevent the bone stress shielding and help with better bone ingrowth [5].

Mg-Nd-Zn system is well known for its good mechanical properties and good corrosion resistance. Those properties depend highly on the preparation method and structure [10]. Mechanical properties of Mg-Nd-Zn alloys can be improved by suitable heat treatment. It is associated with the formation of Guinier-Preston (GP) zones during precipitation. The precipitation process occurs as follows: solid solution, GP zones, β″ Mg_3Nd cph phase, β′ Mg_3Nd (fcc), β $Mg_{12}Nd$ phase [7]. Another way how to improve mechanical properties is an extreme plastic deformation associated with recrystallization and grain refinement. One of the processes how to attain fine structure is extrusion [11]. During extrusion process the material is extremely plastically deformed. Based on the material and procedure conditions, the final product is composed of fine recrystallized grains, but also some elongated grains which were not recrystallized [12]. However, magnesium alloys are prone to the texture formation in the extruded material. After extrusion process, basal planes of hexagonal structure are preferably oriented parallel to the extrusion direction for many magnesium-based alloys [13]. This has great impact on the anisotropy of mechanical properties. Rare earth elements can partially suppress formation of such a specific texture [14].

2 Materials and methods

2.1 Preparation of samples

Mg-3Nd-0.5Zn (composition in wt. %) alloy was prepared from pure Mg (99.9 wt.%), Nd (99.8 wt.%) and Zn (99.9 wt.%) in induction furnace under inert argon (99.996 wt.%) atmosphere. The melt was homogenized at 750 °C for 15 minutes followed by casting into cylindrical brass mold with 50 mm in diameter and 150 mm in height (labeled as as-cast). Three cylindrical samples were prepared by this way. One of them was then heat treated at 540 °C for 16 hours followed by quenching in water (labeled as T4). The third as-cast cylinder was extruded at 400 °C with extrusion ratio of 16 and 0.2 mm/s extrusion speed. Final products were rods with 7.5 mm in diameter (labeled as extruded). Impurities were determined by AAS method and the results are summarized in table 1.

Table 1: Impurities in samples [ppm].

Label	Co	Cu	Ni	Fe
As-cast	0.72	16.80	6.50	64.40
T4	0.24	6.00	6.50	128.20
Extruded	0.47	14.70	4.50	72.90

2.2 Microstructure

Samples were grinded on SiC grinding papers (P80-P4000) and subsequently polished on diamond paste. The final polishing was done on Topol 2 with fine particles of Al_2O_3. Subsequently samples were etched in solution containing 10 ml of acetic acid, 4.2 g of picric acid, 10 ml of distilled water and 70 ml of ethanol. The microstructure was studied by light microscopy and by electron scanning microscope SEM (TescanVEGA3 with energy dispersion spectrometry and EBSD detector). Phase analyses were performed using X-ray diffraction (XRD, X'Pert Philips, 30 mA, 40 kV, CuK_α X-ray radiation)

2.3 Mechanical properties

Compressive tests were performed on LabTest 5.250SP1-VM at room temperature on cylindrical samples with 6 mm in diameter and 9 mm high. Constant deformation speed of 0.001 s^{-1} was used. Compressive yield strength (CYS), ultimate compressive strength (UCS) and total deformation were determined. Tensile properties were measured on the same machine at room temperature on samples with 3.5 mm in diameter in constricted area and 25 mm in length. Constant deformation speed of 0.001 s^{-1} was used. Vickers hardness according to the EN ISO 6507-1 was measured at loading corresponding to 1 kg.

2.4 *Corrosion behaviour*

Immersion tests were carried out in simulated body fluid (SBF) at 37 °C for 14 days. Table 2 shows SBF composition similar to human plasma. Corrosion rate was determined from weight loss and also from dissolved magnesium by atomic absorption spectroscopy.

Table 2: Composition of SBF [mmol/l].

Ions	Na^+	K^+	Mg^{2+}	Ca^{2+}	Cl^-	HCO_3^-	HPO_4^{2-}	SO_4^{2-}
SBF	142.0	5.0	1.5	2.5	147.8	4.2	1.0	0.5

3 Results

3.1 *Microstructure*

Figure 1 shows microstructures of prepared samples. As-cast ingot has typical dendritic structure, with distance between dendritic branches of 8.3 ± 0.7μm. Alloying elements segregated around dendrites, so there was higher concentration of them. Dendrites consist of α-Mg-matrix and eutectic phase $Mg_{12}Nd$ (20.4 ± 3.2 wt.% Nd), which was indicated by XRD diffraction. Mg_3Nd phase was also found in a minor amount. Neodymium and zinc are homogenously dissolved in Mg-matrix (1.3 ± 0.4 wt.% Nd, 0.4 ± 0.1 wt.% Zn). The rest of the neodymium is placed primarily in the intermetallic phases and zinc forms intermetallic phases with neodymium too (31.5 ± 3.9 wt.% Nd, 6.7 ± 0.5 wt.% Zn). Those phases still seem to be binary phases, where Nd in the structure is partially replaced by Zn.

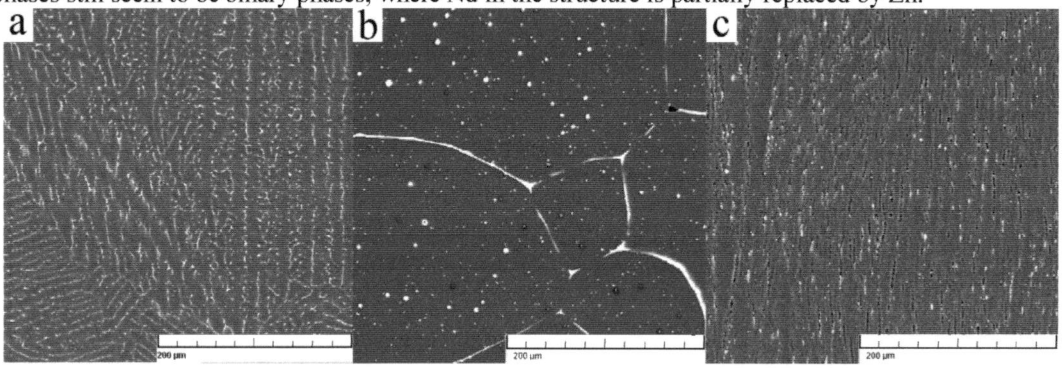

Figure 1: Microstructure: a) as-cast, b) T4, c) extruded.

The heat treatment was performed according to the very well-known approach used for similar alloys. After the heat treatment the original phases dissolved in the Mg-matrix (2.5 ± 0.9 wt.% Nd, 0.5 ± 0.2 wt.% Zn). However, some eutectic phases precipitated at grain boundaries and inside the grains as well. It denotes that during the thermal treatment material had to be partially melted at some areas. The grain size of resulted material ranged from 100 μm to 500 μm.

In figure 1 structure of extruded material is displayed. Intermetallic phases such as $Mg_{12}Nd$, Mg_3Nd and $Mg_{41}Nd_5$ were identified in this state. The amount of neodymium in the Mg-matrix was slightly lower than in the as cast material (1.1 ± 0.3 wt.% Nd, 0.4 ± 0.1 wt.% Zn), which might be caused by precipitation of fine intermetallic phases $Mg_{41}Nd_5$ in the area of original dendritic segregation. Extrusion led to the partial recrystallization of the material. The size of recrystallized equiaxed grains was about 2.3 ± 1.4 μm, which was confirmed by EBSD analysis. A large part of the structure consisted of elongated strongly deformed grains with the nuclei of new grains separated by small angle boundaries with misorientation between 2 and 15 ° (white lines in Figure 2). Performed analysis also confirmed strong texture in the material after extrusion. It is a common phenomenon for magnesium alloys that cause the anisotropy of mechanical properties.

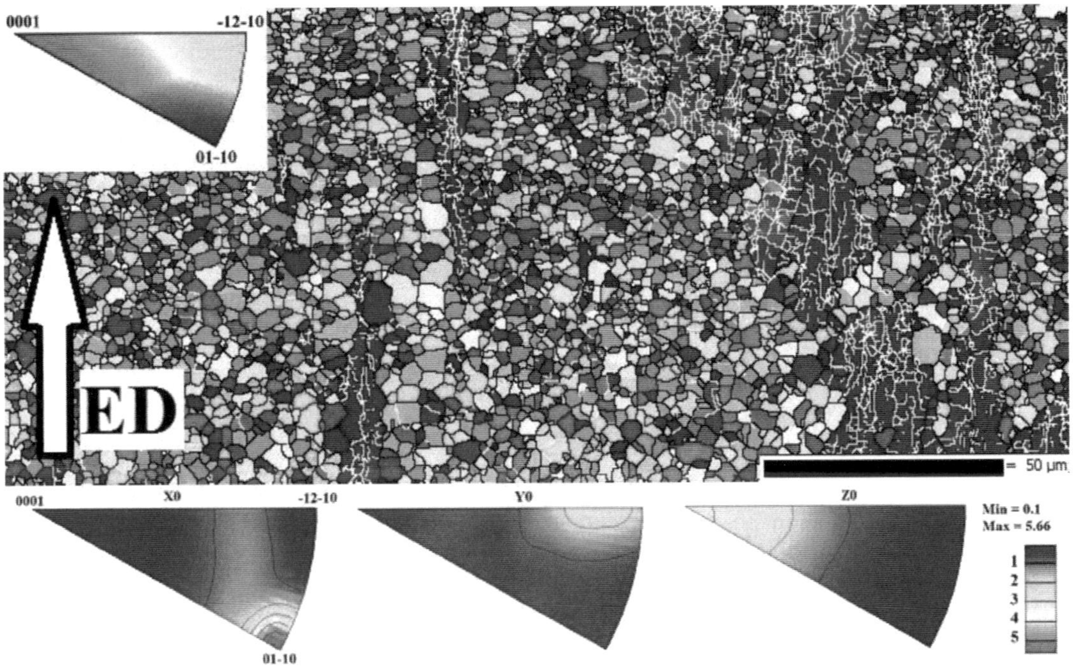

Figure 2: Grain orientation and inverse pole figures of extruded sample. (X0 reflects the direction of the extrusion, ED = extrusion direction.)

3.2 Mechanical properties

From mechanical properties, hardness, compressive and tensile tests were carried out (results are summarized in Table 3). After the heat treatment hardness remains almost the same as for the as-cast material. Compressive stress-strain curves (Figure 3) show strong increase in yield strength of the extruded material compared to the as-cast ingot. This enhancement is caused especially by finer grains but also by existence of fine precipitates which were formed during the extrusion. The same reasons affect the observed value of hardness (Table 3). Thermal treatment of the as-cast ingot leads to increase of ultimate compressive strength and ductility. The reason is connected with the lower amount of brittle intermetallic phases in the structure and probably higher grain size, which support twinning, and therefore, contribute to plasticity by this mechanism.

The tensile yield strength of extruded sample is much higher than the compressive yield strength (Table 3). This results from a strong texture that was confirmed by EBSD analysis. Materials crystalizing in hexagonal structure deforms not only by dislocation slip mechanism but also by specific twinning mechanism. Twinning requires much less energy for plastic deformation, however, it strongly depends on grain orientation [15]. After extrusion the basal planes are oriented parallel to the extrusion direction, which is also the direction of applied compressive force. Such orientation of hexagons towards the compressive force is strongly favourable for twinning mechanism, so it noticeably decreases compressive yield strength of this material. On the other hand, this orientation towards tensile force direction is unfavourable for deformation by twinning, therefore, tensile yield strength is increased [16].

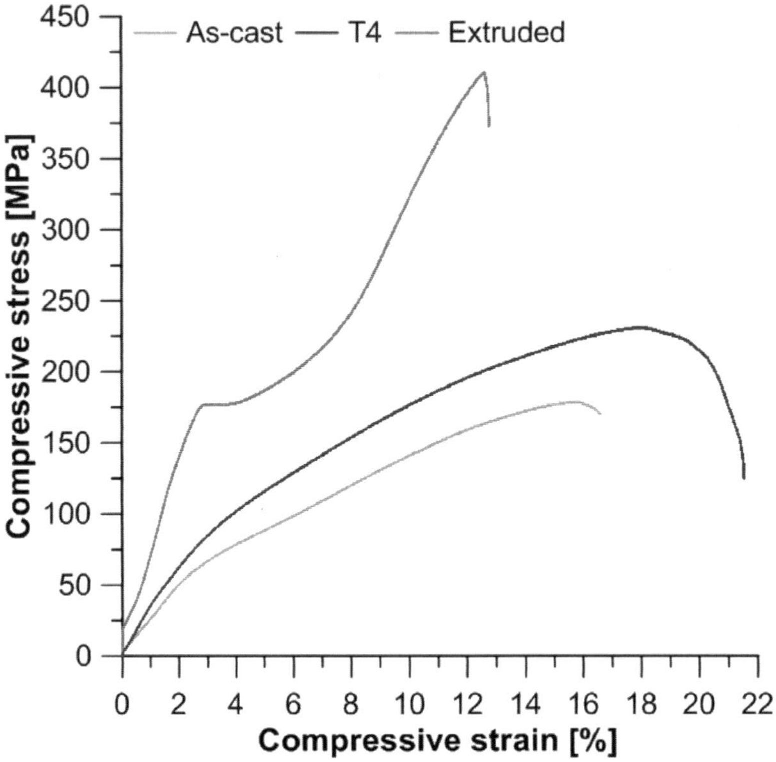

Figure 3: Stress-strain curves of studied materials in compressive conditions.

Table 3: Mechanical properties; CYS = compressive yield strength, UCS = ultimate compressive strength, D = relative deformation, TYS = tensile yield strength, UTS = ultimate tensile strength, E = elongation.

Sample	CYS [MPa]	UCS [MPa]	D [%]	TYS [MPa]	UTS [MPa]	E [%]	HV1
As-cast	58±5	182±5	15.2±0.8	-	-	-	53.1±3.1
T4	51±5	252±30	22.0±2.5	-	-	-	54.2±3.9
Extruded	174±5	423±12	12.7±0.4	251±6	271±8	2.7±0.7	61.1±2.3

3.3 Corrosion behaviour

Corrosion behaviour was studied in SBF solution for 14 days. Weight loss and concentration of magnesium ions in solution were applied as two methods for the estimation of corrosion rate (Table 4). Among studied materials as-cast ingot had the highest corrosion rate.

Thermal treatment led to a rapid decrease of corrosion rate. This is partially caused by more neodymium in the solid solution which has good impact on corrosion rate, although even higher corrosion resistance could be expected if lower amount of intermetallic phases is presented in the structure due to the absence of galvanic cell between magnesium matrix and MgNd phases in the structure. On the other hand, standard reduction potential of MgNd phases is relatively close to Mg, so the Anode-to-Cathode potential difference is low [10]. Therefore, the effect of galvanic corrosion in Mg-Nd alloys is low and the corrosion is more uniform [17].

The extrusion of as-cast ingot also lowered the corrosion rate. This improvement is caused predominantly by finer structure of this material, where grain boundaries accelerate passivation kinetics. Small grains reduce intensity of galvanic corrosion due to formation of close situated anodic-cathodic regions, which should lead to lower corrosion rate and more uniform corrosion [18].

Table 4: Corrosion rate [mg/cm^2/day].

Sample	Weight loss	Ions
As-cast	1.04 ± 0.18	0.83 ± 0.16
T4	0.43 ± 0.05	0.31 ± 0.07
Extruded	0.67 ± 0.07	0.59 ± 0.04

4 Conclusion

Magnesium alloy designated as Mg-3Nd-0.5Zn was prepared by melting of pure metals. Thermal treatment (T4) was applied on one as-cast ingot to investigate changes in microstructure and mechanical properties. Another as-cast ingot was also extruded. The huge changes were observed between these states in both mechanical and corrosion properties. Significantly improved mechanical properties of the extruded material were connected with finer structure, precipitation strengthening, but also strong texture causing anisotropy of properties in tension and compression. Corrosion rates after the heat treatment or extrusion were about half the value of the as-cast sample. Such improvement was related to the increased Nd concentration in solid solution or fine grained structure, respectively.

Acknowledgement
We wish to thank for the financial support of this research by project no. GA16-08963S.

References

[1] Wu D, Chen R S and Ke W 2014 *Materials & Design* **58** pp 324-331
[2] Yang J, Wang J, Wang L, Wu Y, Wang L and Zhang H 2008 *Materials Science and Engineering: A* **479** pp 339-344
[3] Rzychoń T and Kiełbus A 2007 *Journal of achievements in materials and manufacturing engineering* **21** pp 31-34
[4] Zhang X, Yuan G, Mao L, Niu J, Fu P and Ding W 2012 *Journal of the Mechanical Behavior of Biomedical Materials* **7** pp 77-86
[5] Li J, Tan L, Wan P, Yu X and Yang K 2015 *Materials Science and Engineering: C* **49** pp 422-429
[6] Qin H, Zhao Y, An Z, Cheng M, Wang Q, Cheng T, Wang Q, Wang J, Jiang Y, Zhang X and Yuan G 2015 *Biomaterials* **53** pp 211-220
[7] Gärtnerová V, Trojanová Z, Jäger A and Palček P 2004 *Journal of Alloys and Compounds* **378** pp 180-183
[8] Ding Y, Wen C, Hodgson P and Li Y 2014 *Journal of Materials Chemistry B* **2** pp 1912-1933
[9] Zhao X, Shi L and Xu J 2013 *Journal of Materials Science & Technology* **29** pp 781-787
[10] Zhang X, Yuan G, Niu J, Fu P and Ding W 2012 *Journal of the Mechanical Behavior of Biomedical Materials* **9** pp 153-162
[11] Wu Q, Zhu S, Wang L, Liu Q, Yue G, Wang J and Guan S 2012 *Journal of the Mechanical Behavior of Biomedical Materials* **8** pp 1-7.
[12] Dvorsky D, Kubasek J, Vojtěch D, Prusa F and Nova K 2016 *Manufacturing Technology* **16** pp 680-687.
[13] Yin D L, Wang J T, Liu J Q and Zhao X 2009 *Journal of Alloys and Compounds* **478** pp 789-795
[14] Wu Y P, Zhang X M, Deng Y L, Tang C P and Zhong Y Y 2015 *Materials Science and Engineering: A* **644** pp 152-158
[15] Jiang M G, Yan H and Chen R S 2015 *Journal of Alloys and Compounds* **650** pp 399-409
[16] Jiang M G, Xu C, Nakata T, Yan H, Chen R S and Kamado S 2016 *Materials Science and Engineering: A* **667** pp 233-239
[17] Zong Y, Yuan G, Zhang X, Mao L, Niu J and Ding W 2012 *Materials Science and Engineering: B* **177** pp 395-401
[18] Mao L, Yuan G, Wang S, Niu J, Wu G and Ding W 2012 *Materials Letters* **88** pp 1-4

IOP Conf. Series: Materials Science and Engineering **179** (2017) 012019 doi:10.1088/1757-899X/179/1/012019

Mechanical properties determination of AM components

J Dzugan[1], M Sibr[1], P Konopík[1], R Procházka[1], M. Rund[1]
COMTES FHT a.s., Prumyslova 995, 334 41 Dobrany, Czech Republic

E-mail: jan.dzugan@comtesfht.cz

Abstract. Characterisation of engineering materials and components is a crucial part for design and save service life utilization. Due to components processing technologies and exploitation conditions local properties can significantly vary from location to location over larger components as well as over small material volumes with gradual material changes such as welds, coatings or additively manufactured parts. The current paper is dealing with local properties characterisation for additively manufacture (AM) components by micro tensile test (M-TT). Components produced by additive manufacturing techniques yield properties variation in dependence of the considered location within the component regarding to direction in relation to deposition process. Properties vary over the thickness, length, angle or contacts with the supporting structures necessary for a successful components production by additive manufacturing techniques. The properties differences are mainly related to varying heating/reheating and cooling conditions at various locations of usually very complex parts produced mainly by these technologies. The standard testing procedures fail to characterize such local properties of complex shaped objects due to large size requirements on specimens. Therefore, new techniques have to be established for such detailed local characterizations. Results of miniaturized tensile tests application for local properties and orientations are shown here.

1 Introduction

Additive manufacturing (AM) undergoes dramatic development within last few years thanks to big development leap in metal printing technology development and dissemination of a large number of machines for AM among customers. This leads to large application of AM and increased demands on AM produce parts characterization.

Materials produced by AM exhibits some peculiarities in comparison to standardly used semi-products. Standard material products available on market are usually provided, in forged state when almost homogeneous material properties are attained with rather small anisotropy. However, AM products exhibit strong properties dependence on the process parameters, e.g. position within depositing chamber, building orientation (longitudinal, vertical), building direction (initial wall, final outer wall..), specimen/component volume and so on. Therefore it is crucial to be able to measure local properties, which is for most cases impossible with the use of standard sized specimens, due to the experimental material volume requirements. In many cases it is not possible to determine properties with standard sized specimens as the component is smaller than standard specimens. When standard specimen is produced separately, it has different properties, due to different build procedure, heat transfer that results in slightly different properties from a real component. Therefore, specimens with appropriate dimensions fitting into the components produced by AM shall be used. Mini-specimen methods were developed especially for residual service life assessment, such as Automated Ball Indentation (ABI), Small Punch Test (SPT) [1-2] or recently methods using miniaturized standard sized specimens [3-11]. ABI can provide quite variable performance for various materials and thus it not generally widely used and in the case of complex thin walled structures produced by AM it would be also in many cases hardly applicable. Much often is used nowadays SPT. Its drawback is that there is need for known correlation between the SPT specimens and standard sized specimens, which is quite demanding on materials and testing in order to have reliable correlation for each specific material considered. The problem with AM parts is that its properties partly depend on the specific part dimensions and thus it is rather impossible to obtain correlation between standard sized specimens and SPT for AM components. Moreover, in the case of AM parts the anisotropy is also important issue but as SPT is using disc specimens biaxial loaded, SPT provides information on behaviour in one plane, but not direction as standard test such as tensile test, notch toughness test, fatigue test.... All these facts lead to necessity of miniaturized standard specimens' application for AM parts properties characterisation.

This paper demonstrates small specimen's technique application for local tensile properties assessment. Mini-Tensile Tests (M-TT) developed and verified in in [3-11] are applied here for part of the total hip replacement set produced by AM. The material investigated is Ti6Al4V (also known as Ti grade5). Local properties in different orientations are investigated for several sampling locations across the component of interest.

2 Experimental material

The component investigated here is made of Ti6Al4V produced by Selective Laser Melting (SLM) powder bed based AM technology. The standard set of printing parameters was used for the part building. Commercially available Ti-powder was used. Detail of powder particles can be seen in figure 1. There is clearly seen nice globular shape of the particles. The size of the particles is ranging from 10 to 75μm.

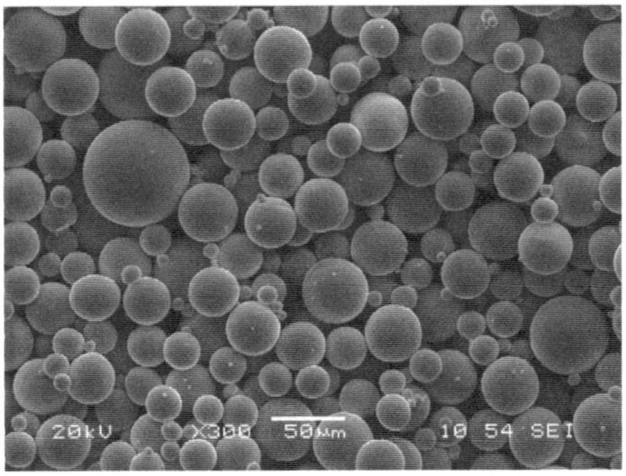

Figure 1. Micrograph of the powder particles.

In order to assess local properties variation across the component investigated, several sampling locations and two sampling orientations were chosen. Four sampling locations evenly distributed over the component length were chosen. At each of the locations longitudinal and transversal directions were investigated. Sampling scheme can be seen in figure 2 together with sampling locations designation.

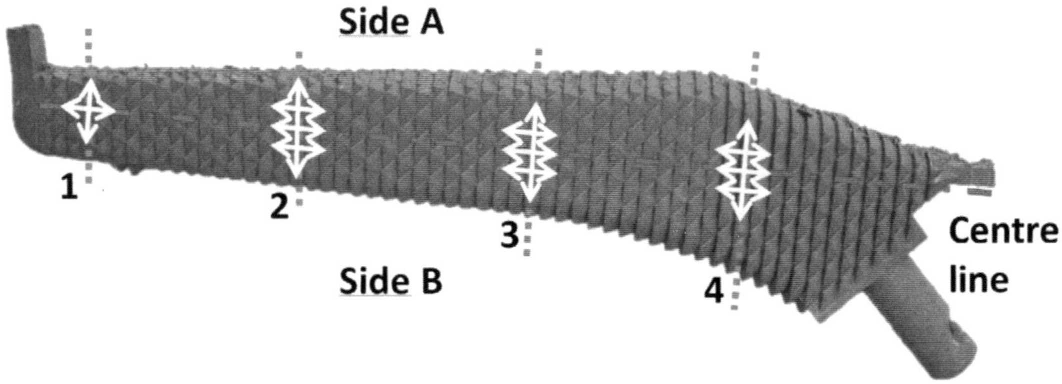

Figure 2. Sampling locations and designation.

3 Tensile tests

Mechanical properties characterization of the material investigated was carried out with the use of Mini-Tensile Tests (M-TT). M-TT testing procedure was verified and performance was shown for example in [7-11]. In the current study, proportional specimens with the geometry according to figure 3 were applied.

Specimens were machined from locations depicted in figure 2. Blocks of the experimental material for M-TT specimens were cut from the component with the use of metallographic saw at first. Subsequently, specimens machining was done by milling following procedures assuring the samples machining influence minimization. Samples plane orientation was perpendicular to the centre line.

Testing was performed under quasi-static loading conditions at room temperature with the use of linear drive based testing system with load capacity of 5kN. Strain was measured with the use of digital image correlation system ARAMIS by GOM. Three specimens per location were tested in the longitudinal direction, except the location 1, where small material volume prevented to have more samples in different orientations. In the case of the transversal direction single specimen per location was investigated. Prior to and after testing specimens dimension were measured for standard parameters determination. Offset yield stress YS, ultimate tensile strength UTS, uniform elongation at maximum force UE, elongation EL and cross section reduction CSR were evaluated. Examples of the obtained tensile records can be found in figure 4. Summarized averaged resulting values of tensile test can be seen in table 1.

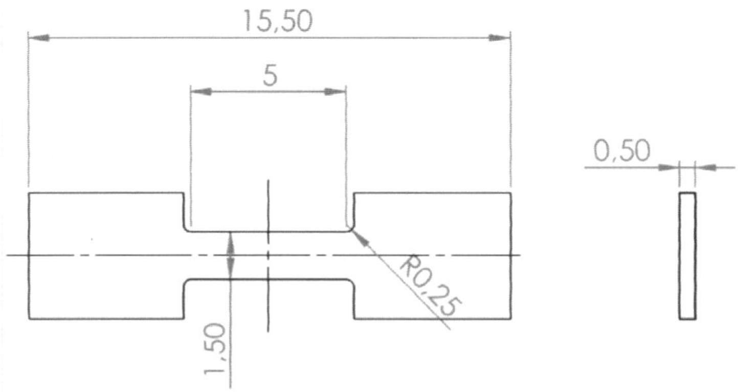

Figure 3. M-TT sample geometry.

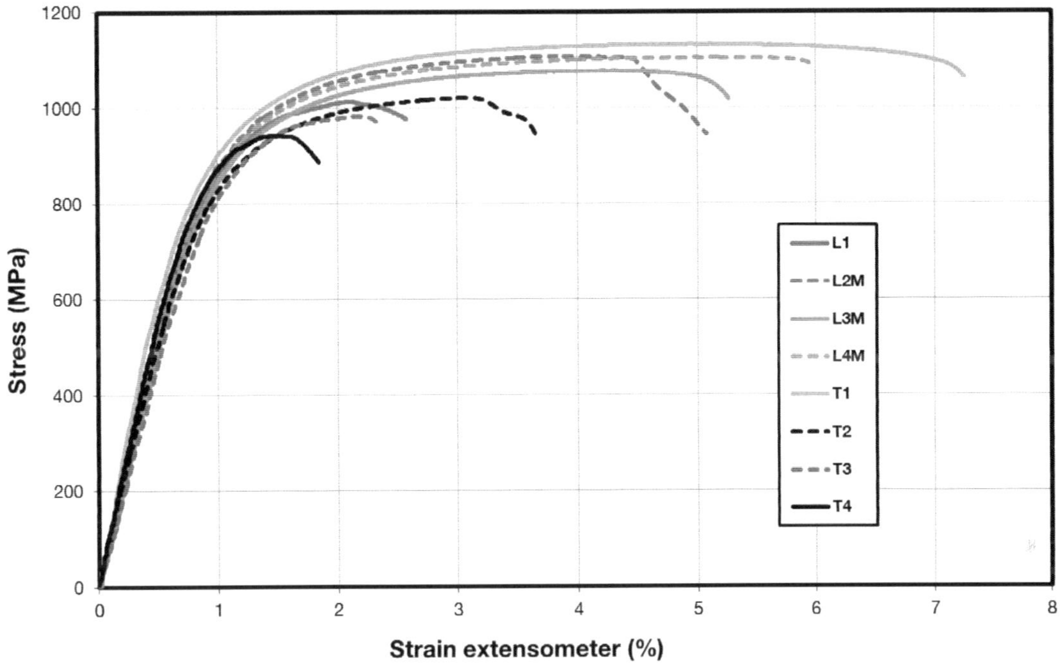

Figure 4. Tensile test records for considered locations and sampling directions.

Table 1. Tensile test results.

Specimen	YS MPa	UTS MPa	UE %	EL %	CSR %
T1	844,3	1131,7	4,1	7,2	28,0
T2	827,4	1020,6	2,1	3,7	13,8
T3	841,0	982,3	1,2	2,4	14,6
T4	848,8	942,5	0,6	1,9	16,4
L1M	880,2	1012,9	1,1	2,6	12,4
L2A	1044,4	1101,7	0,8	2,8	20,3
L2M	904,2	1106,7	2,7	5,1	13,5
L2B	904,0	1130,7	3,6	6,1	12,3
L2 average	950,8	1113,0	2,4	4,7	15,4
L3A	902,0	1099,3	3,0	4,6	23,1
L3M	881,1	1076,5	3,2	5,3	13,1
L3B	892,7	1090,6	4,3	6,2	15,2
L3 average	891,9	1088,8	3,5	5,4	17,1
L4A	908,2	1121,1	4,7	7,6	21,7
L4M	891,0	1104,6	4,1	5,9	27,1
L4B	890,5	1092,8	3,7	6,2	19,7
L4 average	896,6	1106,1	4,1	6,6	22,8

4 Results discussion

The results summarized in table 1 show variation of over 20% in yield stress and tensile strength values and even higher in elongation values across the component investigated.

The transversal tensile properties exhibit clearly monotonic decreasing trend of the ultimate strength values from the location 1 towards the location 4 with the same decreasing trend of the plastic behaviour described by the uniform elongation and elongation. Noticeable embrittlement can be seen here where the original UE 4,1% at the location 1is decreasing up to the final 0,6% at the location 4. However, the yield stress is almost constant for all considered sampling locations.

Observation of the longitudinal results point out rather uniform values of the yield stress across the component, except outlying value for the specimen L2A, that has to undergo some further analysis. The longitudinal tensile strengths show lower value at the location 1 with similar values at the other sampling locations. There can be seen slight ductility increase towards the location 4 when elongation values are observed. Cross section reduction values exhibit higher scatter, but the averaged values follow the same trends as elongation and thus were not discussed separately.

Taking into account longitudinal and transversal test results there seems to be opposite trend in the properties evolution. While the lowest transversal tensile strength with the lowest elongation is found for the location 4, the same location exhibits the best longitudinal values.

5 Conclusions

The paper was dealing with the local properties characterization of AM component made of Ti gr. 5 alloy. Tensile properties were investigated at different component sampling locations in two directions at room temperatures under quasi-static loading conditions. Miniature tensile (M-TT) specimens were employed here for the properties assessment.

The records obtained with the use of M-TT exhibited very good repeatability. Almost identical records in the initial part were obtained; proving well established test methodology from samples preparation up to tests itself execution confirming M-TT methodology application to AM parts characterization.

Concerning the properties assessment across the component of the interest, there can be found certain properties anisotropy when longitudinal and transversal properties are considered. The longitudinal direction is following the build-up layers during printing, while the traversal direction goes across the build-up layers. The longitudinal properties are very similar according to all considered parameters, however the transversal properties exhibit gradual decrease from the thinnest part towards the thickest one. This change of about 200MPa is noticeable and some explanation for this change should be provided by further metallographic investigations.

The study shown here provided detailed investigation of local properties measurement of AM component with the use of M-TT technique. Applicability of this method for AM parts was demonstrated here providing deeper insight into local material behaviour enabling information for AM technology improvement toward tailor made proprieties in different directions. Additionally, local data that can be used as input data for FEM simulation and subsequent optimization of AM component can be consider as another important output of these measurements. Local properties information together with design and AM processes parameters optimization will lead to better components functionality and lower undesired components failure rate.

Acknowlegements

This paper was created by project Development of West-Bohemian Centre of Materials and Metallurgy. No.: LO1412, financed by the MEYS of the Czech Rep. and project TF02000067 - 3D implants printing with high added value – reliability, efficiency, individuality (2016-2019, TA0/TF) sponsored by Technology Agency of The Czech Republic

References

[1] Dymacek P, Seitl S and; Milicka K 2010 *Key Engineering Materials* Influence of friction on stress and strain distributions in small punch creep test models **Vol. 417-418** (Zurich: Trans tech publications ltd) pp 561-4

[2] Dobes F, Dymacek P and Besterci M 2015 *Mat. Sci. and eng. a-struct. mat. prop. Microstr. and proc.* Estimation of the mechanical properties of aluminium and an aluminium composite after equal channel angular pressing by means of the small punch test **Vol. 626** (Amsterdam: Elsevier science SA) pp 313-321

[3] Konopík P and Džugan J 2012 *2nd Int. Conf. SSTT on Det. of Mech. Prop. of Mat. by Small Punch and other Miniature Testing Techn.* Determination of Tensile Properties of Low Carbon Steel and Alloyed Steel 34CrNiMo6 by Small Punch Test and Micro-Tensile Test (Ostrava: Ocelot sro) pp 319-328.

[4] Konopík P, Dzugan J and Prochazka R 2013 *Mat. Sci. and Tech. Conf. and Exh.* Evaluation of local mechanical properties of steel weld by miniature testing technique, pp 2404-11

[5] Konopík P, Dzugan J and Prochazka R 2013 *METAL 2013 – 22nd Int. Conf. on Metallurgy and Mat. (Brno)* Determination of fracture toughness and tensile properties of structural steels by small punch test and micro-tensile test (Ostrava: TANGER Ltd.) pp 722 – 727

[6] Konopik P, Dzugan J and Rund M 2014 *METAL 2014 - 23rd Int. Conf. on Metallurgy and Mat. (Brno)* Dynamic tensile and micro-tensile testing using DIC method (Ostrava: TANGER Ltd.) pp 498-503

[7] Džugan J, Procházka R and Konopík P 2015 *Small Specimen Test Techniques* Micro-tensile test technique development and application to mechanical property determination **6th Volume, ASTM STP 1576** Sokolov M A and Lucon E (West Conshohocken: ASTM International) pp 12-29

[8] Rund M et al 2015 *Procedia Engineering* Investigation of Sample-size Influence on Tensile Test Results at Different Strain Rates, Vol. 114 (Amsterdam: Elsevier science bv) pp 410-15

[9] Dzugan J, Konopik P., Rund M and Prochazka R 2015 *ASME Pressure vessels and piping conf. - 2015* Determination of local tensile and fatigue properties with the use of sub-sized specimens, **Volume 1A: Codes and Standards** (New York: Amer. Soc. Mech. Eng.)

[10] Procházka R, Džugan J and Kövér M 2015 *Arch. of Mat. Sci. and Eng.* Miniature specimen tensile testing of AZ31 alloy processed by ECAP, **Volume 76, Issue 2** (Gliwice: International OCSCO World Press) pp 134-9

[11] Konopik P, Dzugan J and Rund M 2015 *METAL 2015 – 24th Int. Conf. on Metallurgy and Mat. (Brno)* Determination of fracture toughness in the upper shelf region using small sample test techniques (Ostrava: TANGER Ltd.) pp 710-15

Mini-tensile specimen application for sheets characterization

J Džugan[1], M Rund[1], A Prantl[1], P Konopík[1]

[1]COMTES FHT Inc., Průmyslová 995, 334 41, Dobřany, Czech Republic

E-mail: jan.dzugan@comtesfht.cz

Abstract. There are many cases when there is a shortage of the experimental material for detailed analysis and then small size specimens techniques becomes essential. The current paper deals with investigations of mini-tensile tests (MTT) application to metal sheets characterization. In the case of metal sheets assessment the most common are tensile tests for Lankford parameters and strain hardening determination. As most of the processes are not quasi-static and constant strain rate processes, thus assessment of strain rate hardening is also crucial part of the characterization. Previously developed and verified testing procedure of M-TTs for bulk materials is applied here for steel sheet made of DC01 characterization. Tests under quasi-static and dynamic loading conditions are carried out in order to describe above mentioned properties at room temperature. Accurate strain measurement is carried out with digital image correlation systems and results obtained with M-TTs are going to be confronted with standard size specimens' results.

1 Introduction

Metal sheet forming is multifarious process producing components with very complex shapes within very narrow tolerance bounds. In order to achieve desired component shapes, detailed FEM analyses are carried out in order to assure the proposed production process stability and components good quality. Critical component of the FEM simulation are input material characteristics that together with appropriate material model represent the material behaviour in the course of the processing. Metal sheets exhibit certain peculiarities in comparison to bulk materials, due to crystallographic structure and the characteristics of the rolling process sheet metals usually exhibit a certain degree of mechanical properties anisotropy. The properties variation has to be described together with, strain hardening and strain rate sensitivity for precise material behaviour description. There are standard procedures available for these parameters determination that have been used for a many years yet. The current paper shows possibility of determination of all considered properties with the use of miniaturized tensile specimens.

The performance of mini-tensile tests (MTT) for bulk materials characterisation has been successfully shown in [1-6]. The advantage of MTT are very low requirements on the experimental material volume and possibility of local properties determination, even from real components for the local properties determination. Moreover M-TT can be effectively applied for FEM models verification, e.g. by evaluation of local ductility of sheet in the processed component that can be confronted with FEM model prediction for the same location. Additionally, as the forming processes are usually not performed at constant strain rates over whole component, the strain rate sensitivity has to be covered over a wide range of strain rates, from the slowest up to dynamic ones. M-TT allows effectively asses even high strain rate properties with most of the usual testing machines. This is achieved thanks to its small gauge length (3mm) allowing reaching strain rates in the order of several hundreds on testing systems with the actuator velocity of about 1m/s, which is not some extremely sophisticated and thus expensive high strain rate testing machine. Determination of anisotropy, strain-hardening and strain rate sensitivity is presented in this papers for DC01 sheet of thickness 1.5 mm. Strain during all tests are measured with the use of digital image correlation system. Standard size specimens are used for anisotropy and strain hardening determination under a quasi-static loading conditions, where they are compared with MTT results. Strain rate dependency of the parameters consider here are further determined with MTT only.

2 Tensile tests

Tensile tests were performed on flat samples made of steel sheet DC01. The original sheet thickness was 1.5 mm. Standard size specimens with thickness of 1.5 mm and the geometry according to figure 1 were milled as a stock of 10 specimens. M-TT specimens with the geometry according to figure 2 were machined from the middle part of the sheet thickness by spark eroding of the specimen silhouette and grinding to the final thickness. Standard specimens' orientations were considered for both specimen geometries: longitudinal in rolling direction then 45° and 90°from rolling direction. Testing was carried out with the use of servo-hydraulic testing system MTS with capacity of 25 kN and the actuator velocity up to 1.1 m/s for standard sized specimens. MTT specimens were tested on small size testing system with linear drive with the load capacity of 5kN and maximum actuator velocity of 0.3 m/s. All tests were executed at room temperature and the strain measurement was done with the use of ARAMIS Digital Image Correlation system (DIC). Three specimens per condition were tested for all considered cases.

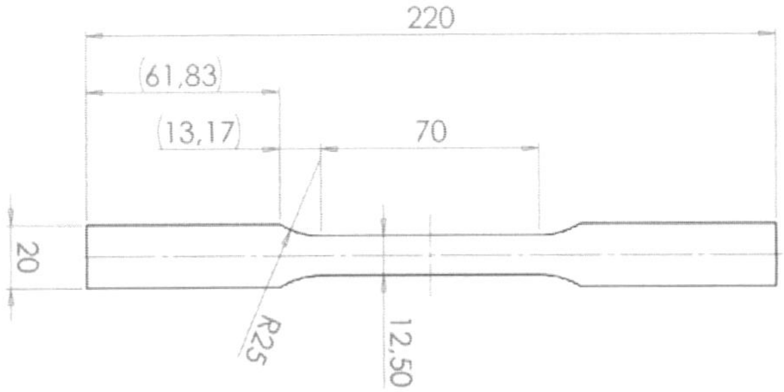

Figure 1. Standard tensile test specimen geometry.

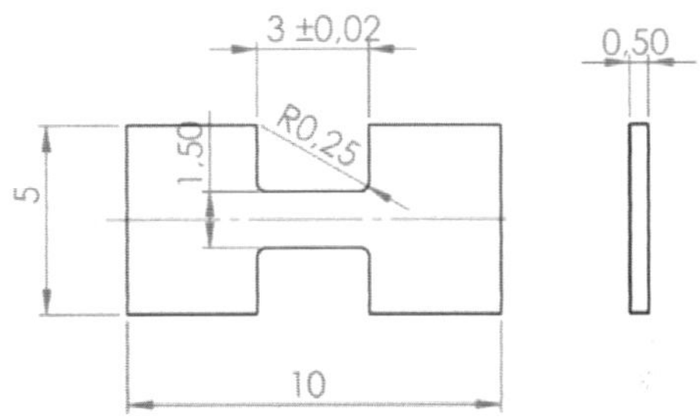

Figure 2. M-TT Specimen geometry.

Quasi-static tensile test were carried out for both specimens geometries and all sampling directions for comparison of the results obtained from standard size and M-TT specimens. On the basis of these tests strain hardening and plastic strain ratio were subsequently evaluated. The strain rate sensitivity determination was determined with the M-TT specimens only. Specimens designation consists of specimen geometry (MTT), specimen orientation and the strain rate. In the case of standard tensile specimens only two latter mentioned parameters are used for designation. Tensile curves obtained for all tested conditions are shown in figure 3-figure 5 and tensile tests results are summarized in table 1, where averaged values from three tests per conditions are shown.

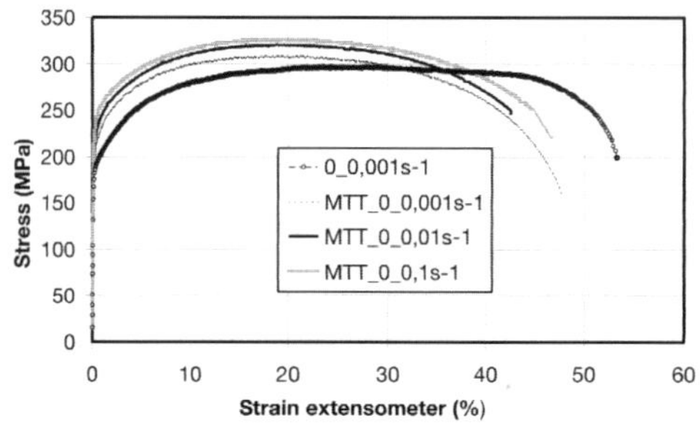

Figure 3. Tensile curves of both specimen geometries in longitudinal direction.

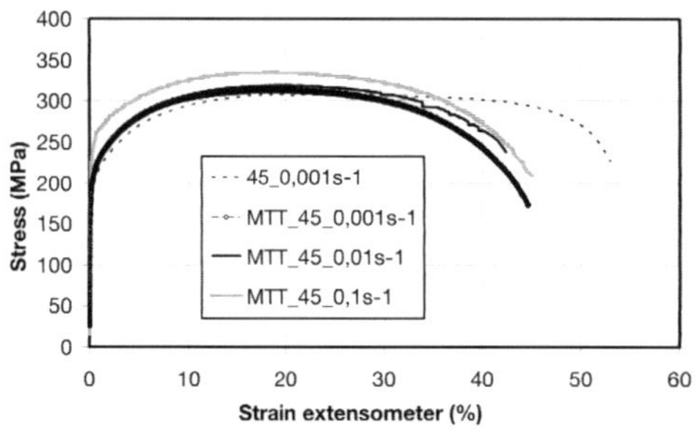

Figure 4. Tensile curves of both specimen geometries 45° orientation.

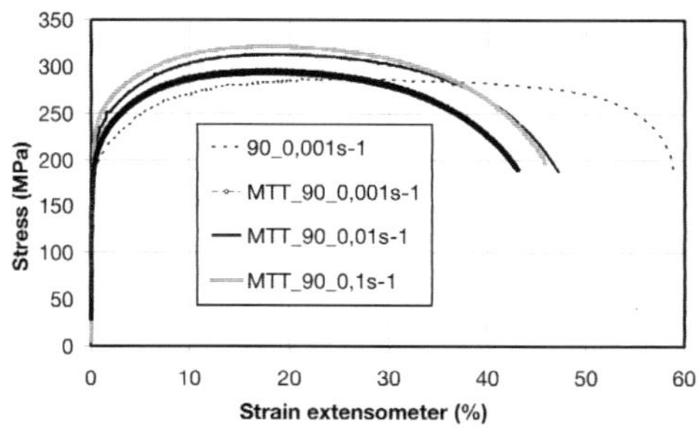

Figure 5. Tensile curves of both specimen geometries 90° orientation.

Table 1. Summary of tensile tests results.

Sample geometry	$\dot{\varepsilon}$ (s^{-1})	Angle (°)	YS (MPa)	UTS (MPa)	A50 (%)
Standard	0,001	0	189,3	297,2	42,1
MTT	0,001	0	197,4	304,4	46,7
MTT	0,01	0	219,8	316,9	42,8
MTT	0,1	0	229,9	326,0	45,9
Standard	0,001	45	199,8	310,8	42,3
MTT	0,001	45	199,9	312,8	43,2
MTT	0,01	45	217,4	324,6	42,6
MTT	0,1	45	226,9	334,2	45,6
Standard	0,001	90	190,2	292,7	45,4
MTT	0,001	90	185,8	292,2	43,8
MTT	0,01	90	211,5	314,8	48,0
MTT	0,1	90	220,5	321,8	45,9

Where:

$\dot{\varepsilon}$	Strain rate (s^{-1})
YS	Yield strength (MPa)
US	Ultimate strength (MPa)
A50	Elongation for initial length 50mm

3 Strain hardening

The evaluation of strain hardening parameters was carried out based on Hollomons equation (1) where the power law relationship between the stress and the amount of plastic strain is used.

$$\sigma = C.\varepsilon^n \tag{1}$$

The fitting range considered for the Hollomons law parameters is between 2-5% of the plastic strain. The example of the evaluation is shown in figure 6. Results are summarized in the figure 7-figure 9.

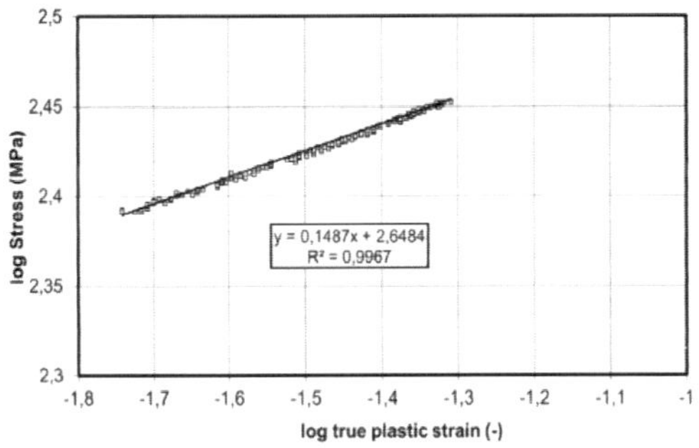

Figure 6. Example of Hollomons law parameters fitting.

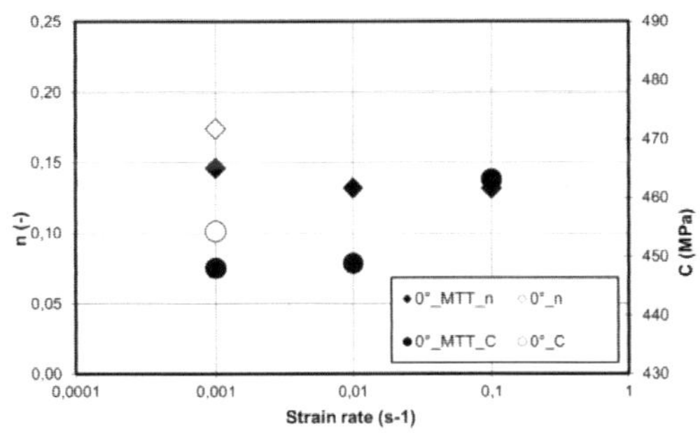

Figure 7. Strain hardening parameters – longitudinal.

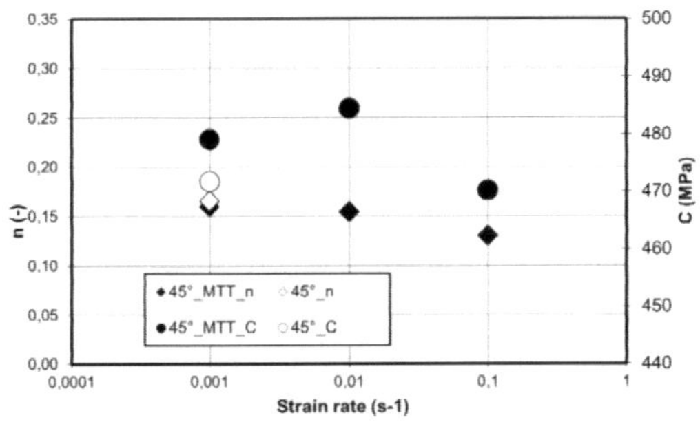

Figure 8. Strain hardening parameters - 45° orientation.

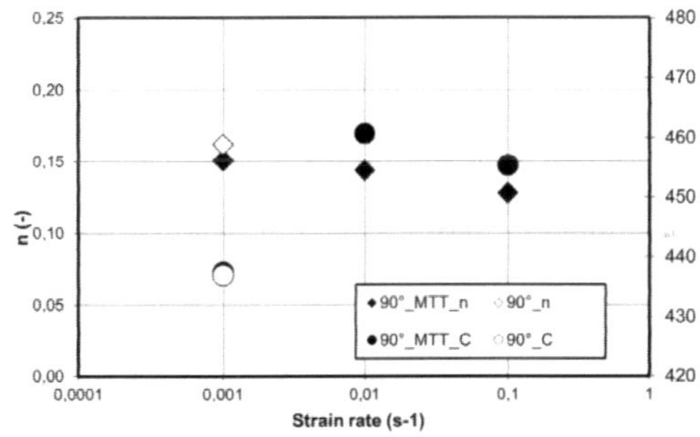

Figure 9. Strain hardening parameters - 90° orientation.

4 Strain rate sensitivity

The strain rate sensitivity of yield stress and ultimate tensile strength was determined within the range of 0.001 up to 0.1 s^{-1} for all considered specimens orientations with the use of MTT only. The evaluated strain rate sensitivity of yield stress and ultimate tensile stress are summarized for specific specimen orientations in figure 10-figure 12.

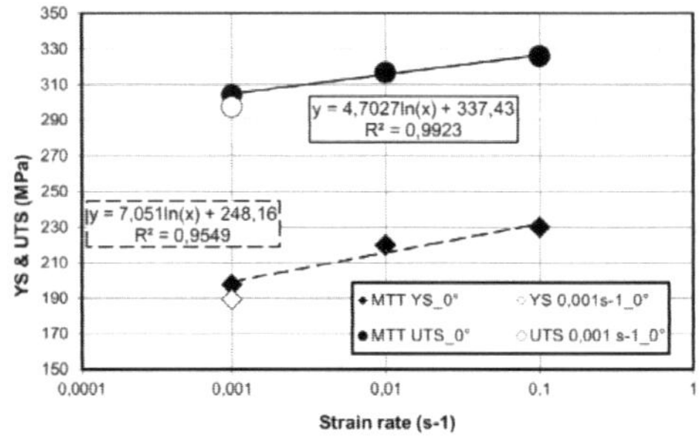

Figure 10. Strain rate sensitivity -longitudinal direction.

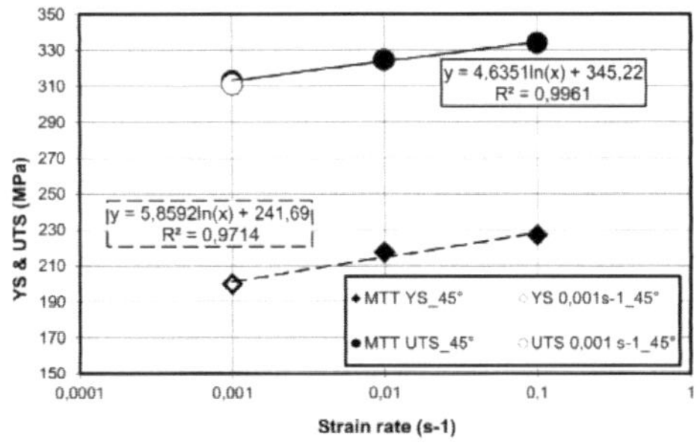

Figure 11. Strain rate sensitivity - 45°orientation.

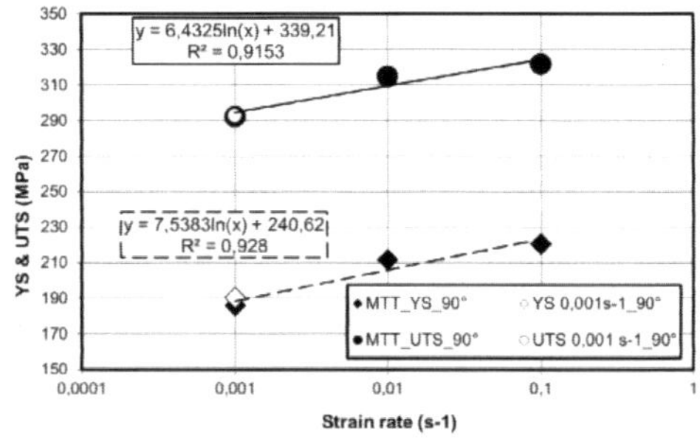

Figure 12. Strain rate sensitivity - 90°orientation.

5 Plastic strain ratio

The plastic strain ration is one of the key parameters in the metal sheet forming [7-11]. The measurement for the plastic strain ratio determination was done with the use of continuous strain measurement by DIC system ARAMIS. The plastic strain ratio was determined from the measured data with the use of equation (2), where mr is determined from the linear regression fit between the lower limit (2%plastic strain) and the upper limit (5% plastic strain) through the origin [12].

$$r = \frac{-m_r}{1+m_r} \tag{2}$$

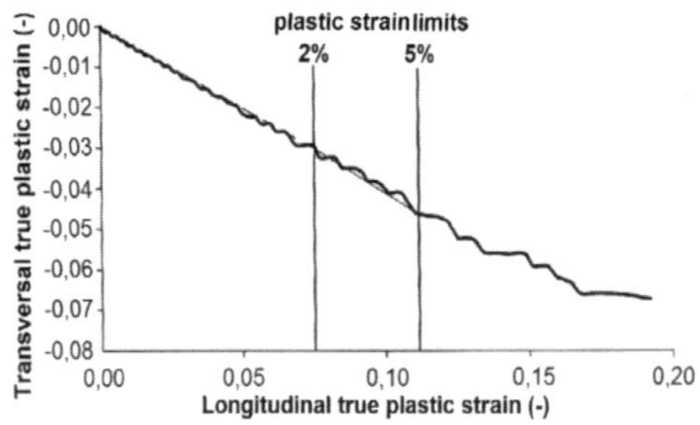

Figure 13. Example of plastic strain determination.

Longitudinal (ε_l) and transversal (ε_b) true plastic strains are calculated according to equations (3) and (4) [12].

$$\varepsilon_l = ln\left[\frac{L_e-\Delta L}{L_e} - \frac{F}{S_0.m_E}\right] \tag{3}$$

IOP Conf. Series: Materials Science and Engineering **179** (2017) 012020 doi:10.1088/1757-899X/179/1/012020

$$\varepsilon_b = ln\left[\frac{b_0 - \Delta b + \frac{b_0 . \upsilon . F}{S_0 . m_E}}{b_0}\right] \tag{4}$$

where:

L_e	Extensometer gauge length	mm
b_0	Original gauge width	mm
ΔL	Instantaneous elongation/ extension of the measurement base	mm
Δb	Instantaneous width elongation	mm
F	Force	N
S_0	Original cross-section area	mm^2
υ	Poisson constant	
m_E	Young modulus	MPa

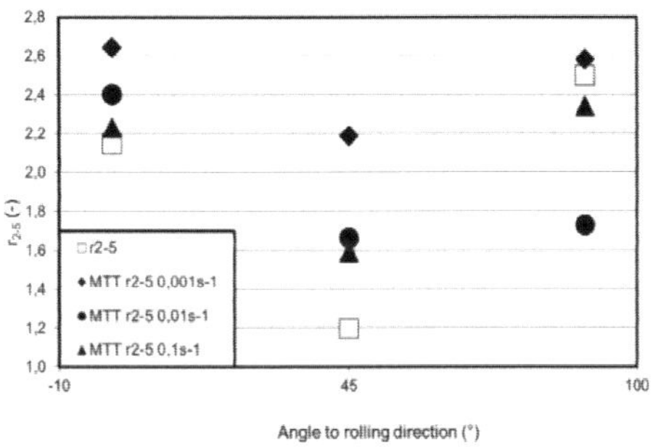

Figure 14. Plastic strain ratio for considered orientations.

6 Conclusions

The paper presented here is dealing with mechanical characteristics determination of sheet materials. Namely strain hardening, strain rate sensitivity and plastic strain ratio for 1.5 mm thick steel sheet made of DC01 were assessed. The properties assessment was done on the basis of the miniaturized tensile tests and standard tests. Results of the MTT specimens are compared with standard size specimens for longitudinal, 45° and 90° sampling orientations. The assessment was done at room temperature. In the case of the strain rate sensitivity MTT specimens were used only.

Obtained tensile tests parameters exhibit excellent agreement between values measured with MTT and standard size specimens for all considered sampling directions. This finding confirms previous results in [1-6].

Strain hardening coefficients assessment points out excellent agreement within a few MPa between C coefficients of MTT and standard sized specimens for all considered sampling directions. In the case of the n parameter, difference of about 0.02 can be found for the longitudinal direction, while the other directions values agree together very well.

The strain rate sensitivity was evaluated with the application of MTT specimens only. Clear consistent trends were achieved for all three directions investigated. The strain rate sensitivity of the ultimate tensile strength was similar for all considered directions. The trend in the case of yield stress is not uniform across the directions investigated. The longitudinal and 90° directions exhibit very similar strain rate hardening behaviour, while the 45° direction yields slightly lower strain rate hardening, however differences are rather small.

Determination of the plastic strain ratio with the use of both specimens geometries points very good agreement between MTT and standard samples for longitudinal and 90°directions. Trends obtained here for this parameter are in agreement with those published in [8-10]. Slightly bigger discrepancy can be found for 45°direction. Investigation of the plastic strain ratio dependency on strain rate shows its increase with increasing velocity as can be expected. The measurement with

MTT specimens for 90°orientation and strain rate 0.01s^{-1} yields rather outlying value. This point was scrutinized, but there was not found any error in the measurement of evaluation and thus it was kept among the results population and was not discarded.

The paper presented here successfully shows possibility of the metal sheet characterisation for forming processes with the use of miniaturize tensile specimens. Generally, very good agreement was found for all considered parameters and conditions between the results attained with the use of standard specimens and MTT. Slight discrepancies between results from MTT and standard size specimens can be assigned to sampling location of the MTT specimens in the middle of the sheet thickness, where small material behaviour deviation from near surface properties can be expected. Further investigations are planned to assess the influence of the MTT specimen localisation within the sheet thickness.

Miniaturized tensile specimens offer very powerful tool for local properties assessment that can be used either for material basic characterisation or as input data for FEM simulations.

Acknowledgements

This paper was created by project TF02000072 - Methods development for formability assessment of thin sheets considering anisotropy and non-linear loading path (2016-2019, TA0/TF) sponsored by Czech Technology Agency and project "Development of West-Bohemian Centre of Materials and Metallurgy" No.: LO1412, financed by the MEYS of the Czech Republic.

References

[1] Rund M et al 2015 Investigation of Sample-size Influence on Tensile Test Results at DifferentStrain Rates Proc. Eng. 114 pp 410-15 doi:10.1016/j.proeng.2015.08.086.

[2] Dzugan J et al 2015 Determination of Local Tensile and Fatigue Properties With the Use of Sub-Sized Specimens Press. Vessels and Piping Conf. PVP2015-45958, doi:10.1115/PVP2015-45958"

[3] Procházka R, Džugan J and Kövér M 2015 Arch. of Mat. Sci. and Eng. Miniature specimen tensile testing of AZ31 alloy processed by ECAP, Volume 76, Issue 2 (Gliwice: International OCSCO World Press) pp 134-9

[4] Konopik P, Dzugan J and Rund M 2015 METAL 2015 – 24th Int. Conf. on Metallurgy and Mat. (Brno) Determination of fracture toughness in the upper shelf region using small sample test techniques (Ostrava: TANGER Ltd.) pp 710-15

[5] Konopik P Dzugan J and Rund M 2014 Dynamic Tensile And Micro-Tensile Testing Using DIC Method Metal 2014 Brno ISBN 978-80-87294-52-9

[6] Konopík P Dzugan J and Prochazka R METAL 2013– 22nd Int. Conf. on Metallurgy and Mat. (Brno) Determination of fracture toughness and tensile properties of structural steels by small punch test and micro-tensile test (Ostrava: TANGER Ltd.) pp 722 – 27

[7] Nasri W et al 2015 Experimental and numerical study concerning the anisotropic behavior of an AA2024-T351 thick sheet using the unified formalism, a multi-mechanism model and a polycrystalline approach 22ème Congrès Français de Mécanique Lyon

[8] Gavrus A and Francillette H 2011 An Anisotropic Behaviour Analysis of AA2024 Aluminium Alloy Undergoing Large Plastic Deformations Aluminium Alloys Theory and Applications ISBN 978-953-307-244-9

[9] Ramos et al 2010 Study of a drawing-quality sheet steel. I: Stress/strain behaviors and Lankfordcoefficients by experiments and micromechanical simulations International Journal of Solids and Structures 47(17) pp 2285-93

[10] Banbic D 2010 Sheet metal forming process Springer-Verlag Berlin Heidelberg doi:10.1007/978-3-540-88113-1

[11] Danckert J and Nielsen K B 1998 Determination of the plastic anisotropy r in sheet metal using automatic tensile test equipment Jour. of Mat. Proc. Tech. 73(1-3) pp 276-80

[12] ISO 10113:2006: Metallic materials Sheet and strip determination of plastic strain ratio

A study of carbide dissolution in bearing steels using computational thermodynamics and kinetics

E Erişir[1], O G Bilir[1], A E Gezmişoğlu[1]
Kocaeli University

eerisir@kocaeli.edu.tr

Abstract. This paper provides a study of carbide dissolution during heating of bearing steel using computational thermodynamics and kinetics. 100Cr6 steel was used for the experimental work and calculations using the Thermo-Calc and DICTRA. Carbide dissolution was calculated at different full and intercritical austenization temperatures from 840 to 1200 °C. The calculations also give the distribution of the alloying elements at carbide and austenite phase boundaries. In experimental work, SEM was used to study microstructures. SEM micrographs were evaluated with line intercept technique and carbide size and quantity were determined. The experimental results were compared with calculations providing good correlation.

1 Introduction

Traditional production cycle of bearing steels include intercritical austenitization to avoid grain coarsening and cracks formed during quenching. However, intercritical austenitization causes to remaining coarse spherical carbides in the final microstructure of bearing parts. It is well known that carbide size and dispersion determine the lifetime of bearings, because coarse carbides providing appropriate nucleation sites for crack propagation and growth [1-3].

In the last decade, a number of researchers have studied computational thermodynamics and kinetics mainly concerned with carbide formation and dissolution in bearing steels. Luzginova et al. [4] studied cementite spheroidization under occurrence of divorced eutectoid transformation reaction. They calculated the time dependence of the cementite fraction and the cementite particle spacing at the different austenitization temperatures with DICTRA. Following the spheroidization, growth of cementite in ferrite was also investigated by Song et al. [5]. According to their DICTRA simulations during cementite precipitation in ferrite, Cr and Mn were partitioned under non-partitioning local equilibrium (NPLE) mode. Si cannot be dissolved in cementite and resulted spikes in Si profile due to local equilibrium (LE). Recently, the carbide dissolution of a bearing steel during the intercritical austenitization was studied by Zhang et al. [6]. They have reported that austenite growth controlled by the diffusion of C during the initial step without redistribution of Cr. In second step, when the carbon content reaches the iso-activity in austenite, the dissolution mechanism becomes partitioning local equilibrium (PLE) mode controlled by the diffusion of Cr. In work by Zhao et al. [7], the cementite dissolution was simulated using MTData software. A cementite radius of 1.88 μm was assumed. Volume fraction of cementite was calculated as a function of time at 860 °C and compared to results from dilatometry and microscopy. It was found that the cementite dissolution is rapid at the initial time and then slow down depending on C and Cr at the interface between austenite and cementite phases. Epp et al. [8] also tried to simulate cementite dissolution using ThermoCalc calculation and in situ X-ray measurements. Cementite content was calculated as 17 wt-% and

measured as 16.3 wt-% using XRD. However, the complete dissolution was not reached at 900 °C after 42 min as predicted at the equilibrium state.
In the present investigation, the carbide dissolution was investigated at different austenitizing temperatures below and above the A_{cm} temperature. Spheroidizing heat treatment has been applied to obtain the initial microstructure. The spheroidizing annealed samples were then austenitized by holding at different temperatures between 840 and 1200 °C. Carbide dissolution has been also calculated using Thermo-Calc and DICTRA software in order to evaluate microstructural variations resulting from different heat-treatment conditions.

2 Experimental

The chemical composition of bearing steel is presented in Table 1. Heat treatments, computational simulation, microstructural characterization and image analyses were done to investigate carbide dissolving behavior.

Table 1. Chemical Composition (in Mass Percent)

C	Si	Mn	P	S	Cr	Ni	Al	Cu	Ti
0,965	0,261	0,29	0,0167	0,0128	1,47	0,0474	0,0261	0,0755	0,0063

In order to establish intercritical and full austenization temperatures, the phase diagrams were calculated using ThermoCalc software [9] and TCFE6 database [10] for steel alloys. According to calculations in Fig 1, cementite phase should be dissolved at temperatures above 901.72 °C. Thus, temperatures of 915, 1050, 1100 and 1200 °C were selected in experimental work.

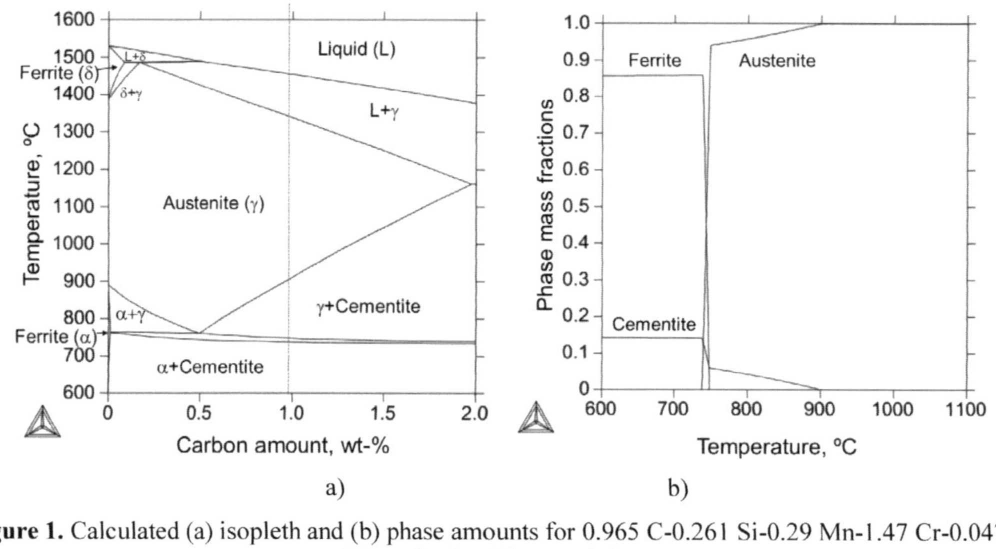

Figure 1. Calculated (a) isopleth and (b) phase amounts for 0.965 C-0.261 Si-0.29 Mn-1.47 Cr-0.0474 Ni steel using Thermo-Calc.

Dissolution behavior of carbides was simulated using DICTRA software [11] and MOBFE2 kinetic database [12] for different temperatures. To create a model for simulation initial structure with spherical carbides was evaluated to determine carbide size and quantity as shown in Fig 2. SEM micrographs of spheroidized specimens were analyzed with line intercept method, average radius was determined. To determine the quantity of carbides in matrix at the equilibrium condition thermodynamic calculations was used. Calculations offer 14.3 volume-% of cementite phase at 720 °C for spheroidization.

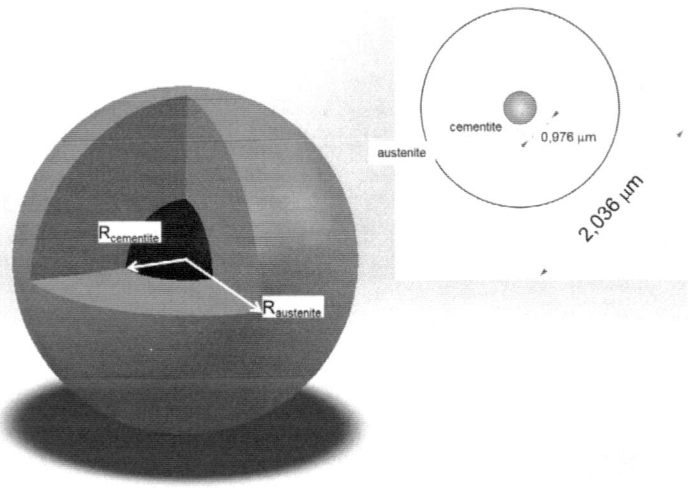

Figure 2. Model for DICTRA calculations.

The microstructures of the samples were examined by scanning electron microscopy (SEM) using secondary electron imaging on a JSM-6060 SEM. The samples were first mechanically polished using 3 μm diamond solution, and then etched using 3% Nital at room temperature. SEM images were evaluated with the line intercept method to determine average radius of carbides and size distribution.

3 Results

Commercially production cycle of bearing steels includes spheroidization for reducing hardness before cold rolling. In spheroidization treatment, specimens were kept at 800 °C for 120 minutes and cooled to 720 °C with cooling rate of 15 K/h and followed by air cooling to room temperature. Fig 3 presents typical microstructure of the sample after spheroidization treatment. Microstructure is consisted of spheroidal carbides in ferritic matrix. An image analysis software was employed which measured diameter of each particle. Due to the particles of bimodal size distribution with overlapped peaks, two different average particle sizes can be determined. The histogram shows a rather broad particle size distribution of 700–1700 nm with an average value of 976 nm. During the dissolution of carbides, coarser carbides are dissolved at the final stage of heat treatment. Thus, in our calculations, only coarser carbides were taken into account.

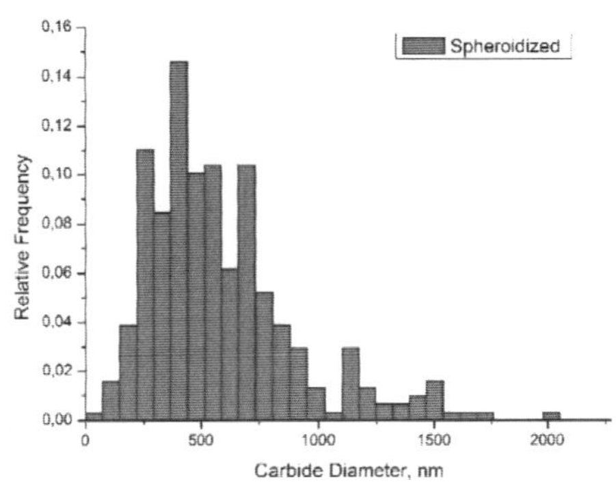

Figure 3. SEM micrograph of spheroidal carbides in ferritic matrix and particle size distribution.

According to the study on carbide dissolution in bearing steels by Kang et al, smaller precipitates tend to dissolve faster than larger particles and the dissolution is diffusion controlled. Thus, DICTRA calculations were applied to larger particles of carbides at different austenitizing temperatures. The simulation begins with a cementite particle with radius of 488 nm surrounded by an austenite grain with radius of 1018 nm. The chemical composition of cementite and austenite calculated at spheroidization temperature of 720 °C using ThermoCalc, as given in Table 2. The radius of cementite obtained from DICTRA simulations were plotted in Figure 4. The results indicate that intercritical austenization treatments should give a small decrease of particle size at 840 and 915 °C. However, at temperatures above 1050 °C, all particles can be dissolved completely.

Table 2. Calculated chemical compositions of phases at 720 °C using ThermoCalc.

Phase	C	Mn	Si	Cr
Ferrite	0.0132143	0.165904	0.290496	0.245577
Cementite	6.73324	1.4147	4.72341×10^{-11}	8.52682
	but not asked by DICTRA			

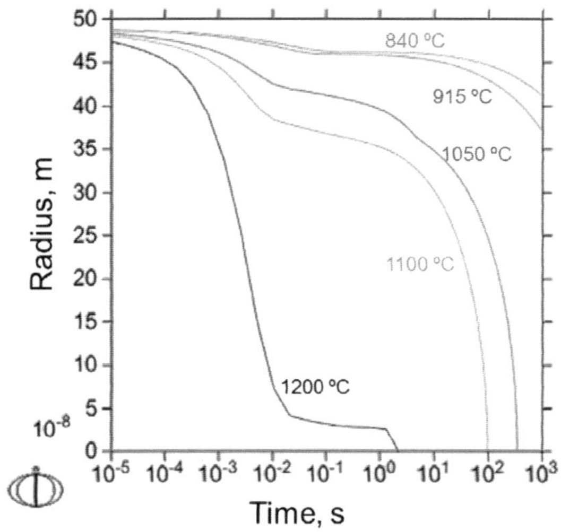

Figure 4. Dissolution of cementite particles at different austenitizing temperatures.

Following the spheroidization treatment, specimens were reheated to 915 °C and 1050 °C. In Fig 5a, it seems that an annealing at 915 °C for 10 min was not enough for complete dissolution of cementite particles. Mostly finer carbides with diameter <500 nm were dissolved compare to the carbide distribution of conventional spheroidization as given in Fig 3. The complete dissolution of carbides was observed after heat treatment at 1050 °C for 15 min as shown in Fig 5b.

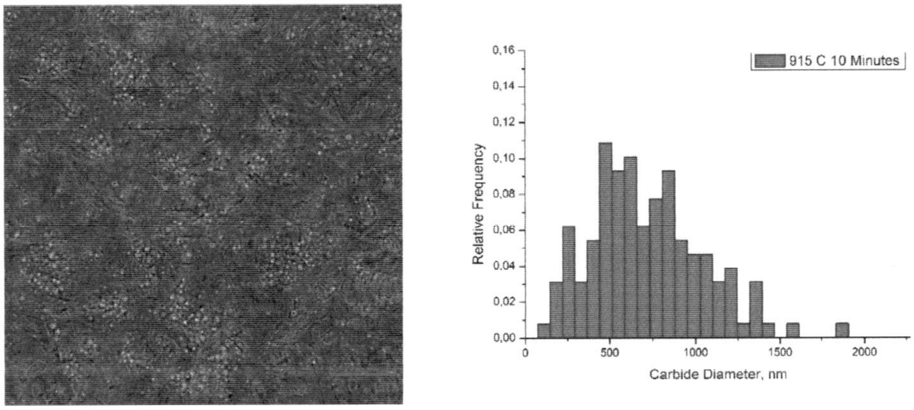

a)

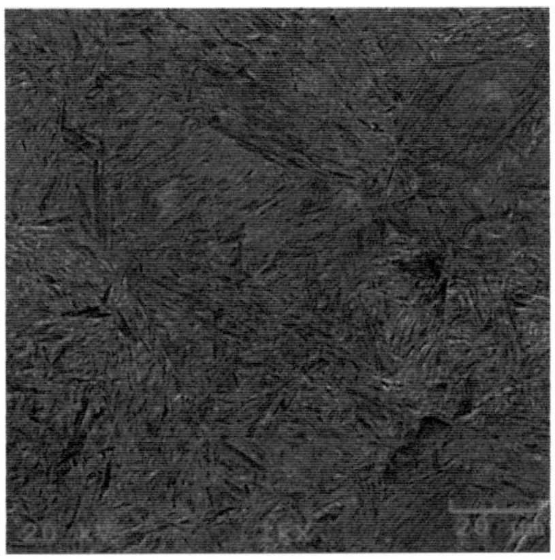

b)

Figure 5. SEM micrograph of samples annealed at a) 915 °C/10 min (with particle size distribution) and b) 1050 °C /15 min prior to a water quench.

4 Discussions

In this work, cementite dissolution at full and intercritical austenitizing temperatures in 100Cr6 steel is experimentally measured using scanning electron microscopy of heat treated samples and simulated by ThermoCalc and DICTRA software. According to calculations and experimental studies, following conclusions are as follows.

Both experimental measurements and simulations reveal that the cementite dissolution is slow at the intercritical austenitizing temperatures. The complete dissolution of cementite can only be attained at higher austenitizing temperatures.

The kinetic simulations indicated that the dissolution of cementite at temperatures above 1050 °C should be finished after a couple of minutes.

According to thermodynamical calculations, cementite should be dissolved temperatures above 901.72 °C. However, SEM observations revealed that holding at 915 °C for 10 min was not enough for complete dissolution of cementite. Kinetic calculations also showed that a longer duration than 1000 s should be needed for complete dissolution of cementite.

Acknowledgments
This research was financially supported by The Scientific and Technological Research Council of Turkey (TÜBİTAK) under research project number 115M445.

References

[1] Stickels, C. A. Carbide refining heat treatments for 52100 bearing steel. Metallurgical Transactions, 1974, 5.4: 865-874.
[2] Bhadeshia, H. K. D. H. Steels for bearings. Progress in materials Science, 2012, 57.2: 268-435.
[3] Kang, J.-H.; Hosseinkhani, Babak; Rivera-Díaz-del-Castillo, Pedro EJ. Rolling contact fatigue in bearings: multiscale overview. Materials Science and Technology, 2012, 28.1: 44-49.
[4] Luzginova, N. V.; Zhao, L.; Sietsma, J. The cementite spheroidization process in high-carbon steels with different chromium contents. Metallurgical and Materials Transactions A, 2008, 39.3: 513-521.
[5] Song, W., Choi, P. P., Inden, G., Prahl, U., Raabe, D., & Bleck, W. On the spheroidized carbide dissolution and elemental partitioning in high carbon bearing steel 100Cr6. Metallurgical and Materials Transactions A, 2014, 45.2: 595-606.
[6] Zhang, X., Tang, D., & Wang, Y. Carbide dissolution during intercritical austenitization in bearing steel. Journal of Wuhan University of Technology-Mater. Sci. Ed., 2014, 29.6: 1242-1245.
[7] Zhao, L., Vermolen, F. J., Sietsma, J., & Wauthier, A. (2006). Cementite dissolution at 860 C in an Fe-Cr-C steel. Metallurgical and Materials Transactions A, 37(6), 1841-1850.
[8] Epp, J., Surm, H., Kessler, O., & Hirsch, T. (2007). In situ X-ray phase analysis and computer simulation of carbide dissolution of ball bearing steel at different austenitizing temperatures. Acta Materialia, 55(17), 5959-5967.
[9] Thermo-Calc TCC Software Users Guide, Version S, Stockholm Technology Park, Stockholm, 2008.
[10] Thermodynamic database TCFE6-TCS Steels/Fe-alloys database (v.6.2) for Thermo-Calc, Thermo-Calc Software AB. www.thermocalc.com
[11] A. Borgenstam A., Engstrom A., Hoglund L., Ågren J., DICTRA, a tool for simulation of diffusional transformations in alloys, J Phase Equilib, 21 (3) (2000), pp. 269–280.
[12] Thermo-calc software MOBFE2 steel and Fe-database version 2 http://www.thermocalc.com

Study of martensitic-ferritic dual phase steels produced by hot stamping

E Erişir[1], **O G Bilir**[1]

[1]Department of Metallurgical and Materials Engineering, Kocaeli University, 41380-Kocaeli-Turkey

corresponding author e-mail address: eerisir@gmail.com

Abstract. The effects of heat treatment and initial microstructure on tensile properties of 22MnB5 and 30MnB5 high-strength hot stamping steels with martensite-ferrite matrix were investigated. Hot stamping steels possessed limited elongations of about 5% in a tensile strength ranging from 1300 to 1500 MPa when quenched at temperatures above A3 temperatures. The total elongations were tried to improve by partial austenization between Ac1 and Ac3 temperature and quenching. Ac1 and Ac3 temperatures were calculated via ThermoCalc. Microstructural characterization was made by using Light Microscope and Scanning Electron Microscope. Microstructure is composed of ferrite+martensite. It was seen that annealing temperature affects the volume fraction of phases. It was concluded that initial microstructure is an important parameter for the final microstructure. This method can be used for automobile parts which require higher TE with sufficient yield and tensile strength. Also this process may be a way of using Zn coated steel sheets in hot stamping process.

1 Introduction

Reducing the fuel consumption without sacrificing the safety properties is only possible with using Ultra High Strength Steels (UHSS) sheets in automotive industry. Dual Phase (DP), Transformation Induced Plasticity (TRIP), Twinning Induced Plasticity (TWIP), Complex Phase (CP) Steels are the most commonly used UHSS. Forming of steel sheets in order to produce body-in-white automobile parts is getting harder as a result of using higher strength steels [1, 2, 3]. Thus, Hot Stamping (HS) process gain importance in recent years by enabling the forming easily to complex geometries without springback or breaking of steel sheets. In HS process, the forming is applied at elevated temperature above Ac3 where the steel sheet has austenitic microstructure. The forming step is followed by quenching in press simultaneously Among other UHSS approaches, HS process gives the highest strength level which is attributed to their fully martensitic microstructure at the end of the process [4]. Market share of hot stamping parts are rising exponentially. There are two important downsides of hot stamped parts. One is that their limited total elongation (TE) behavior. However, TE is not a key necessity in safety zone of an automotive that is responsible of passenger safety. Hot stamped parts are used in safety zones which primarily require high strength. Second drawback of HS is that Zn coated sheet is not suitable for process, because of the Liquid Metal Embrittlement [5].

This study evaluated the extension of the application area of hot stamped parts which require TE in addition to strength. Through-hardened martensitic 22MnB5 and 30MnB5 steels were intercritical annealed at 755 and 775 □C resulting in the formation of dual phase microstructures. It is well known that Fe diffusion into Zn coating may result with an increased melting temperature for Zn coating as 782°C. Thus, this process may create a potential for galvanized steel sheets to use for hot stamping.

2 Material and Experimental

2.1 Material and Heat Treatment

Hot rolled uncoated commercial 22MnB5 and 26MnB5 steels are used in experimental studies. Detailed chemical compositions of specimens are given in Table 1. First heat treatment was austenitization applied at 945 °C for 15 min. Then, two groups of samples are created by depend on cooling media as air cooled and water quenched. Air cooled 22MnB5, water quenched 22MnB5, air cooled 30MnB5 and water quenched 30MnB5 are called respectively as A, B, C and D. Second heat treatment is applied by partial austenization and water quenching. Partial austenization temperatures are used as 755 and 775°C for 10 min. Critical temperatures are calculated using ThermoCalc [6]. Microstructural characterization was made by using Light Microscope (LM) and Scanning Electron Microscope (SEM). The volume fractions of the microstructural constituents in the dual phase microstructures were measured by phase analysis on a selection of Region Of Interest (ROI) using an image analyzer (OLYMPUS Stream) attached to an LM (OLYMPUS, model: BX41RF-LED) [7].

Table 1. Chemical composition of alloys.

Specimens	Chemical Composition (weight %)							
	C	Si	Mn	P	S	Cr	Mo	Ni
22MnB5	0,237	0,292	1,18	0,0315	0,0064	0,121	<0,0050	0,0370
30MnB5	0,317	0,287	1,15	0,0286	0,0096	0,162	<0,0050	0,0509

In Fig.1, an isopleth diagram showing stability of phases as a function of temperature was given for 0.292 Si, 1.18 Mn, 0.121 Cr, and 0.037 Ni. Volume fraction of austenite and ferrite can be calculated via applying lever rule for related intercritical temperature. The intercritical annealing temperature and the carbon content of alloy both determine the volume fraction of austenite at intercritical range. However, at the same intercritical annealing temperature, carbon content of austenite is the same for steels having different C content. In Fig.1 22MnB5 and 30MnB5 were shown as blue (dotted) and red line respectively. Carbon content of partial austenite for each alloy is also shown as black point. The only difference between 22MnB5 and 30MnB by means of chemical composition is the carbon content. Thus, using the same intercritical temperature for 22MnB5 and 30MnB5, 2 different volume fractions of austenite with same carbon content can be achieved. This could help us to understand th effect of austenite volume fraction for a similar austenite chemical composition.

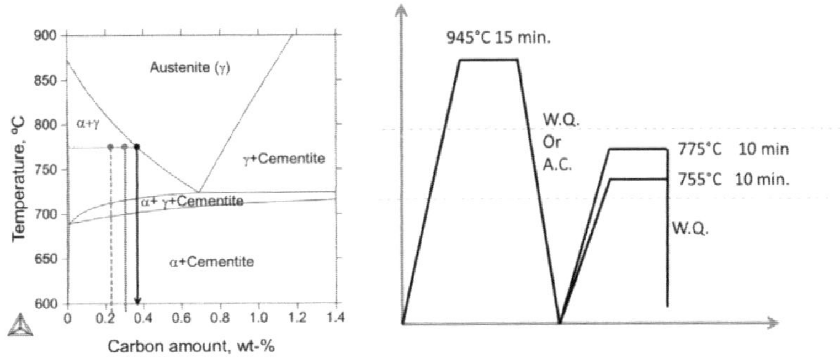

Figure 1. Isopleth diagram and heat treatment steps applied in this study (right). (IA: ıntercritical annealing, W.Q.: Water Quenched, A.C: Air cooled.

2.2 *Modeling of Mechanical Properties By Microstructure*

It is reasonable to apply rule of mixtures to multiphase steels since each of the micro constituents of steel microstructure show different mechanical properties [8]. According to rule of mixtures, mechanical properties of dual phase steels can be predicted by **equation 1** and **2** given below. Where V_M and V_F are volume fraction of martensite and ferrite. UTS_M, UTS_F and UTS_{DP} are the Ultimate Tensile Strength of martensite and ferrite. TE_M, TE_F and TE_{DP} are the Total Elongation of martensite and ferrite.

$$UTS_{DP}= \{(V_M) \text{ x } (UTS_M)\}+\{(V_F) \text{ x } (UTS_F)\} \quad (1)$$

$$TE_{DP}= \{(V_M) \text{ x } (TE_M)\}+\{(V_F) \text{ x } (TE_F)\} \quad (2)$$

Thus, to predict the overall mechanical properties of any dual phase steel one must know the mechanical properties of martensite and ferrite alone. Hot stamped steels are known as fully martensitic steels after the HS operation. Therefore, for low alloyed low carbon steels one can use the mechanical properties of hot stamped steels to predict mechanical properties of martensite alone. If one ignores the additional effect of solid solution strengthening and grain size on mechanical properties for low carbon steel, UTS and TE of martensite may be assumed as 1600MPa and 8% (for optimum values). IF (Interstitial Free) steels are known as single phase ferrite steels. IF steels yield approximately 300MPa of tensile strength with 30% total elongation. As a result, we used mechanical properties of typical IF steel for ferrite. These are the upper limits for commonly used low carbon low alloyed steels without additional treatments. According to Hoofnagel et al. [9] commercial DP600 steels have a volume fraction of 25% martensite while DP800 steels have 41%. Applying proposed model for 25% and 41% volume fraction of martensite resulted with 625 MPa-25% and 833 MPa-21% respectively. These results can be found in mechanical properties of commercial DP600 and DP800 steels.

3 Results and Discussion

In order to evaluate the effect of initial microstructure on dual phase (DP) microstructure, microstructural investigations were performed using SEM and LM. As it can be seen from micrographs in **Fig. 2**, morphology of martensite and ferrite differs depend on initial microstructure. In the intercritical annealing process, austenite may nucleate and grow at different locations depends on initial microstructure [1,2, 10-15]. According to Wei et al. [11] austenite may nucleate at the lath boundaries (LB), packet boundaries (PB) and prior austenite grain boundary (PAGB) during the annealing at the temperature between Ac1 and Ac3. However, a ferrit+pearlite initial microstructure promotes the nucleation of austenite at pearlite colony boundary, ferrite/cementite interface, and ferrite/ferrite interface [10-15].

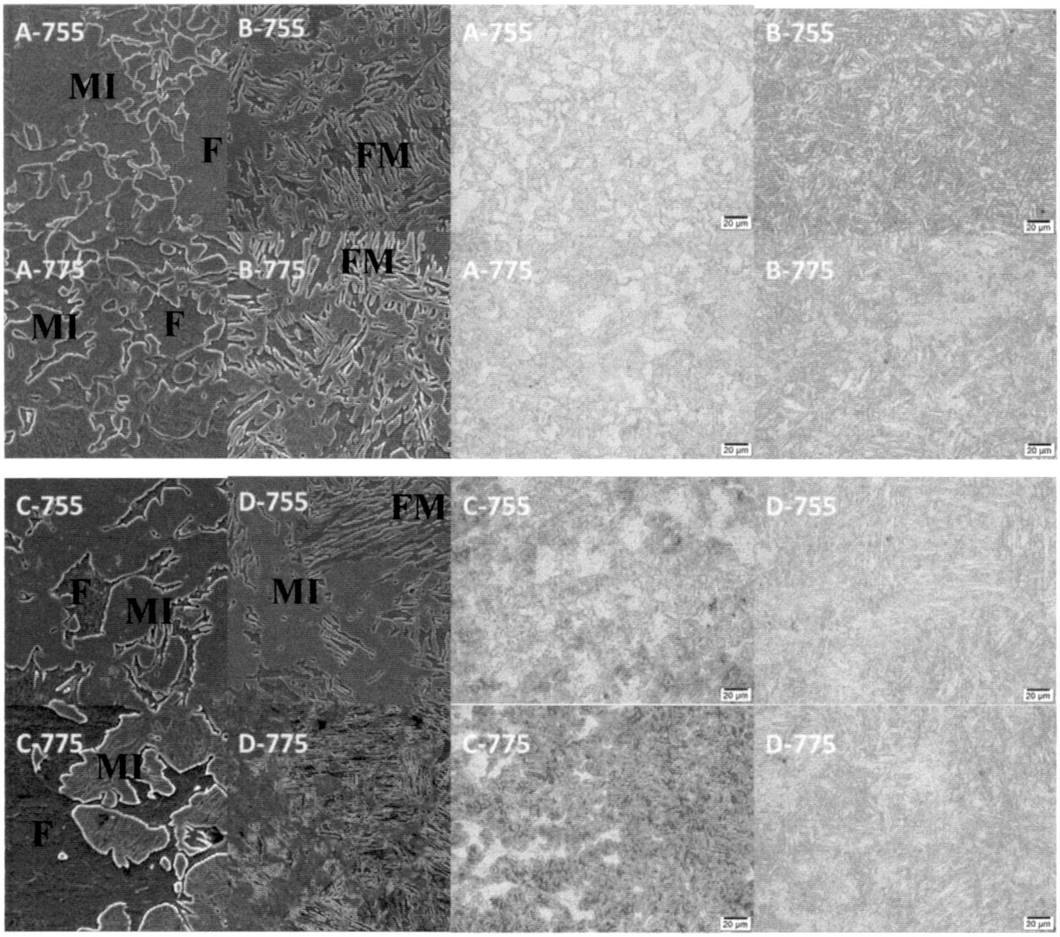

Figure 2. 22MnB5 and 30MnB5 steel with different initial microstructure and partial austenization temperatures. (Air cooled 22MnB5, water quenched 22MnB5, air cooled 30MnB5 and water quenched 30MnB5 are called as A, B, C and D, 755, 775 are intercritical annealing temperature, Darker areas are martensite in light microscope micrographs, MI:Martensite island, FM:Fibrous martensite, F:Ferrite)

Samples B-755, B-775, D-755, D-775 have an initial martensitic microstructure. Thus, DP microstructure mixed of fibrous martensite and ferrite was observed. Fibrous martensite was marked as FM in **Fig2**. In addition to acicular martensite, there are also martensite islands which are coarser and equiaxed. It could be concluded that the formation of different martensite morphologies were due to different nucleation behavior of austenite from initial martensitic microstructure. When an austenite grain nucleates at PAGB or PB, coarse martensite island may be formed. On the other hand, if the nucleation of an austenite grain occurs in LB or PB, it may have finer morphology as martensite fiber. If the initial microstructure is martensite, the transformation of as-quenched martensite to austenite and ferrite can be assumed as a transformation with a single parent phase and two product phases. However, martensite has relatively thin laths and high density of interfaces compare to coarser ferrite and pearlite constituents [16]. Interfaces are usually considered to be preferred nucleation sites. Moreover, thin laths of martensite represent shorter diffusion distances for carbon, which may control the austenitization process. Samples, A-755, A-775, C-755, C-775 have an initial microstructure of ferrite+pearlite. As shown in **Fig 2**, martensite has granular, coarser and equiaxed morphology. Ferrite was surrounded by martensite islands. Surrounding of ferrite by Blocky type of martensite was also seen by other authors and called as networked morphology [17, 18, 19].Turkmen et al [20] indicated that during quenching from austenite and ferrite region, phase transformation of austenite to martensite may cause deformation of neighbor ferrite phase. Since the networked morphology has less contact surfaces between individual phases than fibrous mixed morphology, it is expected to yield lower strength. Ahmad et al [19] investigated the effect ıf of 3 different initial microstructures as austenite, martensite, ferrite+pearlite on tensile strength. They claimed that coarser martensite and ferrite yielded better tensile and yield strength. It was seen that both raising the carbon content of alloy and intercritical annealing temperature resulted with higher volume fraction of martensite as it is expected. It was also seen that raising the annealing temperature also changed the morphology from fibrous to blocky shaped martensite for the samples have initial microstructure of martensite. Lower volume fraction of fibrous morphology was seen in higher temperatures. It was concluded that at higher annealing temperatures grain growth takes place rather than nucleation of fresh phases due to duration of annealing. To compare the predicted volume fractions of martensite and ferrite phases using Thermocalc, image analysis was conducted in LM images. **Table 2** presents the results acquired from image analysis and thermodynamic calculations. It is apparent that an agreement was obtained between calculated and measured values. Image analyses only applied to samples have initial microstructure of ferrite+pearlite. Proposed model was used to predict mechanical properties of aforementioned samples using the image analysis results.

Table 2. Image Analysis Result and comparison with ThermoCalc

	Measured Values (%) via Image Analysis			Calculated Values (%) via ThermoCalc	
Sample	**Ferrite**	**Martensite**	**Deviation**	**Ferrite**	**Martensite**
A-755	49,1	50,9	3,4	44,5	55,5
A-775	25,8	74,2	0,9	29,8	70,2
C-755	26,5	73,5	1,2	29,3	70,7
C-775	11,8	88,2	0,3	11,2	88,8

Sample	Calculated Mechanical Properties via Proposed Model			
	Phase Fractions (%)		**Mechanical Properties**	
	Ferrite	**Martensite**	**UTS**	**TE**
A-755	49,1	50,9	961,7	18,802
A-775	25,8	74,2	1264,6	13,676
C-755	26,5	73,5	1255,5	13,83
C-775	11,8	88,2	1446,6	10,596

4 Conclusion

In this study, an alternative to traditional hot stamping process was investigated. Adding ferrite to microstructure beside martensite can be beneficial to have higher elongation. Thus, partial austenitization was carried out rather than full austanization and followed by quenching. It was seen that annealing temperature is proportional to austenite fraction. In partial austenitization process, austenite fraction can be predicted by ThermoCalc. It was seen that thermodynamical calculations were in good agreement with the results of image analysis. According to results, it is possible to have 1000MPa tensile strength with minimum %10 elongation using the heat treatment steps applied in this study. This process may also pave the way for using Zn coated steels in hot stamping.

References

[1] Erişir E and Bilir O G 2016 Effect of Intercritical Annealing Temperature on Martensite and Bainite Start Temperatures After Partial Austenitization. *JOM*, **68(1)** 203-209. doi:10.1007/s11837-015-1673-4

[2] Erişir E Bilir, O G (2014) Effect of intercritical annealing temperature on phase transformations in medium carbon dual phase steels *Journal of materials engineering and performance* ***23*(3)** 1055-1061.

[3] Calcagnotto M. Adachi Y Pong D Raabe D (2011) Deformation and fracture mechanisms in fine-and ultrafine-grained ferrite/martensite dual-phase steels and the effect of aging. *Acta Materialia* **59(2)** 658-670

[4] Naganathan L. Penter Sheet Metal Forming—Processes and Applications, Chapter 7: Hot Stamping, 2012.

[5] Lee C W Fan D W Sohn I R et al (2012) *Metall and Mat Trans A* **43**: 5122. doi:10.1007/s11661-012-1316-0

[6] Thermodynamic database TCFE6-TCS Steels/Fe-alloys database (v.6.2) for Thermo-Calc, Thermo-Calc SoftwareAB. www.thermocalc.com

[7] http://www.olympus-ims.com/en/microscope/stream2/machinery/

[8] Ahmad E. (2013). Modified Law of Mixture to Describe the Tensile Deformation Behavior of Thermomechanically Processed Dual-Phase Steel. Journal of materials engineering and performance, **22**(8) 2161-2167.

[9] Hoefnagel J P M. et al Experimental and Applied Mechanics, Volume 4: Proceedings of the 2012 Annual Conference on Experimental and Applied Mechanics, Chapter:4

[10] Li Z et al Nucleation of austenite from pearlitic structure in an Fe–0.6C–1Cr alloy (2009) *Scripta Materialia* **60** 485–488,.

[11] Wei R et al (2013) Growth of austenite from as-quenched martensite during intercritical annealing in an Fe–0.1C–3Mn–1.5Si alloy *Acta Materialia* **61** 697–707,

[12] Nakada N et al (2011) Temperature dependence of austenite nucleation behavior from lath martensite *ISIJ International* **51** 299–304.

[13] Shtansky D.V Nakai K Ohmori Y (1999) Pearlite to austenite transformation in an Fe-2.6Cr-1C alloy *Acta Materialia* **47** 2619-2632,.

[14] Savran V.I Leeuwen Y V Hanlon D N Kwakernaak C Sloof W G Sietsma (2007) J Microstructural features of austenite formation in C35 and C45 alloys *Metallurgical and Materials Transactions A* **38** 946–95,.

[15] Savran V I Offerman S E Sietsma J (2010)Austenite nucleation and growth observed on the level of individual grains by three-dimensional x-ray diffraction microscopy, *Metallurgical and Materials Transactions A* **41** 583-591

[16] Esin V. A. et al., In situ synchrotron X-ray diffraction and dilatometric study of austenite formation in a multi-component steel: Influence of initial microstructure and heating rate, Acta Materialia, Vol. 80, 118–131, 2014.

[17] Sarkar, P. P., Kumar, P., Manna, M. K., & Chakraborti, P. C. (2005). Microstructural influence on the electrochemical corrosion behaviour of dual-phase steels in 3.5% NaCl solution. Materials Letters, **59**(19), 2488-2491.

[18] Ahmad, E., Manzoor, T., Ziai, M. M. A., & Hussain, N. (2012). Effect of martensite morphology on tensile deformation of dual-phase steel. Journal of materials engineering and performance, **21**(3), 382-387.

[19] Keleştemur, O., & Yıldız, S. (2009). Effect of various dual-phase heat treatments on the corrosion behavior of reinforcing steel used in the reinforced concrete structures. Construction and building materials, **23**(1), 78-84.

[20] Türkmen, M., & Gündüz, S. (2011). Martensite morphology and strain aging behaviours in intercritically treated low carbon steel. Ironmaking & Steelmaking, **38**(5), 346-352.

IOP Conf. Series: Materials Science and Engineering **179** (2017) 012023 doi:10.1088/1757-899X/179/1/012023

Sheet necking prediction in forming limit diagrams with the anisotropy influence incorporation

P Farahnak, A Prantl, J Dzugan, P Konopik and R Prochazka

COMTES FHT a.s., Průmyslová 995, 334 41 Dobřany, Czech Republic

E-mail: pedram.farahnak@comtesfht.cz

Abstract. A well-known method of materials formability description is the Forming Limit Diagram (FLD). In FLDs/FLC (Forming Limit Curve) a plot of major and minor principal strains in the plane of deformed sheet where necking takes place is represented. The current paper is dealing with anisotropy influence on FLC. This effect is going to be studied with the use of mild steel DC01. Hills anisotropic criterion is considered here and for its application experimental investigation has to be carried out in order to provide input data for the model applied. Uniaxial tensile tests were performed for basic characterisation of sheet samples with the use of contact-less measuring devices. In order to assess the anisotropy experimentally, the specimens were extracted and tested in Rolling Direction (RD), Transverse Direction (TD) and 45° Skew Direction (SD) for the Lankford ratios determination. Detailed stress state description of the material investigated was carried out with the use the stress triaxiality and Lode angle parameter. Therefore, transformations from conventional space into stress state provide a profound insight into the necking phenomenon. According to the experimental and numerical results it was found that it is necessary to take anisotropy effects into account for the material investigated.

1 Introduction

A well-known method of materials formability description is the FLD. The concept was firstly introduced by Keeler [1] and stands as the first safety criterion in deep drawing operations and it is still a conventional approach to assess sheets formability. Based on a deeper analysis of explanatory power of FLD diagram, one can see that it shows the representation of the deformability from a practical-technical point of view rather than from the perspective of ductile damage. In FLDs/FLC a plot of major and minor principal strains in the plane of deformed sheet where necking takes place is represented. Although this concept is very simple and well understood for proportional loading, its experimental determination is rather complex demanding a wide range of sheet metal forming tests. A large number of theories for predicting of sheet necking have been proposed, which can be differentiated in two main theoretical frameworks: the necking theory in homogeneous sheets, based on the original work developed in [2, 3] and the theory of sheet non-homogeneity proposed by Marciniak and Kuczynski (M-K) [4].

From technical and practical points of view, fracture experiments are evaluated in the space of equivalent plastic strain, stress triaxiality and Lode angle parameter that can be obtained by conversion from conventional space. It can be defined and calibrated under the proportional loading using an assumption that the ratio of two principal strain increments is constant. In equation (1),

$$\alpha = \frac{d\varepsilon_2}{d\varepsilon_1} \tag{1}$$

Subsequently, the transformation into another spaces based on plane stress plasticity formulas is considered.

2 Theoretical fundamentals

2.1 Transformation of conventional FLD into the stress state space under plane stress

For proportional loading path, the increments of principal plastic strains follow the equation:

$$d\varepsilon_1 : d\varepsilon_2 : d\varepsilon_3 = 1 : \alpha : -(1+\alpha) \tag{2}$$

Where the material is considered to be incompressible and isotropic. The material is assumed to obey equivalent stress and the associated flow rule,

$$\bar{\sigma} = \sqrt{\frac{1}{2}\left[(\sigma_1 - \sigma_2)^2 + (\sigma_2 - \sigma_3)^2 + (\sigma_3 - \sigma_1)^2\right]}$$
$$d\varepsilon_{ij} = d\Lambda \frac{\partial \bar{\sigma}}{\partial \sigma_{ij}} = \frac{3}{2}\frac{d\Lambda}{\bar{\sigma}} s_{ij} \tag{3}$$

Where σ_{ij} are stress components, σ_1, σ_2 and σ_3 are principal stresses, s_{ij} are deviatoric stress components, $d\Lambda$ is the plastic multiplier, and $d\varepsilon_{ij}$ is the plastic strain increments. For simplicity, the equation (3) is assumed to hold for both positive and negative α values. The stress ratios can be defined as,

$$\frac{\sigma_2}{\sigma_1} = \beta \; ; \; \frac{\sigma_3}{\sigma_1} = \gamma \tag{4}$$

The second term of equation (3) can also be written in the principal stress and strain space, and after some substitutions, the stress triaxiality λ, the Lode angle parameter $\bar{\theta}$ and necking strain $\bar{\varepsilon}$ can be expressed in terms of the stress ratio β and α:

$$\lambda = \frac{\sigma_m}{\bar{\sigma}} = \frac{\beta + 1}{3\sqrt{\beta^2 - \beta + 1}} \; ; \; \bar{\theta} = 1 - \frac{2}{\pi}\arccos\ \xi = 1 - \frac{2}{\pi}\arccos\left[\frac{(\beta+1)(\beta-2)(2\beta-1)}{2(\beta^2 - \beta + 1)^{3/2}}\right]$$
$$\bar{\varepsilon} = \frac{2\varepsilon_1}{\sqrt{3}}\sqrt{1 + \alpha + \alpha^2} \tag{5}$$

Where ζ defined by equation (6) is the normalized third deviatoric stress invariant of the stress tensor and σ_m is the mean stress (for more details see [5-11]).

$$\xi = \frac{27}{2}\frac{(\sigma_1 - \sigma_m)(\sigma_2 - \sigma_m)(\sigma_3 - \sigma_m)}{\bar{\sigma}^3} \tag{6}$$

2.2 *Plasticity model parameter identification*

Plasticity behaviour of isotropic material or von Mises yield criterion has been studied vastly in numerous researches. In this case, elastic properties (Young's modulus and Poisson's ratio) as well as plastic properties such as yield stress and hardening behaviour have been assumed independent of material orientation. However, in many circumstances of practical interest, material behaviour is truly anisotropic with substantial discrepancy among phenomenological properties observed in different material directions. From technical and practical points of view, the isotropy assumption may lead to a poor representation of the actual behaviour. Therefore the formulation and use of appropriate anisotropic plasticity model could raise the accuracy in finite element predictions. A well-known orthotropic model proposed by Hill (1948) is introduced here [12].

2.2.1 *Hill orthotropic model*

The Hill criterion can be considered as an orthotropic extension of von Mises criterion which provides model with the material anisotropic behaviour. Stress tensor components are denoted by σ_{ij}. The rolling, transverse and thickness directions of the sheet are defined on an orthonormal basis and designated as x, y and z. The yield function associated with the Hill criterion can be expressed by following equation (7):

$$\varphi(\sigma, \bar{\sigma}) = F(\sigma_{xx} - \sigma_{yy})^2 + G(\sigma_{zz} - \sigma_{xx})^2 + H(\sigma_{xx} - \sigma_{yy})^2 + 2L\sigma_{yz}^2 + 2M\sigma_{zx}^2 + 2N\sigma_{xy}^2 - \bar{\sigma}^2 \tag{7}$$

Here, F, G, H, L, M and N are anisotropic coefficients, which were calibrated using three R-values (r_0, r_{45}, r_{90}) or Lankford ratios.

$$F = \frac{r_0}{r_{90}(1 + r_0)};\ G = \frac{1}{1 + r_0};\ H = \frac{r_0}{1 + r_0};\ N = \frac{(r_0 + r_{90})(2r_{45} + 1)}{2r_{90}(1 + r_0)} \tag{8}$$

In section 3, measurement of Hill's coefficients from uniaxial tensile test will be discussed.

3 Experiment description

The material investigated is a mild steel DC01 and tensile experiments were performed on different specimen geometries. Figure 1(a) depicts uniaxial tensile test specimen with thickness 1.5 mm that is used for necking prediction. However, in order to characterize the Lankford ratios and subsequently Hill's coefficients in equation (8), it is necessary to perform uniaxial tests on the sample of the same thickness which are extracted in RD, TD and SD. Figure 1 (b) indicates the extractions of samples along aforementioned sampling directions. All the experiments were conducted by constant velocity 4 mm/min and through the experiments, displacement fields are monitored using Digital Image Correlation (DIC) system. In particular, the evolution of width strain is determined by using virtual extensometer. After computing the logarithmic plastic strains in the width and thickness directions (assuming plastic incompressibility), the Lankford ratios were determined from the average slopes,

$$r = \frac{d\varepsilon_w^p}{d\varepsilon_{th}^p} \tag{9}$$

Tables 1 and 2 summarize the identified anisotropy's constants and material properties including Swift Hardening Law (SHL), respectively.

Table 1. Lankford ratios and Hill's coefficients.

r_0	r_{45}	r_{90}	F	G	H	L	M	N
1.66	1.24	2.02	0.30	0.37	0.62	1.5	1.5	1.19

Table 2. Material Properties.

E	ν	σ_y	SHL	A	ε_0	n
200(GPa)	0.3	180(MPa)	$K = A(\bar{\varepsilon}_p + \varepsilon_0)^n$	625(MPa)	0.62	1.19

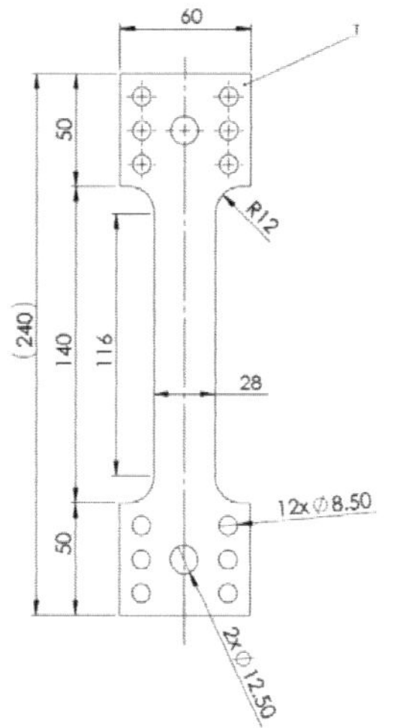

(a)

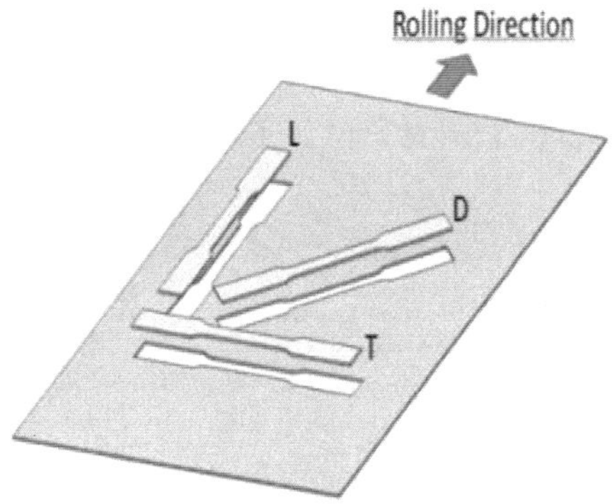

(b)

Figure 1. (a) Uniaxial tensile test specimen; (b). Sampling directions for Lankford ratios identification.

4 Numerical simulation

The aim of numerical simulation is to predict the necking by using isotropic and anisotropic material modelling. For this sake, numerical reaction force-displacement curves are obtained and compared with the experimental data. Figure 2(a) represents the behaviour of the material along the boundary condition. It is noticeable that during the experiment, reaction force and displacement were measured by testing machine in both heads of the specimen and the same boundary condition is imposed for the numerical simulations. In addition, it is shown in figure 2(b) in the necking zone the major and minor strains were measured with the use of the DIC. The same variable is evaluated in the numerical simulations for comparison. Moreover, figure 2(c) indicates the obtained result from the Nakajima test for DC01. Finally, figure 2(d) is the transformation of the Nakajima based on the formulations in section *2.1*, however, the range of interest for this paper is only between points **c** and **d** in which stress triaxiality and Lode angle parameter vary from $0.33 < \lambda < 0.57$ and $0 < \bar{\theta} < 1$, respectively [13-17]. The obtained numerical results from anisotropic material modelling reflect the actual behaviour of the material, while the results of isotropic simulation significantly deviate from the experimental results. Therefore, for material DC01 anisotropy effects must be considered even for uniaxial tensile test.

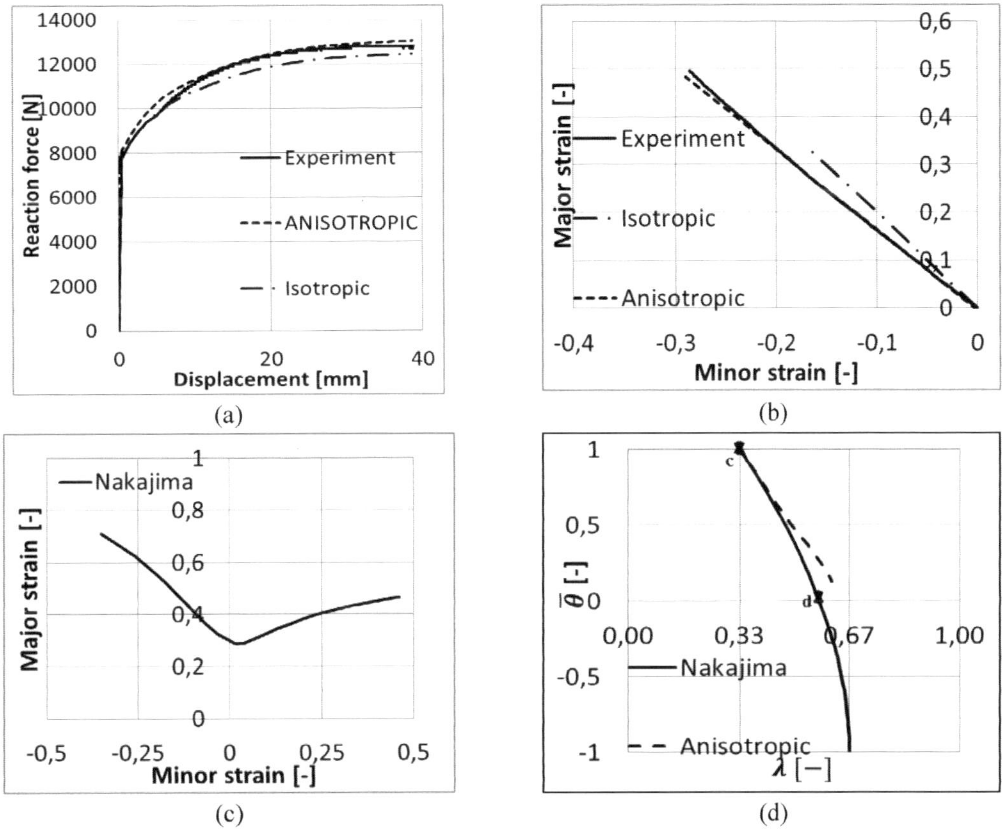

Figure 2. (a) and (b) evaluation along the boundary condition and necking zone, respectively; (c) FLC from Nakajima test; (d) transformation of the FLD into the stress state space.

5 Conclusion

In order to characterize Hill's anisotropy coefficients, it was a prerequisite to extract samples from the sheet of mild steel DC01 with the same thickness 1.5 mm. The obtained results for reaction force-displacement as well as conventional FLD for a critical node inside the necking zone demonstrate that anisotropic FEM simulation represents the actual behavior of the material. Moreover, for obtaining the whole forming curve of mild steel DC01, many Nakajima test with a variety of widths were performed which can only be represented in the conventional space. Therefore, it is possible to transform it into stress state spaces which comprise necking strain, stress triaxiality and Lode angle parameter. Furthermore, in order to cover the whole FLC, plane strain tension and equi-biaxial tension have to be performed with another alternative, like a cruciform specimen in the future research.

Acknowledgements

This paper was created by project TF02000072 - Methods development for formability assessment of thin sheets considering anisotropy and non-linear loading path (2016-2019, TA0/TF) financed by the Technology Agency of the Czech Rep. and project Development of West-Bohemian Centre of Mat.and Metal. No.: LO1412, financed by the MEYS of the Czech Rep.

References

[1] Keeler S P and Backofen W 1963, Plastic instability and fracture in sheets streched over rigid punches *Transactions of American Society for Metals,* vol. **56**, pp. 25-48,.

[2] Hill R 1952 On discontinious plastic states, with special reference to localized necking in thin sheets *mechanics and Physics of Solids,* vol. **1**, no. 1, pp. 19-30.

[3] Swift W 1952 Plastic instability under plane stress *Mechanics and Physics of Solids,* vol. **1**, no. 1, pp. 1-18.

[4] Marciniak Z and Kuczyski K 1967 Limit strains in the processes of stretch-forming sheet metal *Mechanical Sciences,* vol. **9**, no. 9, pp. 609-620,.

[5] Stoughton TB 2000 A general forming limit criterion for sheet metal forming *International Journal of Mechanical Sciences,* vol. **42**, no. 1, pp. 1-27.

[6] Lee YW 2005 Fracture predictions in metal sheets PhD thesis, Massachusetts Institute of technology,.

[7] Bai Y and Wierzbicki T 2008 Forming severity concept for predicting sheet necking under complex loading histories *International Journal of Mechanical Science,* vol. **50**, pp.1012-22

[8] Bai Y and Wierzbicki T 2008 A new model of metal plasticity and fracture with pressure and Lode dependence *International Journal of Plasticity,* vol. **24**, no. 6, pp. 1071-96,.

[9] Bai Y and Wierzbicki T Application of Mohr-Coulomb criterion to ductile fracture *Report*

[10] Hooputra H Gese H Dell H Werner H and Health A 2003 A new comperhensive failure model for crashworthiness simulation-validation for aluminum extrusions *In: 13th European Conference on Digital Simulation for Virtual Prototyping, Virtual Manufacturing and Virtual Environment,* Mainz, Germany, October.

[11] Hooputra H Gese H Dell H and Werner H 2007 A comperhensive failure model for crashworthiness of aluminum extrusions *International Journal of Crashworthiness,* vol. **9**, pp. 449-64.

[12] Souza E A Perić D Owen D R J 2008 Computational Methods for Plasticity: Theory and Applications Swansea University, *John Wiley and Sons Publications, ISBN: 9780470694527,* pp. 414-16,.

[13] Hari C Gholipour J Bardelick A and Worswick MJ 2007 Prediction of necking in tabular hydroforming using an extended stress-based forming limit curve *Journal of Engineering Materials and Technology,* vol. **129**, pp. 36-47,.

[14] Mohr D and Marcadet J 2015 Micromechanically-motivated phenomenological Hosford-Coulom model for predicting ductile fracture initiation at low stress triaxialities *International Journal of Solids and Structures,* vol. **67**, pp. 40-55.

[15] Kubík P, Šebek F, Petruška J, Hůlka J, Růžička J, Španiel M, Džugan J, Prantl A 2013 Calibration of Selected Ductile Fracture Criteria Using Two Types of Specimens, Key Engineering Materials, ISSN: 1662-9795, Vol. **592-593**, pp.258-261.

[16] Španiel M, Prantl A, Džugan J, Růžička J, Moravec M, Kuželka J 2014 Calibration of fracture locus in scope of uncoupled elastic–plastic-ductile fracture material models, Advances in Engineering Software, Vol. **72**, pp.95-108

[17] Dzugan J, et al 2015 Identification of ductile damage parameters for pressure vessel steel. Nucl. Eng. Des., http://dx.doi.org/10.1016/j.nucengdes.2015.12.014

Prediction of thinning of the sheet metal in the program AutoForm and its experimental verification

M Fedorko [1], M Urbánek [1], M Rund [1]

[1] COMTES FHT a.s, Dobřany 334 41, Czech Republic

E-mail: mikulas.fedorko@comtesfht.cz, miroslav.urbanek@comtesfht.cz, marti.rund@comtesfht.cz

Abstract. The manufacture of press-formed parts often involves deep-drawing operations. Deep drawing, however, can be deemed an industrial branch in its own right. Today, many experimental as well as numerical methods are available for designing and optimizing deep drawing operations. The best option, however, is to combine both approaches. The present paper describes one such investigation. Here, measurements and numerical simulation were used for mapping the impact of anisotropy on thickness variation in a spherical-shaped drawn part of DC01 steel. Variation in sheet thickness was measured on spherical-shaped drawn parts of various geometries by means of two cameras, and evaluated with digital image correlation using the ARAMIS software from the company GOM. The forming experiment was carried out on an INOVA 200 kN servohydraulic testing machine in which the force vs. piston displacement curve was recorded. The same experiment was then numerically simulated and analyzed using the AUTOFORM software. Various parameters were monitored, such as thinning, strain magnitude, formability, and others. For the purpose of this simulation, a series of mechanical tests was conducted to obtain descriptions of the experimental material of 1.5 mm thickness. A material model was constructed from the tests data involving the work-hardening curve, the impact of anisotropy, and the forming limit diagram. Specifically, these tests included tensile tests, the Nakajima test, and the stacked test, which were carried out to determine materials data for the model. The actual sheet thickness was measured on a sectioned spherical-shaped drawn part using a NIKON optical microscope.The variations in thickness along defined lines on the sectioned drawn part were compared with the numerical simulations data using digital image correlation. The above-described experimental programme is suitable for calibrating a material model for any computational software and can correctly solve deep-drawing problems.

1 Introduction

Sheet metal forming is the process of converting flat sheet of metal into a part of desired shape without fracture or excessive localized thinning. The application of sheet metal forming includes automotive industry and many more. The automotive industry is a sector with the strictest demands on lightweighting and highest pressure on prices. Consequently, various tools are employed by producers of deep-drawn parts for predicting the part's behaviour before the actual manufacturing operation. Generally, the available options are either simple trials on small-size parts or computer simulations. With intricate press-formed parts, the best choice is to combine both of these options. Forming Limit Diagram (FLD) is used during the design stage of any new sheet metal component for tooling shape & optimizing variables. It is nothing but a combination of major & minor strain. The FLD is a widely

IOP Conf. Series: Materials Science and Engineering **179** (2017) 012024 doi:10.1088/1757-899X/179/1/012024

used concept to represent the formability of thin metallic sheets. To control the operation of sheet metal forming without failure, a diagram is used in which the safe, critical and failure forming regions are shown.

The paper brings new information on material behaviour of chosen steel grade for deep drawing purposes. Authors of this paper have measured fundamental properties of DC01 sheet, constructed a material model for computation with the aid of the AutoForm software, and ran a computer simulation of drawing of a spherical cup. The final steps of this investigation involved validation of the simulation by an actual test of drawing and measuring a spherical cup in a mechanical testing laboratory. The test was performed on a servohydraulic machine and the cup deformation was measured using digital image correlation.

Describing the deep drawing process by means of numerical simulation poses a complex non-linear problem. Deep-drawing grades of sheet tend to be anisotropic, exhibit a defined work-hardening behaviour, and their mathematical description requires up to 10 parameters. Using AutoForm, a special deep-drawing simulation software, parameters of a real-world test can be easily input into the simulation for evaluation. In this study, the amount of thinning in the sheet at the end of the process was compared between a computer simulation and an actual test.

2 Experimental material and method

In order to construct the material model of the DC01 steel, a set of mechanical tests had to be carried out. Anisotropy of the material was first mapped using a series of tensile tests on specimens oriented in various directions (0°, 45°, and 90°) with respect to the sheet rolling direction. Additional material parameters were found by conducting and evaluating stack tests. The outcome of these tests was a curve which describes the anisotropy of material properties.

2.1 Material model

The material model is highly dependent on accurate descriptions of the material´s mechanical behaviour, the selected constitutive model as well as the input data used for its calibration. Therefore, trying to create the appropriate description for the DC01 steel sheet (Table 1.), the following material model was selected to describe as accurately as possible the material behaviour based on the hardening law and the yield criteria.

Table 1. Chemical composition steel DC01.

element	C	Mn	P	S	N	Al
%	0.12	0.6	0.045	0.045	0.003	0.04

2.1.1 Yield criteria. In order to obtain a balance between the accuracy of the model and parameter identification costs, most of the advanced yield criteria are using a set of seven or eight experimental values. Usually, six input values are used for parameter identification: the three yield stresses (σ0, σ45, σ90) and the three r-values (ratio between the strain in the width and the thickness) obtained from uniaxial tensile tests (performed with orientations of 0°, 45° and 90° to the rolling direction), calculated using the formula:

$$r = \frac{\varepsilon_w}{\varepsilon_t} = \frac{ln(\frac{w}{w_0})}{ln(\frac{t}{t_0})} \tag{1}$$

For the definition of our material model, we will use the balanced biaxial strain ratio (rb) obtained from the stack test and the biaxial test. The balanced biaxial strain ratio rb-value can be calculated comparing the strain in two directions (rolling direction and transverse to rolling direction) and using the formula:

$$r_b = \frac{\varepsilon_{TD}}{\varepsilon_{RD}} \tag{2}$$

2.1.2 Hardening law. In order to define the hardening behaviour of the metallic sheet we decided to use the Swift hardening law. In the model of Swift law, three material constants, strength coefficient (K), strain-hardening exponent (n) and initial strain (ε0) should be estimated by curve fitting the measured true stress-strain data before necking to the following equation:

$$\sigma = K(\varepsilon_0 + \varepsilon)^n \tag{3}$$

2.1.3 Uniaxial tensile test. The mechanical properties of the material has been measured on the tensile test in three directions (0°, 45° and 90°) related to the rolling direction. The strain measurements of the tests have been performed by Digital Image Correlation (DIC) method [1] with the use of ARAMIS system [1]. The system is using two cameras that enable optical 3D strain measurements.

The mechanical properties evaluated from the uniaxial tensile test are the engineering strain (e) and engineering stress (s), these properties are obtained directly from measurements:

$$e = \frac{\Delta L}{L_0} \tag{4}$$

$$s = \frac{F}{A_0} \tag{5}$$

Finally the flow stress behaviour has been obtained rcalculating the true strain (ε) and true stress (σ). These parameters calculated directly from the engineering parameters before the necking and with an extrapolation after it.

$$\varepsilon = ln(1 + e) \tag{6}$$

$$\sigma = s(1 + e) \tag{7}$$

The flow stress curves are shown in the Figure1:

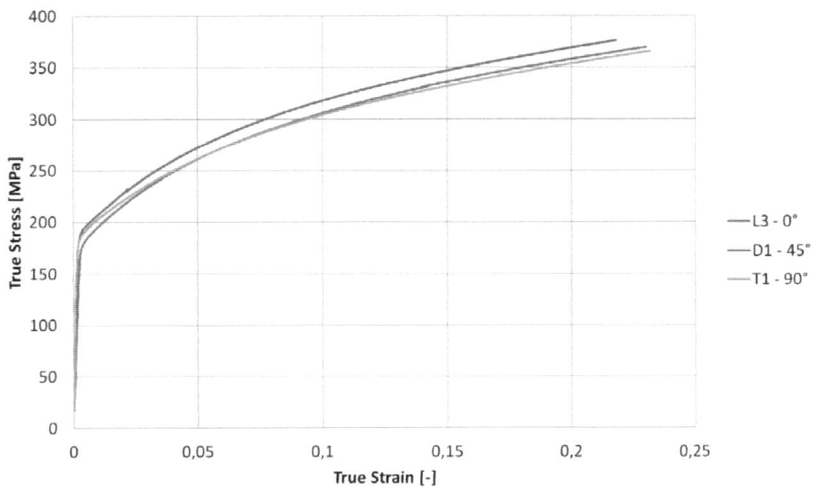

Figure 1. Flow stress curves for each direction.

After the calculation of the true stress-strain we fitted the constants for the Swift hardening law for the rolling direction curve (0°), obtaining the following parameters:

Table 2. Swift model parameters.

K	ε0	n
625.1	0.005	0.412

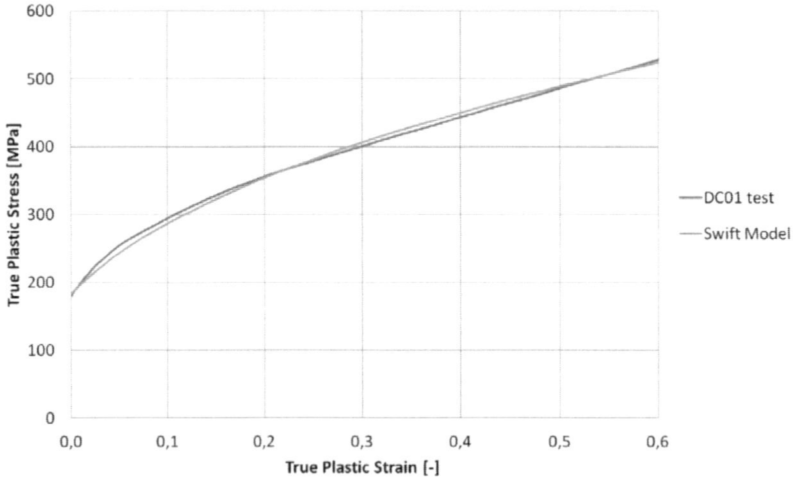

Figure 2. Swift model fitted to the tensile test.

The anisotropy parameters have been calculated after the initialization of the straining and before the necking of the specimen. The results obtained after the evaluation of the uniaxial tensile tests are:

Table 3. Anisotropy parameters gained from uniaxial tensile tests

σ_0[MPa]	σ_{45}[MPa]	σ_{90}[MPa]	r_0[-]	r_{45}[-]	r_{90}[-]
178.7	192.7	184.8	1.78	1.25	2.15

2.1.4 Stack test

To complete the characterization of the anisotropy model it is necessary to obtain another parameter in order to create a material model more reliable for advanced yield criteria. This parameter is the balanced biaxial strain ratio r_b-value that is obtained from the disk compression test (stack test), the bulge test or the biaxial test.

Figure 3. Compression stack test sample.

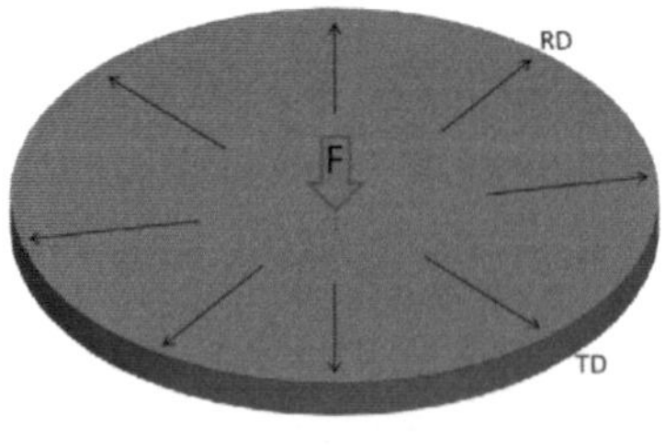

Figure 4. Compression and force and strain distribution during stack test.

To obtain this parameter the stack test has been performed, using disks of 25mm diameter and stacking a total of 10 disks [2]. To conserve the anisotropy properties, the disks were piled considering the rolling direction. To reduce the friction between the stacked specimen and the compression plates of the tool, a Teflon-foil acting as a solid lubricant is placed at each end of the specimen. During the test, the force was gained from the universal testing machine Zwick Roell 250kN and the strain and displacement values were gathered by another DIC method using MERCURY system [3]. Two independent cameras were arranged with an angle of 90° to each other, one of them in the rolling direction and the other one perpendicular to the rolling direction. With this setup it was possible to cover more than the half lateral surface and to measure the strain in the rolling direction and in the transversal direction. Using the formula previously described we have obtained the value $r_b = 0.951$ gathering the strain values of the central disks of the pile, where the strain field is homogeneous.
In order to calculate the yield locus for the DC01 steel sheet, the BBC2008 yield criterion was used [4]. The material parameters used as in put in the BBC2008 identification procedure [5] are the previously calculated:

- Three Yield stresses for each direction (σ_0, σ_{45}, σ_{90})
- Three r-values (r_0,r_{45},r_{90})
- Balanced biaxial stress (r_b)

The calculated BBC2008 yield locus is shown in the figure 5a. In our case was used 7 parameters BBC2008 yield criterion plotted on the yield locus graph (Figure 5a) [6]. Based on the results, a forming limit curve as a function of ε_1 and ε_2, true strain values on the sheet surface, was constructed, as illustrated in Figure 5b [6]. Functions obtained in this manner were input into numerical simulations of the spherical cut drawing process in the AUTOFORM software.

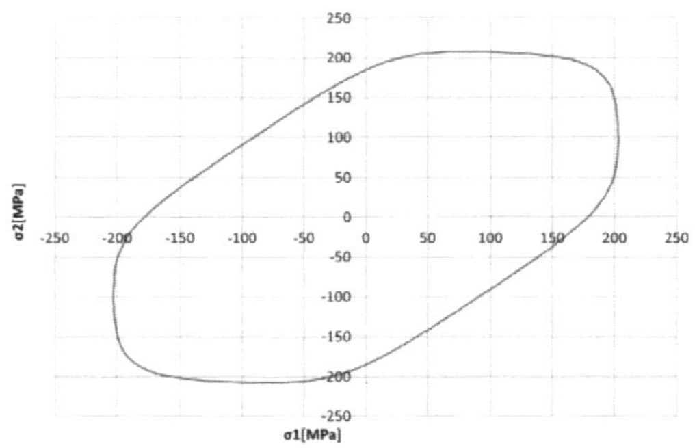

a) Yield surface

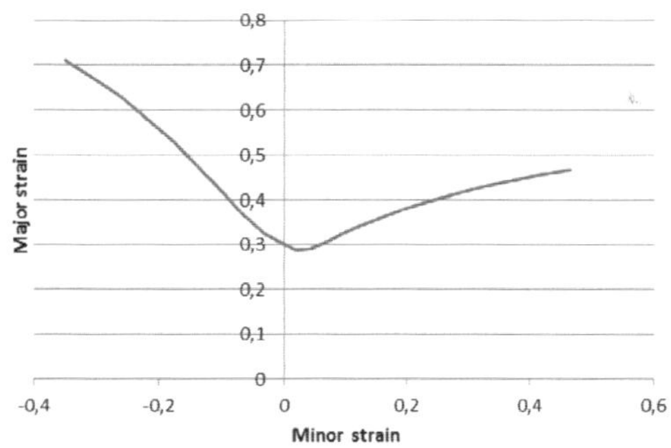

b) Forming limit diagram

Figure. 5 Material properties of DC01.

3 FEM simulation

The AutoForm software was developed for modelling deep drawing processes. It relies on the finite element method and uses no other than shell elements. In the program, the mathematical descriptions of materials properties are tailored to deep drawing simulations. The data and the graphs listed above (Figures 1, 2 and 5) were used as inputs into the numerical simulation of the sheet behaviour.

Drawing on hands-on experience and expertise from sheet-forming industry, the program is intended mainly for the automotive sector. Every step in the sequence can be analyzed and optimized separately as an independent solution. The program features a built-in function which provides a full-scale solution for the entire manufacturing process. Default variables evaluated by the program include the sheet thickness, % thinning, wrinkling, and other values associated with deep drawing. The input into the AutoForm simulation was a CAD file with the geometry of the Nakajima test fixture shown in Figure 6. The parameters of the Nakazima test are shown in Table 4 Only a single-action draw was simulated, where the positions of the punch, the die and the blank holder are precisely defined. In this test, the blank was bolted down firmly. Therefore, the blank holder in the model was fixed. The velocity of punch was set in accordance with the completed validation test. Results of the computer simulation show the thinning of the spherical cup, as well as the indication of thinning in the graph.

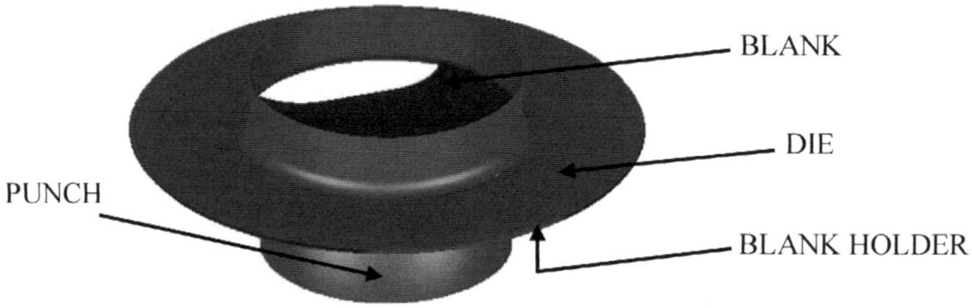

Figure 6. Nakajima test.

Table 4. Proces parameters.

Proces parameters	
Hemispherical punch radius:	60
Die profile radius:	5mm
Die inner radius:	100 mm
Blank material	DC01
Blank holder force:	1000kN
Friction coefficient:	0.05
Element type:	Elastic plastic shell
Element size of blank:	1.5mm

The key outcomes of these simulations were the shape of the drawn part and thickness variations along defined cross-sections. Other variables which were compared between the actual test and the numerical simulations included the behavior of the drawing process, the force required for drawing the part and the impact of anisotropy on strain distribution.

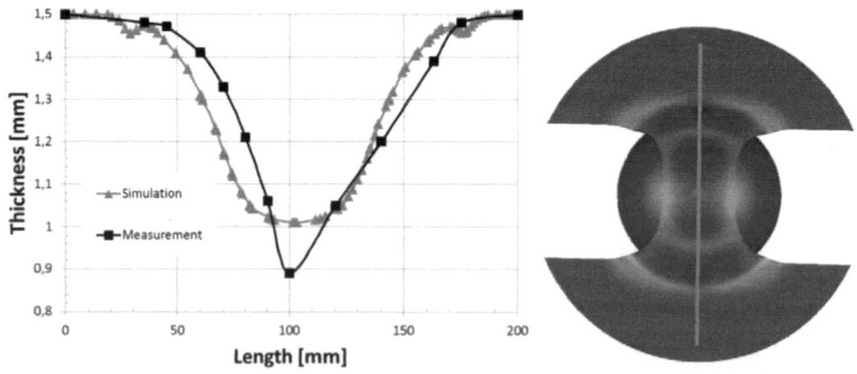

Figure 7. Comparison between simulation data and readings from verification measurement. Data were measured on the red line (right part of the figure).

Figure 7 compares the thickness variation in the drawn part along a line across the tear. The simulation indicates uniform thinning in the sheet, whereas the validation test shows led to a non-symmetric profile caused by tearing.

4 Conclusion

This paper evaluates sheet thickness variation in a spherical drawn part of DC01 deep-drawing steel. Based on mechanical testing data, a material model was constructed and input into numerical simulations in AUTOFORM software. Real-part measurements and numerical simulation data were compared. As emerged from the experimental and numerical comparison , the achieved sheet thickness for both the cases coresponds. Based on the experimentally conducted measurement and simulation, it can be stated that the created material model describes sheet anisotropic behaviour and sheet tendency to thinning. Thus, the study produced a material model for DC01 steel which describes the anisotropy of properties and its effect on sheet thickness variation in Nakajima drawing test.

Acknowledgement

This study was created by project Development of West-Bohemian Centre of Materials and Metallurgy No.: LO1412, financed by the MEYS of the Czech Republic.

References

[1] Sutton, M., Digital and Image Correlation: Principle Developments and Applications for Parameter Estimation, University of South Carolina. http://www.gdr2519.cnrs.fr/ecole2011/ET_2011/M_Sutton.pdf

[2] Merklein M., Kuppert A., A method for the layer compression test considering the anisotropic material behaviour. Int J Mater Form, (2009) Vol.2 Suppl 1: 483-486. DOI 10.1007/s12289-009-0592-8

[3] Folgar Ribadas H., Urbánek M., Procházka R., Džugan J., Experiences with cruciform sample to characterize anisotropy plasticity and hardening for steel sheet, IRF2016, Porto, Portugal

[4] Comsa D.S., Banabic D., Plane/stress zield criterion for highly-anisotropic sheet metals, Numisheet 2008, ed. P. Hora, Interlaken, 2008, p.43-48.

[5] Jurco P., Banabic D., A user-friendly programme for calculating Forming Limit Diagrams, Techinical University of Cluj-Napoca, Daicoviciu 15, Cluj-Napoca, Romania

[6] Hill R., A theory of the yielding flow of anisotropic metals, Proc. Roy. Soc. **A193** (1948) 281-297

IOP Conf. Series: Materials Science and Engineering **179** (2017) 012025 doi:10.1088/1757-899X/179/1/012025

3D printed porous stainless steel for potential use in medicine

M Fousová[1], J Kubásek[1], D Vojtěch[1], J Fojt[1], J Čapek[1, 2]

[1]Department of Metals and Corrosion Engineering, University of Chemistry and Technology Prague, Technická 5, 166 28 Prague 6, Czech Republic

[2]Institute of physics, Academy of Science of the Czech Republic (AS CR), Na Slovance 1999/2, 182 21 Prague 8, Czech Republic

E-mail: fousovam@vscht.cz

Abstract. 3D printing technologies like Selective Laser Melting (SLM) or Electron Beam Melting (EBM) produce components of very complicated shapes from various kinds of materials. In this work a highly porous (porosity of almost 90 vol. %) stainless steel component was manufactured by SLM. The material was characterized in terms of structure, surface chemistry and mechanical properties. It was observed that mechanical properties of the material were similar to those of trabecular human bone. The tests realized in this work confirmed suitability of the porous material prepared by SLM for the use in medicine, for example, for scaffolds designed to repair bone defects.

1 Introduction

With the constant progress in development of biomaterials and availability of new technologies, regenerative medicine keeps achieving greater success. However, there are still many challenges that are hoped being overcome in the near future. One of them is the design of bone scaffolds and implants which would replicate the biomechanical properties of the host bones perfectly.

For the bone applications, especially metals have been used for many years. Among them, stainless steel, cobalt alloys and titanium along with its alloys are well established for permanent implants. These materials are biocompatible, have good mechanical properties and express high corrosion resistance in the body environment [1]. Such characteristics are very important and even fundamental for the use of biomaterials, but do not guarantee flawless operation of the final implant. Although mechanical strength of over-mentioned metals is high enough to bear physiological loading, much higher stiffness over natural bones is the main source of bone tissue loss and eventual implant failure. The extent of such negative influence depends especially on the difference in elastic moduli between natural bone and the implant material. When a bone is coupled with an implant, it begins to share its load-carrying capacity. As a consequence, the bone is subjected to reduced stresses, which is perceived as a stimulus for bone resorption. Such a phenomenon, termed 'stress shielding', was observed decades ago and since then scientists have been searching for possible solutions [2-4].

An effective method is to introduce porosity into bulk materials and change their relative density [5]. When damaged bone must be repaired or even replaced, porous metals present suitable candidates as their stiffness and porosity can be adjusted on demands. In case when pores are opened towards the implant surface and interconnected, metallic replacements also allow the bone tissue to grow into and, thus, accelerate the osseointegration process [6,7]. The fabrication of

porous metals has been studied intensively and many methods have been developed. Four main manufacturing routes - solid state processing, liquid state processing, vapor deposition and electron deposition – can be classified according to the state of matter in which the metal is processed. In addition, first two approaches cover tens of various methods, e.g. use of blowing agents, gas injection, powder metallurgy techniques, sintering of fibers etc. [8]. Despite a high number of available traditional techniques, only limited control over the internal architecture is provided. Although pore size and shape can be adjusted to some extent by the set-up of process parameters, pores distribution remains rather random [9]. However, current boom in additive manufacturing (AM) opens new horizons. Thanks to a layer-by-layer additive principle, complex structures that were impossible to be prepared by conventional techniques can be nowadays achieved [10].
New era in manufacturing is highly appreciated especially in the field of regenerative medicine, where newly prepared customized implants may already meet the increasing demands on implant performance [11]. Metallic materials satisfying requirements for mechanical load in the human body can be processed by several AM techniques among which selective laser melting (SLM) and electron beam melting (EBM) are the most widespread [12,13]. This paper is, therefore, focused on characterization of stainless steel 316L porous scaffold prepared by SLM which could potentially serve as a bone scaffold or an augmentation in the vicinity of joint replacements.

2 Experimental

Studied scaffold was prepared by selective laser melting technology using a M2 Cusing machine (ConceptLaser, Germany) equipped with 200W Yb:YAG fiber laser. As an input material a gas-atomized powder of stainless steel AISI 316L (Misan GmbH) with particle size in the range of 15-50 μm was used. An initial powder layer thickness of 30 μm was selected to be processed in a layer-by-layer manufacturing process. In each manufacturing step processing parameters were kept constant: laser power of 200 W, scanning speed of 200 mm s^{-1} and hatching distance of 1 mm. With hatching distance set to 1 mm and cross-scanning strategy, a geometrically defined lattice structure was obtained. Final product (shown in figure 1) was in the form of 50 x 30 x 20 mm^3 block and possessed square-shaped pores of an average size of 750 μm separated by 250 μm thick struts. The total porosity was determined to be 87 vol.%.

Figure 1. AISI 316L scaffold manufactured by means of 3D printing.

Macrostructure and surface morphology of the produced scaffold were studied by scanning electron microscope TESCAN VEGA-3LMU equipped with an Oxford instruments INCA 350 EDX analyzer (SEM-EDX). Material microstructure was observed on metallographic cross-sections using an Olympus PME3 light microscope and SEM-EDX. Cross-sections were grinded up to P4000, polished on a diamond paste D2, final polished with colloid alumina and etched electrochemically in 10 wt.% oxalic acid (6 V, 80 s).
X-ray photoelectron spectroscopy (ESCAprobe P, Omicron Nanotechnology Ltd. equipped with an Al Kα (λ = 1486.7 eV) X-ray source) was used to study the surface chemistry. To assess the influence of SLM technology, obtained surface was compared with a reference sample of AISI 316L steel produced conventionally by means of hot forging. The spectra were measured with an energy step of 0.05 eV and normalized to the adventitious carbon (C1s peak, 285.0 eV). NIST X-

ray Photoelectron Spectroscopy Database was used to evaluate chemical states of detected elements. Prior the spectroscopy measurement, the specimens were thoroughly cleaned with distilled water, ethanol and acetone.

Compressive tests were carried out on a LabTest 5.250SP1-VM universal loading machine to determine basic mechanical properties. Tests were performed at room temperature, with a strain rate of 0.001 s^{-1}, on three rectangular samples cut from the manufactured scaffold.

3 Results and Discussion

3.1 Scaffold macrostructure

Macrostructure of the manufactured scaffold is shown in figure 2. Struts thickness of about 250 μm corresponds to the laser beam spot diameter. Struts are inclined by 45° from the building direction. Every four struts enclose a square-shaped pore, altogether forming a structure with fully interconnected open porosity. In higher magnification (figure 2b) a joint where four struts converge is displayed. One can also observe the surface is not completely smooth, but is roughened by adhering spherical particles. These are indicated by white arrows in figure 2b. As the diameter of these particles falls into the range of the initial powder diameter (15-50 μm), it can be supposed it concerns unmelted particles which were sintered to the final product due to the thermal diffusion effects. As the inclined struts were partially built on the loose powder, particles below each layer were partially melted and bonded to its bottom [14]. Therefore, it can be noticed that the adhering particles prevail on the bottom part of the struts. Figure 2c brings a strut surface in detail. Arrows are pointing at arcs representing overlapping layers.

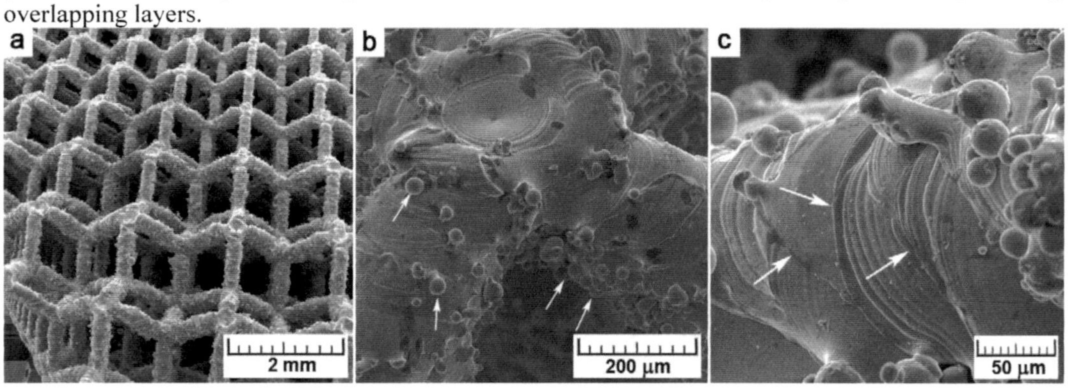

Figure 2. Macrostructure and surface morphology of the SLM scaffold. Arrows in b) show unmelted powder particles and in c) overlapping layers.

3.2 Microstructure of SLM AISI316L stainless steel

Figure 3a represents a cross-section through 4 scaffold struts. Here, dark arcs can be also observed. The distance between these is about 80-150 μm, which significantly exceeds the powder layer thickness setting indicating that the material is melted also in several precedent layers and a strong interlayer bonding is formed. In agreement with other works [15-17], material exhibits microstructure typical for laser-treated austenitic steels that undergo rapid cooling. Austenitic grains are elongated in the building direction as they grew through several melted layers and possess very fine cellular-dendritic substructure (shown in figure 3b).

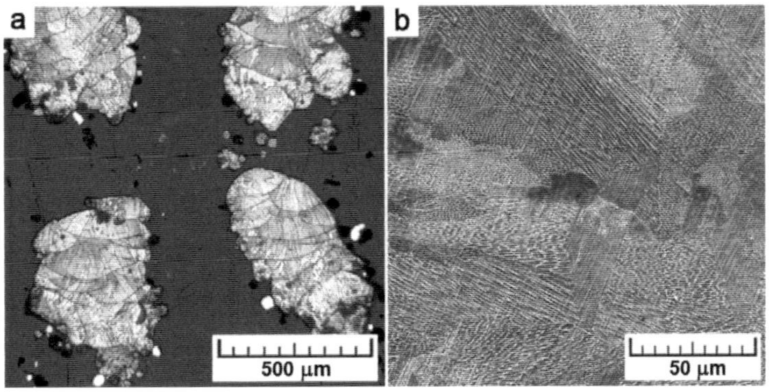

Figure 3. a) Microstructure of the scaffold struts and b) cellular-dendritic substructure.

3.3 Surface chemistry

XPS analysis was carried out on SLM scaffold as well as hot forged reference for which the 316L stainless steel biocompatibility is well known. The aim was to assess changes brought about by the laser melting process and estimate if the observed changes may have negative effect on interactions with cells. Survey spectra for both materials are given in figure 4. Except from main constituting elements (Fe, Cr, Ni, Mo) adventitious carbon, oxygen and nitrogen were also detected. Binding energies declare that, in both cases, the surface layers are formed by a mixture of iron and chromium oxides particularly (Fe_2O_3, Fe_3O_4, $FeCr_2O_4$ and Cr_2O_3). The changes in surface composition induced by SLM process are shown in figure 5. Nickel and molybdenum were detected only in the passive oxidic layer of hot forged reference (in the form of NiO and MoO_3, respectively). Within SLM surface a strong enrichment in chromium was observed. Such an enrichment occurred due to a high affinity of chromium to oxygen at high temperatures induced by laser melting, as some oxygen residues remained in the protective argon atmosphere.

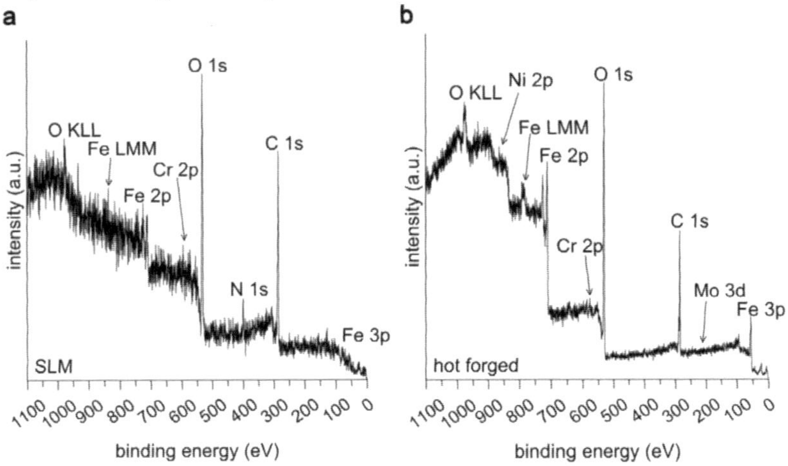

Figure 4. Survey spectra for a) SLM scaffold and b) hot forged reference obtained by XPS.

From observed facts, it can be concluded that SLM should not exert any negative effect on biocompatibility. Although high Cr content could represent a toxic hazard, all the Cr is present in the form of chromium (III) oxide or chromite which are almost insoluble in the body environment and, conversely, play the key role in the stainless steels corrosion resistance.

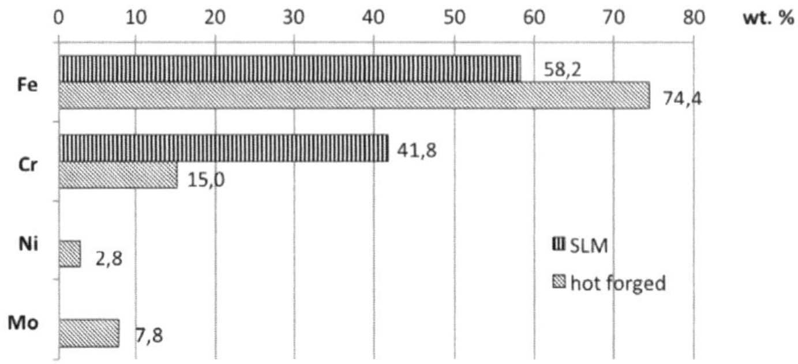

Figure 5. Surface chemical composition of SLM AISI 316L stainless steel in comparison with hot forged reference (only metallic elements included).

3.4 Mechanical behaviour in compression

Compressive tests yielded stress-strain curves, one of which is given in figure 6. It exhibits a high level of plasticity, with compressive strain reaching 56.8±0.1%. A plateau in between 5 and 35% can be observed which is related to a high plastic deformation with only low increase in stress. Afterwards, a densification stage occurs, where the stress increases sharply. Such a phenomenon is often observed in case of highly porous metallic materials. It is caused by the collapse of struts followed by complete collapse of cells accompanied with compaction of the porous structure [18,19]. Table 1 summarizes mechanical properties subtracted from the curve. To illustrate the impact of porosity introduced into the material on its mechanical properties, table 1 also lists values achieved in our previous study [20] with nonporous AISI 316L stainless steel prepared equivalently by SLM technology under comparable conditions. Since the aim of this paper is to show if the studied scaffold is suitable for medical use, also a comparison with human bone is given.

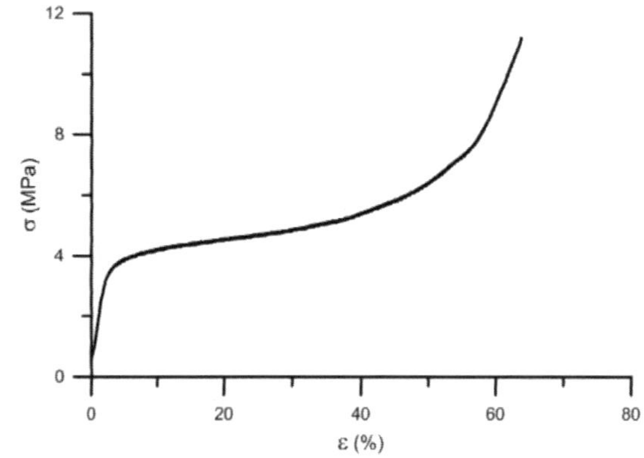

Figure 6. Compressive stress-strain curve for SLM scaffold.

Table 1. Mechanical properties in compression for SLM AISI 316L scaffold, compact material and human bone (CYS = compressive yield strength, UCS = ultimate compressive strength, ε = maximal compressive strain, E = Young´s modulus in compression).

	CYS (MPa)	UCS (MPa)	ε (%)	E (GPa)	Ref.
Scaffold	3.01±0.13	10.6±0.6	56.8±0.1	0.15±0.03	this study
Compact	497±24	981±20	18.1±6.1	187±5	[20]
Cortical bone		~170-193	~1.85-2.60	7-27	[21]
Trabecular bone		~2.4-11.6		~0.1-0.4	[22]

4 Conclusion

By SLM technology AISI 316L stainless steel scaffold was successfully prepared. In this study, basic characterization was carried out. As mechanical properties in compression approached those for human trabecular bone and the surface chemistry seems not to negatively influence material biocompatibility, prepared scaffold appears to be a promising tool for bone defects repair.

Acknowledgements

Authors want to acknowledge the Czech Science Foundation (project no. P108/12/G043) for the financial support of this research.

References

[1] Park J and Lakes R S 2007 Biomaterials: An Introduction (New York: Springer Science+Business Media) chapter 5 p 100-136
[2] Huiskes R, Weinans H and Van Rietbergen B 1992 *Clin. Orthop. Relat. Res.* **274** pp 124-134
[3] Long M and Rack H J 1998 *Biomaterials* **19 (18)** pp. 1621-39
[4] Cheal E J, Spector M. and Hayes W C 1992 *J. Orthop. Res.* **10 (3)** pp. 405-422
[5] Gibson L J and Ashby M F 1997 Cellular solids: structure and properties (Cambridge: Cambridge Univ. Press)
[6] Zhang S, Wei Q, Cheng L, Li S and Shi Y 2014 *Materials & Design* **63** pp. 185-193
[7] Ryan G, Pandit A and Apatsidis D P 2006 *Biomaterials* **27 (13)** pp. 2651-70
[8] Banhart J 2001 *Prog. Mater. Sci.* **46 (6)** p p. 559-632
[9] Colombo P and Degischer H P 2010 *Mater. Sci. Tech. Ser.* **26 (10)** pp. 1145-58
[10] Murr L E 2016 *J. Mater. Sci. Technol.* In press. http://dx.doi.org/10.1016/j.jmst.2016.08.011
[11] Emelogu A, Marufuzzaman M, Thompson S M, Shamsaei N and Bian L 2016 *Additive Manufacturing* **11** pp. 97-113
[12] Heinl P, Müller L, Körner C, Singer R F and Müller F A 2008 *Acta Biomat.* **4 (5)** pp. 1536-44
[13] Pattanayak S K, Fukuda A, Matsushita T, Takemoto T, Fujibayashi S, Sasaki K, Nishida N, Nakamura N and Kokubo T 2011 *Acta Biomat.* 7 **(3)** pp. 1398-1406
[14] Yan C, Hao L, Hussein A and Young P 2015 *J. Mech. Behav. Biomed. Mater.* **51** pp. 61-73
[15] Cherry J A, Davies H M, Mehmood S, Lavery N P, Brown S G R and Sienz J *2015 Int. J. Adv. Manuf. Tech.* **76 (5)** pp. 869-879
[16] Yasa E and Kruth J P 2011 *Procedia Eng.* **19** pp. 389-395
[17] Sun Z, Tan X, Tor S B and Yeong W Y 2016 *Materials & Design* **104** pp. 197-204
[18] Esen Z and Bor Ş 2007 *Scripta Mater.* **56 (5)** pp. 341-344
[19] Wang X, Li Y, Xiong J, Hodgson P D and Wen C 2009 *Acta Biomat.* **5 (9)** pp. 3616-24
[20] Fousová M, Vojtěch D, Kubásek J, Dvorský D, Machová M 2015 *Manufecturing Technology* **15 (5)** pp. 809-814
[21] Wu S, Liu X, Yeung K W K, Liu Ch and Yang X 2014 *Mater. Sci. Eng. R* **80** pp. 1-36
[22] Keaveny T M, Morgan E F, Niebur G L and Yeh O C 2001 *Annu. Rev. Biomed. Eng.* **3** pp. 307-333

Strain-amplitude dependent cyclic hardening of 08Ch18N10T austenitic stainless steel

J Fumfera[1], J Džugan[2], J Kuželka[1], R Procházka[2] and M Španiel[1]

[1] Czech Technical University in Prague, Faculty of Mechanical Engineering, Department of Mechanics, Biomechanics and Mechatronics, Technická 4, 166 07 Praha 6, Czech Republic

[2] COMTES FHT a.s., Průmyslová 995, 334 41 Dobřany, Czech Republic

E-mail: jaromir.fumfera@fs.cvut.cz

Abstract. In this paper, cyclic plasticity response of 08Ch10N10T austenitic steel under large strain amplitude loading is presented. Non-Masing behaviour and strain-amplitude dependent cyclic hardening is observed and analysed. Impact on testing and modelling associated with this phenomenon are described.

1 Cyclic plasticity response of 08Ch18N10T austenitic stainless steel

Basic material models of cyclic plasticity assume so called Masing behavior. Illustration of this behavior is in figure 1, where all hysteresis loops are shifted in stress-strain space, so their lower-left points coincide. For every loading amplitude applied, hardening path is the same. Problem is, that construction materials usually do not meet this simplification, as can be seen in figure 2 for 08Ch18N10T austenitic stainless steel.

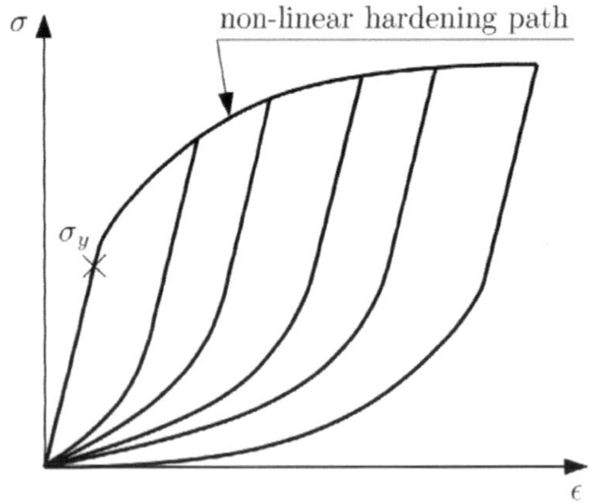

Figure 1. Example of Masing behavior.

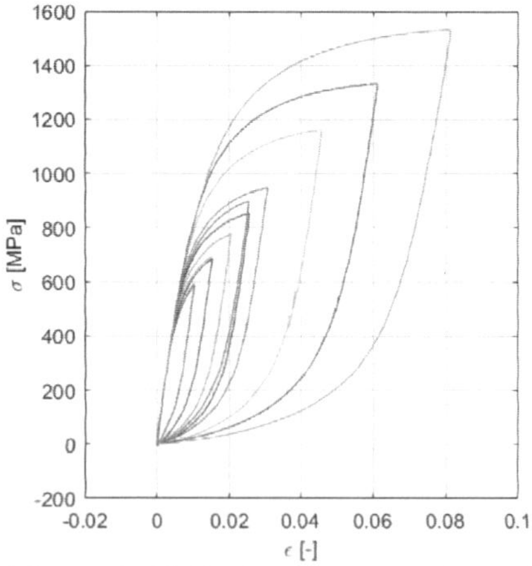

Figure 2. 08Ch18N10T austenitic steel Non-Masing behavior.

For higher loading amplitudes, the deviation is significant and material model assuming Masing behavior should be used.

The primal cause of non-Masing behavior is material strain-amplitude (also known as strain-range) dependent cyclic hardening. Physical nature of this phenomenon is on the microstructure level of material and resides in different dislocation mechanism activation depending on loading level applied. Electron microscopy of austenitic stainless steel SUS304-HP performed in [1] observed that for larger amplitude loading material slip band density increased and the dislocation structure changed from a planar array for lower load levels to a more cellular-like structure for higher load levels.

As can be seen in figure 3, for lower strain-amplitude loading levels, there is no cyclic hardening. Hysteresis loop saturates after few initial cycles and remains the same up to final softening stage caused by crack growth, which leads to final failure. For higher amplitudes, there is no saturation and material cyclically hardens up to final softening stage. Absolute number of cycles to failure are in table 1.

Table 1. Number of cycles to failure

ϵ_a [%]	N_f	ϵ_a [%]	N_f	ϵ_a [%]	N_f	ϵ_a [%]	N_f
0.5	16172	1	2013	1.5	785	3	223
0.75	4172	1.25	1212	2.25	351	4	135

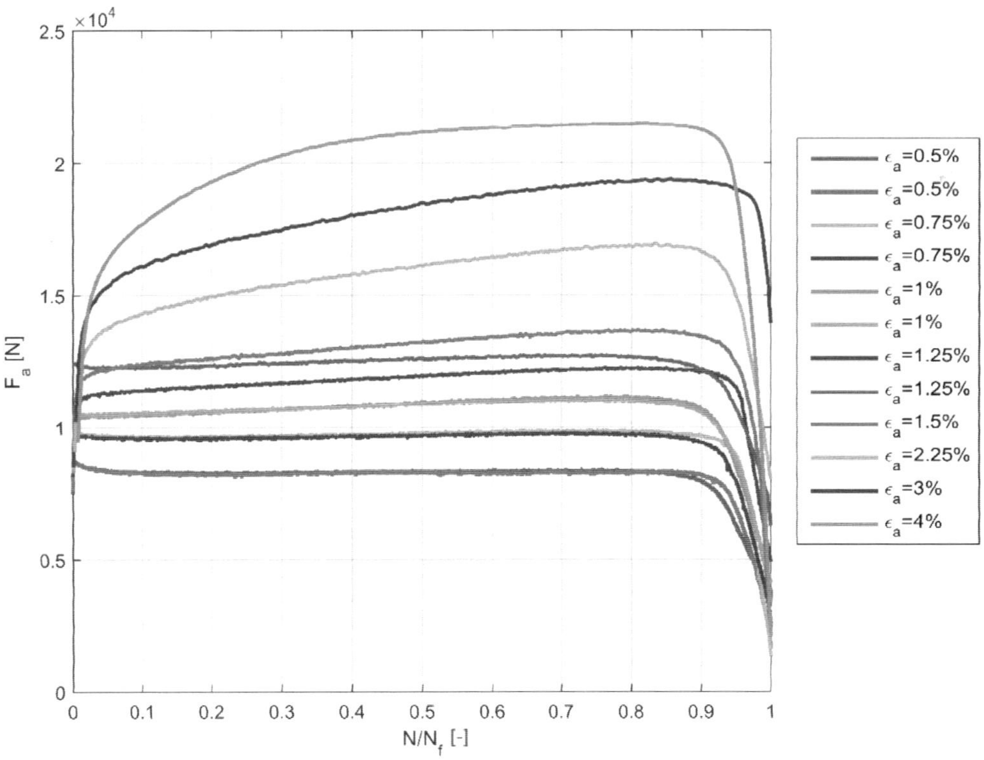

Figure 3. Force evolution during fatigue life test for various strain-amplitude levels.

Figure 2 illustrates and experimental data analysis proved, that there is a difference in yield stress value and non-linear hardening path between different strain amplitudes. The difference in Yield stress is from about 200MPa in monotonic tensile test [2] up to about 300MPa for $\epsilon_a = 3\%$.
The tendency of increasing yield stress with increasing strain amplitude is supported by performed material hardness test. More on correlation of material hardness and yield stress for austenitic steels can be found for example in [3], but briefly summarized: material hardness increases with increasing yield stress. Hardness test proved increasing material hardness on specimen with variable geometry in axial direction as can be seen in figure 4. Strain amplitude in narrowest part of specimen (distance 0 mm in figure 4) is $\epsilon_a = 3\%$ and corresponds with highest material hardness measured. As specimen get thicker along the axis (distance > 0 mm in figure 4), strain amplitude gets lower as well as material hardness.

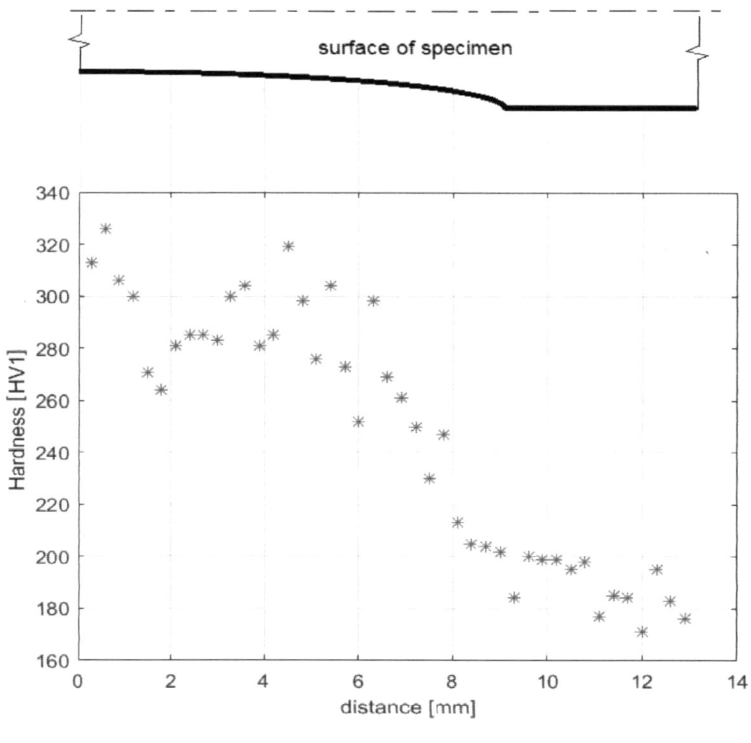

Figure 4. Material hardness in various points of variable geometry specimen surface.

For the given amplitude level, actual yield stress changes during few initial cycles (see figure 5 for illustration) and then does not change during the fatigue life even for higher strain-amplitude levels (see figure 6). Cyclic hardening is then represented only by different non-linear hardening path for different stage of the fatigue life. Young's modulus is not affected by cyclic hardening.

2 Impacts on experiment

2.1 Classical experiment scheme issue

This phenomenon brings some complications starting with low cycle fatigue tests. According to [4], for strain amplitudes higher than $\epsilon_a = 1\%$, specimen with variable longitudinal geometry should be used to prevent undesirable phenomena like loss of stability control etc. Classical scheme of strain-controlled experimental tests looks like this one: deformation is measured by the extensometer, whose tips are attached to the specimen's body outside the curved area in the cylindrical part of the specimen. Deformation measured by the extensometer controls the loading of the specimen.

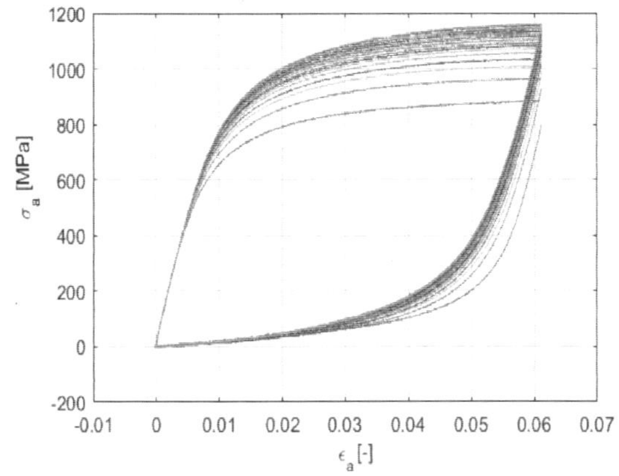

Figure 5. The first 10 hysteresis loops for $\boldsymbol{\epsilon_a} = \mathbf{3\%}$ (shifted so their lower-left points coincide).

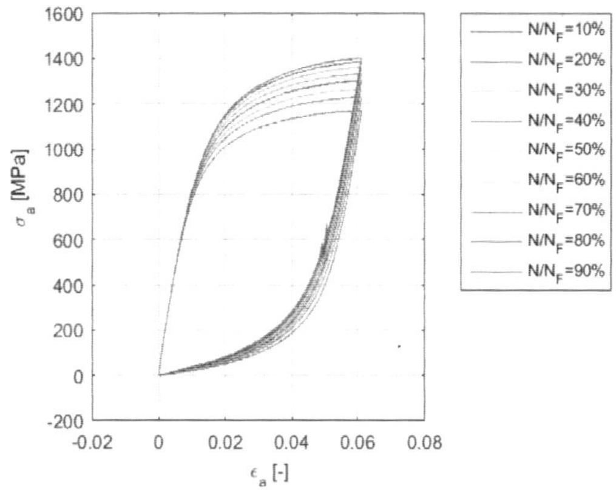

Figure 6. Hysteresis loops evolution during fatigue life for $\boldsymbol{\epsilon_a} = \mathbf{3\%}$ (shifted so their lower-left points coincide).

Problem is, that strain-amplitude dependent cyclic hardening causes significant hardening of higher loaded areas in the narrow parts of the specimen, but lower or no cyclic hardening in the thicker parts of the specimen, where strain amplitudes are lower. This phenomenon along with this classical scheme of experiment control leads to decrease of strain amplitude in the area of interest during the cycles, so experiments are not performed under the constant strain amplitude, strain amplitude is known only approximately and number of cycles to failure is not obtained for constant value of strain amplitude.

2.2 Proposed solution

There are two possible ways of performing experiments. The first one is to measure deformation and its evolution during the fatigue life by an optical device (DIC for example) and ex-post evaluation results. The evaluation procedure scheme is following. Manson-Coffin law [5], [6] determines how many cycles to failure N material lasts under plastic strain amplitude ϵ_a^p as

$$\epsilon_a^p = \epsilon_f'(2N)^c. \tag{1}$$

ϵ_a^p is known in every cycle from optical deformation measurement and ϵ_f' and c are material constants. Equation (1) can be modified into form

$$\mathrm{N} = 0.5 \cdot \exp\left(\frac{1}{\mathrm{c}} \cdot \ln\frac{\epsilon_a^p}{\epsilon_f'}\right). \tag{2}$$

Palmgren-Miner linear damage hypothesis [7] determines how much damage D cause number of cycles n_i as

$$\sum_i \frac{n_i}{N_i} = D \tag{3}$$

where N_i is number of cycles to failure on loading level i. Loading level can be represented for example by particular value of strain-amplitude applied. When $D = 1$, material fails. Combining equation (2) into equation (3) results in

$$\sum_i \frac{n_i}{0.5 \cdot \exp\left(\frac{1}{c} \cdot \ln \frac{\epsilon^p_{a\,i}}{\epsilon'_f}\right)} = D \tag{4}$$

where $\epsilon^p_{a\,i}$ is plastic strain amplitude of load level i. Final summation of each D from equation (4) during the fatigue test must meet the sum

$$\sum_{j=1}^{N_f} D_j = 1 \tag{5}$$

where $\epsilon^p_{a\,i}$ is known in each particular cycle and N_f is number of cycles to failure during the fatigue test. The only unknowns are material parameters ϵ'_f and c, which can be determined from applying this procedure to at least two sets of fatigue tests under at least two various load levels. For more than two set of tests, least square method should be used to determine only two unknown parameters.

This procedure can be tricky, because it presupposes validity of the Manson-Coffin law even for large strain amplitude loading and validity of Palmgren-Miner linear damage hypothesis.

The second way is to control the experiment by strain measured directly in selected area, for example by an optical (DIC) device. But there are few complications. First, deformation measured by DIC can be less accurate depending on camera resolution and size of measured area (accuracy of common extensometer is about $\pm 1{,}5\mu m$ vs about $\pm 8\mu m$ for 4Mpx camera chip and $350 mm^2$ measured area). Second, there must be some clear markers in the specimen's body of which the deformation is determined. But experiments show, that small cracks appear on the specimen's surface already in the half of the fatigue life, became larger and a new initiate as number of cycles increases, which leads to destruction of markers. This approach can allow holding constant strain amplitude value for certain number of cycles, but certainly does not allow to perform whole fatigue life test up to the final failure. Attaching extensometer tips to the selected area would probably encountered the same problem.

3 Material model outline

First attempt to include strain-amplitude dependency into material model was made in [8] by introducing strain memory surface concept, later modified in [9] introducing non-hardening region. These classic models had been modified to include various phenomena like ratcheting or non-proportional loading, one of the most recent modification by [10] is being tested and seems promising. The author's team has not chosen final form of material model yet, so only brief general outline will follow to remind the main issues.

Generally speaking, there are two basic mechanisms of modeling cyclic plasticity. Isotropic hardening model describes change of the yield surface size (value of yield stress) while kinematic hardening describes shift of the yield surface center point in space of principal stresses (determines non-linear hardening path). Including strain-amplitude dependent material behavior into material model usually needs combination of both mechanisms mentioned above. Using von-Mises yield criterion, the plasticity condition can be described by equation

$$f = \sqrt{\frac{2}{3}(\underline{S} - \underline{\alpha}):(\underline{S} - \underline{\alpha})} - \sigma_y = 0 \tag{6}$$

where $\underline{S}$ is deviatoric part of the stress tensor, $\underline{\alpha}$ is deviatoric backstress tensor (responsible for kinematic hardening mechanism), σ_y is the current yield stress (responsible for isotropic hardening mechanism).

3.1 Isotropic hardening part

As it was mentioned above, material shows increase of yield stress depending on strain amplitude applied, but after few cycles, the yield stress value stays constant for the constant strain amplitude applied (see figures 6 and 7). So in general, actual yield stress σ_y from equation (1) must depend on applied load level function, let's call it L. This dependency should have fast saturation to the target value, so there is no significant cumulative variable dependency function, let's call it C.

3.2 Kinematic hardening part

Material hardening on the high amplitude level loading seems to be reflected in non-linear hardening path only as number of cycles increases (see figure 6). So in general, this phenomenon clearly has to be modelled using kinematic hardening mechanism and backstress $\underline{\alpha}$ must be function of applied load level L and cumulative variable C.

So the final plasticity condition equation should be in a form

$$f = \sqrt{\frac{2}{3}\left(\underline{S} - \underline{\alpha}(L,C)\right):\left(\underline{S} - \underline{\alpha}(L,C)\right)} - \sigma_y(L) = 0 \quad (2)$$

Specification of function L and C will be the subject of a future article.

4 Conclusion

Cyclic loading response of 08Ch18N10T austenitic stainless steel shows strain-amplitude dependent cyclic hardening and significant non-Masing behaviour. Analysis of cyclic hardening mechanisms shows change of yield stress value dependent on applied load level in first few cycles, but no dependency during the rest of fatigue life. Cyclic hardening is then coupled with change in non-linear hardening part only and should be modelled using kinematic hardening mechanism only. Some consequences of strain-amplitude dependent cyclic hardening on testing are mentioned along with proposed solution. Some impacts on material modelling are reminded.

Acknowledgements

The authors would like to acknowledge support from the Technology Agency of the Czech Republic, grant No. TA04020806, and support from the Grant Agency of the Czech Technical University in Prague, grant No. SGS15/187/OHK2/3T/12.

References

[1] Ye D 2006 The low-cycle fatigue, deformation and final fracture behavior of an austenitic stainless steel *Mat. Scienece and Engineering* 415 p 104-17
[2] Džugan J et al 2016 Identification of ductile damage parameters for pressure vessel steel *Nucl. Eng. Des.(2015),* http://dx.doi.org/10.1016/j.nucengdes.2015.12.014
[3] Busby J T 2005 The relationship between hardness and yield stress in irradiated austenitic and ferritic steels *J. Nuclear Mat.* 336 p 267–78
[4] American Society for Testing and Materials 2004 Standard practice for strain-controlled fatigue testing, E606-04 *ASM International: The Materials Information Society*
[5] Coffin L F 1953 A study of the effects of cyclic thermal stresses on a ductile metal *ASME 1953;53-A76:931–50*
[6] Manson S S 1954 Behavior of materials under conditions of thermal stress. *National advisory committee for aeronautics report 1170, Cleveland*
[7] Miner M A 1945 Cumulative Damage In Fatigue *J. Applied Mechanics* 12 p 159–64
[8] Chaboche J L, Dang-Van K and Cordier G 1979 Modelization of strain memory effect on the cyclic hardening of 316 stainless steel *Transactions of the 5th International Conference on Structural Mechanics in Reactor Technology, Berlin, No. Div. L in 11/*3 p 1-10
[9] Ohno N 1982 A constitutive model of cyclic plasticity with a nonhardening strain region *J. Appl. Mech.* 49 p 721-727
[10] Khutia N, Dey P P, Sivaprasad S and Tarafder S 2014. Development of new cyclic plasticity model for 304LN stainless steel through simulation and experimental investigation *Mech. Mater.* 78 p 85–101

Sintering and heat treatment of $Al_{15}Ti_5Co_{35}Ni_{25}Fe_{20}$ high-entropy alloy

K Górecki[1,*], J Zýka[2], J Malek[2,3], J Horvát[3], J Čapek[4], P Bała[1,5]

[1]AGH University of Science and Technology, Faculty of Metals Engineering and Industrial Computer Science, Department of Physical & Powder Metallurgy, Krakow, Poland, EU,*
[2]UJP PRAHA a.s., 15610 Prague 5, Zbraslav, Czech Republic, EU
[3]Czech Technical University in Prague, Faculty of Mechanical Engineering, Department of Materials Engineering, 12000 Prague 2, Czech Republic, EU
[4]Czech Technical University in Prague, Faculty of Nuclear Sciences and Physical Engineering, Prague 2 12000, Czech Republic, EU
[5]AGH University of Science and Technology, Academic Centre for Materials and Nanotechnology, Krakow, Poland, EU

[1]E-mail: kgorecki@agh.edu.pl

Abstract. The new $Al_{15}Ti_5Co_{35}Ni_{25}Fe_{20}$ high entropy alloy was obtained by the powder metallurgy from the elemental powders (-325 mesh). The obtained alloy should be characterized by the dual BCC + FCC phase microstructure according to the valence electron concentration calculations (VEC=7.9). The main aim of the research was to determine the optimal conditions for sintering and heat treatment of $Al_{15}Ti_5Co_{35}Ni_{25}Fe_{20}$ high entropy alloy. The alloy was sintered for 40 minutes at 600 °C for the preliminary consolidation of the powders and initial diffusion of the material. Then samples were annealed at various temperatures: 700, 800, 900, 1000 °C for 1 hour. Samples after annealing were characterized by the XRD and SEM. For the chosen temperature the annealing time was elongated to 2, 3, 5, 10 and 20 hours. The chemical homogeneity was determined.

1 Introduction

The rapid development of modern technology includes high-entropy alloys (HEAs) which could be used instead of conventional metallic materials, which are mostly composed from one or two principal elements. HEAs are composed of at least five principal elements with concentrations of each between 5-35 at.%. Number of elements in alloys could be explained by the dependence between mixing entropy and enthalpy. It is considered that formation of solid solution phases in HEAs is possible when the mixing entropy is high enough to counterbalance the mixing enthalpy, what is observed when HEAs consist at least five principal elements [1–5]. HEAs usually form simple solid solution phases with FCC and/or BCC structures. The number of obtained phases is much lower than calculated by Gibbs phase rule, which directly results from effects of high entropy [6, 7]. As atoms of each principal element can be located in all lattice positions with equal probability, the multielement lattices are highly disordered. Furthermore as metallic diameters of the elements are different, this intensifies the disorder effect. At least HEAs could be considered like an atomic-scale composite, where interactions between principal elements indicate indirect synergistic effect of all elements. It has been reported that HEAs have very good thermal stability and high-temperature strength [8–10].

As an alternative for mostly used methods of synthesis HEAs such like arc melting [11–13] and mechanical alloying or spark plasma sintering [14–17] investigated in this paper $Al_{15}Ti_5Co_{35}Ni_{25}Fe_{20}$ alloy was synthesized by the use powder metallurgy.

2 Experimental

$Al_{15}Ti_5Co_{35}Ni_{25}Fe_{20}$ alloy (indexes represents atomic percent of elements) have been prepared from the metal base fine powders (~325 mesh; purity 99.9 at. pct.). Metal base powders have been mixed by 10 hours in the Turbula Shaker-Mixer under argon atmosphere. In order to intensify the mixing of powders the metal balls were used (ball to powder ratio 1:12). Mixed powders were pressed hydrostatically under pressure 400 MPa and sintered in vacuum furnace CLASSIC 5015T under pressure 10^{-3} – 10^{-5} Pa (temperature 600 °C for 40 minutes). Sintering parameters have been chosen to avoid liquid phase (aluminum melting point 660 °C) and formation of intermetallic phases. However, these parameters were enough for the preliminary diffusion of elements.

Due to finding proper heat treatment temperature samples obtained after sintering have been annealed at varied temperatures: 700, 800, 900 and 1000 °C for 1 hour. After treatment samples have been investigated by Scanning Electron Microscope JEOL JSM 7600F (back-scattered electrons imaging;15 kV) and by the XRD measurements (X'Pert PRO MPD in Bragg-Brentano geometry using cobalt X-ray tube 45kV, 30 mA with pinhole 4x1 mm2). Analysis of obtained data allows to choose one temperature of annealing (1000 °C) for further investigation where annealing time was elongated to 2, 3, 5, 10 and 20 hours.

3 Results and discussion

Parameter VEC and thermodynamic parameters calculated for obtained samples are presented in Table 1. Parameter VEC (Valence Electron Concentration) corresponds to the phase composition of alloy. For investigated sample calculated VEC is 7.9 what represents dual phase structure of obtained sample – BCC+FCC. ΔH_{mix} allows to predict the formation of solid solutions, calculated value fulfils criteria of the High-Entropy Alloys (-14.58 kJ/mol), even if the value of mixing entropy is high while the ΔH_{mix} is to low or to high the solid solution would not be formed. Rest of parameters δ (ratio between atomic radius of single alloy element to average atomic radius of alloy) and Ω ($\Omega = T_m \Delta S_{mix}/|\Delta H_{mix}|$ where $T_m = \sum c_i (T_m)_i$, $(T_m)_i$ – melting point of each element) combined with ΔH_{mix} indicates formation of solid state. Calculated values fulfil the criteria – δ is 6.01 and Ω is 1.21.

Table 1. Thermodynamic parameters calculated on the basis of chemical composition for the obtained sample and comparison with the values for model High-Entropy Alloys (HEA).

	VEC	ΔH_{mix} [kJ/mol]	δ [%]	Ω
values for model HEAs	BCC≤6.88; 8≤FCC [18]	-15 ÷ 5 [19, 20]	<6.6 [19, 20]	≥1.1 [19, 20]
values calculated for obtained sample	BCC+FCC 7.9	-14.58	6.01	1.21

3.1 XRD analysis

Obtained XRD patterns were presented in Figure 1. Part A of the Figure presents the influence of varied temperatures of annealing on the phase composition while part B presents the influence of the time of heat treatment on the phase composition. In both cases the BCC and FCC phases were investigated. BCC phases were recognized as iron base solid solution (BCC1 – space group number 229) and as CoFe phase (*BCC2* – space group number 221). FCC phase were recognized as nickel base solid solution (space group number 225).

3.1.1 Varied temperatures of annealing. For the sample that was only sintered and have not been annealing, the BCC and FCC phases were observed. However, the patterns assigned as BCC1 and BCC2 phases were combined together. During the heat treatment with the increase of the temperature of annealing the aforementioned patterns became separated. Perhaps the additional patterns (CoFe phase) could be formed by the diffusion of cobalt in to Fe base solid solutions. In the case of FCC phase it should be noted that the intensity of patterns near 50, 60 and 70 °2θ increase with the temperature of annealing. Moreover, for the XRD patterns obtained for 1000 °C the intensity of patterns assigned to FCC phase is much higher than for the BCC phase patterns. It could be assumed that with the increase of temperature the samples are more homogenized. This fact determined choosing temperature 1000 °C for further investigations.

3.1.2 Varied time of thermal treatment. Extend of the time of heat treatment also affects the homogenization of material. For the 20 h of annealing the intensity of FCC patterns is almost two times higher than for the samples annealed by 1 h. Obtained patterns are sharp and well separated. It could be assumed that increase of the time of heat treatment causes more advanced homogenization. As it was aforementioned before in the case of the varied temperatures of annealing also in this case patterns combined of BCC phases separates.

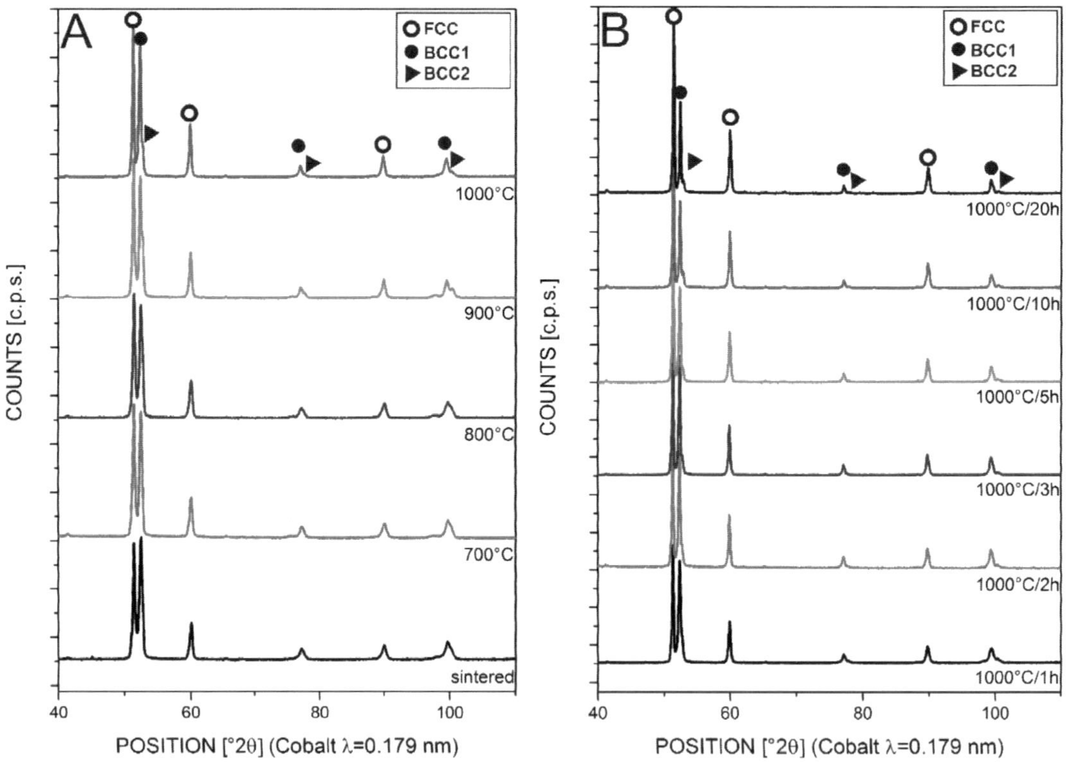

Figure 1. XRD patterns of investigated alloy: A – varied temperatures of annealing, B – varied time of annealing at 1000 °C.

3.2 Scanning Electron Microscopy

Microstructures of samples obtained after heat treatment are presented in Figure 2. At the presented figures the bright and dark areas could be noted, that represents phases described by the XRD measurements. Obtained materials are characterized by the huge porosity. The mechanical pore (MP) and diffusion pore (DP) could be observed. First type of pores is created during the hydrostatic pressing and second type during the diffusion of elements atoms. It should be noted that with the increase of the temperature the amount of darker areas and amount of diffusions pores increase. Both phenomena could be explained by the easier diffusion activated by increase of temperature. Not only temperature influences the diffusion activation, the time of annealing also indicate better chemical homogeneity of sintered materials.

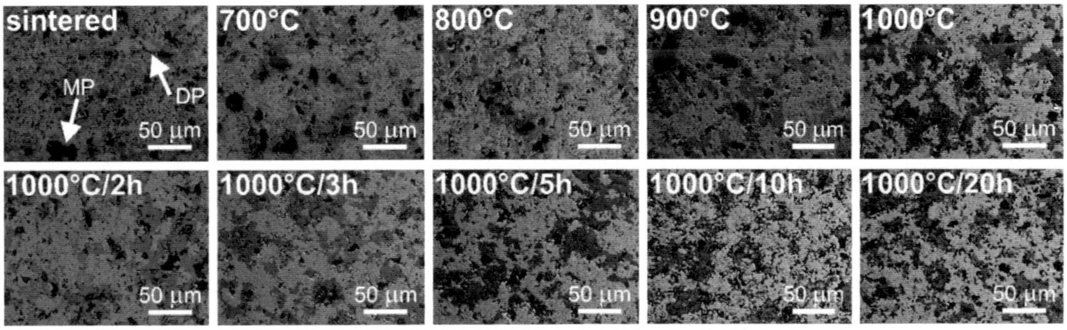

Figure 2. Microstructures of investigated alloy after different heat treatment (SEM:BSE). Top line of figures represents the various temperatures of annealing, time of treatment 1h. Bottom line of figures presents the various time of treatment in the same temperature (1000 °C); MP - mechanical pore, DP - diffusion pore.

The EDS map of elements distribution for sintered material (Figure 3) shows that the obtained material is highly heterogeneous. It should be noted that the areas enriched by nickel are also enriched by aluminium and partially by titanium.

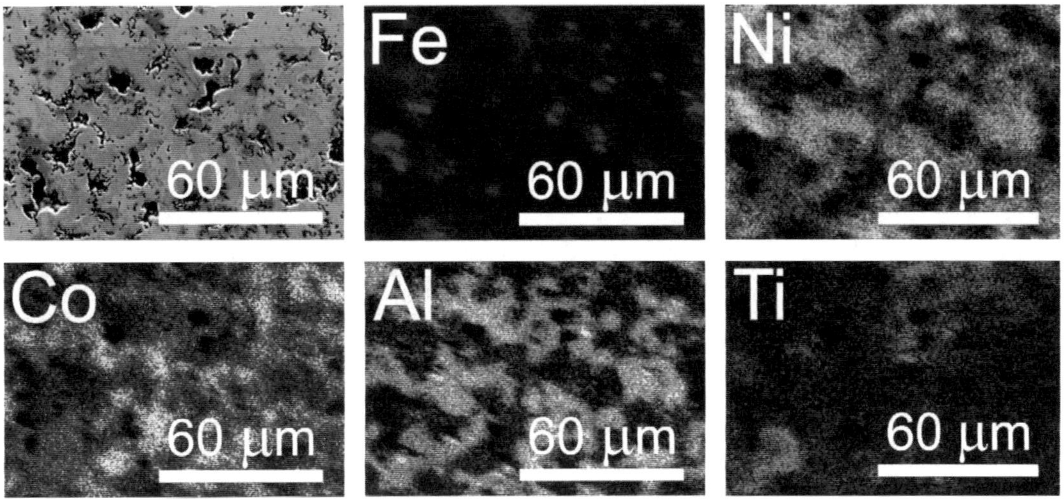

Figure 3. EDS map of elements distribution for sample after sintering.

The EDS map of element distribution obtained for the sample annealed at temperature 1000 °C for 1 hour is presented in Figure 4. Analysis of obtained data allows observing much better homogenization of chemical composition than for sample after only sintering. The brighter area is enriched by iron and cobalt while the darker area is enriched by aluminium, titanium and partially by nickel.

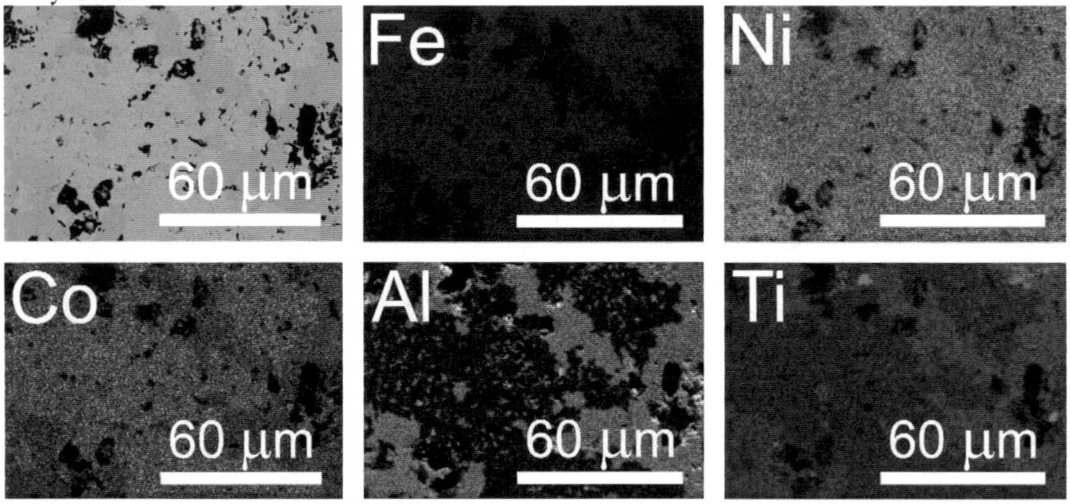

Figure 4. EDS map distribution of elements for sample after sintering and annealing at 1000 °C for 1 hour.

Extend of the time of annealing from 1 hour to 20 hours allows observing the increase of homogeneity of elements distribution, especially titanium and cobalt (Figure 5). Aluminium enriched darker areas while iron enriched brighter areas.

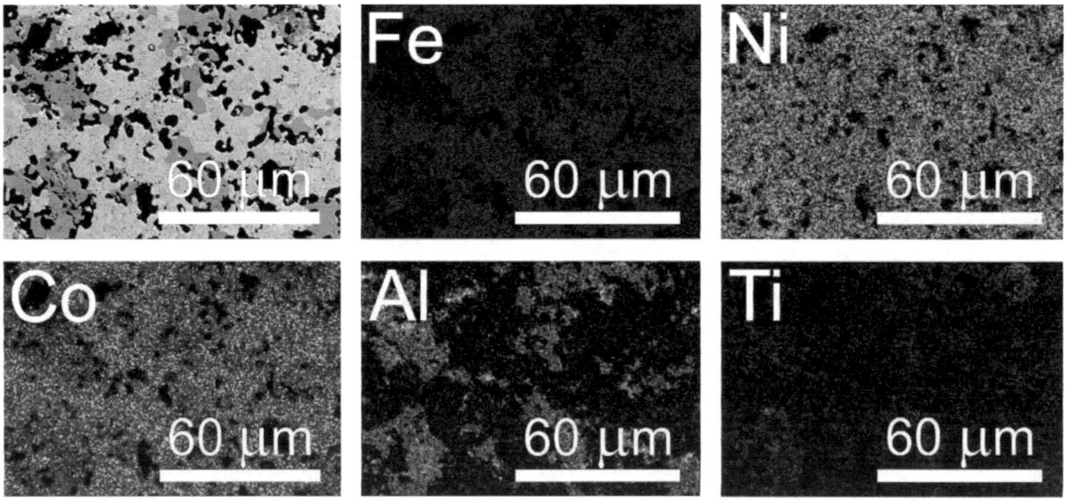

Figure 5. EDS map distribution of elements for sample after sintering and annealing at 1000°C for 20 hours.

On the basis of data obtained by the SEM and XRD it could be said that the darker areas of examined samples are enriched by aluminium and nickel that constitute the FCC phase (nickel-based solid state solution), while brighter areas represents the BCC phase (iron-based solid state solution).

4 Conclusions

According to obtained data and calculated values of thermodynamic parameters it is possible to obtain the High-Entropy Alloys from the elemental powders by the powder metallurgy techniques. Sintering of powders in the temperature 600 °C by 40 minutes it is not enough for fully homogenization of alloy, however in those conditions the diffusion of elements is activated. XRD measurements allow to determine dual phase structure composed from BCC and FCC phases. BCC phase corresponds to the iron-based solid solution and to the FeCo phase while the FCC corresponds to the nickel-based solid solution. XRD data confirms theoretical assumptions. It should be noted that increase of temperature and time of annealing leads to the homogenization of chemical and phase composition – separation of BCC phases with decreasing of BCC2 phase; increasing of FCC phase amount; nickel, titanium and aluminium tend to enrich the darker areas of microstructure while iron and cobalt enrich the brighter areas of microstructure. With the increase of both parameters the amount of diffusion pores increase, what also confirm the homogenization of the material.

Acknowledgements
Research supported by Ministry of Science and Higher Education Republic of Poland (project No. 15.11.110.428)

References

[1] Yeh J W, Chen S K, Lin S J, Gan J Y, Chin T S, Shun T T, Tsau C H and Chang S Y 2004 Nanostructured high-entropy alloys with multiple principal elements: Novel alloy design concepts and outcomes, Adv. Eng. Mater. 6 (2004) 299–303

[2] Ranganathan S 2003 Alloyed pleasures: Multimetallic cocktails, Curr. Sci. 85 (2003) 1404–1406.

[3] Chen T K, Wong M S, Shun T T and Yeh J W 2005 Nanostructured nitride films of multi-element high-entropy alloys by reactive DC sputtering, Surf. Coatings Technol. 200 (2005) 1361–1365

[4] Yeh J W, Lin S, Chin T, Gan J, Chen S, Shun T T, Tsau C H and Chou S Y 2004 Formation of simple crystal structures in Cu-Co-Ni-Cr-Al-Fe-Ti-V alloys with multiprincipal metallic elements, Metall. Mater. Trans. A. 35 (2004) 2533–2536

[5] Hsu C, Yeh J W, Chen S and Shun T 2004 Wear Resistance and High-Temperature Compression Strength of Fcc CuCoNiCrAl0.5Fe Alloy with Boron Addition, Metall. Mater. Trans. A. 35A (2004) 1465–1469.

[6] Cantor B 2014 Multicomponent and High Entropy Alloys, Entropy 16 (2014) 4749–4768.

[7] Cantor B, Chang I T H, Knight P and Vincent A J B 2004 Microstructural development in equiatomic multicomponent alloys, Mater. Sci. Eng. A. 375–377 (2004) 213–218.

[8] Yao M J, Pradeep K G, Tasan C C and Raabe D 2014 A novel, single phase, non-equiatomic FeMnNiCoCr high-entropy alloy with exceptional phase stability and tensile ductility, Scr. Mater. 72–73 (2014) 5–8.

[9] Otto F, Dlouhý A, Somsen C, Bei H, Eggeler G and George E P 2013 The influences of temperature and microstructure on the tensile properties of a CoCrFeMnNi high-entropy alloy, Acta Mater. 61 (2013) 5743–5755.

[10] Senkov O N, Scott J M, Senkova S V, Miracle D and Woodward C F 2011 Microstructure and room temperature properties of a high-entropy TaNbHfZrTi alloy, J. Alloys Compd. 509 (2011) 6043–6048

[11] Gwalani B, Soni V, Choudhuri D, Lee M, Hwang J Y, Nam S J, Ryu H, Hong S H and Banerjee R 2016 Stability of ordered L1 2 and B 2 precipitates in face centered cubic based high entropy alloys - Al 0.3 CoFeCrNi and Al 0.3 CuFeCrNi 2, Scr. Mater. 123 (2016) 130–134.

[12] Ng C, Guo S, Luan J, Wang Q, Lu J, Shi S and Liu C T 2014 Phase stability and tensile properties of Co-free Al0.5CrCuFeNi2 high-entropy alloys, J. Alloys Compd. 584 (2014) 530–537.
[13] Górecki K, Bała P, Cios G, Kozieł T, Stępień M and Wieczerzak K 2016 The influence of cooling rate during crystallization on the effective Partitioning coefficient in High-Entropy Alloys from Al-Ti-Co-Ni-Fe system, Metall. Mater. Trans. A. 47 A (2016) 3257–3262
[14] Mohanty S, Gurao N P and Biswas K 2014 Sinter ageing of equiatomic Al20Co20Cu20Zn20Ni20 high entropy alloy via mechanical alloying, Mater. Sci. Eng. A. 617 (2014) 211–218.
[15] Bo R 2014 Age hardening of AlCrMoNiTi High Entropy Alloy prepared by powder metallurgy, Rare Met. Mater. Eng. 43 (2014) 1286–1290.
[16] Chen W, Fu Z, Fang S, Xiao H and Zhu D 2013 Alloying behavior, microstructure and mechanical properties, J. Mater. 51 (2013) 854–860.
[17] Fang S, Chen W and Fu Z 2014 Microstructure and mechanical properties of twinned mechanical alloying and spark plasma sintering, Mater. Des. 54 (2014) 973–979.
[18] Guo S, Ng C, Lu J and Liu C T 2011 Effect of valence electron concentration on stability of fcc or bcc phase in high entropy alloys, J. Appl. Phys. 109 (2011) 103505-2–103505-5.
[19] Yang X and Zhang Y 2012 Prediction of high-entropy stabilized solid-solution in multi-component alloys, Mater. Chem. Phys. 132 (2012) 233–238.
[20] Miracle D, Miller J, Senkov O, Woodward C, Uchic M and Tiley J 2014 Exploration and development of High Entropy Alloys for structural applications, Entropy 16 (2014) 494–525.

IOP Conf. Series: Materials Science and Engineering **179** (2017) 012028 doi:10.1088/1757-899X/179/1/012028

Mechanical properties and structure of AZ61 magnesium alloy processed by equal channel angular pressing

O Hilšer[1], S Rusz[1], T Tański[2], P Snopiński[2], J Džugan[3] and M Kraus[1]

[1]VSB - Technical University of Ostrava, Ostrava, Czech Republic, EU
[2]Silesian University of Technology, Division of Material Processing Technology , Gliwice, Poland, EU
[3]Comtes FHT a.s., Dobřany, Czech Republic, EU

E-mail: ondrej.hilser@vsb.cz

Abstract. An equal channel angular pressing (ECAP) procedure has been developed to produce a fine-grained AZ61 magnesium alloy. The results show that the microstructure can be effectively refined with increasing equivalent strain during ECAP. For increasing ECAP process efficiency was conventional tool as a helix in the horizontal part of channel built. This fine-grained alloy has an excellent strength accompanied by reasonable good tensile ductility. The success of the development of this ECAP procedure can offer a good opportunity for the development of magnesium alloys with good mechanical properties.

1 Introduction

It is well known that there is relationship between grain size and properties of the materials. The grain Refinement causes a significant increase in strength in bulk metallic materials due to the Hall–Petch relationship which connects the mechanical properties and microstructural features such as the grain size. This relationship illustrates the potential for the mechanical strength increase by decreasing the size of grains and subgrains and this has been main driving force in the expansion of ultrafine grained (UFG) and nanostructured (NS) materials. From the practical point of view, materials with refined structure are most readily manufactured in bulk form by samples subjected to severe plastic deformation (SPD) processes. Over the last few decades, severe plastic deformation techniques have been extensively utilized to fabricate a bulk (NS) materials that exhibit an unusual properties which is very attractive for diverse structural and functional applications [1]. Ultrafine grained metals and its alloys made by severe plastic deformation techniques exhibits in fact excellent mechanical properties which are a result of significant grain refinement connected with dislocation strengthening.

The equal-channel angular pressing method is at present one of the most promising and commonly used of all Severe Plastic Deformation processes. In this technique, a sample with circular, square or rectangular cross section is pressed through a die consisting of two channels. These channels have identical cross-section and intersects at a specific angle of (Φ). The work sample to be processed is prepared to fit tightly the die channel and has to be well lubricated in order to minimalize friction coefficient. The specimen is inserted to the first channel of a die and then pressed through the second channel (with a constant speed) using a plunger. Ideally, severe deformation process occurs by simple shear in the area of intersection of two channels. As the work sample is pressed through this area, new regions of sample are exposed to shearing strain providing a homogeneous shear deformation of the specimen apart from its end regions, which result in strain accumulation and grain/subgrain refinement [1-3, 9].

2 Experimental procedure

2.1 Experimental material

Experiments were conducted on a commercial AZ61 alloy to evaluate the potential for achieving an ultrafine grain size through the use of EX-ECAP two-step processing procedure of hot extrusion plus equal-channel angular pressing.

Tested material was in initial state casted and extruded at 430°C (703 K) to the dimensions 40x40 – 1000 mm (OML Skawina). Extrusion of as-casted Mg alloys has a positive impact on the homogenous grain size distribution [13] and homogenous distribution of the phases Mg_{17} (Al, Zn)$_{12}$ [5]. The chemical composition of the material investigated is shown in **Table 1**.

Table 1. Chemical composition of AZ61 alloy

Weight [%]	Fe	Si	Mn	Ni	Cu	Mg	Zn	Al
AZ61	0.003	0.030	0.230	0.001	0.003	92.210	1.092	6.430

2.2 ECAP process

Samples were ECAPed at 200°C (473 K) on the hydraulic press DP 2000. Deformation route B_c (rotation about longitudinal axis by 90° in the same direction) was applied.

The ECAP process is based on extrusion of the sample through the tool with an internal L-shaped channel, without any change of cross-section of the sample [1]. The sample is inserted from above into the vertical channel and then extruded through the tool. This operation is then repeated in order to achieve the required degree of deformation of the material leading to a refinement of the structure. It is possible to use various types of deformation route changes for the ECAP process [1-3].

Equal channel angular pressing that was carried out using die with modifying tool geometry. ECAP + TE, i.e. Equal-Channel Angular Pressing combined with Twist Extrusion) combination was used for process efficiency increase. The tool contains an in-built helix with spiral angle of 30°. Advantage of the in-built helix consists in creation of counter-pressure, which enables an increase of the deformation degree and thus achievement of substantial grain refinement at smaller number of passes through the ECAP tool in comparison with the previously used tool geometries [3, 6]. Material of the ECAP tool is HOTVAR, that exhibits better strength stability during forming processes at higher temperatures [4].

2.3 Investigation procedure

Detailed studies of plastically deformed grains, distribution and types of phases have been done by transmission electron microscopy (TEM) using Tecnai G2 F20 (200 kV) microscope fitted with high-angle annular dark field scanning transmission electron microscopy detector (HAADF-STEM) and energy dispersive X-ray (EDX) EDAX microanalysis. TEM specimens have been prepared by twinjet electro-polishing (Tenupol-5) using a solution of 10.6 g lithium chloride LiCl, 22.32 g magnesium perchlorate $Mg(ClO_4)_2$, 1000 ml methanol, and 200 ml 2-butoxy-ethanol at - 40°C and 80 V. EBSD data was collected using a high resolution FEI FEGSEM Quanta 3D equipped with TSL software.

From the reason of small sample dimension the Micro-Tensile Test Technique was developed in [10-12], that was applied for current material characterization.

3 Investigation results

3.1 Microstructure

TEM investigations of samples after 1 and 3 ECAP passes were performed in order to examine microstructural features, especially the change of grain size and distribution but also size and chemical composition of precipitates. **Figure 1** presents a set of bright field (BF) images taken at different magnifications and corresponding selected area diffraction patterns (SADP) of sample after 1 pass of ECAP. High density of dislocations, twins and formation of shear bands can be observed in BF images. The grains possess slightly elongated shape and mean size of about 5.4 μm. SADP was indexed to Mg with a [123] zone axis exhibiting a diffused character of reflects associated with a strongly deformed structure. Also an additional reflects can be recognized on SADP image, most probably originated from precipitates. BF and dark field (DF) microstructures and corresponding SADP of sample after 3 passes of ECAP are presented in **Figure 2.** In this case more refinement microstructure can be seen as is demonstrated on DF image performed from reflect of (110) plane of Mg structure. The mean grain size is about 1.9 μm, what indicates that two additional passes of ECAP caused almost 2.5 times reduction of grain size. Higher magnification image shows individual grain of magnesium matrix with [001] zone axis having equiaxed shape indicating that the ECAP process causes loss of texture of grains before present in the samples after 1 pass of ECAP. Although, reduction of grain size was observed after 3 passes of ECAP, shape of the grains and lower dislocation density may indicate that the dynamic recrystallization occurred, since the ECAP process was conducted at 200°C [7-8].

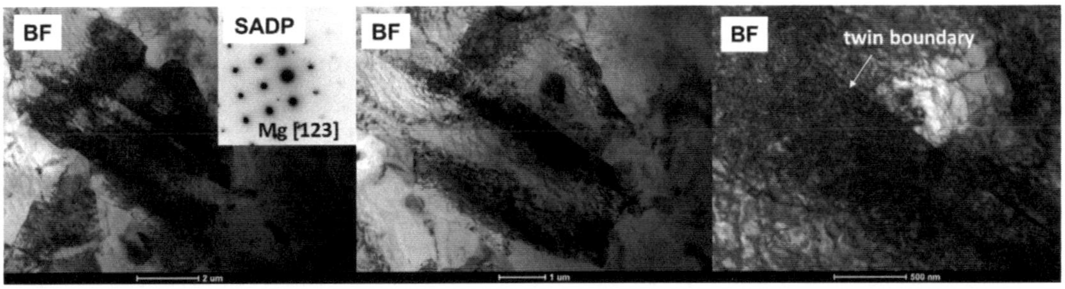

Figure 1. Set of bright field (BF) images and corresponding selected area diffraction pattern (SADP) of sample after 1 pass of ECAP

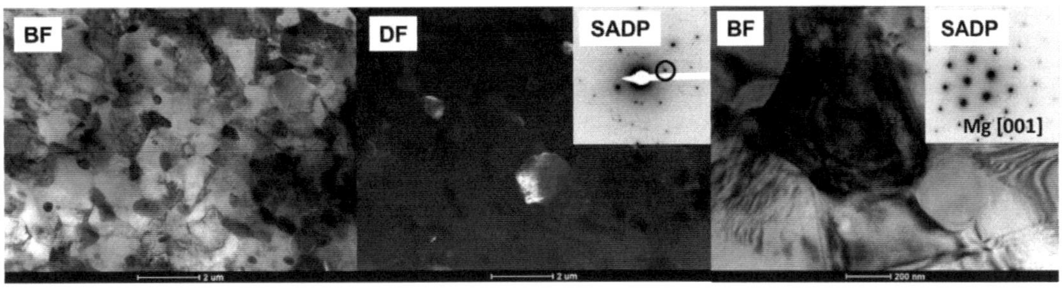

Figure 2. BF and dark field (DF) microstructures and corresponding SADPs of sample after 3 passes of ECAP

Figure 3 presents BF images taken at different magnifications and corresponding SADP of samples after 1 (upper) and 3 (lower) passes, respectively. One can see that in case of sample after

1 pass of ECAP, irregular precipitate of $Mg_{17}Al_{12}$ type with the [101] zone axis is located inside grain with high density of dislocations and twins. This type of precipitates is mainly present in AZ61 magnesium alloys. Also it can be assumed that these precipitates contain some amount of Zn. In the case of sample after 3 passes through ECAP performed at higher temperature the $Mg_{17}Al_{12}$ precipitates of size of about 200 nm possess rounded shape and mainly are located at the grain boundaries strongly affecting on adjacent area.

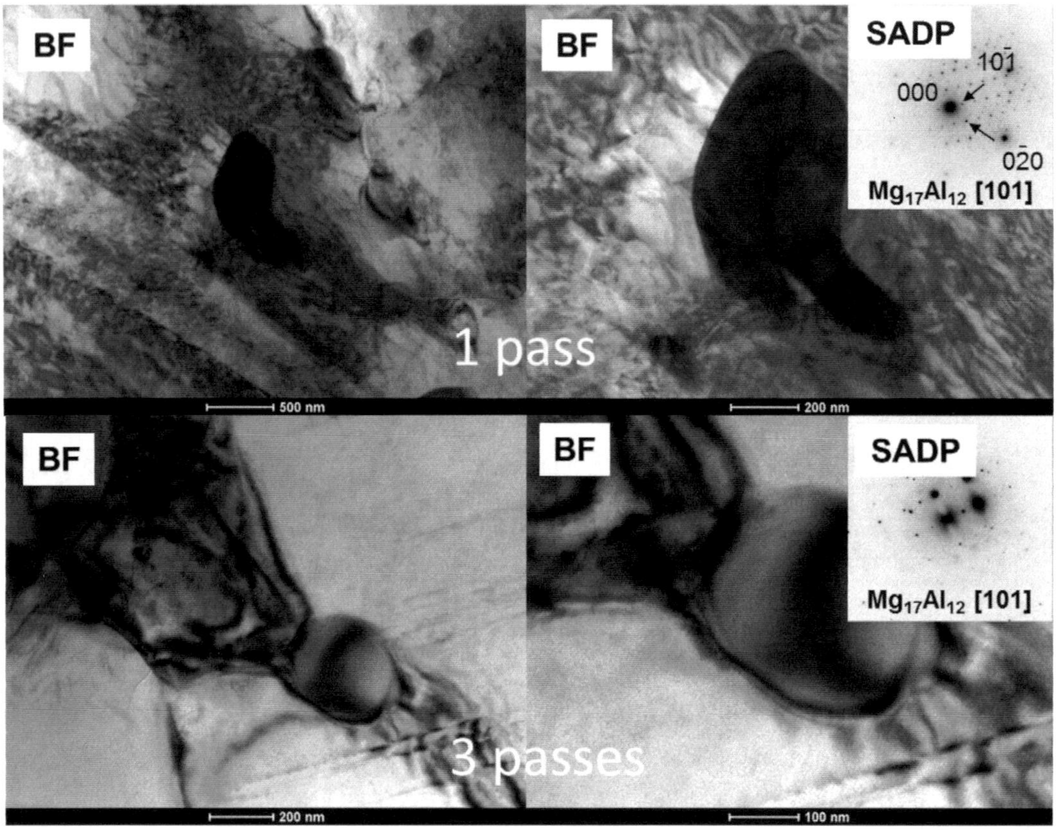

Figure 3. BF images taken at different magnifications and corresponding SADPs of samples after 1 (upper) and 3 (lower) passes, respectively

Figure 4 present BF images at two different magnifications performed in two beam condition for Mg [761] zone axis in sample after 3 passes of ECAP. There are several precipitates of size of about 200 nm at the grain boundaries but no precipitates inside the grain. In the higher magnification image the stacking faults contrast (or dislocation pileups) is visible between two neighboring $Mg_{17}Al_{12}$ precipitates. Therefore, it can conclude that location of small precipitates on grain boundaries strongly effecting these areas due to blocking movement of dislocations during plastic deformation. Extremely fine grains with average grain size of 0.5 μm are formed by dynamic recrystallization being attributable to dislocation pileups that occur in the vicinity of dynamically precipitated γ-$Mg_{17}Al_{12}$ phase particles of less than 100 nm diameter. Such mechanism can be successfully adapted to our case even the grain and particle sizes are small different.

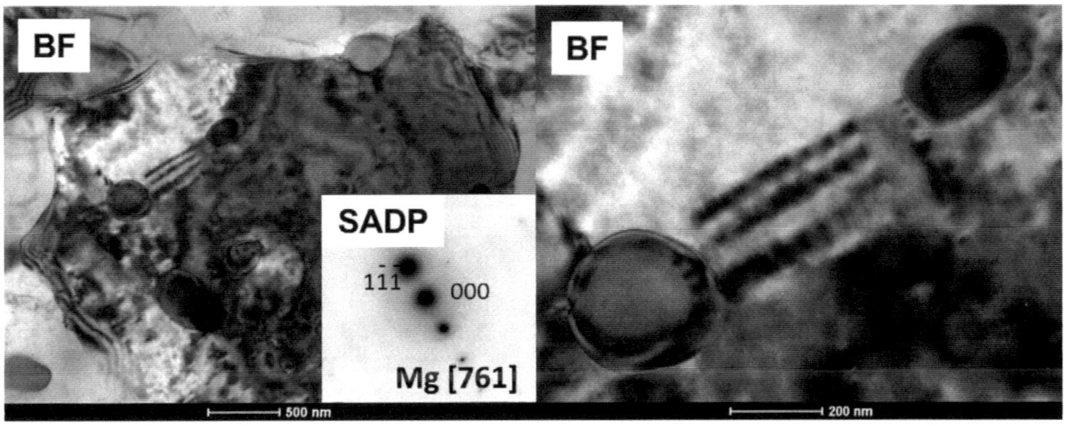

Figure 4. BF images at two different magnifications performed in two beam condition for Mg [761] zone axis in sample after 3 passes of ECAP

3.2 Mechanical properties

Mechanical properties characterization of the material investigated was carried out with the use of Mini-Tensile Tests (M-TT). M-TT specimens geometry and testing procedure was verified in [10-12]. Testing was performed under quasi-static loading conditions at room temperature. Strain was measured with the use of digital image correlation system ARAMIS by GOM. Four specimens' batches were tested: initial state – MPO, state after one pass – MP1, state after two passes – MP2 and the state after three passes MP3. Specimens were extracted in the longitudinal direction from ECAPed bars. Three specimens per condition were tested. Youngs modulus **E**, offset yield stress **Rp0,2**, ultimate tensile strength **Rm**, uniform elongation at maximum force **Ag** and cross section reduction **CR** were evaluated. Examples of obtained tensile curves together with the specimens geometry can be found in **figure 5**. Summarized averaged resulting values of tensile test can be seen in **table 2**. Results clearly show strength parameters increase after the first pass by about 20% together with uniform elongation increase by almost 40%. However subsequent passes do not have any significant influence on the tensile properties that remain almost the same, except the yield stress that is gradually decreasing.

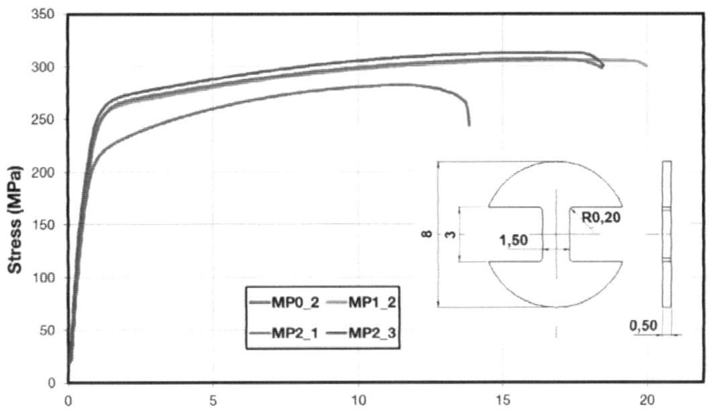

Figure 5. Obtained tensile curves of AZ61 processed by ECAP method

Table 2. Mechanical properties of AZ61 measured by M-TT

Speciment	No. of passes	E [GPa]	$Rp_{0,2}$ [MPa]	Rm [MPa]	Ag [%]	CR [%]
MP0	Initial state	35,8	205,3	285,8	11,3	29,6
MP1	1st pass	35,8	230,0	309,4	15,3	25,3
MP2	2nd pass	38,0	217,5	308,7	15,4	24,7
MP3	3rd pass	37,6	213,7	301,2	16,4	24,7

4 Conclusion

The main aim of the experiments is a refinement of the structure of magnesium alloy AZ61 using the minimum number of ECAP passes with special tools. To increase the degree of distortion, thereby achieving the desired structure is an important factor in appropriate modification tool geometry. Geometric adjustment of instruments is particularly evident in the new instrument ECAP helix is located 30° in the horizontal part of the channel, which shows an overall increase in efficiency of the SPD process.

Structure investigations demonstrate that ECAP process allows obtaining highly deformed grains that contains high density of shearing bands. Further ECAP process also causes subsequent grain refinement. The TEM analysis shows that there are presented in the structure of the severely deformed alloy two different areas, those defects free and with high dislocation density. It was also confirmed the presence of main hardening phase that occurs after artificial ageing. It was also found that the grain refinement occurs on the subgrain level.

Results from M-TT shows an highest increase of strength parameters ($Rp_{0,2}$ and Rm) after first pass, but after further passes decrease $Rp_{0,2}$ and Rm. It is evident that during ECAP process softening process occurs. This hypothesis was confirmed by Ag and CR parameters measurements and microstructure investigations. The Young's modulus of AZ61 studied is introduced in Table 2. E value seems by very low for typical magnesium alloy (prox. 42GPa). In the initial state (casted and hot extruded) is E = 36GPa, lower value of Young's modulus may be caused by hot extrusion after casting.

It was confirmed a positive impact of ECAP tool modification on the grain refinement process. During ECAP occurred due to high degree of deformation softening (polygonization and dynamic recrystallization).

Acknowledgements

International Visegrad Fund Scholarship no. 51600916 granted for period between September 2016 and June 2017. The research works was sponsored Research and development of productions and materials systems SP2016/73. Project "Creation of an international scientific team and incorporation to scientific networks in the area of nanotechnology and unconventional forming material. CZ.1.07/2.3.00/20.0038", which is financed by the European Social Fund and state budget of the Czech Republic.

References

[1] Valiev R Z and Langdon T J 2006 Principles of equal-channel Angular pressing as a processing tool for grain refinement. *Prog. in Mater. Sci.* **51** pp 881 – 981

[2] Rusz S, Tylsar S, Kedron J, Dutkiewicz J and Donic T 2010 Enhancement of efficiency of SPD proces by application of new geometry of ECAP tool. *Act. Met. Slov.* **16** pp 229 – 236

[3] Rusz S, Cizek L, Salajka M, Tylsar S, Kedron J, Michenka V, Donic T, Hadasik E and Klos M 2014 Ultrafine grain Refinement of AlMn1Cu and AZ31 alloys by SPD process. *Arch. of Met. and Mater.* **59** pp 359 – 364

[4] Hruby J, Rentka J, Schindlerova V, Krejci L and Sevcikova X 2013 Possibilities of Prediction of service life of forming tools. *Manuf. tech.* **13** pp 178 – 181

[5] Cizek L, Kocich R, Greger M, Prazmowski M and Tanski T 2006 Study of plastic deformation of magnesium alloys with graduate aluminium content. *Act. Met. Slov.* **12** pp 490 – 496

[6] Rusz S, Cizek L, Tylsar S, Kedron J, Salajka M, Dutkiewicz J, Hadasik E and Klos M 2012 Structure of AZ31 magnesium alloys after ECAP processing. *Jour. of Tre. in the Dev. of Mach. and Assoc. Tech.* **16** pp 51 – 54

[7] Al-Samman T and Gottstein G 2008 Dynamic recrystallization during high temperature deformation of magnesium. *Mater. Sci. and Eng.* A **490**, pp 411 – 420

[8] Qiang CH, Dayu S., Chuankai H, Zude Z and Baoguo Y 2012 Grain Refinement in an-cast AZ61 magnesium alloy processed by multi – axial forging under the multitemperature processing procedure. *Mater. Sci. and Eng.* A **541**, pp 98 – 104

[9] Snopinski P, Tanski T, Hilser O, Matysiak W, Wisniowski M, Krzeminski L and Tillova E 2015 Effect of Equal channel Angular pressing combined with heat Treatment on structure and properties of AlMg3 aluminium alloy. *Jour. of Ach. in Mater. and Man. Eng.* **73** pp 36 – 44

[10] Dzugan J, Konopik P and Prochazka R 2014 Micro-tensile test technique development and application to Mechanical property determination. Int. Symp. on Small Spec. Test Tech. (Houston) (ASTM international) pp 12 – 30

[11] Konopik P, Dzugan J and Prochazka R 2013Determination of fracture toughness and tensile properties of structural steels by small punch test and micro – tensile test. METAL 2013: 22nd Int. Conf. on Metall. and Mater. (Brno) (Ostrava: TANGER) p 2098

[12] Konopik P, Dzugan J and Prochazka R 2014 Dynamic Tensile and Micro-Tensile Testing Using DIC Method. METAL 2014: 23rd Int. Conf. on Metall. and Mater. (Brno) (Ostrava: TANGER) p 2125

[13] Matsubara K, Miyahara Y, Horita Z and Langdon T J 2003 Developing superplasticity in a magnesium alloy through a combination of Extrusion and ECAP. *Act. Mater.***51** pp. 3073 – 3084

IOP Conf. Series: Materials Science and Engineering **179** (2017) 012029 doi:10.1088/1757-899X/179/1/012029

The effect of the material surface oxidation on laser light absorption

M Hruška, J Tesař, M Vostřák, E Smazalová

University of West Bohemia, Pilsen, Czech Republic, EU

maslej@ntc.zcu.cz, tesar@ntc.zcu.cz, mvostrak@ntc.zcu.cz, smazal@ntc.zcu.cz

Abstract. This study examines the influence of surface condition on its interaction with a laser beam. The relation was tested on milled C45 steel samples with a different width of created oxide layer. The milled surface with roughness 1.6 was modified by the creation of oxide layer. The oxide layers were created by solid state pulsed laser with scanning optics. The modified samples were processed by high power diode laser with a wide spread beam. The surface temperature was measured by the thermal camera during this process. The surface hardness and hardness depth profile were measured for all modified samples and the emissivity values were measured. The effect of pretreatment on hardness and microstructure will be discussed. The goal of this paper is to describe the relation between the parameters of created oxide layer and its effect on laser beam absorption.

1 Introduction

Laser hardening [1,2] is the modern technology of improving surface properties of the metallic materials. The material is rapidly heated to a high temperature by guiding the laser beam across the workpiece surface. Short interaction period of the laser beam and the processed surface cause, that the absorbed energy is concentrated only at the irradiated surface area, whereas the core remains cold. The rapid removal of heat from the surface layer to the colder unaffected bulk of a processed part allows the formation of a required microstructure. The used laser hardening technology is explained in [3].

The depth of hardened layer is dependent on absorbed energy. Dependence of hardened depth on various coatings was described in [4]. The relation between measured surface temperature and laser absorption was emphasized in this paper. An infrared camera was used in [5] for process characterization, control, and temperature field monitoring. ThermaCAM SC3000 was applied to temperature fields measurements during hardening of a cylinder in [6]. The view of the camera was at an angle of 30° with respect to the laser beam. The emissivity was set to the value of 0.88, corresponding to the applied surface coating, which was used to improve absorptivity. The same camera but at an angle of 53° was used in [7]. The emissivity of the sample was estimated during the heating with the propane torch in order to see how emissivity changes with temperature.

It is obvious that laser hardening is very closely connected with the surface condition because it can radically influence the absorption of a laser radiation. The absorption of the laser beam can be changed by modification of the optical properties of the irradiated sample surface. The modification can be achieved by pulsed laser processing. This method allows controlled growth of oxide layer. This phenomenon is described in detail in [8].

The incident, reflected, absorbed, and transmitted fluxes are connected together by energy conservation law. From this fact can be deduced that the sum of optical properties reflectivity (ρ), absorptivity (α), and transmissivity (τ) equals one [9]:

$$\rho + \alpha + \tau = 1 \quad (1)$$

For an opaque object without transmission ($\tau = 0$) the relation (1) simplifies

$$\rho + \alpha = 1 \quad (2)$$

Kirchhoff's law provides a relation between the absorption and emission processes and thus between emissivity and absorptivity

$$\varepsilon = \alpha \quad (3)$$

These relations are valid for similar spectral or total spectral condition. In this case, there are three sets of optical properties (three different wavelengths) for that Kirchhoff's law, is valid. The three wavelengths are the wavelength of engraving laser 1064 nm, the wavelength of hardening laser 808 nm, and the measuring range of infrared cameras 7.5 – 14 μm, eventually 1 μm. This means that, for example, the emissivity of the surface in the range of infrared camera ε_{IR} has no direct connection with absorptivity on the wavelength of hardening laser α_{HPDD}.

2 Experimental procedure

2.1 Surface modification process and absorptivity measurement

Tests were performed on a sample from milled steel C45 with roughness 1.6 and with a dimension of 100 x 100 x 25 mm. Oxide layers of square dimension 7 x 7 mm were created by marking laser SPI-G3-SP-20P with MOPA arrangement according to Figure 1. Parameters, to heating but not remelting the surface, were chosen. The scanning speed of the laser beam and the laser power were chosen common for all samples. The power was set to 17.25 W and the value of 800 mm/s was selected for scanning velocity. The depth of the oxide layer depends on energy E_v brought into the treated area. The quantum of the input energy was controlled by hatch density and allows operated growth of oxide layer. It means that it is controlled by the value of l_s distance between laser crossings. The manner of heating the treated surface layer by the laser is featured in Figure 2. 24 different oxide layers were made overall. Absorptivities of these 24 layers were measured in the ultraviolet (UV), visual (VIS), near-infrared (NIR) and mid-infrared (MIR) region. The method of measurement of spectral normal hemispherical reflectivity at room temperature was applied. FTIR spectrometer Nicolet 6700 and UV/Vis spectrophotometer Specord 210 BU were used for absorptivity measurement.

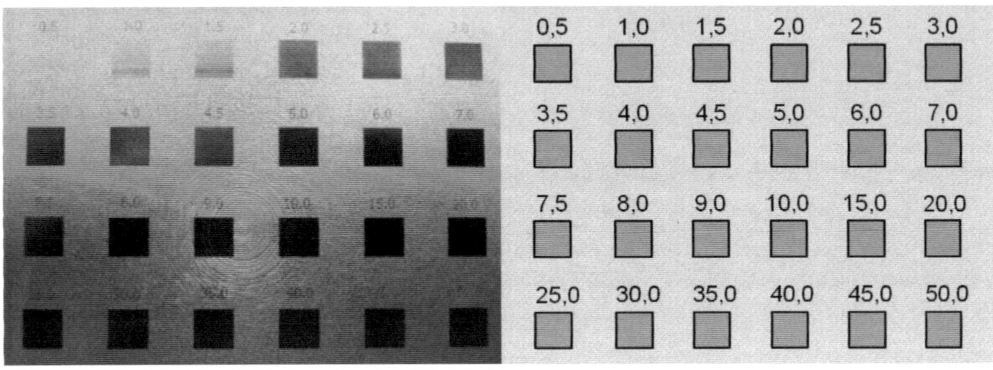

Figure 1. Sample after treating by marking laser (on the left), scheme of created oxide layers with used input energy values (on the right).

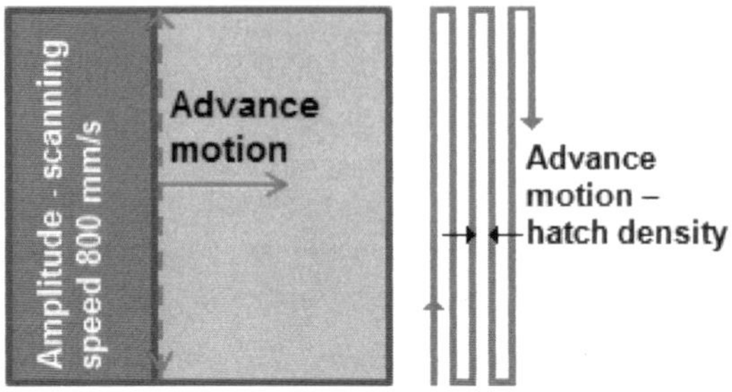

Figure 2. Technology of creating oxide layer by laser beam.

2.2 Laser hardening process

The sample with treated squares undergoes consequently laser hardening with continuous HPDD (High Power Direct Diode) laser Coherent Highlight ISL – 4000L. The laser head is set on industrial manipulator Fanuc M710iC. The laser emits at wavelength 808 nm and its maximal power is 4.3 kW. The laser spot has a rectangular shape with dimension 12 x 6 mm, Gaussian energetic distribution in the 6 mm long side direction and uniform energetic distribution in the 12 mm long side direction. The laser beam with a power of 2 kW and traversing speed of 1.2 m/min was applied for processing of treated sample. The process was observed by three IR cameras (Figure 3).

The thermo-vision MSC640 was mounted on laser head to move with a laser beam. The camera measures emitted infrared radiation on the wavelengths around 1 μm (780 – 1080 nm) and had set temperature measurement range 673 – 1530°C. The thermo-visions Flir A615 and Variocam HD were set on tripods and these measures in the wavelength range 7.5 – 14 μm. The A615 camera has set temperature measurement range 100 – 650°C and the Variocam used 250 – 2000°C. The thermo-cameras measured with recording frequency 10 Hz. Measured values from the MSC640 were used for the evaluation, values from the other cameras have rather an informational character due to high emissivity dependence.

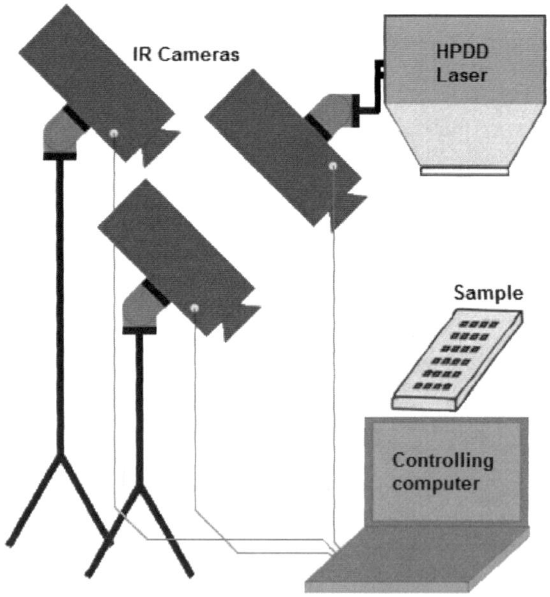

Figure 3. Experimental arrangement.

3 Results and discussion

3.1 Temperature measurement and post-processing

The temperature was measured directly in laser spot during the hardening process. Temperature values were evaluated in LenaSense LumaSpec RT software the values were got by analyzing value at one point placed in the middle of the laser spot and the comparable value was obtained by averaging measured values from a whole laser spot area. The measurement by the MSC640 is significantly less dependent on the emissivity of the measured surface then using LWIR (Long Wave Infra Red) cameras because it measures infrared radiation on the wavelengths around 1 μm. The measurement accuracy of the thermo-vision is ± 0.5 % of measured value.

Table 1. Maximal temperatures – T_{max} and values obtained by averaging – T_{av} in relation to brought energy E_v (width of oxide layer), T_{max}, T_{av} (Figure 4) for non-treated surface flow along 1150–1200°C

E_v [J/mm^2]	0.5	1	1.5	2	2.5	3	3.5	4	4.5	5	6.1	7
T_{max} [°C]	1058	1097	1139	1206	1214	1256	1231	1238	1252	1267	1270	1307
T_{av} [°C]	1034	1076	1118	1191	1199	1253	1216	1229	1235	1260	1255	1305
E_v [J/mm^2]	7.5	8	9	10	15	20	25	30	35	40	45	50
T_{max} [°C]	1238	1303	1306	1327	1327	1389	1471	1442	1458	1451	1565	1643
T_{av} [°C]	1220	1274	1290	1292	1323	1376	1395	1405	1425	1428	1465	1583

The measured surface temperatures are shown in Table 1. One crossing of the sample by the laser beam is noted in Figure 4. Six variants of oxide layers with different width were overridden by the laser beam during this one crossing. The temperatures displayed in Figure 4 were measured for six different oxide layers created by laser heating with brought energy E_v from 25 to 50 J/mm^2. It is evident from Figure 4, that temperature measured on basic non-treated material flow maximally to 1200°C, the temperatures measured on treated areas flow along 1450 – 1650°C. Though the temperatures are above the melting point, the surface is not melted. The phenomenon can be caused by a measurement error of thermo-vision.

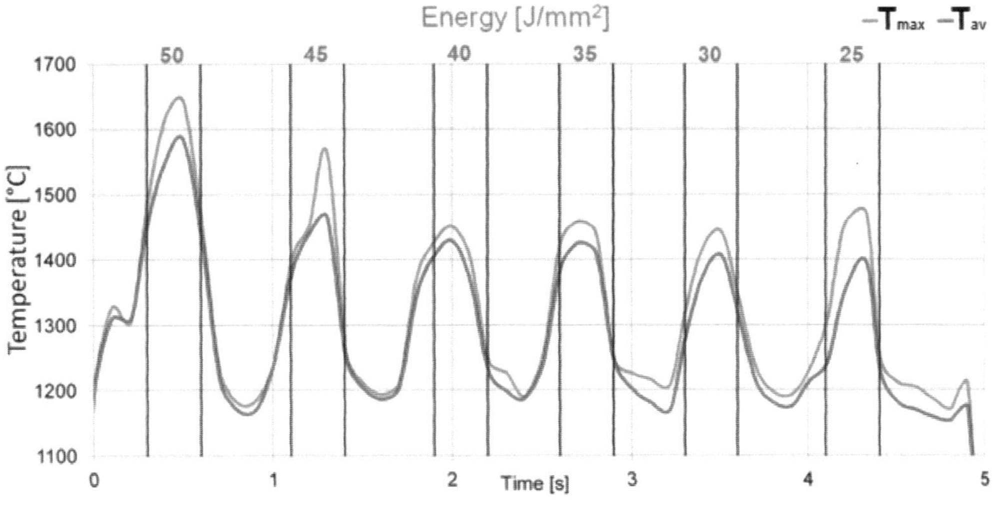

Figure 4. Compare – Temperature value analyzed from one point in the middle of the laser spot (blue line), Temperature value got by averaging measured values from a whole laser spot area (red line).

3.2. Microstructure and surface properties

Surface hardness and hardness depth profile were measured. The results in Tables 1, 2 implicate the relation between measured surface temperature and laser absorption that affects properties like surface hardness and hardness depth. Higher surface temperature means higher laser absorption and higher laser absorption results in higher hardness depth. The surface temperature for laser hardening process has to exceed value 1050°C for C45 material to cause microstructure transformation and to assure the creation of required martensitic structure. Surface hardness decreases with the further gradual increase of surface temperature. It is caused by a decline in heat dissipation from the surface layer.

Table 2 Hardness measured 150 μm under the surface H_{150} and hardness depth profile H_D, for untreated surface H_{150UN}=595 HV1, H_{DUN}=400 μm

E_v [J/mm^2]	0.5	1	1.5	2	2.5	3	3.5	4	4.5	5	6.1	7
H_{150} [HV1]	522	590	510	607	710	721	691	695	559	587	659	623
H_D[μm]	300	350	350	500	400	400	500	500	500	600	550	600
E_v [J/mm^2]	7.5	8	9	10	15	20	25	30	35	40	45	50
H_{150} [HV1]	710	678	649	620	706	637	710	713	675	702	774	683
H_D[μm]	550	600	650	600	500	500	550	500	650	650	600	700

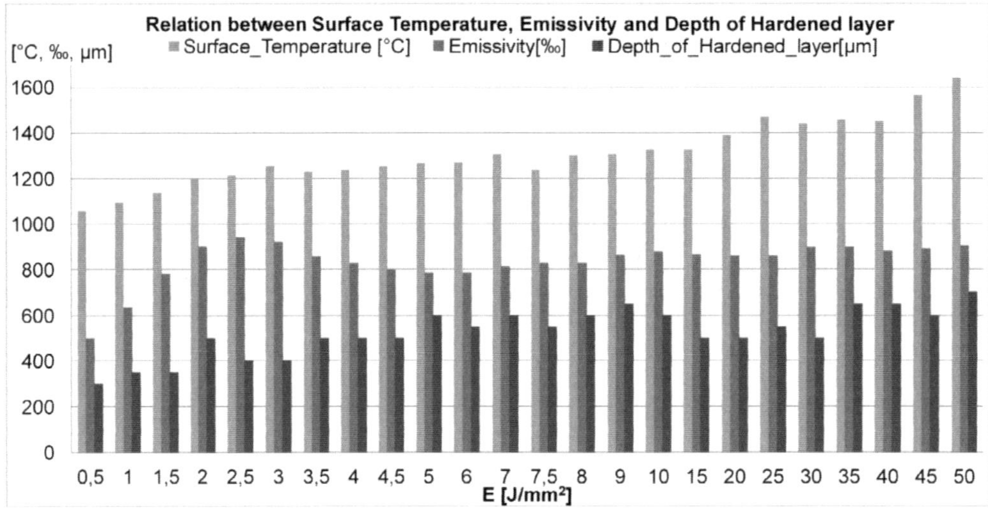

Figure 5. Compare of measured Hardness, Depth of hardened layer and Surface Temperature.

The achieved depth of hardened layer depends on the surface emissivity. Notice that growing surface temperature results in increasing depth of the hardened layer of the irradiated surface. This upward trend can be seen in figure 5. The emissivity displayed in figure 5 was measured for 808 nm wavelength. The difference in the depth of hardened layer between the untreated surface and the treated area with the highest depth is around 300 um. This result proves that the emissivity was seriously increased. The surface hardness of treated area was increased too.

This paper proves that thermal surface modification can seriously change absorptivity, resulting in the influence of the created hardened layer parameters.

4 Conclusion

The paper is focused on characterization of the relation between the surface absorptivity and its effect on the absorption of laser radiation during the hardening process. The absorptivity influences the surface temperature, measured during the process, and consequently parameters of formed hardened layer. The surface absorptivity can be changed by the creation of oxide layer. The width of the oxide layer depends on "brought energy". "Brought energy" is input energy brought into the processed part. This input energy can be very precisely driven into the surface by using the IR laser beam and precious deflection system (scanning head). Higher input energy results in a higher depth of oxide layer [10] and thicker oxide layer causes higher absorption of laser radiation.

Acknowledgements
The result was developed within the CENTEM project, reg. no. CZ.1.05/2.1.00/03.0088, cofunded by the ERDF as part of the Ministry of Education, Youth and Sports OP RDI programme and, in the follow-up sustainability stage, supported through CENTEM PLUS (LO1402) by financial means from the Ministry of Education, Youth and Sports under the "National Sustainability Programme I." and project no. SGS-2016-005.

References

[1] HRUŠKA, Matěj, VOSTŘÁK, Marek, SMAZALOVÁ, Eva, ŠVANTNER, Michal. Standard and scanning laser hardening procedure. *METAL International Conference on Metallurgy and Materials* [online]. 2013, s. 1009–1014. Dostupné z: http://apps.webofknowledge.com/full_record.do?product=UA&search_mode=Refine&qid=3&SID=U1nUP8aht9cZmirQKr2&page=1&doc=3

[2] HRUŠKA, M., VOSTŘÁK, M., SMAZALOVÁ, E., ŠVANTNER, M. 3D scanning laser hardening. *METAL 2014 - 23rd International Conference on Metallurgy and Materials, Conference Proceedings*. 2014, s. 921–926.

[3] KENNEDY, E., BYRNE, G., COLLINS, D. N. A review of the use of high power diode lasers in surface hardening. *Journal of Materials Processing Technology* [online]. 2004, roč. 155-156, č. 1-3, s. 1855–1860. ISSN 09240136. Dostupné z: doi:10.1016/j.jmatprotec.2004.04.276

[4] Ways t o Intensify Laser Hardening Technology. 1998, roč. 47, s. 133–136.

[5] ATTIA, H., TAVAKOLI, S., VARGAS, R., THOMSON, V. Laser-assisted high-speed finish turning of superalloy Inconel 718 under dry conditions. *CIRP Annals - Manufacturing Technology* [online]. 2010, roč. 59, č. 1, s. 83–88. ISSN 00078506. Dostupné z: doi:10.1016/j.cirp.2010.03.093

[6] PATWA, Rahul, SHIN, Yung C. Predictive modeling of laser hardening of AISI5150H steels. *International Journal of Machine Tools and Manufacture* [online]. 2007, roč. 47, č. 2, s. 307–320. ISSN 08906955. Dostupné z: doi:10.1016/j.ijmachtools.2006.03.016

[7] SKVARENINA, Stephen, SHIN, Yung C. Predictive modeling and experimental results for laser hardening of AISI 1536 steel with complex geometric features by a high power diode laser. *Surface and Coatings Technology* [online]. 2006, roč. 201, č. 6, s. 2256–2269. ISSN 02578972. Dostupné z: doi:10.1016/j.surfcoat.2006.03.039

[8] LAAKSO, P, RUOTSALAINEN, S, PANTSAR, H, PENTTILÄ, R. Relation of laser parameters in color marking of stainless steel [online]. 2009, s. 15. Dostupné z: http://www.vtt.fi/files/research/ism/manufacturingsystems/relation_of_laser_parameters_in_color_marking_of_stainless_steel.pdf

[9] ROBERTO RINALDI, BORIS G. VAINER, PETR KUNC, IVANA KNIZKOVA, KLAUS GOTTSCHALK, GIOVANNI M. CARLOMAGNO, CHRISTOPHE ALLOUIS, ROCCO PAGLIARA, RALPH A. ROTOLANTE, ERMANNO GRINZATO, Carosena Meola. *Infrared Thermography (Recent Advances and Future Trends)*. B.m.: Carosena Meola, Department of Aerospace Engineering University of Naples Federico II, Italy, 2012. ISBN 9781608051434.

[10] ADAMS, D. P. et al. Nanosecond pulsed laser irradiation of stainless steel 304L: Oxide growth and effects on underlying metal. *Surface and Coatings Technology* [online]. 2013, roč. 222, s. 1–8. ISSN 02578972. Dostupné z: doi:10.1016/j.surfcoat.2012.12.044

IOP Conf. Series: Materials Science and Engineering **179** (2017) 012030 doi:10.1088/1757-899X/179/1/012030

Testing of T91 steel in heavy liquid metals

M Chocholoušek, Z Fulín, J Janoušek, Z Špirit

Centrum vyzkumu Rez s.r.o., Morseova 1245/6, Plzen, Czech Republic, EU

E-mail: michal.chocholousek@cvrez.cz

Abstract. Tests of candidate construction materials for a heavy liquid metal environment are performed at Centrum Vyzkumu Rez. Tests are focused among other things on the influence of corrosive environments on the mechanical properties of T91 steel. Non-standard environments require special testing devices, which must be able to perform tests in liquid lead or liquid lead bismuth eutectic. An important issue is also the monitoring of the oxygen volume, which has an influence on the production and stability of oxide layers and therefore on crack initiation. This article presents the issue of testing steel T91 and the associated development of a testing device for slow strain rate tests, especially in liquid lead bismuth eutectic environment.

1 Introduction

There is a need to research material properties of candidate materials in reactor coolants for the development of Generation IV reactors. Material research also requires the development of appropriate test equipment for testing in active and inactive radiation environments.

The efforts of Centrum vyzkumu Rez (CVR) include the support of developing lead-cooled fast reactors (LFR). A workplace equipped with chambers for testing in an environment of liquid lead (Pb) and liquid lead bismuth eutectic (LBE) was created within the Sustainable Energy (SUSEN) project for this purpose. This facility was built for testing inactive materials.

The candidate materials for structural steels and alloys include those with known and proven properties in the radioactive environment. The research focuses on their properties in liquid media, along with the influence of surface finishes. Due to the operating temperatures under consideration, the most interesting temperatures for testing are up to 480°C, however investigating the area of accidental conditions extends the temperature range of interest up to 650 °C.

Material research at CVR [1-3] is mainly focused on how a corrosive medium effects the behavior of T91 steel in Pb and LBE environments [4-6]. Tests are performed with the control of the oxygen content in the medium in the form of metal oxides and which has a considerable influence on the life of the structure. High concentrations result in the formation of lead oxides which may, in large quantities, impede heat transfer, or block narrower parts of the device. Conversely, too low oxygen concentration in the medium leads to the dissolution of material and the release of alloying elements [7]. A suitable concentration of oxygen leads to the formation of a thin oxide layer protecting the steel surface from the medium. Optimizing the oxygen concentration is thus a key issue for the operation of LFR.

This article presents the issue of testing steel T91 and the associated development of a testing device for slow strain rate tests (SSRT) in a heavy liquid metal (HLM) environment, especially in LBE, with oxygen regulation. It also presents methodologies for crack initiation of T91 steel in the HLM environment and the basic information about sample preparation before and after SSRT testing.

IOP Conf. Series: Materials Science and Engineering **179** (2017) 012030 doi:10.1088/1757-899X/179/1/012030

2 Methodology and testing devices

The influence of the corrosion media is examined for samples exposed to HLM (Pb or LBE) with monitored and controlled oxygen content. The media is prepared in the melting chamber (Figure 1), separated from the testing chamber (Figure 2) and the testing specimen. It is melted in a protective atmosphere at a pressure of 1.5 bar to 2 bar. Ar is used as the initial protective atmosphere. During melting, the concentration of oxygen in the environment is regulated. A mixture of 98 % Ar + 2% H2 is used for the reduction. The composition of the medium is adjusted prior to the test within the separate melting chamber, so the reduction can be accelerated by increasing the temperature (without thermal effects on the sample). The medium cools to the final temperature after the desired concentration of oxygen is achieved and the medium is then transferred into the test chamber.

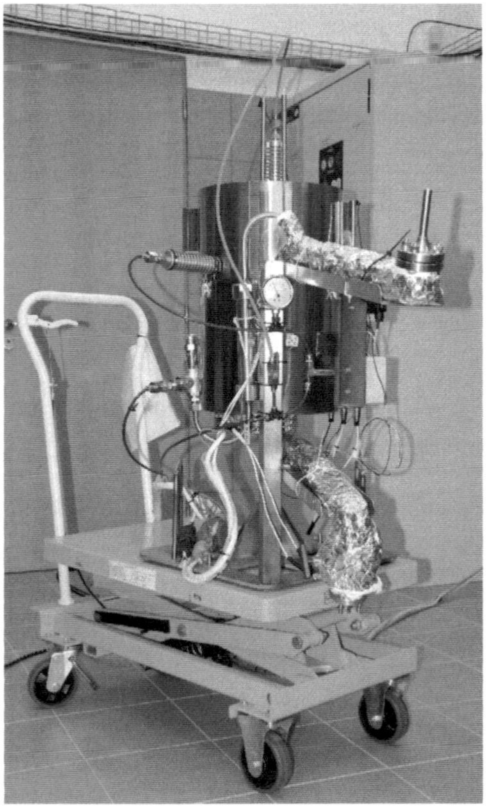

Figure 1. Chamber for melting and preparation of the HLM medium.

Figure 2. Chamber for testing the material's mechanical properties on the electromechanical machine.

The gas control system is electronically controlled based on the oximeter [3] voltage response, which is, according to equation (1), converted to a percentage value of the oxygen content in the media:

$$E(T, c_o) = K_1 + K_2\, T + K_3 T \, ln\, c_o \qquad (1)$$

where E is the voltage measured by the oximeter in Volts, T is temperature in Kelvin, c_o is the weight concentration of oxygen in the medium and $K1$, $K2$, $K3$ are constants derived from thermodynamic relations in the system of the medium (in a system Pb-Bi-O or Pb-O) and depending on the type of oximeter. The oximeter is based on Bi/Bi_2O_3.

The sample temperature is measured at the bottom of the chamber, meaning that the thermocouple is not in direct contact with the medium. Regulated gas is released under the surface of the media and it creates a slight circulation of fluid. The temperature of the vessel wall is then considered as the temperature of the medium.
The testing cell consists of a chamber and a rigid frame attached to the cover of the chamber, which allows the specimen to be manipulated outside the narrow space of the autoclave. The frame is equipped with a system for attaching adapters, which can be changed according to the shape of the samples. The test chamber is designed for a maximum load of 10 kN and a temperature of 650 °C. The system is not equipped with an extensometer for measuring deformation directly on the sample, the deformation is measured through displacement of the crosshead. Due to the requirements for gas-tightness, bellow is welded between the rod for holding the sample and the cover of the chamber through which the rod passes. The bellows enable up to 20 mm deformation of the sample.
During testing, the test chamber is placed in the working space of a Kappa50DS made by Messphysik. The test machine is sufficiently load-sized (± 50 kN) to the load capacity of the test chamber (± 10 kN). The machine allows testing with a slow rate of deformation, which is suitable for observing the influence of the fluid medium on steel embrittlement.
Small standard CT samples, non-standard samples such as tapered samples (Figure 3) and notched tensile samples with a circular cross section are used for testing the influence of the media. The influence of the media on the surface of T91 is often studied at a slow strain rate. During the tensile loading the sample is loaded up to the ultimate tensile strength (UTS) and then unloaded. The formation of cracks is then observed on the sample and the results in LBE (or Pb) environment and in air are compared. The influence of load dependence on crack initiation threshold can be evaluated.

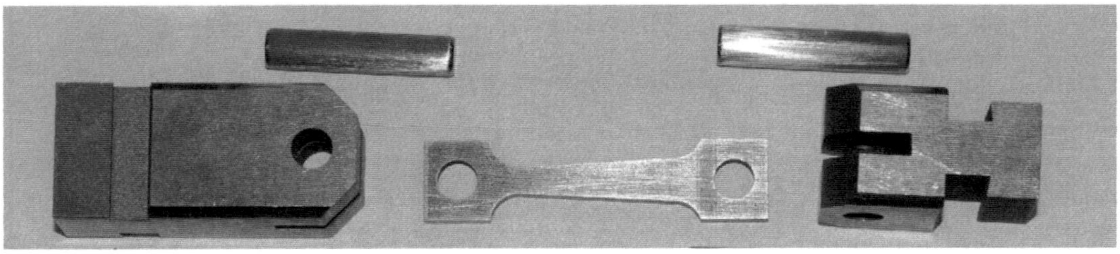

Figure 3. Adapters and tapered specimen of T91 steel for slow strain rate testing for exploring the corrosion influence of heavy liquid metals.

The surface or, if necessary, fracture surfaces are examined using a scanning electron microscope. The fracture surface of the sample is filled with liquid metal, although the medium is transferred from the test chamber after the end of the test. Therefore, at least one of the two counterpart fracture surfaces is cleaned. In the case of LBE and T91 steel, the fracture surface is cleaned from LBE residues using a solution of CH_3COOH: H_2O_2: C_2H_5OH = 1: 1: 1.
Tapered samples of T91 steel are currently being tested. The sample surface is treated differently on different sides (e.g. wire cutting without treatment, grinding, polishing). The concentration of oxygen in the LBE environment is regulated in the range of 10^{-8} to 10^{-7} wt% at a temperature of 300 °C. Tests are performed with a slow strain rate of 10^{-6} s^{-1}. The loading rate is related to the narrowest cross section of the tapered sample. Tests are performed to fracture or up to the maximum tensile stress and then unloaded. The resulting specimens are then compared, the fractured samples are used for evaluating the crack characteristics for plastic deformation and samples without significant plastic narrowing are used for the examination of crack formation due to the LBE embrittlement effect (see figure 4).

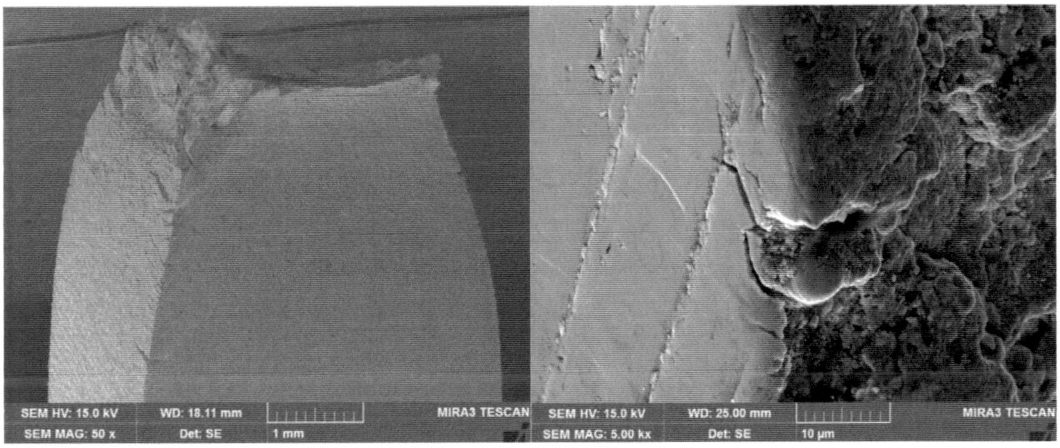

Figure 4. Tapered specimen of T91 steel after SSRT (10^{-6} s^{-1}) at 300 °C in LBE environment. Cracks and fracture area after test up to rupture (left). Crack on the edge of the tapered specimen after test up to ultimate tensile strength (right) on the polished side.

3 Conclusion

CVR developed and manufactured test equipments in cooperation with Škoda JS for testing in an environment of heavy liquid metals with oxygen regulation, with a maximum load of 10 kN and for temperatures up to 650 °C. Up to now the equipment was used primarily for testing the material properties of T91 steel and performing experiments to determine the influence of corrosive media.

At the oxygen concentrations in the range of 10^{-8} to 10^{-7} wt%, strain rate and a temperature of 300 °C, a minor effect on the formation of cracks was observed in T91 steel. The planned experimental activities will be further extended to a wider range of temperatures of up to approximately 500 °C and to a lower strain rate. In addition to the above mentioned surface treatments, tests on samples of T91 steel with a coating based on AlTiN will follow.

The development of an extensometer suitable for accurate measurement of the sample deformation in the HLM environment is considered. The extensometer would mainly be used in the description of fracture characteristics of T91 steel in the mentioned environments.

Acknowledgements

The presented work was financially supported by the Ministry of Education, Youth and Sport Czech Republic Project LQ1603 (Research for SUSEN). This work has been realized within the SUSEN Project (established in the framework of the European Regional Development Fund (ERDF) in project CZ.1.05/2.1.00/03.0108). Thanks also go to Škoda JS a.s, which participated in the development of the device and produced it according to our requirements.

References

[1] Di Gabriele F, Doubková A, Hojná A, Investigation of the sensitivity to EAC of steel T91 in contact with liquid LBE. *J. Nucl. Mater.* 2008, 376 (2008), pp. 307–311.

[2] Hojná A, Di Gabriele F, On the kinetics of LME for the ferritic–martensitic steel T91 immersed in liquid PbBi eutectic. *J. Nucl. Mater.* 2011, 413 (2011), pp. 21-29.

[3] Klecka J, Hojná A, Di Gabriele F, Mechanical properties of the steel T91 in contact with lead. *Nuclear Engineering and Design.* 2015, 283 (2015), pp. 131-138.

[4] Auger T et al., Fracture mechanics behaviour of the T91 martensitic steel in contact with liquid lead-bismuth eutectic for application in an accelerator driven system. *J. Nucl. Mater.* 2011, 415 (2011) 293-301.

[5] Auger T, Hamouche Y, Medina-Almazan L, Gorse D, Liquid metal embrittlement of T91 and

316L steels by heavy liquid metals: A fracture mechanics assessment. *J. Nucl. Mater.* 2008, 377 (2008), pp. 253–260.

[6] Yurechko M, Schroer C, Wedemeyer O, Skrypnik A, Konys J, Creep-to-rupture of 9%Cr steel T91 in air and oxygen-controlled lead at 650 °C. *J. Nucl. Mater.* 2011, *419 (2011)*, pp. 320–328.

[7] Nuclear energy agency organisation for economic co-operation and development http://www.oecd-nea.org/science/pubs/2015/7268-lead-bismuth-2015.pdf (*accessed 7. 10. 2015*).

Reinforcement of the bulk crowns excavator technology by dispersion hardening of titanium carbide to increase wear resistance

I Chumanov[1], A Anikeev[2], V Chumanov [3]

Federal State Autonomous Educational Institution of Higher Education "South Ural State University (national research university)", 454080 Russia, Chelyabinsk, Lenin prospekt, 76

E-mail: chiv71@susu.ac.ru, anikeev-ml@mail.ru, chumanovvi@susu.ru

Abstract. Dispersed particles embedded in the metal, improve the mechanical properties, particularly wear resistance. Introduction of the metal particles in a controlled and their distribution is difficult because of the density difference and the metal particles introduced. This article provides a method for the introduction of particles in the volume of metal, it contains a proposed strengthening technique described experiment course. During the experiment, the volume-reinforced castings from wear-resistant steel, the results of studies of the microstructure by optical and scanning electron microscopy, and mechanical tests of prototypes have been obtained.

1 Introduction

Open development is the most efficient way of mining. The most important link of the open-pit mining technological chain is the process of excavation, continuity of which is largely determined by the durability of replaceable teeth of the bucket quarry excavators that experience direct interaction with the rock. During excavation of the exploded array particularly hard and abrasive rock crowns of bucket excavators are working in extreme conditions of abrasive wear (a large number of favorably oriented cutting edges on the surface of the blasted rock fragments, the dominant role of metal micro cutting, intense shock loads, etc.), which cause their rapid failure [1, 2].

2 Materials and methods

The bucket crowns of excavators are traditionally made of steel grade 110G13L (Hadfield steel). This highly manganese steel combines qualities such as toughness and wear resistance, low thermal conductivity, sufficiently large linear rate of crystallization and shrinkage, has an increased tendency to columnar crystallization, has a coarse-grained structure with carbides that are predominantly located along the grain boundaries of the metal. The analysis of defective parts' for various purposes fall out that are made of this particular grade has shown that, despite the different conditions of the castings, the main reason of their failure/fall out is the deterioration in the abrasion in contact with abrasive material. One of the ways to increase the abrasion resistance of a metal is grinding the metal grain, but to crush the structure of the subsequent of the following with heat treatment is almost impossible, since the steel does not undergo phase transitions. Significant improvement of the primary structure of cast steels of this type can be achieved through integrated operations deoxidation and modification [3]. However, the introduction into a crystallizable melt of dispersed particles (or modifiers) is associated with the problem of uneven distribution of metal particles in the input volume, resulting in unevenness of chemical composition and properties distribution [4-7].

The department of "Technology and materials' technology production" FGAOU VO "South Ural State University (NIU)" branch in Zlatoust city, one of the areas of activity, which is the hardening of existing steels by introduction into the crystallizing melt particles of dispersed carbides, the method of introduction of dispersed particles directly into the Styrofoam model was proposed. The company "Uralpromdetal", which is specializing in the production of molded parts for industrial purposes, a series of experiments on the production of the excavators' crowns with the high wear resistance due to introduction of titanium carbide has been held.

As a reinforcing piece the crown of the tooth "1U-3352" mark (the estimated weight of 5.8 kg) for the dredge "Caterpillar 313", mark adapter "Cat J350" has been chosen (Figure 1).

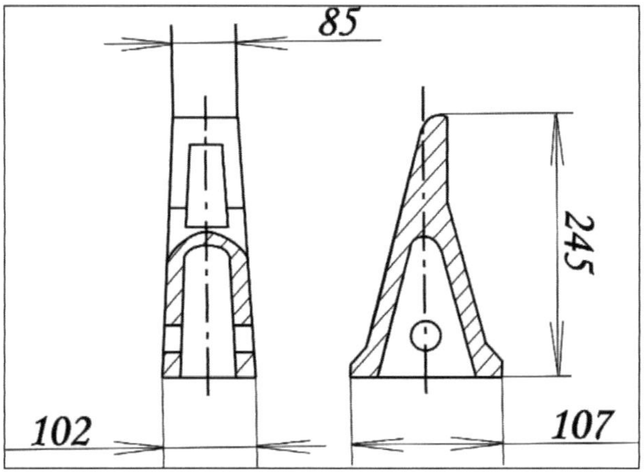

Figure 1. Drawing crown brand «1U3352».

The holes in the sides of the crown are needed for fixing it on the adapter using the "finger" with the horizontal mount. As the material of the tooth the commonly used Hadfield steel has been chosen (C 09-15%; Si 0,3-1%; Mn 11,5-15%; Ni NB 1%; Cr NB 1%; S NB 0,05%; P NB 0.12%).The models were made of polystyrene, painted with nonstick foundry and then dried. The titanium carbide was selected as the hardening phase, the density of which is 4.92 g/cm3 and the dispersity of which is 0,02-0,35 mkm. Inside the fabricated models the cylindrical holes were cut, into which the carbide particulate was filled up in different amounts (Table 1), after which the holes were sealed with polystyrene.

Table 1. Number of dispersed particles in the crowns.

Sample number	The content by weight of carbides casting, %	Number of introduced particles, grams
0	0	0
1	0.5	29
2	0.75	43.5
3	1	58

3 Discussion

One of the challenges, which the researches group had to face, was the task of ensuring cooperation between the titanium carbides and the metal – to ensure wetting of carbide. Many researchers deal with the issues involved in the wetting process of various materials [8-13], since these issues are among the top in the materials interaction. To ensure the interoperability of titanium carbide and metal melt the reducing atmosphere was used, which contributed to an increase in the contact angle [14].

These models with embedded titanium carbide were connected to the gating system and were fitted into the flask at an angle of 45 degrees. This casting arrangement was necessary to disperse particles after melting polystyrene, due to the difference of densities and the metal carbide, surfaced on one of the working surfaces of the crown of the excavator. The dry quartz sand was sued when backfilling flask models, the shape was reinforced with vacuum, and the metal casting was produced on top for 35 seconds. The flask after filling was heated for 3 hours in order to avoid cracking of parts due to the obtained rapid cooling. Then the flask was removed from the cast crowns and was separated from the gating system. The obtained crowns had a satisfactory surface quality, the right geometry, and the absence of surface defects (Figure 2).

Figure 2. Appearance received crowns of the teeth of the excavator.

In order to objectively assess the impact of input carbides on the structure and qualities of cast metal the heat treatment of the received crowns has not been made [15]. All the crowns were subjected to the study of the structural heterogeneity, hardness, and abrasive wear resistance. The microstructure investigation showed that all crowns have austenite-pearlite structure with excess carbides (Mn, Fe) 3C. The assessment of the changes in size of the dendritic cells (and, therefore, grains points) produced with the method of measuring the lengths of chords of circles with an inverted microscope "C. Zeizz Observer. D1m", which is equipped with a "Thixomet. PRO" image analyzer complex [16].

The study revealed that in the crowns №1-3 (with embedded titanium carbide) there is a slight refinement in the working structure of the surface areas of the crown (1-2 points) compared to the crown №0 (without carbides). It is worth noting that the entire volume of crowns №1-3 (in addition to the working surfaces) there are local areas of structural heterogeneity (structure refinement for 1-3 points compared to the original structure), which is probably explained by uneven distribution of embedded carbide.

The study of microstructures with the help of scanning electronic microscope "JEOL JSM – 6460 LV" showed that the implanted titanium carbide has been distributed unevenly over the volume of blanks: the largest number observed on the working surfaces of crowns №1-3 (Figure 3a), but also for crowns volume local area with a "loose" carbides are being present. Spectra taken from the detected particles confirmed that it is the titanium carbide (Figure 3b).

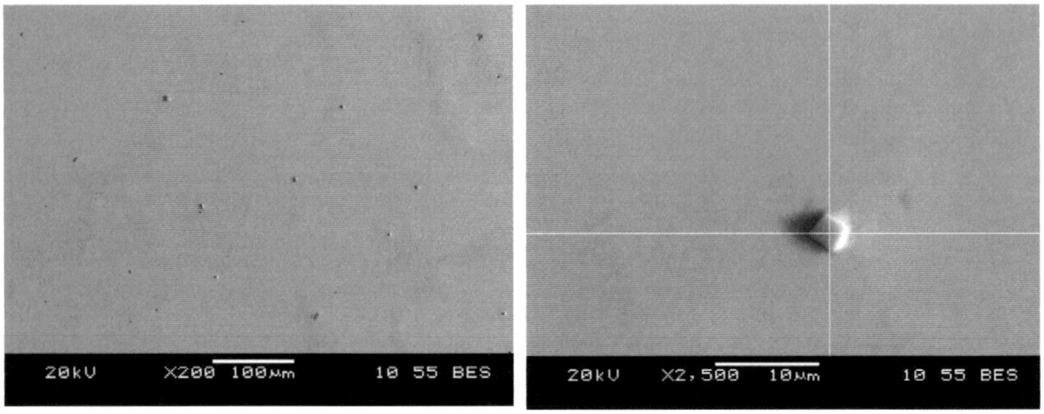

a b

Figure 3. Microstructure received crowns with titanium carbide inserted: and - external working surface, (×200); b - discovered titanium carbide, (× 2500).

4 Conclusions

The hardness testing was performed on Rockwel Hadness Tester (mark "TR-5004") and Brinell (mark "TB-5004"). The study showed that the difference in hardness between samples in the middle of the sample is small and varies from 4-30 HB (1-4 HRC). However, the difference between external and internal workspaces of the same crown can reach 128 HB (13,4 HRC). An increase in the hardness of the internal working area of the inner to outer crown has been observed (Table 2).

Table 2. The study of the hardness obtained crowns.

Place of execution	Internal workspace crown		The middle of the crown		External workspace crown		Average value HB/HRC
Performance number	1	2	3	4	5	6	
	HB/HRC	HB/HRC	HB / HRC	HB/HRC	HB/HRC	HB/HRC	
Sample №0	285/32.3	313/35.7	295/33.6	313/35.7	306/34.9	325/36.8	306.1/34.8
Sample № 1	278/31.5	282/32.1	317/36.2	337/38.4	354/40.0	350/39.3	319,6/ 36.25
Sample №2	313/35.3	329/37.4	269/30.2	325/36.8	354/40.0	350/39.5	323,3/ 36.5
Sample №3	329/37.3	275/31.2	325/40.1	333/37.9	354/40.0	403/44.6	336,5/ 38.5

The determination of abrasive wear resistance was carried out on a laboratory setting by measuring changes in sample weights before and after abrasion of the abrasive cloth under a load. Each sample was performed by three passes through the abrasive pelt 1.5 meters long under a load of 3 kg. The study showed that the sample bits №1 virtually has no difference in the durability of the sample №0 (without carbide), while crown samples №2-3 have abrasive wear to 1.5-1.9 times higher as compared with the sample bits without carbides (Table 3).

Table 3. The measurement results obtained durability crowns.

Sample number	Test sample weight, grams	Sample weight after pass 1, grams	Sample weight after pass 2, grams.	Sample weight after pass 3, grams	The total loss of mass of the sample, grams
0	15.330	15.324	15.320	15.316	0.015
1	15.923	15.917	15.912	15.908	0.014
2	16.015	16.011	16.007	16.006	0.010
3	15.201	15.198	15.196	15.195	0.008

Thus, the developed and approved method of hardening excavator crowns is able to increase its operating time in 1.5-1.9 times, thereby reducing the frequency of teeth replacements and increasing the efficiency of the use of excavation equipment.

Acknowledgements

The work within the framework of PP № 11.1470.2014 / K, as well as supported by the Ministry of Education under the contract 14.Z 56.15.7690-MK.

References

[1] Dragobetskii V, Shapoval A, Mospan D., Trosko O, Lotous V. Excavator bucket teeth strengthening using a plastic explosive deformation. (Komsomolsk: issue 4) pp. 363-368.

[2] Rakhimyanov V, Yanpolskiy V, Yusupov A 2016 Jet electrochemical machining of the steel 110G13L (System. Methods. Tehnologies, Vol. 30, issue 2) pp. 34-34.

[3] Chumanov I, Chumanov V, Anikeev A 2015 Investigating the effect of carbide disperse particles on hardness and wear resistance of experimental materials in cast and deformed conditions(Indian Journal of Science and Technology, Vol.8. issue 34) 7 P.

[4] Chumanov I, Anikeev A 2015 Investigation of Macro- and Microstructure of Experimental Composite Materials Obtained by Disperse Hardening Method (Indian Journal of Science and Technology, Vol.8. issue 34.) pp. 2-9.

[5] Chumanov I.V., Anikeev A.N. 2015 Growth and distribution disperse particles in liquid-phase interaction of the system Fe –TiC (Procedia Engineering ICIE-2015) pp. 809-815.

[6] Chumanov V, Chumanov I, Anikeev A and Garifulin R 20010 Hardening of the surface layers of the formation of a hollow shell by centrifugal casting (Russian Metallurgy (Metally), issue 12) pp. 1125-1128.

[7] Chumanov V, Kareva N, Chumanov I, Anikeev A 2013 Stady and analisis of the structural constituents of billets hardened by fine-grained particles and formed by centrifugal casting (Russian Metallurgy (Metally), Vol. 2012, issue 6) pp. 540-543.

[8] Passerone A, Muolo M, Valenza F 2016 Critical Issues for Producing UHTC-Brazed Joints: Wetting and Reactivity (Journal of Materials Engineering and Performance, issue 6) pp.1-18.

[9]. Gambaro S, Valenza F, Passerone A, Cacciamani G, Muolo M 2016 Brazing transparent YAG to Ti6Al4V: Reactivity and characterization (Journal of the European Ceramic Society, issue 6) pp.1-12.

[10] Zhao S, Zhang X, Liu G, Qiao G, Passerone A 2015 Wetting and interfacial behavior of molten Cu on Co-Si (-Mo) coated SiC (Ceramics International, Vol. 41, issue 10) pp. 13493-13501

[11] Xi L, Kaban I, Nowak R, Stoica M, Eckert J 2016 Investigation of Ni-B Alloys for Joining of TiB_2 Ultra –High Temperature Ceramic (Journal of Materials Engineering and Performance) pp. 1-7

[12] Sobczak N, Purgert R, Asthana R, Siewiorek A, Pirowski Z 2016 Wettability and reactivity of Y2O3 with liquid nickel and its alloys (Ceramic Engineering and Science Proceedings № 8) pp.309-321.

[13] Sobczak N, Asthana R 2014 Wettability in joining of advanced ceramics and composites: Issues and challenges (Ceramic Transactionsб issue 248) pp.591-600.

[14] Chumanov I, Anikeev A 2016 Study of the atmosphere influence the wetting of titanium carbide liquid low carbon steel (Electrometallurgy, issue 7) pp. 9-12.

[15] Chumanov I.V., Anikeev A.N. 2015 Investigation of Macro- and Microstructure of Experimental Composite Materials Obtained by Disperse Hardening Method (Indian Journal of Science and Technology,. Vol 8, issue 34) pp.1–7.

[16] Official website analyzer Thixomet - http://www.thixomet.ru/?src=cal-24-01-2005-rus.html

IOP Conf. Series: Materials Science and Engineering **179** (2017) 012032 doi:10.1088/1757-899X/179/1/012032

Creep test with use of miniaturized specimens

E Chvostová, J Džugan

COMTES FHT a.s., Dobřany, Czech Republic, EU

E-mail: eva.chvostova@comtesfht.cz

Abstract. Application of the mini samples methods is very common especially for residual service life assessment of the components operating in the energy sector. Residual lifetime of operating device can be evaluated using standard tests, but these usually cannot be performed due to limited material amount that can be extracted from the components. It is possible to use in these cases a semi destructive sampling of materials and testing methods employing miniature specimens. This article deals with comparison of the creep results obtained with the use of on standard test bars and sub sized samples such as Small Punch Test and newly proposed miniature samples that are axially loaded. The experiment was performed on steam line steel CSN 15 128 after operation. Emphasis was put to practical use of the test results and therefore residual life was assessed with the use of standard and new mini specimens. The results obtained for standard and mini samples exhibit very good agreement without necessity of any correlation as in the case of SPT and thus much more straight forward approach can be applied even for "unknown materials".

1 Introduction

Residual lifetime estimation and early detection of possible risks of mechanical equipment failure and structure is a key question for reliable operation of industrial plants. Currently, this issue is particularly critical for the power stations and petrochemical plants that are approaching their designed end of life. The remaining service life can be evaluated using standard tests, but also by using semi-destructive techniques using mini samples.

One of used semi-destructive methods is the Small Punch Test (SPT) [1]. SPT is usually based on the conversion of the obtained results into conventional mechanical properties [2]. The correlation parameters must be verified for each new material. There are also other available methods, where a small test sample is subjected to the same loading mode, as is the case of standard samples. For example the micro-tensile test [3-5], the miniature fracture toughness test [6-7], the miniature fatigue test [8] and mini-Charpy test for DBTT determination [9]. These methods maintain minimal material demands without requiring previously established correlations or at least they use a much more reliable type of correlation (without necessity to measure wide range materials). The above mentioned miniaturized testing techniques have been verified. These testing methods using miniaturized specimens maintaining testing loading mode are being developed and successfully applied and this paper shows results of miniaturized standard specimens testing in the field of creep tests.

2 Test samples

As a reference specimens standard creep test samples with diameter of 8mm and gauge length of 40mm were used here. The SPT test were done on disc shaped specimens with diameter of 8mm and thickness of 0.500 mm ± 0.001 mm. The geometry of the newly designed mini specimen is shown in Figure 1. The specimen geometry comes from modified dog bone shape and dimensions are based on the shape of the sample with dimensions of 20x20x4 mm, that can be extracted from real in service components by semi-destructive device EDSE (the Electric Discharge Sampling Equipment) [10]. There can be produced at least 4 specimens of considered geometry out of the extracted experimental material. Figure 2 documents comparison of all specimens investigated in this study.

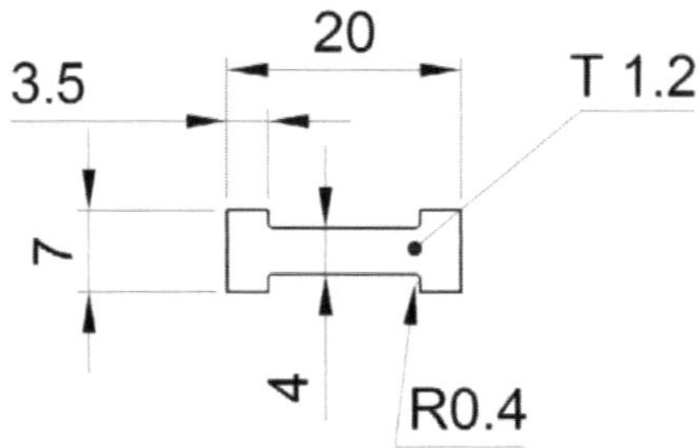

Figure 1. Geometry of miniature creep test specimen.

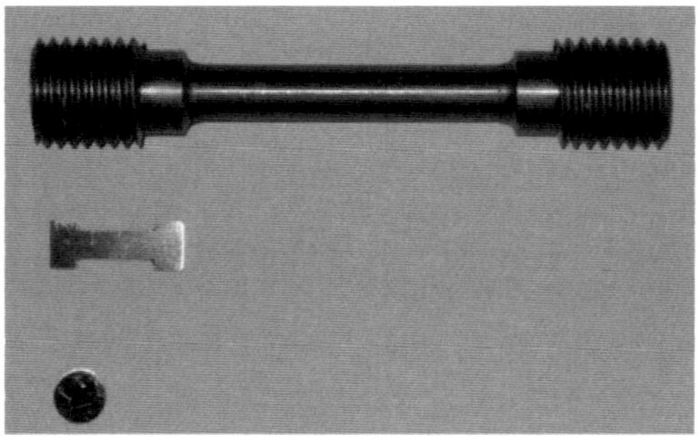

Figure 2. Used specimens for testing creep.

The experimental material was pipe line steel CSN 15 128 after certain service period. Chemical composition of the steel investigated is shown in Table 1.

Table 1. Chemical composition of the experimental material.

C	Mn	Si	Cr	Mo	V	Al	P	S
0.14	0.52	0.36	0.69	0.48	0.27	0.019	0.02	0.02

3 The creep testing and evaluation

Creep tests on standard test bars were executed with the use of conventional creep machines fitted with furnace, strain and temperature measurement. Testing of the SPT creep specimens was carried out on specially adopted machines. In the case of miniaturized standard specimens testing, standard machines with modified grips were engaged.

Testing of SPTs on standard creep machine requires a special testing fixtures converting tension load into compression. The sample is inserted into the testing fixture and consequently positioned into the creep machine [1]. Testing set up can be seen in Figure 3. The test disc is placed on a special die No. 1 and then it is inserted into the fixture bottom part No. 2. The upper part No. 3 is subsequently screwed on the bottom part and the guide No. 4 is afterwards inserted followed by Al_2O_3 ball of diameter 2.5 mm and punch No. 5. The tip of the punch is tapered in order to centre the push ball during the test. After the testing fixture assembly, it is positioned in the creep stand

IOP Conf. Series: Materials Science and Engineering **179** (2017) 012032 doi:10.1088/1757-899X/179/1/012032

and loaded. Load line displacement or strain is not measured in the current study. Time to fracture is evaluated here only.

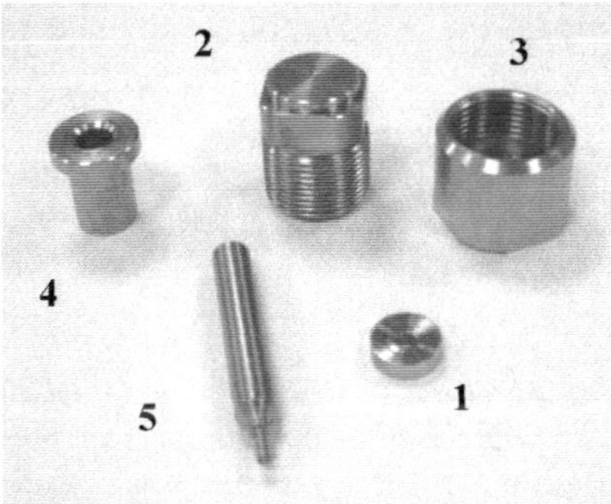

Figure 3. SP Test product disassembled and SP Test preparation assembled [9].

Mini samples testing was done with newly designed and manufactured fixture which is screwed onto the rods of conventional creep machines, the machine itself does not require any alternation. In order to ensure consistent contact force for all tests, the testing fixture is tightened by torque of 25 Nm using a torque wrench. Testing set up for testing of mini specimens is depicted in Figure 4. The proposed fixture is intended only for time to fracture determination. Specimens of all considered geometries after creep tests are displayed in Figure 5.

Figure 4. Test arrangement creep - mini sample.

The creep tests were carried out according to ČSN EN ISO 204 [11], the results of creep tests are evaluated in the form of time to fracture. Tests were performed on 18 standard creep specimens, 8 mini specimens and 16 SPTs. Tests were carried out in temperature range from 500 to 600 °C. The longest time to fracture was 7,699 hours in the case of standard specimens. In the case of miniaturized specimens times to fracture were up to several hundreds of hours in the current study.

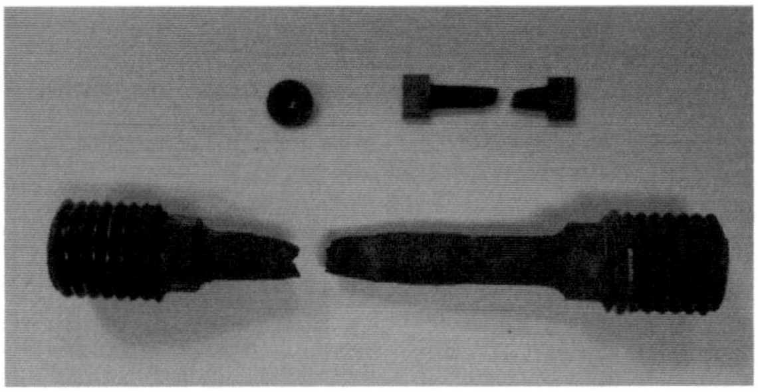

Figure 5. The test specimens after testing creep.

In the case of SPT creep tests there is difficulty with stress definition within the test-piece due to uneven stress distribution. There was derived expression (1) for stress calculation based on the theory of bending thin plates that is widely used in the publications for SPT creep test evaluation [13]. This expression was used here for stress calculation.

$$\frac{F}{\sigma} = 3.33 k_{sp} R^{-0.2} r^{1.2} h_0 \tag{1}$$

where ***R*** is die hole radius (2.2 mm), ***F*** is applied force in N, h_0 is initial specimen thickness (0.5 mm), ***r*** is ball radius (1.25 mm), k_{sp} is strain hardening coefficient (1.0) and σ is evaluated tensile stress.

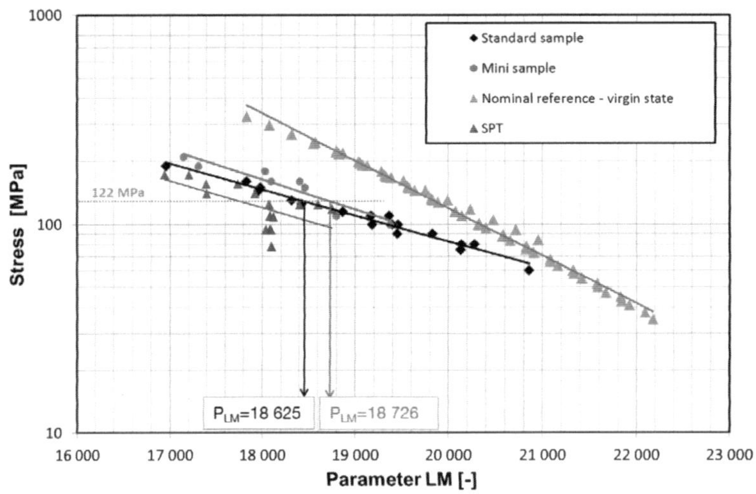

Figure 6. Comparison of creep results for material 15 128 after service.

Standard and new mini-specimens were evaluated with the use of standard formulas such as e.g. stress calculation. In the case of SPT evaluation, a correlation for the material investigated was established on the basis of standard tests as a reference values and results of SPT.

Comparison of results for all three groups of specimens was done with the use of Larson - Miller parameter defined as $P_{LM} = T (20 + \log \tau)$, where ***T*** is temperature in K, and τ is time in hours. The results are summarized in Figure 6. There are shown reference creep properties for the material investigated as specified by materials standard in virgin state [12] and results obtained with full sized specimens, new mini specimens and SPT. The values obtained clearly show material properties degradation due to service on difference between standard reference curve according to CSN 415128 and shift of presently measured values towards shorter times to fracture. The SPT results exhibit significantly higher scatter in comparison with other results and also slightly lower times to fracture in comparison with full size specimens. However, if some of outlaying values around P_{LM} of about 18 500 would be removed better agreement should be achieved, that can be expected as correlation based on current data population is applied. Comparison of full sized and axially loaded mini specimens yields excellent agreement within data scatter. This is very promising result as no correlation was applied here and thus this finding confirms of direct comparability of results obtained on specimens of different sizes.

The results obtained for full size and new mini samples are applied for residual service life estimate of the material that is investigated after period of service operation. The calculation was performed with the use of following parameters: internal gas pressure in pipe 18 MPa, operation temperature of 543 °C and the pipeline dimensions: diameter 324 mm, wall thickness 48 mm. Based on these input data, FEM simulation shown stress in the pipe at level of 122 MPa. These parameters were together with results from graph displayed in Figure 6 inserted into Larson-Miller relation shown below.

$$\tau_Z = 10^{\left(\frac{P}{T} - 20\right)} = 10^{\left(\frac{18625}{543 + 273.16} - 20\right)} = 661\,[h] \quad \textbf{standard sample} \tag{2}$$

$$\tau_Z = 10^{\left(\frac{P}{T} - 20\right)} = 10^{\left(\frac{18726}{543 + 273.16} - 20\right)} = 879\,[h] \quad \textbf{mini sample} \tag{3}$$

Calculation results shown residual service life estimate of 661 h for standard specimens' data and 879 h for mini samples results. Both results are very close confirming applicability of mini-samples utilization for residual service life estimation of in-service components.

4 Conclusions

The paper deals with possibilities of creep properties determination with the use of miniaturized experimental specimens. Standard creep specimen results are compared with results obtained from SPT creep tests and newly proposed mini-creep specimens. The standard sized specimens provide reliable results, however it is usually impossible to machine them out of real in-service components. SPT utilize small size specimens that can be extracted from real components, but the evaluation procedure requiring known correlation relation for considered material and if possible verified also for the state investigated that is also rather difficult and in many cases impossible. Therefore new miniaturized creep specimens were proposed combining the advantage of maintaining the same loading mode as standard creep tests while material demand for specimens machining can be covered by volume that can be extracted from serviced component. The dimensions of creep mini-samples proposed here are based on material volume that can be extracted by available EDSE sampling device.

Creep tests were performed for all considered sample geometries for after service pipeline steel in temperature range between 500 and 600 °C. In the case of SPT tests correlation function was established for the material investigated based on relation between results of standard tests and SPT prior further comparison. Subsequently, all results were summarized for comparison in the form of Larson-Miller plot. There is visible slight offset of SPT results from the other two investigated data sets. This was caused by wider data scatter mainly. If some of the outlying results would be removed, the agreement with other results would be significantly better. This agreement would be expected as the correlation was established on the same data population. The results of SPT point out just inclination to higher data scatter and necessity to establish correlation specifically for the material investigated.

Results of mini creep specimens exhibited very good agreement with standard creep test results. This is very positive finding as the same evaluation was used for both specimens' batches without any further adjustments for mini-specimens. The results confirm possibility of direct comparability of results attained for specimens of different sizes in loaded in the same mode.

The results of standard tests and mini creep tests were also used for residual service life estimation according to Larson-Miller approach. Very close residual service lives were obtained confirming previous statements about the mini specimens' successful application to real components service life assessment.

More data with the miniaturized specimens have to be accumulated for more general conclusions and recommendations. Testing set up allowing also strain measurements is considered for future investigations.

Acknowledgements

This paper was created by project Development of West-Bohemian Centre of Materials and Metallurgy No.: LO1412, financed by the MEYS of the Czech Republic.

References

[1] Small punch test method for metallic materials. 2007 Part A: A Code of Practice for Small Punch Creep Testing, Part B: A code of practice for small punch testing for tensile and fracture behaviour. CWA 15627

[2] Konopík P and Džugan J 2012 *2nd Int. Conf. SSTT on Det. of Mech. Prop. of Mat. by Small Punch and other Miniature Testing Techn.* Determination of Tensile Properties of Low Carbon Steel and Alloyed Steel 34CrNiMo6 by Small Punch Test and Micro-Tensile Test (Ostrava: Ocelot sro) pp 319-328.

[3] Džugan J, Procházka R and Konopík P 2015 *Small Specimen Test Techniques* Micro-tensile test technique development and application to mechanical property determination **6th Volume, ASTM STP 1576** Sokolov M A and Lucon E (West Conshohocken: ASTM Int.) pp 12-29

[4] Rund M et al 2015 *Procedia Engineering* Investigation of Sample-size Influence on Tensile Test Results at Different Strain Rates, **Vol. 114** (Amsterdam: Elsevier science bv) pp 410-15

[5] Konopík P, Dzugan J and Prochazka R 2013 *Mat. Sci. and Tech. Conf. and Exh.* Evaluation of local mechanical properties of steel weld by miniature testing technique, pp 2404-11

[6] Konopik P, Dzugan J and Rund M 2015 *METAL 2015 – 24th Int. Conf. on Metallurgy and Mat. (Brno)* Determination of fracture toughness in the upper shelf region using small sample test techniques (Ostrava: TANGER Ltd.) pp 710-15

[7] Konopík P, Dzugan J and Prochazka R 2013 *METAL 2013 – 22nd Int. Conf. on Metallurgy and Mat. (Brno)* Determination of fracture toughness and tensile properties of structural steels by small punch test and micro-tensile test (Ostrava: TANGER Ltd.) pp 722 – 27

[8] Dzugan J, Konopik P., Rund M and Prochazka R 2015 *ASME Pressure vessels and piping conf. - 2015* Determination of local tensile and fatigue properties with the use of sub-sized specimens, **Volume 1A: Codes and Standards** (New York: Amer. Soc. Mech. Eng.)

[9] Chvostová, E, Chocholoušek, M., Mentl, V., Volák, J.: Construction materials mechanical properties assessment with the use of miniaturized test specimens, Internal report VYZ 1374/11, Research and Testing Institute Plzen

[10] ETD Technology Newsletter, EDSE: Innovation in 'Boat Sampling' [online], [cit. 2015-05-20]: https://static1.squarespace.com/static/5304e62de4b00674c06afd5c/t/555efc36e4b08ef1ebb76245/1432288310982/May+2015+Newsletter.pdf

[11] Standard ČSN EN ISO 204 Metallic materials – Uniaxial tensile creep test – testing method. 1. 10. 2009

[12] ČSN 415128, Steell 15 128 Cr-Mo-V, 4. 12. 1984

[13] Bicego, V.: Suggested parts to be put in the ECCC WG1.1 report on miniature creep testing,CESI 2003

Using of material-technological modelling for designing production of closed die forgings

K Ibrahim, I Vorel, Š Jeníček, J Káňa, D Aišman, V Kotěšovec

University of West Bohemia, Regional Technological Institute, Univerzitní 22, Plzeň, Czech Republic

E-mail: ibrahim@rti.zcu.cz

Abstract. Production of forgings is a complex and demanding process which consists of a number of forging operations and, in many cases, includes post-forge heat treatment. An optimized manufacturing line is a prerequisite for obtaining prime-quality products which in turn are essential to profitable operation of a forging company. Problems may, however, arise from modifications to the manufacturing route due to changing customer needs. As a result, the production may have to be suspended temporarily to enable changeover and optimization. Using material-technological modelling, the required modifications can be tested and optimized under laboratory conditions outside the plant without disrupting the production. Thanks to material-technological modelling, the process parameters can be varied rapidly in response to changes in market requirements. Outcomes of the modelling runs include optimum parameters for the forging part's manufacturing route, values of mechanical properties, and results of microstructure analysis. This article describes the use of material-technological modelling for exploring the impact of the amount of deformation and the rate of cooling of a particular forged part from the finish-forging temperature on its microstructure and related mechanical properties.

1 Introduction

A large number of parameters play their roles in forging processes. The key ones include temperature and strain magnitude and rate. Their values together with cooling conditions impart the resulting properties to forged parts. It is often impossible to map their effects in actual operation, mainly due to financial constraints associated with production.

One available solution is the use of material-technological modelling. This method involves simulating the real-world forging process in a laboratory in order to study the effects of process parameters without interrupting or slowing down production in the forge shop [1]. As a result, countless trials can be run which – if carried out in the factory – would entail a major financial burden. The product of material-technological modelling is a specimen with a processing history identical to the actual manufacturing route for the forged part [2, 3]. Unlike FEM simulations, this approach enables microstructure characterization and mechanical testing. Its design requires concrete data gathered in real-world production. This data, i.e. temperatures, intervals between operations, and cooling rates, as well findings from the FEM simulation of the forming process, including deformation values, are used for developing material-technological modelling specifications [4, 5,6,7].

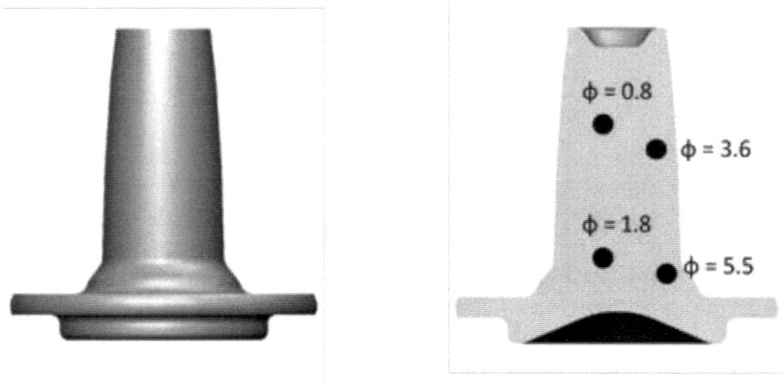

Figure. 1 General view of the forged part, a cross-section with the points selected for modelling, and their total true strain levels.

2 Experimental

This material-technological modelling study was conducted to correlate true strain levels and the rate of cooling of a particular forged part from the finish-forging temperature with resulting microstructure and mechanical properties. This forging is used for making part of the drive train of a lorry "Figure 1". Its final microstructure was specified as a mixture of ferrite and pearlite. Material-technological models were constructed for several points on the forging's cross section. For these points, FEM simulations using the DEFORM software reported the following true strain levels: φ = 0.8; 1.8; 3.6; 5.5. Cooling curves starting at the finish-forging temperature were plotted for these points "Figure 2". The curves, i.e. the associated cooling rates, were designed to enable a particular limit cooling rate to be identified: the one at which no bainite forms in the material within the critical cooling interval of 950/300 °C.

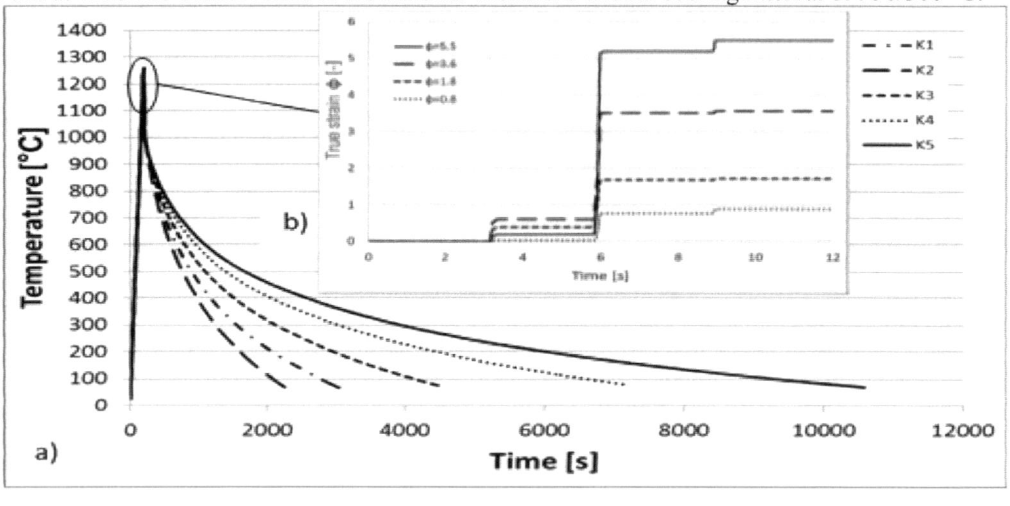

Figure. 2: a) Cooling rates from the finish-forging temperature in the material-technological models of the chosen points, b) Profiles of true strain introduced during forging operation.

Modelling was carried out in a thermomechanical simulator provided with an electrical induction-resistance heating system which offers heating rates of up to 100 °C/sec. The specimens obtained by material-technological modelling were sectioned to prepare metallographic sections and provide stock for static tensile testing and notch toughness testing. The experimental material was 30MnVS6 steel "Figure 3".

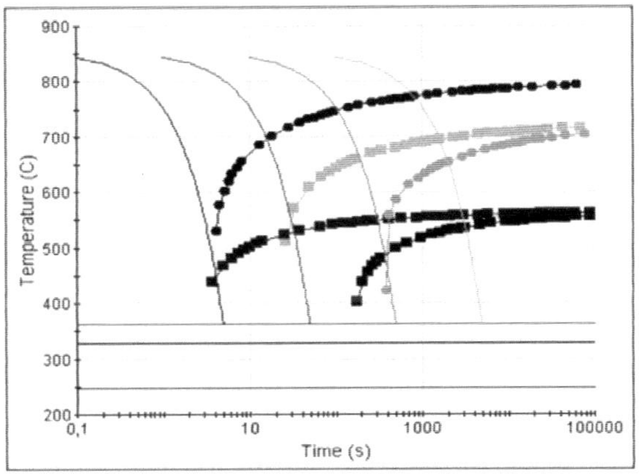

Figure 3. CCT diagram for the 30MnVS6 steel as computed and constructed using the JMAtPro program.

3 Results and discussion

3.1 Mechanical properties

The values of mechanical properties confirmed the expectation that the strength of material increases with the amount of deformation "Figure 4" and the tensile test was applied using small flat specimens machined from the deformed parts with dimensions of 5 mm reduced section and a cross section 2 mm × 1.2 mm. The strength level was also affected by the rate of cooling from the finish-forging temperature "Figure 5". The highest strength was obtained for K1 cooling curve. By contrast, the lowest strength – at an identical amount of strain – was obtained upon cooling according to K5 curve. This effect of cooling rate is probably down to the resulting fractions of phases (ferrite, pearlite, and bainite), and to the presumed precipitation of vanadium carbides within proeutectoid ferrite and within the ferrite in pearlite.

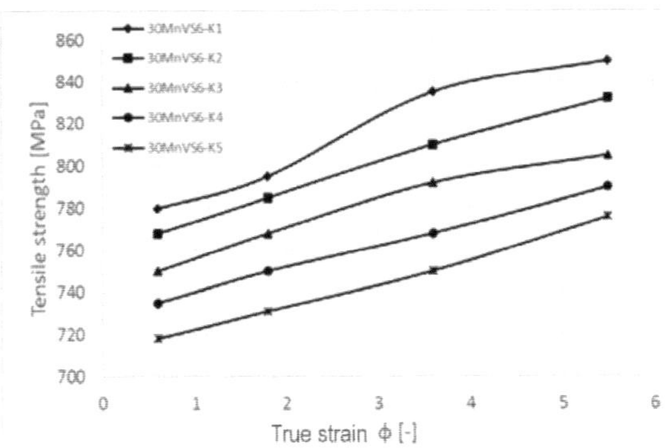

Figure 4. Tensile strengths of specimens representing material-technological models with various total strain levels and cooling rates.

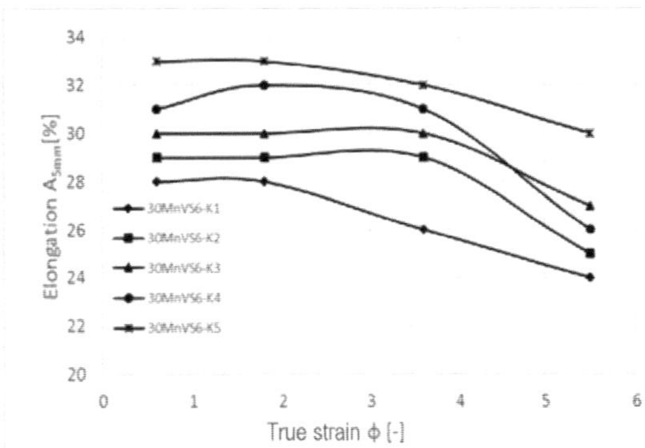

Figure 5. Elongation A5mm of specimens representing material-technological models with various total strain levels and cooling rates.

The highest elongations, 30–33 %, were attained upon cooling according to K5 curve "Figure 5". By contrast, the lowest elongations, A5mm = 28–24 %, were obtained upon cooling according to K1 curve. This relationship between the cooling rate and the A5mm elongation is attributed to ferrite fraction, which increases with decreasing cooling rate. On the other hand, A5mm elongation decreased with increasing amount of strain, as expected "Figure 5". This is in agreement with the general principles of deformation behaviour of metals.

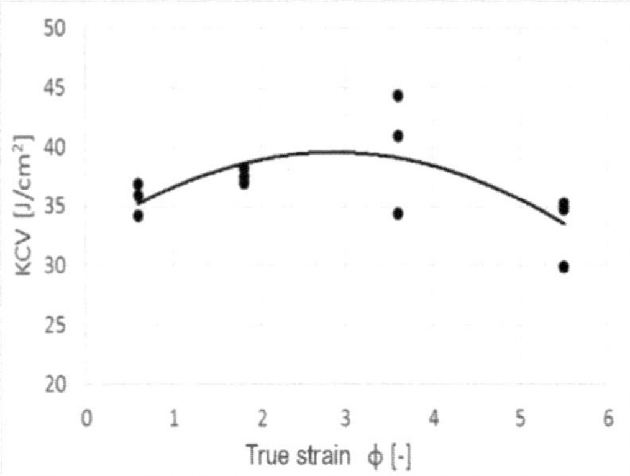

Figure 6. Notch toughness KCV of specimens representing material-technological models with various total strain levels – K5 cooling curve.

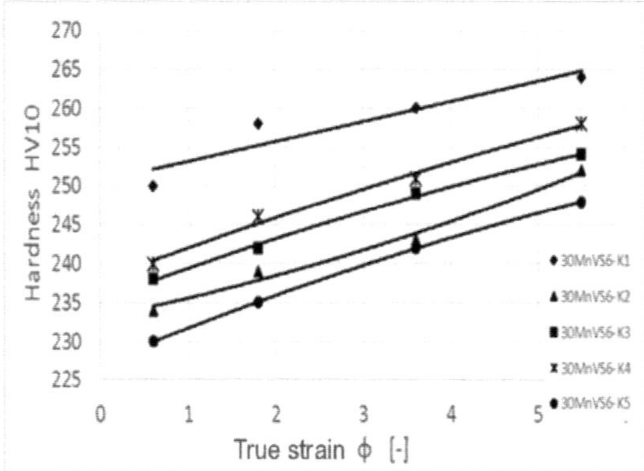

Figure 7. Hardness number HV10 of specimens representing material-technological models with various total strain levels and cooling rates.

Impact toughness tests using charpy impact samples were carried out on specimens which had cooled according to K5 curve "Figure 6" due to the final microstructure of its samples. The charpy samples were also machined from the deformed parts with dimensions of 2 mm thickness, 4 mm × 27.5 mm as a cross section, the geometry of the notch is 1 mm deep with angle of 60° and maximum radius 0.1 mm. This curve was chosen because of the requirement for the material-technological modelling specimens to contain fully ferritic-pearlitic microstructure. Notch toughness values (KCV) showed a flat peak at approx 40 J/cm2. The lowest notch toughness levels were found in specimens with true strains of φ= 5.5. Interesting information about the impact of deformation on fracture behavior was thus obtained. The expectation that at higher strains, and therefore higher work-hardening levels, the notch toughness will decrease considerably, was not confirmed entirely. Macrohardness was measured in the specimens "Figure 7". It increased linearly with true strain, with highest levels at $\varphi = 5.5$, and the lowest values at $\varphi = 0.8$. This correlation with deformation is probably related to the essential principles of work hardening and precipitation hardening. The lowest hardness was found in specimens which had cooled according to K5 curve. The highest hardness was obtained with the K1 curve. This can be explained by the various ferrite, pearlite, and bainite fractions. The lowest amounts of pearlite (and correspondingly higher amounts of ferrite) were found in specimens which had cooled according to K5 curve, as opposed to specimens cooling according to K1 curve.

3.2 Metallographic examination

The microstructure of the specimens was revealed with 3 % nital and examined under optical microscope. All specimens which had cooled according to K1 curve consisted of ferrite, pearlite, and a small amount of bainite "Figure 8". Bainite was also found in specimens cooling according to K2 curve. K3–K5 curves led to ferritic-pearlitic microstructures "Figure 9". With increasing true strain, the microstructure became notably finer, as the size of ferrite, pearlite and bainite grains and particles decreased "Figure 10, 11".

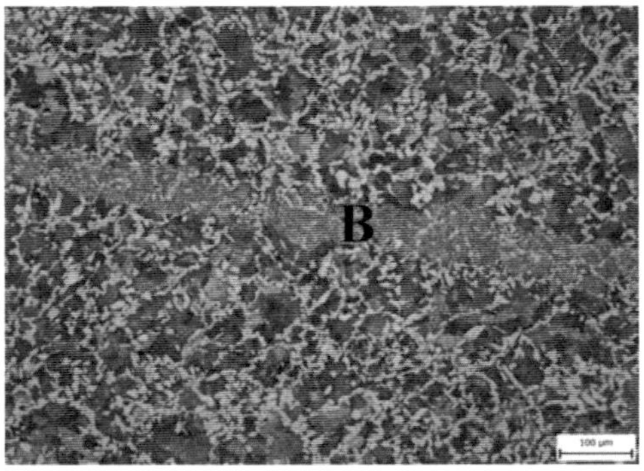

Figure 8. Micrograph of specimen cooling according to K1 curve, total strain φ = 0.8.

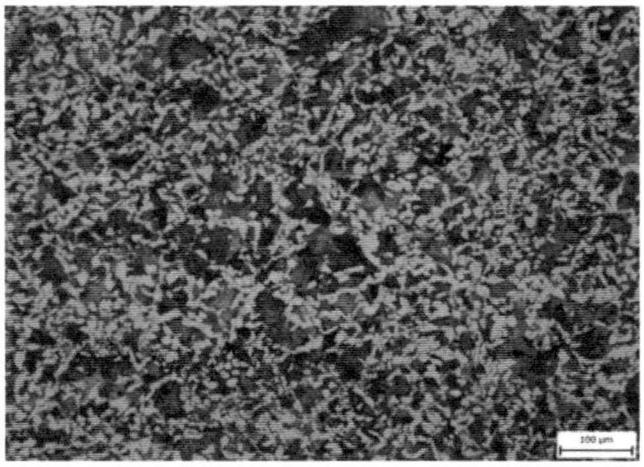

Figure 9. Micrograph of specimen cooling according to K5 curve, total strain φ = 0.8.

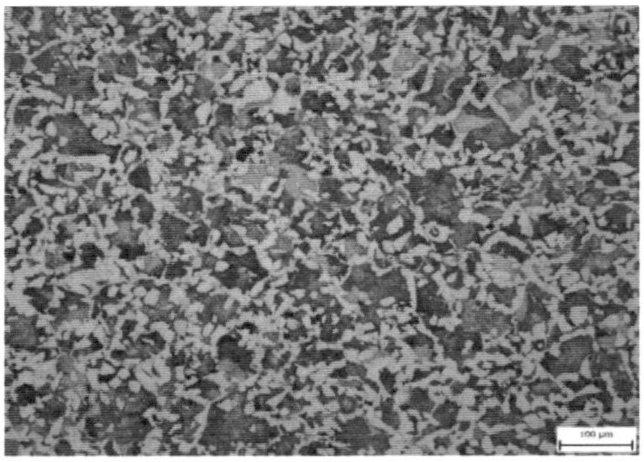

Figure 10. Micrograph of specimen cooling according to K5 curve, total strain φ = 1.8.

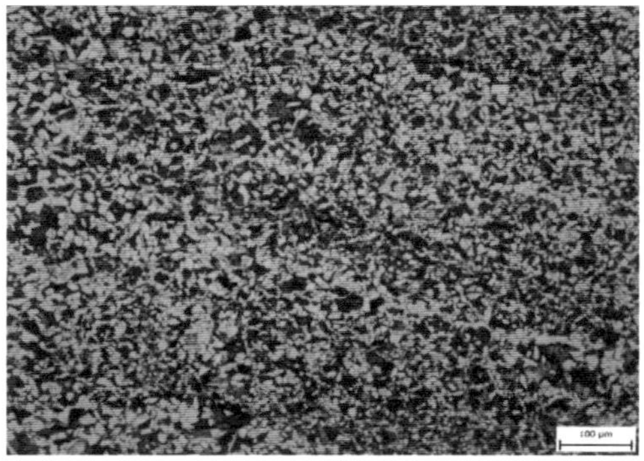

Figure 11. Micrograph of specimen cooling according to K5 curve, total strain φ = 5.5.

4 Conclusion

Using material-technological modelling, correlation between true strain levels, the rate of cooling of a particular forged part from the finish-forging temperature and the resulting microstructure and mechanical properties was examined. Several points were identified on the forging's cross section. For these points, FEM simulations using the DEFORM software reported the following true strains: φ = 0.8; 1.8; 3.6; 5.5. Material-technological models were constructed for these points to describe the forging process in thermophysical terms. One boundary condition for this study required the final specimens to contain exclusively ferritic-pearlitic microstructure. Important conclusions have been drawn from the experiment and verified. Tensile strength increased with the amount of strain in individual specimens, each of which represented selected points on the cross-section through a forged part. This effect can be strengthened by accelerated cooling through the critical interval of 950/300 °C. Elongation decreases with increasing amount of strain and cooling rate. Notch toughness shows a flat peat at 40 J/cm2. Hardness decreases with cooling rate. The hardness of the forged part can be optimized for machinability by appropriate cooling. The prescribed ferritic-pearlitic microstructure can be obtained by cooling through the 950/300 °C interval in more than 1600 seconds.

Acknowledgements

This paper includes results created within the project LO1502 Development of Regional Technological Institute. The project is subsidised by the Ministry of Education, Youth and Sports from specific resources of the state budget for research and development.

References

[1] MAŠEK B, JIRKOVÁ H, MALINA J, SKÁLOVÁ L and MEYER L W 2007 Physical Modelling of Microstructure Development During Technological Processes with Intensive Incremental Deformation Vol. 345-346, No. 1-2 Key Engineering Materials p 934-946

[2] PILEČEK V, VANČURA F, JIRKOVÁ H and MAŠEK B 2014 Material-Technological Modelling of Die Forging of 42CrMoS4 Steel Vol. 48, Issue 6 Materiali in technologije p 869-873

[3] VOREL I, PILEČEK V, VANČURA F, JIRKOVÁ H and MAŠEK B 2015 Material-Technological Modelling of C45 Steel Die Forgings. *Procedia Engineering*100(2015)714 –721

[4] VOREL I, VANČURA F and MAŠEK B 2015 Material-Technological Modelling of Controlled Cooling of Closed die Forgings from Finish Forging Temperature. 24th International Conference on Metallurgy and Materials. Ostrava: TANGER Ltd pp 202-208

[5] MAŠEK B, JIRKOVÁ H, KUČEROVÁ L, RONEŠOVÁ A and MALINA J 2011 Material-Technological Modelling of Real Thin Sheet Rolling Process, 20th Anniversary International Conference on Metallurgy and Materials, pp 216-220

[6] MAŠEK B, PILE4EK V, VANČURA F and JIRKOVÁ H 2014 Modelling a Process chain of die Forging Process. XXXIII. Verformungskundliches Kolloquium. Leoben, pp 72-76

[7] MALEČEK L, FEDORKO M, VANČURA F, JIRKOVÁ H and MAŠEK B 2015 Development of Numerical Models for the Heat-Treatment- process Optimisation in a Closed-Die Forging Production. MATERIALI IN TECHNOLOGIJE 3rd year, pp 471-475

Fatigue performance of blade steel T552 in a corrosive environment

J Janoušek[1, 2], S Hřeben[3], Z Špirit[1, 2], J Strejcius[1] and J Kasl[4]

[1]Research Centre Rez, Hlavní 130, 250 68 Husinec-Řež, Czech Republic

[2]University of West Bohemia, Faculty of Mechanical Engineering, Department of Material Science and Technology, Univerzitní 22, 306 14 Plzeň, Czech Republic

[3]Doosan Škoda Power, Tylova 1/57, 301 28 Plzeň, Czech Republic

[4]Research and Testing Institute Plzen, Tylova 1581/46, 301 00 Plzeň, Czech Republic

jas@cvrez.cz

Abstract. This contribution is based on an experimental programme which deals with the issue of blade steel T552 corrosion fatigue damage. The specimens were taken from a reference material delivered for blade production and tested in a corrosion cell under pre-stress of 300 MPa. A chloride solution of 35 ppm was chosen as the environment at a temperature of 80 °C. Such an environment can be considered as the limit state that should not occur during the operating regime. The results are summarized and compared with experiments conducted at another workplace.

1 Introduction

Fatigue failure of steam turbine blades caused by pitting corrosion is one of the most dangerous issues of power plants. It occurs preferentially on blades in the low pressure (LP) rotors (especially on the last L-1 and L-0 stages) where early condensate develops [1, 2]. It is often reported that damage in LP steam turbines rotors initiates in highly localized areas, most commonly at corrosion pits that act as stress raisers [3, 4]. The corrosion events develop and pass through distinct stages; namely, initiation of metastable pits, the survival of few metastable pits to form stable pits, the growth of stable pits, the transition of pits into cracks, the growth of subcritical cracks, and finally unstable fracture [3]. It is possible to keep the total amount of impurities in steam below concentrations of 50 ppb; 2 ppb for chlorides, and oxygen below a threshold of 10 ppb during standard operational conditions [5]. Thermodynamic calculations predict that the concentration of nonvolatile substances (such as chlorides, sulfates and carbonates) in a liquid phase might be up to 100 times higher than in steam. [6] The concentration might be up to several ppm, especially in the transitional operational state during putting the block in and out of operation, or due to condenser leakage, etc. [7]. An environment with a chloride solution of 35 ppm can be considered as the limit state that should not occur during the operating regime [8].

The most applied materials are 12 % of chromium modified martensitic steels, then 12-15 % chromium precipitation hardened steels and, in sporadic cases, titanium alloys [9]. In terms of static

loading, the blade is exposed to centrifugal force as well as the pressure force of medium. The effect of the flow medium which flows around the blades is one of the sources of dynamic excitation. An unequal pressure field and rotor shaking could also result in excitation [10, 11]. The dynamic stresses are low compared to centrifugal forces of rotation, and the acting stress ratios can exceed values of $R = 0.9$ [1]. The number of cycles corresponds with very high cycle fatigue (VHCF), therefore the value of 10^8 cycles is established for the fatigue limit in accordance with standards [12].
Linear elastic fracture mechanics is used for evaluating the influence of corrosion pits on fatigue. Investigations show that the pits can be treated as semicircular surface cracks with the width at the surface 2c and the pit depth a [13]. The results are summarized in the Fatigue Crack Growth Rate (FCGR) diagram and Kitagawa-Takahashi (K-T) diagram. As an output, K-T diagrams can be used for assessment of the fatigue limit and life-time of corroded parts of steam turbines [13].

2 Description of similar tests for the same material

In the framework [1] the test environments were air and two aqueous solutions at 90 °C. De-aerated 300 ppb Cl^- solution - in such an oxygen-free environment, corrosion pits repassivate and stop growing even after exposure to more corrosive conditions such as an aerated 6 ppm Cl^- solution. The specimens were smooth with an hourglass shape and a cylindrical gauge length and they were prepared from dual certified 403/410 12% Cr martensitic steel. The specimens were ground and polished after machining. They were stress-relief annealed in high vacuum at 10^{-6} Pa (heating from room temperature to 600 °C for 1 h, holding for 2 h, cooling from 600 °C to 400 °C for 2 h and to room temperature in approximately 12 h) to eliminate residual stresses. The material was hardened at 913 °C and tempered. The mean grain size was 6 µm and 44 µm (2 batches). The number of inclusions per observed area was higher by a factor of 2.7 between batches. Mainly Al_2O_3, MnO, Cr_2O_3 and MnS inclusions were found. The probability of crack initiation is enhanced with an increased number of inclusions that act as stress raisers [1]. Enhanced fatigue strength of the material is, among other things, related to the lower number of inclusions. The specimens were pre-pitted in a droplet cell for studying pit-to-crack transition. Pit depths were 50 µm, 100 µm and 250 µm. The tests were conducted on ultrasonic fatigue testing equipment (dynamic component up to 20 kHz) in combination with a servo-hydraulic machine (static component). The fatigue limit was defined as the stress range where two specimens survived at least 1 x 10^9 cycles without failure.

In [1] a 35 l reservoir with flow rate 3 l/h was used for creating the environment. Most contact parts were made of 316 stainless steel. The corrosion potential was measured using an Ag/AgCl reference electrode, but the potentials quoted refer to the saturated calomel electrode (SCE) at 25 °C. Immersion pre-tests showed that corrosion pits can form in aerated 6 ppm Cl- solution but the pit growth rate is low, therefore it does not compete with the artificially generated pit. The maximum depth of such formed pits was 20 µm after an immersion time of 14 hours which was the time taken to reach 109 cycles. The pH of the solutions was between 5.2 and 7.2. Ultrapure water with an initial conductivity of 0.06 µS/cm and AR grade NaCl was used. The solution was exchanged when their conductivity increased by 10 % of the initial value.

Although environmental effects are usually less significant at high loading frequency, it is known that close to threshold, and for small cracks, dissolution of microstructural barriers can lead to a reduced threshold. This was not observed in [1] due to crack closure by oxide debris and enhanced roughness. A closure-free situation was confirmed for stress ratios $R \geq 0.72$ for tests in air, $R \geq 0.76$ for de-aerated solution and $R \geq 0.72$ for aerated solution. Specimens tested in air and de-aerated 300 ppb Cl- solution at $R = 0.8$ near fatigue crack growth threshold ΔK_{th} show no evidence of fretting in contrast to aerated 6 ppm Cl- solution. No satisfactory explanation can be given for lower crack growth rates, and a higher near ΔK_{th} was found in de-aerated 300 ppb Cl- than in aerated 6 ppm Cl- solution.

In [1] for S-N tests on smooth specimens, a decrease of the fatigue limit for increasing R and corrosiveness of environment was found. For high $R = 0.8$ was observed only a slight decrease of the stress range by 5 MPa compared to air. The reason for this is a change in the failure mechanism.

Characteristic fatigue fracture surfaces with crack initiation at inclusions or small defects were found for small R = 0.05 and for R = 0.5 in aerated 6 ppm Cl- solution. In contrast, a cup-and-cone like fracture typical for tensile-testing was observed for R = 0.5 and 0.8 although fatigue lives above 108 cycles were achieved. Pre-pitted specimens with a pit depth of 100 µm did not cause final failure at R = 0.8 in air because it occurred elsewhere.

3 Experimental programme

The martensitic steel X12CrNiMoV12-3 Böhler T552 (also known as 403/410 12% Cr steel designated 1.4938 or 1.4939) was used as the reference material. The chemical composition is given in Tab. 1. The hardness of the material was between 309-323 HB (confirmed 325 - 339 HV0.5), yield strength R_{p02} between 954 MPa and 961 MPa and ultimate (tensile) strength R_m between 1058 MPa and 1061 MPa. The following processes were used for heat treatment: hardening 1040 °C/1 h/oil, tempering 620 °C/5 h/air and stress annealing 580 °C/4 h/air.

Table 1. Chemical composition of the steel T552 [14].

Cr	Ni	Mo	Mn	V	Si	C	N2	P	S
11.70	2.71	1.70	0.75	0.30	0.21	0.12	0.038	0.016	0.002

Pre-stress was chosen for all subsequent tests up to 300 MPa. The resonant testing machine TESTRONIC 250 made by the Swiss firm RUMUL was used for all the following experiments. It is a dynamic testing machine, which works in full resonance for loads up to 250 kN. The maximum oscillating stroke extension available is up to 4 mm. The resonant operating frequency is given in a range from 40 Hz up to 250 Hz by the oscillating masses and also by the stiffness of the specimen.
The corrosion measuring cell was designed, engineered and manufactured for tests in a corrosive environment. The cell is made of an inert material, Teflon, glass and silicone rubber (Fig. 1). The cell has a flow volume of 1 litre, the entire electrolyte solution is about 1 000 litres per hour and the electrolyte is tempered to the required temperature with an accuracy of 1 °C. The bubbling inert gas enables control of the content of the dissolved oxygen in the electrolyte. It is also possible to adjust the pH. The control of electrical conductivity is performed continuously. During the test of corrosion fatigue the electrode potential of the specimen is also recorded. Specimens 240 mm long with a straight working part and a diameter of 8 mm were tested in the flow cell. (see dimensions in Fig. 2). The NaCl solution with a concentration of 35 ppm of chlorides was chosen for testing in a corrosive environment at a temperature of 80 °C. The content of oxygen was not controlled.

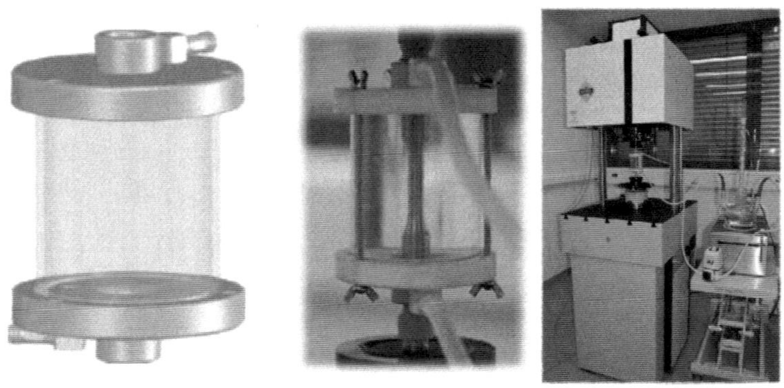

Figure 1. Corrosion cell with testing equipment.

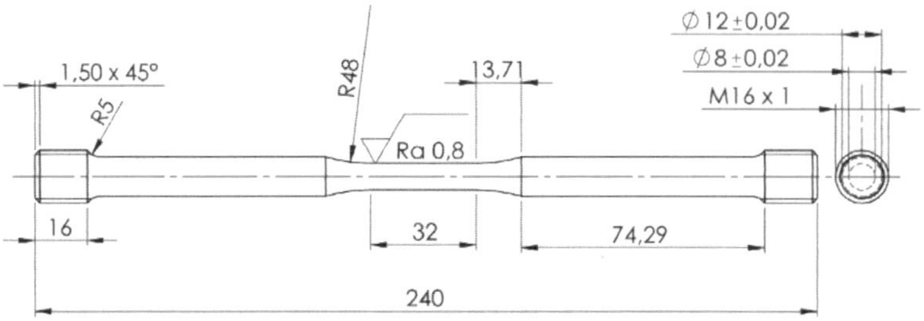

Figure 2. Specimen used for testing in corrosion cell with dimensions.

4 Evaluation and comparison of results

A considerable scatter of results was found for tests in air (see Fig. 4). This behaviour is related to the globular inclusion (oxides) content with a size of about 10 μm (see Fig. 3), and many inner initiations in particular were proved on inclusions with a size of up to 40 μm. The microstructure of the heat treatment state must be composed of fine-grained tempered martensite without hard lines or a coarse carbide network with a maximum grain size of 5.0 according to DIN EN ISO 643. In our case the mean grain size was G4 i.e. 88.4 μm.

Figure 3. 100x and 500x microscopic magnification of cross centre and longitudinal centre. Two mechanisms of fatigue rupture on surface inclusions and inner inclusion (so-called 'fish eye').

The tests were performed for pre-stress 300 MPa. The slanted branch of selected points of the S-N curve was fitted by Fletcher's version of the Levenberg-Maquardt algorithm for the minimization of the sum of the squares of the equation residuals. The program was created in a numerical computing environment and the programming language MATLAB. This regression line curve is defined by an equation that was found by a probabilistic approach as

$$\log N_a = 23.4 - 6.9 \log \sigma_a \tag{1}$$

The S-N curve is depicted in Fig. 4. The red points are valid, (i.e. sample failed) and green arrows indicate points without rupture (or a fracture in the threaded attachment) and they were not included in the linear regression. The frequency was chosen intentionally low, approximately 66 Hz, to prevent heating of the specimen. The fatigue limit was determined on the basis of two undamaged specimens to a value 360 MPa.

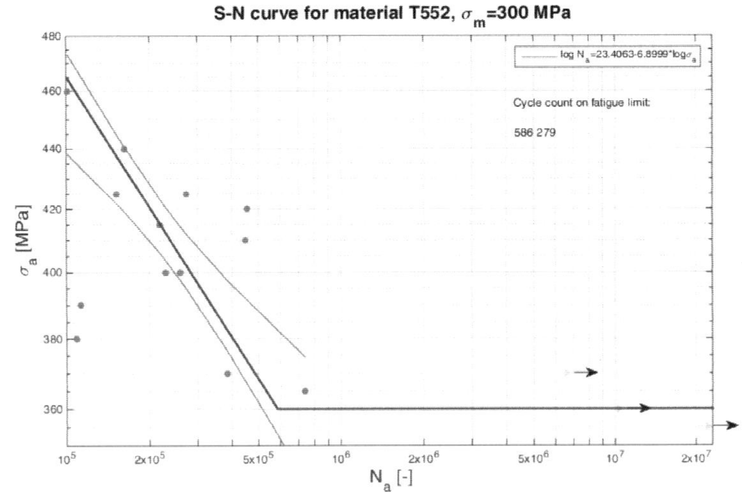

Figure 4. S-N curve for pre-stress in normal component σ_m = 300 MPa tested in air.

The NaCl solution with a concentration of 35 ppm chlorides was chosen for testing in a corrosive environment at a temperature of 80 °C. The content of oxygen was not controlled. The S-N curve is depicted in Fig. 5. The fatigue limit was determined on the basis of four undamaged specimens to a value of 160 MPa. The formula for the slanted branch of the S-N curve for a corrosive environment was determined by the linear regression in the form:

$$\log N_a = 15.6 - 4.1 \log \sigma_a \tag{2}$$

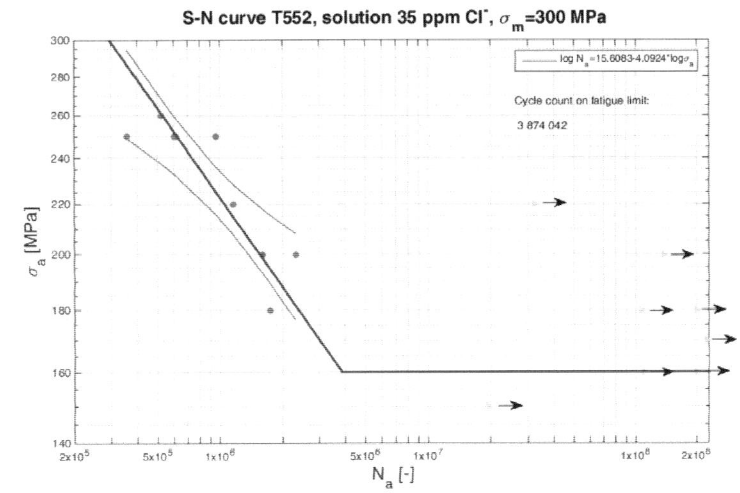

Figure 5. S-N curve for pre-stress in normal component σ_m = 300 MPa tested in a corrosion cell for an aggressive environment (solution of 35 ppm chlorides) at a temperature of 80°C.

5 Conclusions

The results of high cycle fatigue tests on blade steel Böhler T552 are summarized. The tests were performed in air and in a corrosive environment of a solution with a concentration of 35 ppm chlorides at a temperature of 80 °C. The results can be compared with experiments published by another workplace in [1] where a fatigue limit of 315 MPa was obtained for pre-stress 348 MPa. This is a very similar result because newly published experiments showed a fatigue limit of 360 MPa for pre-stress 300 MPa.

A fatigue limit of 265 MPa was obtained in [1] for pre-stress 292 MPa for tests in aerated 6 ppm Cl- solution at 90 °C. In this article the fatigue limit is about 100 MPa lower for almost the same pre-stress. The more aggressive solution is obviously the reason. To sum it up, the solution of 35 ppm chlorides decreased the fatigue limit by about 200 MPa compared to the air condition.

Another experimental programme for this steel includes corrosion fatigue tests on specimens with improved surface corrosion resistance.

Acknowledgements

This work was financially supported by the Ministry of Education, Youth and Sport Czech Republic Project LQ1603 (Research for SUSEN). This work has been performed within the SUSEN Project in the framework of the European Regional Development Fund (ERDF) in the project CZ.1.05/2.1.00/03.0108 and by the Technology Agency of the Czech Republic (TA CR) under project TA04020685.

References

[1] Schönbauer B M, Stanzl-Tschegg S E, Perlega A, Salzman R N, Rieger N F, Zhou S, Turnbull A, Gandy D, Fatigue life estimation of pitted 12% Cr steam turbine blade steel in different environments and at different stress ration, International Journal of Fatigue **65** (2014) 33-43.

[2] Steam, Chemistry, and Corrosion in the Phase Transition Zone of Steam Turbines, EPRI, Palo Alto, CA: 1999.

[3] Černý M, Alloy Corrosion in Steam Turbines, Experimental Stress Analysis 2014.

[4] Černý M, Alloy Localized Corrosion in Steam Turbines, Experimental Stress Analysis 2015.

[5] Strejcius J, Folková E, Kasl J, Řehořek J, Špirit Z, Corrosion fatigue strength of precipitation hardened martensitic steels MLX17 & T671 in a chloride ionts - contaminated steam condensate, Increase of Lifetime for Components of Energy Equipment in Power Plants, Srní, Czech Republic, 2015, (in Czech).

[6] Janoušek J, Strejcius J, Hřeben S, The Environment Influence On Bladed Steel During Operational Conditions, Applied Mechanics and Materials Vol. **827** (2016) pp 185-188.

[7] Low-Pressure Steam Turbine Corrosion Mechanisms and Interactions: State of Knowledge 2010, Report, EPRI.

[8] Janoušek J, Hřeben S, Špirit Z, Strejcius J, Kasl J, Fatigue performance of blade steel T671 for different kinds of loading, Baltica X, Finland, 2016.

[9] Development of a Corrosion-Fatigue Prediction Methodology for Steam Turbine Blades: AISI 403/410 (12% Cr) and 17-4PH Blade Steels, EPRI, Palo Alto, CA: 2015.

[10] Míšek T, Dynamic Analysis of Bladed Disks of Axial Turbomachinery, doctor thesis, UWB Pilsen, 2011, (in Czech).

[11] Balda M, Červená O, Monitoring Residual Fatigue Lives of Steam Turbines Blades, Experimental Stress Analysis 2015.

[12] ČSN 420363 – Fatigue Testing of Metals – Methods of fatigue testing of metals.

[13] Černý M J, Allowed Stress Range Prediction of Fatigue Loading for Steel Alloy for Steam Turbines, Experimental Stress Analysis 2016.

[14] Atest – Inspection Certifikacate of producer and provider METAL RAVNE (Slovenia).

IOP Conf. Series: Materials Science and Engineering **179** (2017) 012035 doi:10.1088/1757-899X/179/1/012035

Effect of silicon on stability of austenite during isothermal annealing of low-alloy steel with medium carbon content in the transition region between pearlitic and bainitic transformation

Š Jeníček, I Vorel, J Káňa, K Ibrahim, V Kotěšovec

University of West Bohemia, Regional Technological Institute, Univerzitní 22, Plzeň, Czech Republic

E-mail: jeniceks@vctt.zcu.cz

Abstract. In a vast majority of steels, a prerequisite to successful heat treatment is the phase transformation of initial austenite to the desired type of microstructure which may consist of ferrite, pearlite, bainite, martensite or their combinations. Diffusion plays an important role in this phase transformation. Together with enthalpy and entropy, two thermodynamic quantities, diffusion represents the decisive mechanism for the formation of the particular phase. The basis of diffusion is the thermally-activated movement of ions of alloying and residual elements. It is generally known that austenite becomes more stable during isothermal treatment in the transitional region between pearlitic and bainitic transformation. This is due to thermodynamic processes which arise from the chemical composition of the steel. The transformation of austenite to pearlite or bainite is generally accompanied by formation of cementite. The latter can be suppressed by adding silicon to the steel because this element does not dissolve in cementite, and therefore prevents its formation. The strength of this effect of silicon depends mainly on the temperature of isothermal treatment. If a steel with a sufficient silicon content is annealed at a temperature, at which silicon cannot migrate by diffusion, cementite cannot form and austenite becomes stable for hours.

1 Introduction

Phase transformation from austenite to pearlite that takes place during continuous cooling or during isothermal treatment is controlled by diffusion. It is a reconstructive phase transformation based on thermally activated migration of interstitial atoms, substitutional ions of alloying addition, and iron ions. The control mechanism of this transformation is diffusion. The rate of transformation is mainly dictated by the temperature at which the treatment and the austenite-pearlite transformation are taking place. Bainitic transformation is semi-diffusional in nature, as it is accompanied by diffusion of only the interstitial atoms. The austenite-bainite transformation itself is based on the shear mechanism.

The transition region between pearlitic and bainitic transformations is characterized by increased stability of austenite. This is due to the thermodynamic conditions, which arise from the chemical composition of the steel . The reason is the fact that pearlitic transformation takes place through formation and growth of nuclei which are highly sensitive to the temperature of transformation [1]. At lower temperatures, the nucleation and growth slow down, whereas the free enthalpy of supercooled austenite increases. The amount of supercooling plays a key role in the bainitic transformation [2]. This results in a situation where the amount of supercooling of austenite is sufficient to prevent heterogeneous nucleation of pearlite but at the same time, it is inadequate to induce the shear mechanism of bainitic transformation [3,4,5]. By this means, and predominantly during isothermal treatment, a region of higher stability of austenite arises, as known from TTT diagrams. Silicon is generally understood to prevent cementite precipitation [6,7,8]. Its effect depends on the temperature of isothermal treatment. When steel is annealed at a temperature, which impedes silicon diffusion, cementite cannot precipitate within a short timeframe [9].

2 Experimental programme

The objective of this experimental programme was to examine the possible effects of the silicon content and the temperature of isothermal treatment in the region between pearlitic and bainitic transformations on austenite stability and on the resulting microstructure in a low-alloy steel with 0.56 weight percent carbon. (Table 1).

Table 1. Chemical composition of experimental steel.

C	Si	Mn	Cr	P	S	M_s [°C]	M_f [°C]
0.56	2.0	0.59	1.33	0.005	0.003	252	126

Thermal treatment sequences were proposed for experimental isothermal annealing in the transition region between pearlitic and bainitic transformations. The specimens were heated to 950 °C for 300 seconds to obtain homogenous austenite, and then cooled to various temperatures in the region between the pearlite and bainite noses of the diagram. These temperatures were chosen so that the various annealing sequences took place in pearlitic transformation and bainitic transformation regions, as well as in the transition region between them, as seen in the TTT diagram of the experimental steel which was constructed using calculations completed in the JMatPro program (Figure. 2)

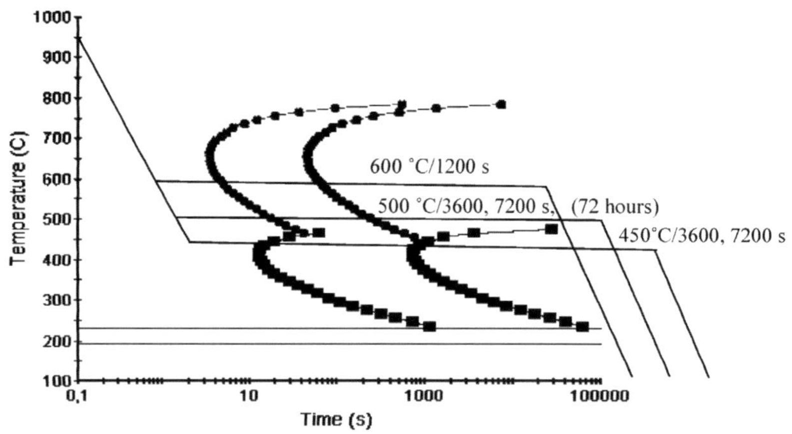

Figure 2. Experimental isothermal annealing sequences in the transition region between pearlitic and bainitic transformations.

The isothermal holding times were 1200, 3600, and 7200 seconds. After the annealing, the specimens were quenched in water. The specimens were threaded round bars with a diameter of 8 mm. Prior to this experiment, they had been annealed at 1050 °C for 15 minutes. The heating and cooling was carried out in a thermomechanical simulator with a induction-resistance heating capability.

3 Results and discussion

Microstructures of the specimens were examined using scanning electron microscopy. The etchant was 3 % nital. For reference, hardness was measured on the metallographic sections.

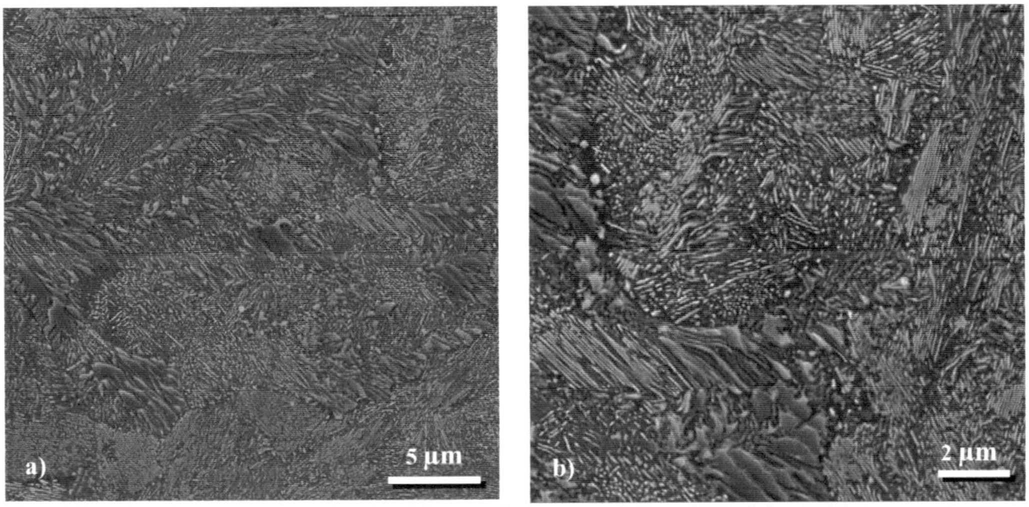

Figure 3. Micrograph of the specimen annealed at 600 °C for 1200 s
a) pearlite – 5000×, SEM, b) pearlite – 10000×, SEM.

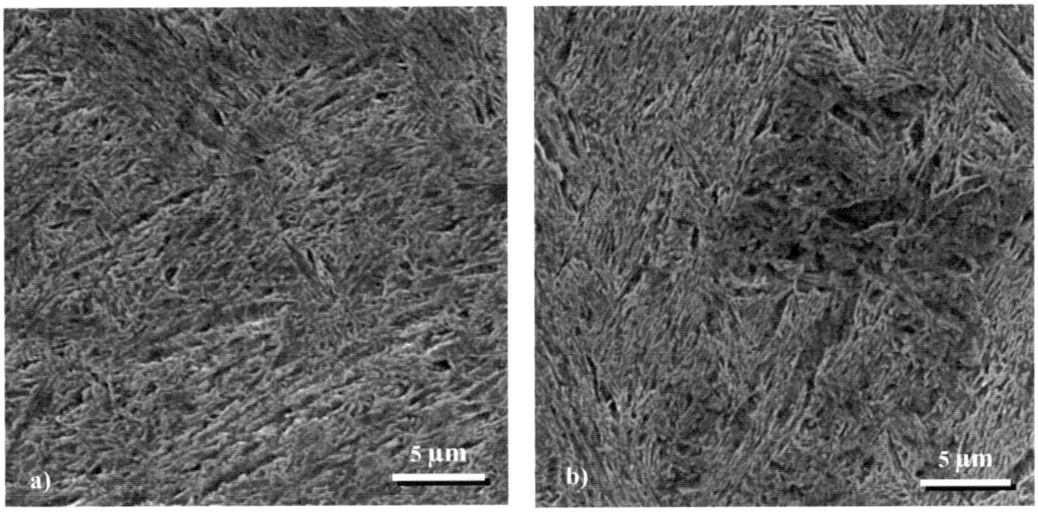

Figure 4. Microstructures of specimens annealed at 500 °C
a) specimen 500 °C/3600 s – martensite – 5000×, SEM, b) specimen 500 °C/7200 s – martensite – 5000×, SEM.

In specimens treated at 600 °C/1200 s, the microstructure consisted entirely of lamellar pearlite with a hardness of 344 HV10 (Figure 3). Specimens 500 °C/3600 s and 500 °C/7200 s contained martensite with a hardness of 720 and 728 HV10, respectively (Figure 4). These findings suggest that the austenite-pearlite transformation has not occurred, although it should have taken place in full, according to the theoretical diagram. For this reason, an additional sequence was carried out: annealing at 500 °C for 72 hours (referred to as 500 °C/72 hrs). It was completed in a salt bath placed in the heat treatment furnace. Specimen 500 °C/72 hrs contained a mixture of lamellar pearlite and martensite (Figure 5). The hardness of this specimen was 430 HV10.

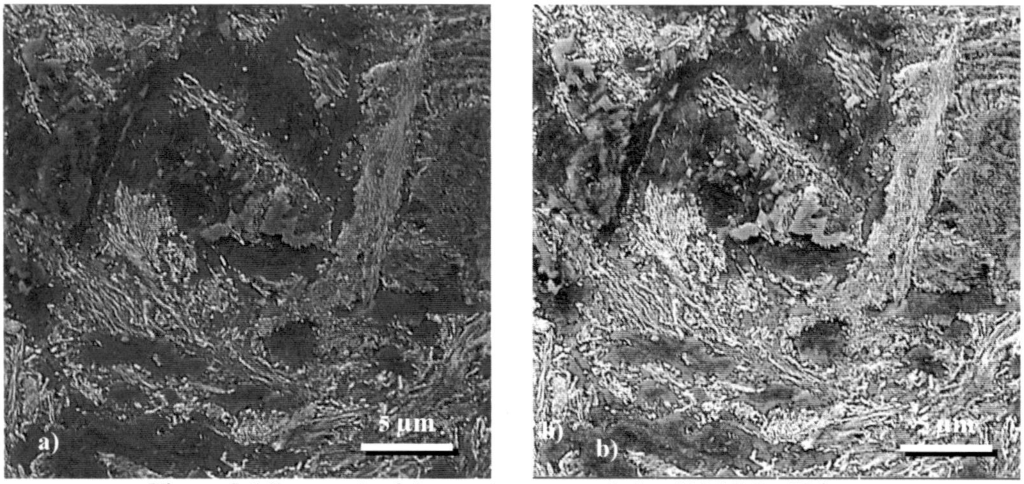

Figure 5. Micrograph of the specimen annealed at 500 °C for 72 hours
a) Pearlite – martensite – 5000×, SEM, b) Pearlite– martensite – phase contrast micrograph – 5000×, SEM.

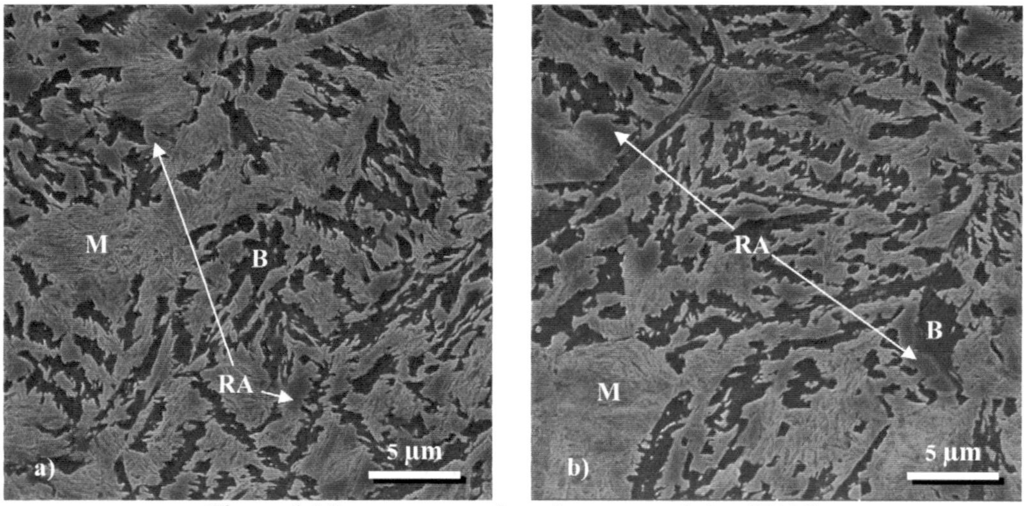

Figure 6. Microstructures of specimens annealed at 450 °C
a) specimen 450 °C/3600 s – bainite – martensite – retained austenite – 5000×, SEM, b) specimen 450 °C/7200 s – bainite – martensite – retained austenite – 5000×, SEM.

In specimen 450 °C/3600 s, microstructure consisted of a mixture of martensite, retained austenite, and upper bainite with a hardness of 520 HV10 (Figure 6a). Isothermal annealing for an extended period led to a slight increase in the bainitic phase and retained austenite fractions at the expense of the martensitic phase fraction (Figure 6b). The hardness of specimen 450 °C/7200 s was 525 HV10 (Table 2).

Table 2. Comparison of microstructures and hardness for different isothermal annealing temperatures and holding times.

Isothermal annealing temperature [°C]	Holding time [hours]	Microstructure	Hardness HV10
600	1/3	Pearlite	344
500	1	Martensite	720
500	2	Martensite	728
500	72	Martensite, pearlite	430
450	1	Martensite,bainite, retained austenitie	520
450	2	Martensite,bainite, retained austenitie	525

The effect of silicon on microstructural evolution in specimens 600 °C/1200 s can be explained with reference to the diffusional nature of the austenite-pearlite transformation. The isothermal annealing temperature was sufficient for the diffusion of substitutional elements including silicon – which therefore could not prevent the formation of cementite. The formation of cementite is one of the preconditions for the pearlitic reaction. The situation in specimens 500 °C/3600 s and 500 °C/7200 s was, in all likelihood, such that the undercooling of austenite was large enough to enable heterogeneous nucleation of pearlite but, at the same time, the intensity of silicon diffusion reached the critical limit for formation of pearlitic cementite. This means that the carbon-enriched lamellae which would at higher temperatures form cementite, contained such an amount of silicon which in fact prevented cementite from forming. The potential for reduction of silicon level in the lamellae which were super-saturated with carbon was, however, severely limited by the low temperature. This hypothesis could explain the long times of phase transformation in the sequences denoted as 500 °C/3600 s, 500 °C/7200 s, and 500 °C/72 hrs because the austenite-pearlite transformation would strongly depend on the diffusion rate of silicon.

In specimens 450 °C/3600 s and 450 °C/7200 s, the effect of silicon was manifested predominantly in the mechanism of bainitic transformation, more precisely in suppressing the carbide precipitation along the boundaries of ferrite laths which had formed by shear.

Given the above, it should be noted that besides the influence of silicon is necessary to take into consideration the effect of manganese and carbon, due to their effect on the stabilization of austenite. This step will be further aim of the research, which will be simultaneously carried out with regard to the distribution of manganese, carbon and silicon between observed phases. Further, detailed analysis will be made of the observed microstructures with respect to carbide precipitation. The comparison of the results with the results of an experiment on steel of similar carbon content and manganese but different silicon content. will be carried out

4 Conclusion

A low-alloy steel containing 0.56 % carbon, 2 % silicon, 0.56 % manganese, and 1.33 % chromium was used for a study of the stability of austenite during isothermal annealing in the transition region between pearlitic and bainitic transformations, in relation to high silicon content in the steel. It was proven that in this steel austenite can remain stable during isothermal annealing at 500 °C for periods of several hours. This finding could be explained by the effect of silicon on the mechanism of pearlite formation in steels. At sufficiently low temperatures which nevertheless still enable heterogeneous nucleation of pearlite on the boundaries of austenite grains and its growth by diffusion, higher silicon level can considerably slow down the austenite-pearlite transformation. It is very likely that this is due to very low but non-zero diffusivity of silicon which can ultimately result in a very slow austenite-pearlite transformation.

Next stage of the research will be focused on the distribution of manganese, carbon and silicon between the present phases, the influence of these elements on the course of phase transformations of the austenite and detailed microstructure analysis.

Acknowledgements
This paper includes results of the project LO1502 Development of Regional Technological Institute. The project is subsidised by the Ministry of Education of the Czech Republic from specific resources of the state budget for research and development.

References

[1] Pandit A 2011 Theory of the Pearlite Transformation in Steels Ph.D. Thesis, University of Cambridge
[2] Jaramillo R A and Babu S S 2005 Austenite Decomposition in Fe-C-Si-Mn Steel under 30 Tesla Magnetic Field *Solid→Solid Phase Transformations in Inorganic Materials* Volume 1, pp. 873-878.
[3] Bhadeshia H K D H 2001 Bainite in Steels 460p
[4] Caballero F G and Bhadeshia H K D H 2002 Very Strong Low Temperature Bainite *Materials Science and Technology*, 18 279-284
[5] Speer J G and Edmonds D V and Rizzo F C and Matlock D K 2004 Partioning of carbon from supersaturated plates of ferrite, with application to steel processing and fundamentals of the bainite transformation Current Opinion in Solid State and Materials Science 8:219–237
[6] Kucerova L 2011 Comparison of Microstructures and Properties Obtained After Different Heat Treatment Strategies of High Strength Low Alloyed Steel Journal of Iron and Steel Research International vol. 18, pp. 427-431
[7] Jirkova H, Kucerova L, Masek B 2012 Effect of Quenching and Partitioning Temperatures in the Q-P Process on the Properties of AHSS with Various Amounts of Manganese and Silicon Materials Science Forum, vols. 706-709, pp. 2734-2739
[8] Masek B et al. 2011 Improvement of Mechanical Properties of automotive Components Using Hot Stamping with Integrated Q-P Process Journal of Iron and Steel Research International, vol. 18 (1-2), pp. 730-734.
[9] Fielding L C D The Bainite Controversy 2013 *Materials Science and Technology*, 9, 383-399

IOP Conf. Series: Materials Science and Engineering **179** (2017) 012036 doi:10.1088/1757-899X/179/1/012036

Semi-solid processing of high-chromium tool steel to obtain microstructures without carbide network

H Jirková[1], D Aišman[1], K Rubešová[1], K Opatová[2] and B Mašek[1]

University of West Bohemia, Research Centre of Forming Technology, Univerzitní 8, 306 14 Pilsen, Czech Republic

University of West Bohemia, Regional Technological Institute, Univerzitní 8, 306 14 Pilsen, Czech Republic

E-mail: hstankov@kmm.zcu.cz

Abstract. Treatment of high-alloy tool steels that involves transition to the semi-solid state can transform the sharp-edged primary carbides which usually form during solidification. These carbides severely impair toughness and are virtually impossible to eliminate by conventional treatment routes. Upon classical semi-solid processing which dissolves these carbides, the resulting microstructure consists of polyhedral and super-saturated austenite embedded in lamellar austenite-carbide network. This type of microstructure reflects in the mechanical properties, predominantly in material behaviour under tensile loading. Such a network, however, can be removed by appropriate thermomechanical treatment.

In the present experiment, various procedures involving heating to the semi-solid state were tested on X210Cr12 tool steel. The feedstock was heated to the temperature range of 1220 – 1280 °C. The heating was followed by procedures involving either water quenching to the forming temperature, room temperature or temperature from the range from 500 °C to 1000 °C followed by reheating to the forming temperature. It was found that the development of the lamellar network strongly depends on the temperature of heating to semi-solid state. Thermomechanical treatment produced microstructures in which the matrix consisted of a mixture of polyhedral austenite grains and the M-A constituent. In addition, the initial lamellar eutectic network was partially or even completely melted and substituted with a mixture of very fine recrystallized austenite grains and precipitates of chromium carbides. Some fine M_7C_3 carbides were present in the austenitic-martensitic matrix as well. When appropriate processing parameters were chosen, very good mechanical properties were obtained, among them a hardness of 860 HV10.

1 Introduction

The high-alloy tool steels of the X210Cr12-type produced by conventional metallurgy contain large sharp-edged M_7C_3 carbides which remain stable even at high temperatures [1]. Consequently, there is virtually no way of to modify these carbides by means of conventional heat treatment to a more favourable morphology, i. e. to the microstructure with finer and more uniformly dispersed carbides. Since large sharp-edged primary carbides impair toughness, this kind of steels must be produced by powder metallurgy which can eliminate the risk of formation of large chromium carbides.

Primary M_7C_3 chromium carbides can dissolve at temperatures above the solidus. Therefore, a method based on semi-solid-processing has been proposed for transition of these carbides leading to the microstructure refinement of high-alloyed tool steels. During semi-solid processing, the material exists as a mixture of liquid and solid phases. If the material in the semi-solid state exhibits thixotropy, it can be shaped by thixoforming. Thixoforming is a technique that can produce intricate-shape parts using a single forming cycle [4]. During thixoforming, microstructures characterized by polyhedral islands of super-saturated austenite surrounded by lamellar austenite-carbide-eutectics are developed [2, 3]. Hence, no large sharp-edged primary carbides remain in the structure. In these structures, austenite exhibits extraordinary thermal stability. Its thermal decomposition begins only at temperatures higher than 500 °C. The decomposition finishes when the material is annealed at 550–600 °C [5]. Austenite itself is ductile but the carbide network cannot sustain sufficient plastic deformation at room temperature. Nevertheless, these structures can be formed successfully by compressive deformation, with the achievable plastic deformation being no less than 35 %. It has been verified experimentally that they can also be formed at high temperatures between 1000 °C and the solidus temperature [6, 7]. At an appropriate temperature, the austenite-carbide network can be broken up by forming. With the right amount and intensity of deformation, carbides can be dispersed uniformly throughout the austenitic matrix. After cooling, these carbides can remain dispersed, and contribute to the strength of the resulting structure. In addition, they prevent austenite grain growth upon high-temperature deformation. Applied deformation and temperature can also cause partial dissolution of these carbides. After reprecipitation, they can contribute to the strengthening of the matrix. To achieve optimal properties, the matrix can be altered by additional heat treatment, such as quenching and tempering or even quenching and partitioning of carbon. If mechanical working finishes under appropriate conditions, a structure with fine martensite is obtained.

2 Experimental programme

To obtain carbide precipitates with an optimal distribution, sizes, and morphology, and microstructures with optimal austenite grain sizes and martensite volume fractions, it is necessary to optimize a whole range of treatment parameters, starting with the temperature of heating into the semi-solid state. The X210Cr12 tool steel was selected for this pilot experimental programme.

2.1 Experimental material

X210Cr12 is a high-carbon and high-chromium steel (table 1). It has been developed for applications in punching machines and presses, mainly for heavy-duty punches and highly-complex progressive and combination tools. The initial annealed microstructure contained large sharp-edged primary chromium carbides and very fine cementite embedded in a ferritic matrix.

Table 1. Chemical composition of X210Cr12 steel [wt. %].

C	Cr	Mn	Si	Ni	P	S
1.8	11	0.2	0.2	0.5	0.03	0.035

Before definition of processing parameters, it was necessary to identify the freezing range and the dissolution temperature of chromium carbides. To this end, an approximate calculation using the JMatPro software was carried out [8]. It was found that up to 758 °C, the material retains a stable ferrite-cementite microstructure. It begins to melt at 1225 °C and finishes melting at 1373 °C. According to computation, the primary chromium carbides finish to dissolve at 1255 °C.
When the material is heated into the semi-solid state and subsequently cooled, its microstructure changes strongly. The resulting microstructure consists of grains of super-saturated austenite embedded in a lamellar ledeburitic network. This network comprises austenite and secondary M_7C_3 carbides.

2.2 Thermomechanical treatment

To break up the lamellar network and initiate recrystallization, the material had to be worked using a large amount of deformation. In the present case, open-die forging in a hydraulic press was used. As the material was to be heated into the semi-solid state, the workpiece was enclosed in a case made of low-carbon steel. This simplified the handling of the partially melted material and made its temperature field more homogeneous. The case of 30 mm diameter, 6 mm wall thickness, and 55 mm length was made of the SJ355 low-carbon steel with the melting temperature above 1400 °C. The workpieces were heated in an air furnace. Flat dies were used for the forming operation. A total of 10 different treatment procedures were carried out (table 2).

Table 2. Parameters of thermomechanical treatment procedures.

Procedure no.	Heating temperature [°C]	Holding time [min]	Cooling temperature [°C]	Reheating temperature [°C]	Holding time at reheating temp.[min]	No. of forming steps [-]	HV10 [-]
1	1265	15	500	1050	5	1	520
2	1265	15	500	1100	5	1	487
3	1265	15	RT	1050	12	1	520
4	1220	15	600	1050	6	3	788
5	1220	15	1100	–	–	4	803
6	1225	60	900	1080	1.5	5	836
7	1200	15	1000	1070	2	4	848
8	1240	15	900	1080	1.5	5	864
9	1240	60	900	1080	1.5	5	855
10	1280	16	900	1080	1.5	5	866

Procedures 1 through 3 involved the heating temperature of 1265 °C with a holding time of 15 minutes. This temperature had been tested in earlier experiments conducted in a thermomechanical simulator [3]. According to the calculations carried out using the JMatPro software, all primary chromium carbides should be dissolved, and the structure should consist of the melt and austenite at this temperature. The calculated proportion of the melt was 30 %. The treatment variants included quenching in water to the temperature of 500 °C (procedures 1 and 2) and to the room temperature (procedure 3), followed by material reheating to the forming temperature, either 1050 °C or 1110 °C, and holding for 5 minutes. The specimens were upset to a half height in a single operation.

Subsequently, a reduced heating temperature of 1220 °C was tested. It is just below the calculated solidus temperature. At this temperature, the microstructure still contains about 8 % of M_7C_3 carbides. In procedure 4, quenching to the temperature of 600 °C was followed by reheating in a furnace to the forming temperature of 1050 °C. The specimen was first upset to a half height, then drawn-out to 50 mm and then upset again all the way to the height of 20 mm. Another variant (procedure 5) had the same heating temperature but involved cooling to no less than 1100 °C followed by forming: upsetting – drawing-out – upsetting – drawing-out. The forming procedure finished at the temperature below 800 °C. In order to explore the effects of the holding time, the procedure 6 had the holding time extended to 60 minutes. The aim was to ascertain whether the austenite grains coarsen, whether coarser grains affect the morphology of recrystallized grains after forming, and whether primary chromium carbides dissolve in any greater extent.

In another variant (procedure 7), the heating temperature was reduced even more, i. e. to 1200 °C. At this temperature, the material is not expected to contain any liquid phase, and therefore the effects of liquid fraction on microstructural evolution from previous experiments can be compared. Upon heating, the microstructure consisted of a mixture of austenite and 9 wt. % of carbides according to JMatPro calculation. The heating temperature of 1240 °C was combined with the holding times of 15 min (procedure 8), and 60 min (procedure 9). This temperature approaches the limit, at which primary chromium carbide are fully dissolved. Nevertheless, their fraction is still approximately 7 wt. % at this temperature. The temperature at the end of cooling was 900 °C, after which the material was reheated to the forming temperature of 1080 °C.

In order to be able to describe the impact of a larger liquid fraction on the microstructure evolution, procedure 10 was designed with the highest heating temperature of 1280 °C. It was expected to lead to complete dissolution of carbides and even to partial melting of austenite. According to calculations, the liquid fraction is 35 %.

The microstructures were examined for their homogeneity using optical microscopy. More detailed analysis of microstructure was performed in a scanning electron microscope. The distribution of chromium was mapped by EDX, and microstructure constituents identified using EBSD. Mechanical properties were measured by HV10 hardness testing.

3 Results and discussion

Procedure 1 that involved the heating temperature of 1265 °C and a single upsetting operation led to a strongly non-uniform microstructure with a distinct diagonal cross pattern on the cross-section through the specimen. As expected, in the locations which had not been strongly affected by deformation a structure of polyhedral austenite grains embedded in a lamellar ledeburitic network was developed (figure 1). The location of the largest strain exhibited a heavily distorted structure in which austenite grains were elongated, and the austenite-carbide network was somewhat fragmented but still enveloping isolated austenite grains. There were very fine recrystallized austenite grains within the network (figure 2). Lamellar structure was found in some locations. The hardness was 520 HV10. Another slight increase in the forming temperature, from 1050 °C to 1100 °C, caused no considerable changes in the microstructure evolution.

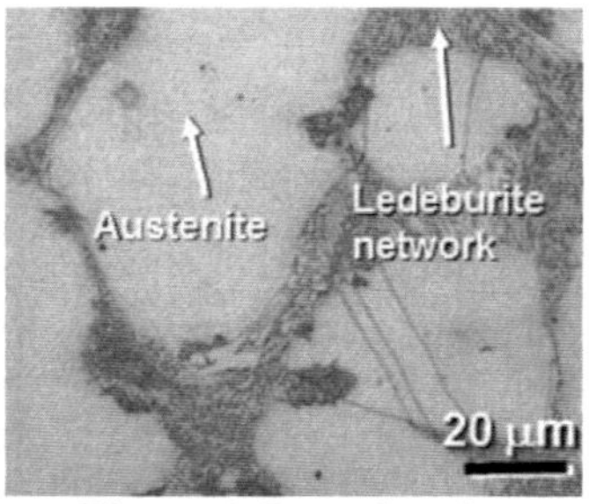

Figure 1. Procedure 1 – Deformation-free microstructure.

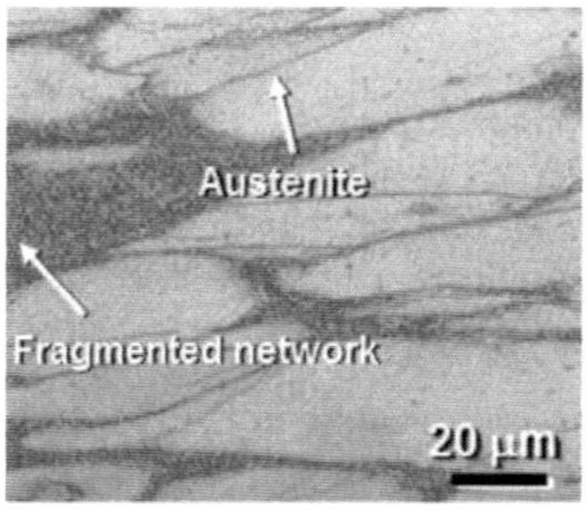

Figure 2. Procedure 1 – Microstructure upon the heaviest deformation.

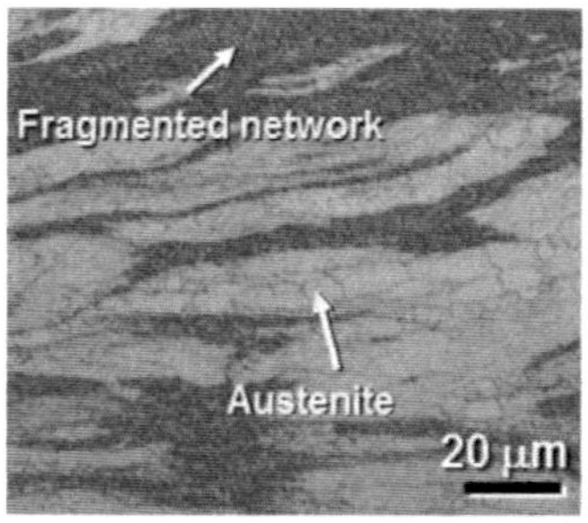

Figure 3. Procedure 3 – Microstructure upon the heaviest deformation – process of formation of slip bands in austenite grains.

After cooling all the way down to room temperature and reheating to the forming temperature of 1050 °C, the region along the axis of the specimen exhibited effects of recrystallization in austenite grains and formation of slip bands and subgrains (figure 3). The character of lamellar network was preserved in some areas.

If the heating temperature was reduced to 1220 °C and 1225 °C, i. e. just below the solidus curve, the primary chromium carbides did not dissolve (figure 4). The character of the microstructure depended on the subsequent thermomechanical treatment procedure. Unlike in the first three variants, there was a greater deformation intensity because the forming involved between 3 and 5 reductions. Quenching in water to the temperature of 600 °C and subsequent reheating to the forming temperature of 1050 °C (procedure 4) had produced a matrix consisting of a mixture of fine recrystallized austenite grains which transformed to martensite, resulting in the M-A constituent. This was only visible using large magnifications which confirmed the presence of martensite (figure 5). The units of the M-A constituent had an approximate size of 1 μm, with some units sized up to 3 μm. This matrix contained not only primary chromium carbides but also secondary chromium carbides present between the units of the M-A constituent. Since no melting had occurred during heating, no ledeburite network formed. All these factors were reflected in the increase in hardness to 788 HV10 (table 2). When cooling took place to no less than 1100 °C and immediately followed by forming without any reheating, the character was similar to variant 5. The larger amount of deformation was another factor why the fraction of unrecrystallized austenite grains was smaller. The hardness was 803 HV10. When the heating time was extended from 15 minutes to 60 minutes (procedure 6), there was no considerable change in the morphology of primary chromium carbides. The microstructure configuration was therefore very similar. Nevertheless, the hardness increased to 836 HV10. A further reduction in the heating temperature to 1200 °C (procedure 7) led to no considerable microstructural changes. The hardness was 848 HV10.

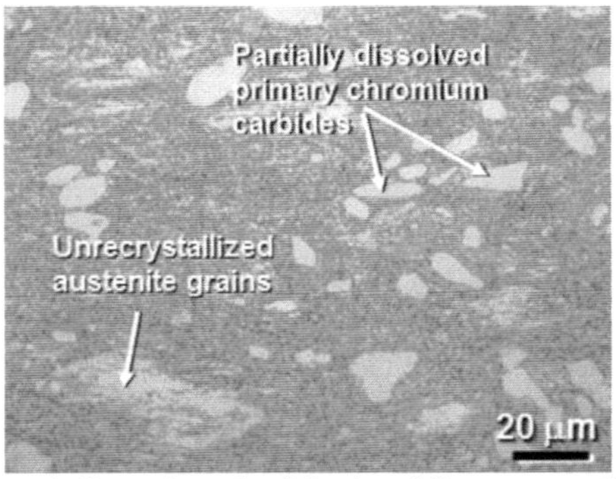

Figure 4. Procedure 4 – Matrix of fine recrystallized units of the M-A constituent with chromium carbide precipitates and primary chromium carbides that had dissolved to various level.

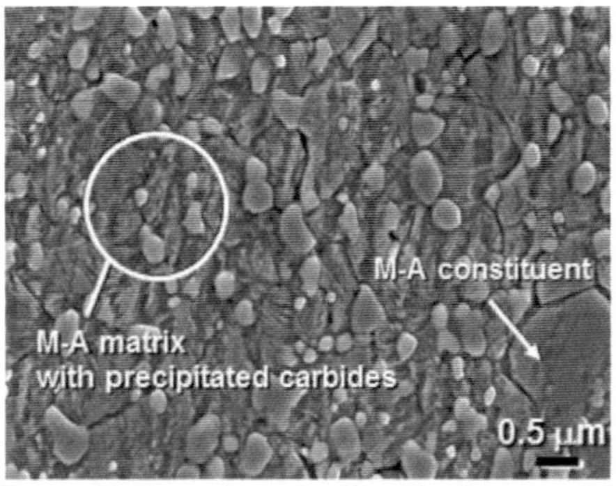

Figure 5. Procedure 4 – Detail of martensitic matrix with a small amount of austenite and secondary chromium carbide precipitates between units of M-A constituent.

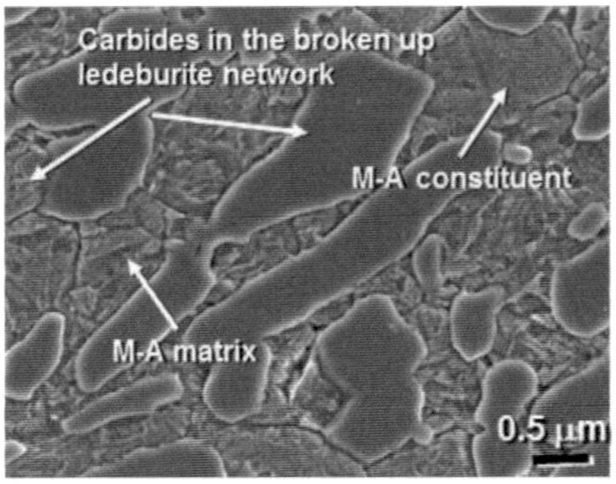

Figure 6. Procedure 8 – Detail of carbides in the broken up ledeburite network which is embedded in fine recrystallized austenite grains that transformed to martensite.

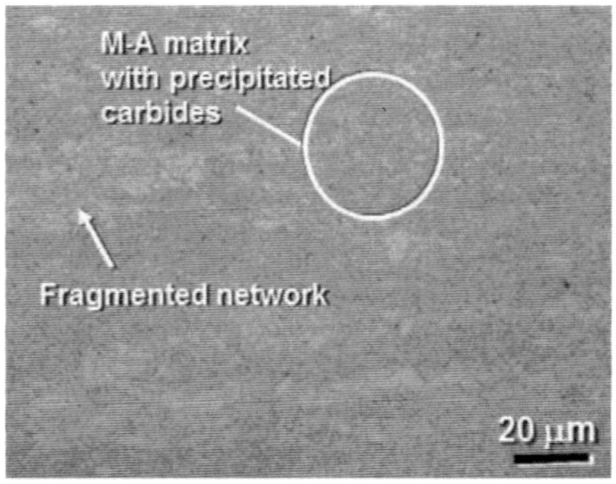

Figure 7. Procedure 10 – Martensitic-austenitic matrix with remnants of ledeburite network fragmented by deformation.

The experimental research continued by increasing the heating temperature to 1240 °C. Two procedures were tested with the only difference in the heating time (procedures 8 and 9). Thanks to five-step deformation, the resulting microstructure was very fine. The matrix consisted of recrystallized austenite grains which had mostly transformed to martensite and had the form of the M-A constituent. Their mean size was between 0.5 and 1 μm (figure 6). Chromium carbide precipitates were found in the matrix. Thanks to the presence of liquid phase after heating, the microstructure contained distinct remnants of carbide network along prior austenite grain boundaries. This network was mostly fragmented. EBSD analysis confirmed that the matrix consists predominantly of martensite and Cr_7C_3 carbides. The hardness was 864 HV10. The extension the holding time from 15 minutes to 60 minutes caused no considerable changes in the microstructure. The hardness was 855 HV10.

Thanks to a higher amount of melted phase present at the heating temperature of 1280 °C (procedure 10), there was a larger fraction of the ledeburite network. Its removal by forming was not as successful as in the previous cases (figure 7). Despite that, a very high hardness of 866 HV10 was obtained, thanks to a very fine austenitic-martensitic matrix with chromium carbide precipitates.

4 Conclusion

An unconventional thermomechanical treatment route was used to remove large sharp-edged primary chromium carbides in the X210Cr12 steel. These normally form during solidification at the metallurgical stage of production, and it is impossible to remove them by conventional heat treatment. The underlying principle was to use material heating to the semi-solid state to transform the initial microstructure to polyhedral austenite embedded in carbide-austenite network, and then use forming to break up this network and thus produce a fine microstructure of the M-A constituent and fine chromium carbide precipitates. The semi-solid processing stage was necessary in order to achieve the temperature at which the primary sharp-edged chromium carbides dissolve. By this means, the carbides can be converted to an austenitic-carbidic structure which can be hot-formed. Using plastic deformation, the carbide network can be broken up, and the carbides can be uniformly dispersed.

The results showed that the temperature of 1240 °C must be exceeded in order to obtain sufficient amount of dissolved primary chromium carbides. When lower heating temperatures were used, i.e. 1220 °C and 1200 °C, the microstructure became much finer thanks to dynamic recrystallization of austenite, high hardness was achieved, but the primary chromium carbides remained. They can be melted completely at heating temperatures higher than 1255 °C followed by multi-step deformation below the solidus temperature. Such a procedure can lead to microstructures in which a matrix of the M-A constituent prevails, the prior austenite grains are smaller than 1 μm, and fine chromium carbide precipitates are present. The hardness of such structures exceeds 850 HV10.

Acknowledgement

This paper includes results achieved within the project SGS-2015-028 Semi-solid Processing and New Structures without Carbide Net. The project is subsidised from specific resources of the state budget for research and development.

References

[1] Wang J, Guo W, Sun H, Li H, Gou H, Zhang J 2016 Plastic deformation behaviours and hardening mechanism of M7C3 carbide *Materials Science and Engineering* A **662** 88-94

[2] Rassili A, Atkinson H V 2010 A review on steel thixoforning *Trans. Nonferrous Met. Soc. China* **20** 1048-54

[3] Mašek B, Aišman D, Behulová M, Jirková H 2010 Structure of miniature components from steel produced by forming in semi-solid state *Trans. Nonferrous Met. Soc. China* **20** 1037-41

[4] Rogal L, Dutkiewicz J, Szklarz Z, Krawiec H, Kot M, Zimowski S 2014 Mechanical properties and corrosion resistance of steel X210CrW12 after semi-solid processing and heat treatment *Materials Characterization* **88** 100-110

[5] Jirková H, Aišman D, Mašek B 2010 Unconventional structure of X210Cr12 steel obtained by thixoforming *Journal of Alloys and Compounds* **504** S500–S3

[6] Aišman D, Jirková H, Kučerová L, Mašek B 2011 Metastabulka structure of austenite base obtained by rapid solidification in a semi-solid state *Journal of Alloys and Compounds***509** S312–S15

[7] Aišman D, Jirková H, Mašek B 2012 The influence of deformation and cooling parameters after transition through semi-solid state on structure development of ledeburite steel *Journal of Alloys and Compounds* **536** S204-S7

[8] JMatPro, Release 9.0, Sente Software Ltd., 2016

High-strength state of ultrafine-grained martensitic steel produced by high pressure torsion

M V Karavaeva[1], M A Nikitina[2], A V Ganeev, R K Islamgaliev

Ufa State Aviation Technical University, Institute of Physics of Advanced Materials
12 K. Marx str., Ufa 450008 Russia

E-mail: [1]karma11@mail.ru, [2]nik.marina.al@gmail.com

Abstract The paper presents the study results on the effect of severe plastic deformation (SPD) via high pressure torsion (HPT) on the structure and properties of martensitic steel. The contribution of different strengthening mechanisms in the strength of steel has been analyzed. It is shown that independently of the deformation temperature the main contribution in hardening belongs to grain boundaries (about 50 %), whereas the dislocation and solid solution components achieve 15 and 25 %, respectively.

1 Introduction

High strength in martensitic steels at room and elevated temperatures is traditionally ensured by two-stage thermal treatment. At the first stage as a result of quenching strengthening takes place due to phase hardening during martensitic transformation. At the second stage during heating of the quenched steel, the strength grows according to the mechanism of dispersion hardening at the expense of formation of precipitates [1-4].

It has been recently shown that the strength of steels can be enhanced using severe plastic deformation, for example by high pressure torsion (HPT) [5-8]. During HPT an ultrafine-grained microstructure with grain sizes less than 1 μm can be observed, which allows additional strengthening in accordance with the Hall-Petch ratio [9,10]:

$$\sigma_y = \sigma_0 + K_y d^{-1/2}, \qquad (1)$$

where σ_0 – the friction stress of a crystalline lattice, K_y – the coefficient of strengthening at the expense of grain boundaries, d – the grain size.

A grain size depends on the parameters of deformation and thermal treatment. Usually when a deformation temperature increases, grain sizes grow [11]. Therefore, in order to achieve the maximum strengthening effect due to microstructure refinement, one should reduce the deformation temperature. However, several publications have appeared recently [7,8,12], which demonstrate that the dependence of strength on the deformation temperature is not always decreasing for steels. The maximum strength for carbon steels 10 and 45 was observed after severe plastic deformation at 350 °C [7, 8]. The observed strength in the austenite steel after deformation at 300 °C is higher than that after deformation at room temperature [12]. This is explained by a complex microstructure observed in steels as a result of severe plastic deformation. The deformation temperature affects not only grain sizes, but also such structural features as sizes and distribution of secondary particles, formation of segregations, phase composition, formation of twins [7-8, 11-12]. The influence of temperature on the microstructure and properties of complex-alloyed steels and alloys, for which dispersed particles are one of the main factors for strength enhancement, is especially ambiguous.

This work deals with the possibility of strength enhancement in martensitic steel as a result of severe plastic deformation.

2 Material and procedure

Martensitic steel in the shape of a hot-rolled rod is taken as research material. Table 1 displays the chemical composition of the steel under study.

Table 1. Chemical composition of the research steel, wt.%.

C	Si	Mn	Ni	S	P	Cr	Mo	W	V
0.1 - 0.16	to 0.6	to 0.6	1.5 - 1.8	to 0.03	to 0.03	10.5 - 12	0.35 - 0.5	1.6 - 2	0.18 - 0.3

The steel samples with a diameter of 20 mm and a thickness of 1 mm in the quenched and tempered states are subjected to HPT at room and elevated T=300 °C temperatures. The pressure is 6 GPa, the number of rotations is N=10.

Thin foils for TEM investigations were cut at a distance of ½ of the radius from the sample center. After mechanical polishing of foils to a thickness of 0.1 mm, a disc with a diameter of 3mm is cut out. It is thinned on a jet polishing set "Tenupol-5" in the electrolyte based on n-butyl alcohol. The thinning is performed in the regime: U=60V at the current stabilization till holing. The fine structure of foils is studied using a transmission electron microscope JEM-2100 at an accelerating voltage of 200 kV and a scanning electron microscope JSM-6390.

Diffraction patterns were recorded on X-ray diffractometer DRON-4 with Co Kα emission using a monochromator, 0.02 ° step and a holding time of 5 seconds for each step. Analysis of diffraction patterns was performed using «MAUD» program [13] which is in the public domain[14]

The dislocation density ρ_{XRD} is determined according to the results of X-ray studies via the formula:

$$\rho_{XRD} = \frac{2\sqrt{3}\langle\varepsilon^2\rangle^{1/2}}{bd_{XRD}}, \qquad (2)$$

where $\langle\varepsilon^2\rangle^{1/2}$ – the level of elastic microdistorsions of the crystalline lattice; b – the Burgers vector of dislocations; d_{XRD} – the coherent scattering domain sizes.

The average grain size is measured by Random Intercept method.

The microhardness is measured on a Micromet-5101 set at a load of 100 g for 10 seconds.

3 Results

The microstructure of the steel in the initial state is a ferrite matrix with non-uniformly distributed particles of strengthening phases, which can be clearly seen on the image taken with the SEM (Fig.1a). The average grain size of the matrix phase is 1400±100 nm. The majority of particles is localized on grain boundaries of the ferrite matrix. The average size of particles is 171±30 nm, the volume fraction is 5 %. Clear grain boundaries can be observed in the fine structure (Fig. 1b). The majority of grains are free of dislocations. Pile-ups of dislocations can be seen in separate grains.

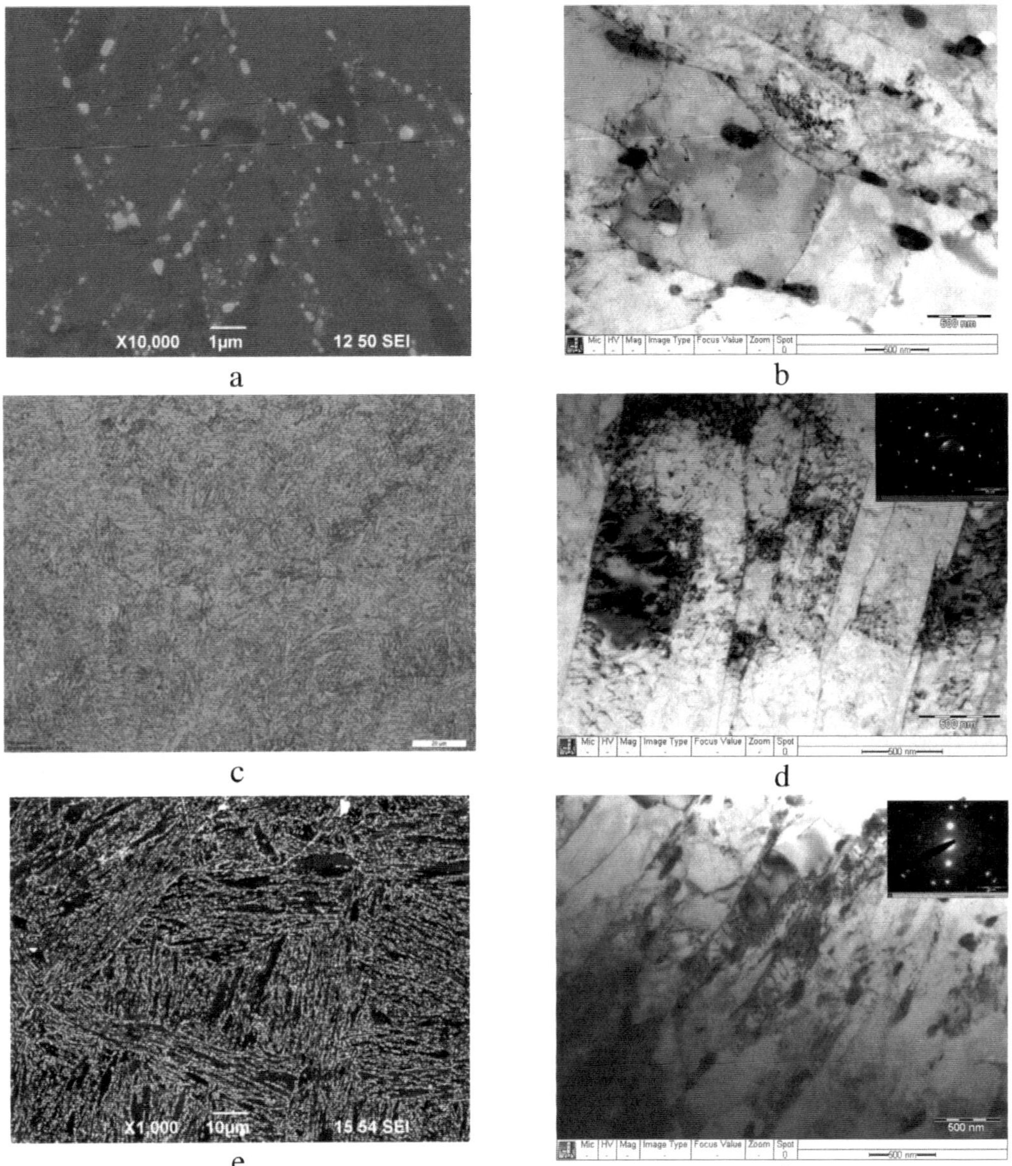

Figure 1. Microstructure of steel in the initial (a, b); quenched (c, d) and tempered (e, f) states.

Non-uniform distribution of particles in the volume, as a rule, decreases the deformability of the material. Therefore, prior to deformation the samples are quenched in oil after heating and holding at 1050°C, for 1 hour, then they are tempered at 800 °C for 1 hour.

After quenching martensitic microstructure can be observed (Fig.1 c, d). The average thickness of martensite plates is 518 nm. The martensite is characterized by high dislocation density within a plate. Separate dislocations as well as wide dislocation walls are observed within martensite plates. There are no secondary phase particles, which are typical of the initial state, in the martensitic structure.

After tempering uniform distribution of secondary phase particles in the volume is observed (Fig. 1 e, f). The particles precipitate in the boundaries of prior austenite grains and along the boundaries of martensite plates. Only about 25 % of particles are located in martensite plate bodies. The average size of particles is 250 nm, the volume fraction achieves 19 %. The plate structure of the martensite retains. The average size of plates is 556 nm.

As a result of HPT the microstructure changes significantly. The sizes of structural components are determined by the deformation temperature. Thus, after HPT at room temperature an ultrafine-grained microstructure with an average grain size of about 140 nm forms (Fig. 2 a, c). After torsion under HPT at an elevated temperature T=300°C the average grain size is 200 nm (Fig. 2 d, f). The diffraction pattern with spots located around the circle (Fig. 2c) testifies to high-angle misorientations. The dark-field images display (Fig. 2e) that after HPT at room temperature grains have the shape close to the equi-axed one.

After HPT at an elevated temperature the spots on the diffraction pattern are located around the circle, but they have azimuthal blurring indicating to high internal stresses (Fig. 2d).

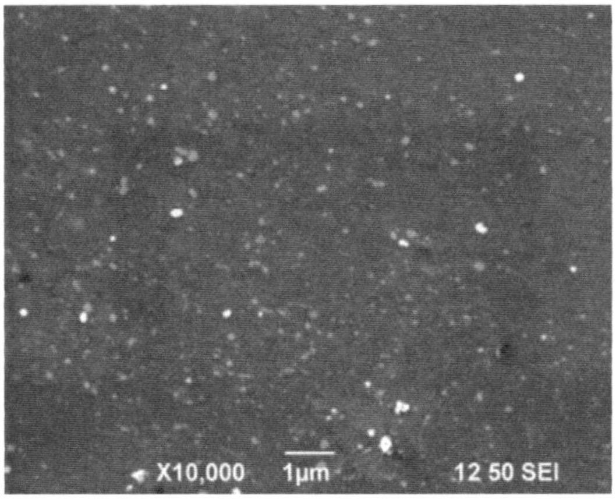

a

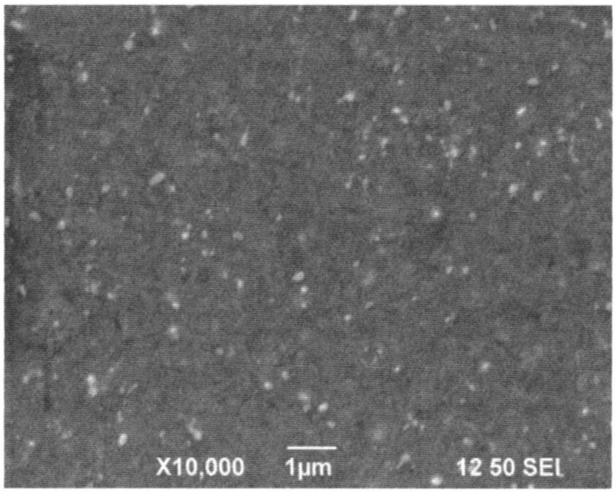

b

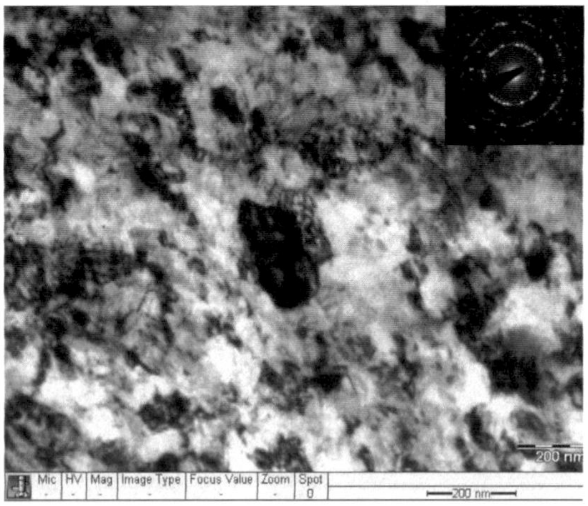

c

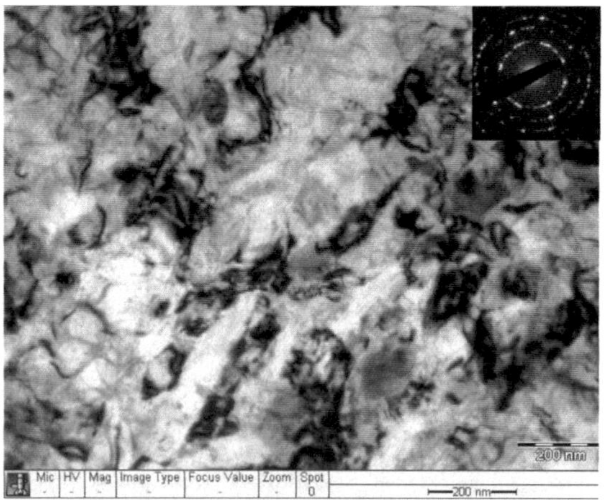

d

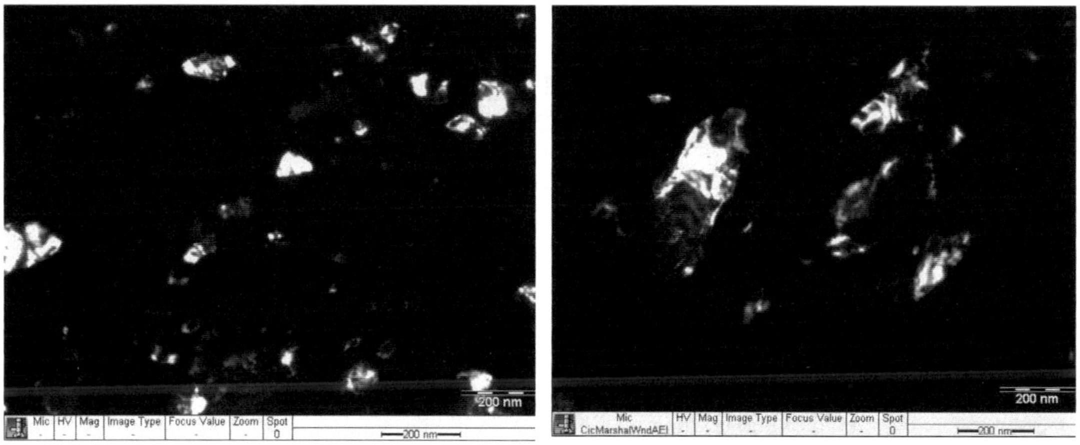

e f

Figure 2. Microstructure of steel 13Kh11N2V2MF-Sh after torsion: a, c, e – at room temperature; b, d, f – at T=300 °C; a, b –SEM; c-f – TEM: c, d – bright-field image, e, f – dark-field image.

Examination of the microstructure in SEM at a lower magnification (Fig. 2 a, b) demonstrates a significant reduction of the volume fraction of secondary phase particles in the deformed samples as compared to the quenched and tempered states (Fig.1 e). The volume fraction of particles is 4.5 % after HPT at 20 °C and 3.6 % after HPT at 300 °C. The average particle size is 109 nm after HPT at room temperature and 160 nm after HPT at 300 ° (Fig. 2 a, b). This is confirmed by the data of X-ray analysis of the samples., which are listed in Table 2.

Table 2. X-ray analysis results.

State	Lattice parameter, A	CSD size, nm	Microdistorsions, %
Quenching 1050°C+tempering 800 °C	2.87818(3)	1500	0,039
HPT at 20°C, 10 rot, 6 GPa	2.88043(5)	35	0,27
HPT at 300°C, 10 rot, 6 GPa	2.88021(8)	40	0,32

The results of X-ray analysis by the Rietveld method using the MAUD software package are listed in Table 2. These results demonstrate that in samples after HPT the level of microdistorsions of the crystalline lattice increased by an order as compared to the quenched and tempered steel. A much higher level of microdistorsions is observed after SPD at an elevated temperature: 0.32 % against 0.27 % for steel after SPD at room temperature. Besides, HPT results in increase of the lattice parameter as compared to the state after quenching and tempering. A significant reduction of CSD dimensions is observed (Table 2) as well.

Thus, the deformation temperature impacts not only the grain sizes, but also the quantity and sizes of secondary phase particles. In particular, after the deformation at an elevated temperature the particle sizes are larger, the volume fraction is lower than those after the deformation at room temperature.

The microstructure change as a result of severe plastic deformation affects considerably the steel microhardness (Fig. 3). The microhardness in the initial state is 3200 MPa, after quenching the microhardness increases up to 5100 MPa, and after tempering it returns practically to the value typical of the initial state – 3500 MPa.

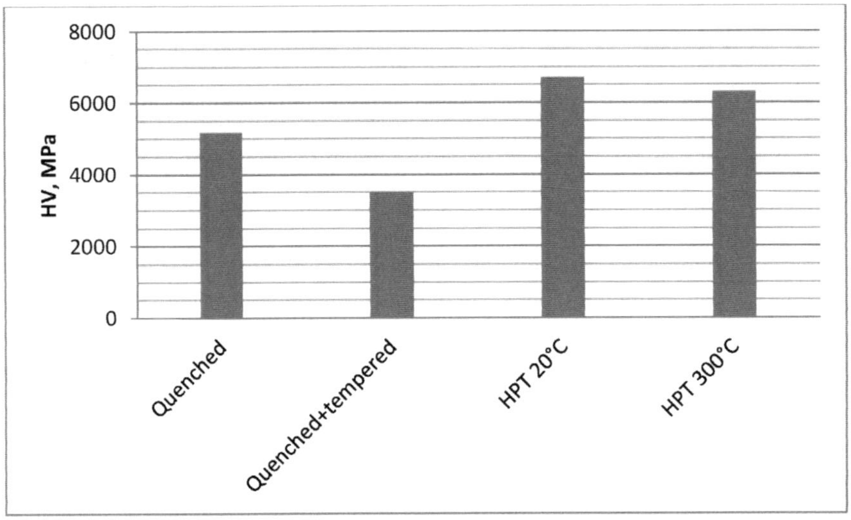

Figure 3. Microhardness of steel in different states.

As a result of SPD the microhardness increases practically two times as compared to the initial state and by 25 % as compared to the quenched state of steel (Fig.3). The deformation temperature insignificantly effects the microhardness. Thus, the microhardness value after HPT at room temperature is 6700 MPa, it is 6300 MPa after HPT at an elevated temperature.

4 Discussion

Achievement of a high-strength state in steels as a result of severe plastic deformation is ensured by formation of a unique microstructure, in which a number of strengthening factors is combined. First of all, it is an ultrafine-grained matrix that ensures grain boundary hardening in accordance with the Hall-Petch ratio [9, 10]. Besides, solid solution hardening of an alloyed ferrite and dispersion hardening at the expense of high-dispersion particles of the strengthening phases take place for the complex-alloyed steel.

Solid solution hardening ($\Delta\sigma_{SS}$) is determined by the content of the alloying element and its strengthening effect [15]:

$$\Delta\sigma_{SS} = k_i c_i, \tag{3}$$

where k_i – the strengthening coefficient for ferrite, which is an increment of the yield stress when 1 % (wt) of the i-th alloying element dissolves in it; (for carbon k = 4670 MPa/ % [16]) c_i – the concentration, % (wt), i-th alloying element.

Hardening at the expense of dispersion particles can be determined in accordance with the Orowan's ratio [17]:

$$\Delta\sigma_{or} = 1.2Gbf^{1/3}d^{-1} \quad , \tag{4}$$

where G=78 GPa – the shear modulus of the ferrite, b=0.25 nm – the Burgers vector for dislocations ½<111> of the ferrite, f, d – the volume fraction and size of particles.
The dislocation hardening can be calculated via the formula:

$$\Delta\sigma_d = \alpha M b G \rho^{1/2} \quad (5)$$

where $\Delta\sigma_d$ – the dislocation hardening, where α is the coefficient depending on the character of dislocation interaction in the course of work hardening; M is the orientation multiplier: for α -iron M = 2.75, and the product αM = 0.5, b – the Burgers vector, ρ - the dislocation density[16].
In general at first approximation the contribution of different mechanisms in the hardening is additive:

$$\sigma_y = \Delta\sigma_0 + \Delta\sigma_{ss} + \Delta\sigma_{or} + \Delta\sigma_{gb} + \Delta\sigma_d \quad (6)$$

where σ_0 - the friction stress of the α-iron lattice; $\Delta\sigma_{SS}$ - solid solution hardening; $\Delta\sigma_{or}$ - dispersion hardening; $\Delta\sigma_{gb}$ - grain boundary hardening; $\Delta\sigma_d$ - dislocation hardening.

The friction stress of the α-iron lattice (σ_0) is determined by the Peierls–Nabarro stress [18]:

$$\sigma_0 = 2 \cdot 10^{-4} G \quad (7)$$

Table 3 lists the results of calculation of the contribution of different strengthening mechanisms in the yield stress of steel in accordance with equations (1),(3-5), (7). The comparison with the experimental results is made with account of the empirically set correlation between strength and hardness: $\sigma_y = \frac{1}{3}HV$ [19]. This relation can be used to analyze the yield stress, as the yield stress of steels with an ultrafine-grained microstructure produced by HPT is, as a rule, close to the ultimate strength value.
The following assumptions were accepted during calculation of solid solution hardening. One-phase structure, namely solid solution of all the alloying elements, is taken for the quenched structure. The carbon content, which most significantly hardens α-Fe, in the tempered structure is taken as equilibrium (0.006 % according to diagram "iron-cementite"). The results of X-ray analysis are taken into account for steel after HPT (Table 2), which showed the increase of the lattice parameter as a result of deformation by 0.002Å, which corresponds to the growth of the carbon content in the solid solution by 0.02 %.

Table 3. Contribution of different strengthening mechanisms in the yield stress of steel.

state	Estimate values					Experimental values	
	$\Delta\sigma_{gb}$, MPa	$\Delta\sigma_d$, MPa	$\Delta\sigma_{SS}$, MPa	$\Delta\sigma_{or}$, MPa	σ_y, MPa	1/3 HV, MPa	HV, MPa
quenched	555	185	1049	-	1789	1700	5100
tempered	536	20	448	356	1360	1166	3500
HPT 20°C	1069	319	541	166	2095	2233	6700
HPT 300°C	894	325	541	51	1811	2100	6300

The results listed in the table show that the estimated values are very close to the experimental ones, which proves the possibility of considering the principle of linear superposition when evaluating the contribution of different mechanisms in the yield stress in accordance with (4). Thus, the strength of the quenched steel is ensured mainly by solid solution hardening (58 %), and the contribution of grain boundaries is significant (30 %). After tempering the solid solution hardening reduces to 38 % with the dispersion hardening increase to 30 %.

Another situation is observed during HPT. Independently of the deformation temperature the main contribution belongs to grain boundaries. Nevertheless, this contribution is only about 50 %. The dislocation (to 15 %) and solid solution (to 25 %) components increase significantly.

An interesting fact is that the estimate values of the yield stress for the quenched steel and quenched and tempered steel exceed the experimental value of HV/3, whereas the values are reduced by 10% for the ultrafine-grained steel after HPT. This can be caused by the inaccuracy connected with evaluation of microstructural parameters as well as the content of alloying elements in the solid solution. However, one can suppose synergetic effect or some new strengthening mechanism, for example, at the expense of segregations formation on grain boundaries [20, 21].

The results obtained counts in favor of the complex structure formed during SPD that gives possibility of simultaneous activation of several mechanisms of hardening. As a result of such collective effect, the superstrength is achieved in steels with the ultrafine-grained structure.

5 Conclusions

1. As a result of HPT the ultrafine-grained microstructure has been produced in martensitic steel. Significant reduction of the volume fraction of secondary phases from 19 % in the tempered state to 3.6-4.5 % in the deformed state is observed. When the deformation temperature increases from the room one to 300°C, the grain sizes grow from 140 nm to 200 nm, the sizes of particles increase, and the volume fraction of secondary phases decreases.
2. Evaluation of the contribution of different strengthening mechanisms in the strength properties showed that grain boundary hardening ensures only about 50% of strength. Only the combination of various strengthening mechanisms of the complex ultrafine-grained structure produced by HPT makes it possible producing superstrong steels.

Acknowledgements

The authors are grateful to the Russian Science Foundation for the support of this work through project No. 15-19-00144

References

[1] Song M, Zhu R, Foley D C, Sun C, Chen Y, T.Hartwig K, Zhang X 2013 Enhancement of strength and ductility in ultrafine-grained T91 steel through thermomechanical treatments *J Mater Sci* **48** pp 7360–73

[2] Song M, Wub Y D, Chen D, Wang X M, Sun C, Yu K Y, Chen Y, Shao L, Yang Y, Hartwig K T, Zhang X 2014 Response of equal channel angular extrusion processed ultrafine-grained T91 steel subjected to high temperature heavy ion irradiation *Acta Materialia* **74** pp 285–295

[3] Fan Z Q, Hao T, Zhao S X, Luo G N, Liu C S, Fang Q F 2013The microstructure and mechanical properties of T91 steel processed by ECAP at room temperature *Journal of Nuclear Materials* **434** pp 417–421

[4] Yang G, Huang C X, Wang C, Zhang L Y, Hu C, Zhang Z F, Wu S D 2009 Enhancement of mechanical properties of heat-resistant martensitic steel processed by equal channel angular pressing *Materials Science and Engineering* A **515** pp 199–206

[5] Valiev R Z, Islamgaliev R K, Alexandrov I V 2000 Bulk nanostructured materials from severe plastic deformation *Prog Mater Sci* **45** pp 103

[6] Zhilyaev A P, Langdon T G 2008 Using high-pressure torsion for metal processing: Fundamentals and applications *Prog Mater Sci* **53** pp 893–979

[7] Ganeev A V, Karavaeva M V, Sauvage X, Courtois-Manara E, Ivanisenko Y, Valiev R Z 2014 On the nature of high-strength state of carbon steel produced by severe plastic deformation IOP Conf. Series. *Mater. Sci. Eng.* **63** 012128 pp 1-7
[8] Karavaeva M V, Kiseleva S K, Ganeev A V, Protasova E O, Valiev R Z, Ganiev M M, Simonova L A 2015 Superior strength of carbon steel with an ultrafine-grained microstructure and its enhanced thermal stability *J Mater Sci* **50** pp 6730-38
[9] Hall E I 1951The deformation and ageing of mild steel: III Discussion of results. *Proc Phus Soc London* **B64** pp 747-753
[10] Petch N J 1953 The cleavage strength of polycrystals *J Iron and Steel Inst.* **174** pp 25-28
[11] Huang X, Yangb G, Gao Y L, Wu S D, Zhang Z F 2008 Influence of processing temperature on the microstructures and tensile properties of 304L stainless steel by ECAPC *Mat Sci Eng* A. **485** pp 643–650
[12] Abramova M M, Enikeev N A, Valiev R Z, Etienne A, Radiguet B, Ivanisenko Y, Sauvage X 2014 Grain boundary segregation induced strengthening of an ultrafine-grained austenitic stainless steel. Materials Letters **136** pp 349-352 (in Russian)
[13] M. Ferrari and L. Lutterotti. J. Appl. Phys. 76, 7246 (1994).
[14] http://www.ing.unitn.it/~maud/
[15] Schmauder S, Kohler C 2011 Atomistic simulations of solid solution strengthening of a-iron. *Computational Materials Science* **50, 4** pp1238 – 43
[16] Goldstein MI, Litvinov VS, Bronfin BM (1986) Metal physics of the high-strength state. Metallurgiya, Moscow (in Russian)
[17] Ning J l, Courtois-Manara E, Kurmanaeva L, Ganeev AV, Valiev R Z , Kübel C, Ivanisenko Y 2013 Tensile properties and work hardening behaviors of ultrafine grained carbon steel and pure iron processed by warm high pressure torsion *Mat Sci Eng* A **581** pp 8–15
[18] Nabarro F R N 1997 Fifty-year study of the Peierls-Nabarro stress *Mat Sci Eng* A **234-236** pp 67-76
[19] Pavlina E J, Van Tyne C J 2008 Correlation of yield strength and tensile strength with hardness for steels *Journal of Materials Engineering and Performance* **17, 6** pp 888-893
[20] Sabirov I, Murashkin M Yu, Valiev R Z 2013 Nanostructured aluminium alloys produced by severe plastic deformation: New horizons in development *Mat Sc. Eng* A **560** pp 1-24
[21] Nurislamova G, Sauvage X, Murashkin M, Islamgaliev R, Valiev R 2008 Nanostructure and related mechanical properties of an Al-Mg-Si alloy processed by severe plastic deformation *Phil Mag Lett* **88, 6** pp 459-466

IOP Conf. Series: Materials Science and Engineering **179** (2017) 012038 doi:10.1088/1757-899X/179/1/012038

Characterization of TiAl based alloys with various content of carbon

A Klimová, J Lapin and T Pelachová

Institute of Materials and Machine Mechanics, Slovak Academy of Sciences, Dúbravská cesta 9, 845 13 Bratislava 3, Slovak Republic

E-mail: ummsakli@savba.sk

Abstract. TiAl based alloys prepared by induction melting with nominal compositions Ti-(37.5-41)Al-(3.5-4)Nb-1Mo-0.1B-xC (at.%), where carbon content x is ranging from 3 to 7 at.%, were characterised by optical microscopy (OM), scanning electron microscopy (SEM), energy dispersive spectroscopy (EDS), X-ray diffraction analysis (XRD) and elemental LECO analysers of light elements (O, N and C). Morphology and volume fraction of coexisting phases are quantitatively analysed using computerised image analyser. The effect of carbon content on Vickers microhardness and hardness of the alloys is evaluated and discussed. The microstructure of the alloys consists of fine-grained lamellar matrix with elongated Ti_2AlC particles dispersed in the whole volume of the ingots. The volume fraction of the carbide particles increases with the carbon content of the studied alloys from 8 to 18 vol.%. The average Vickers hardness of the ingots doesn't show the significant variations with their carbon content, but the hardness considerably exceeds the average values in the surface regions where the small rod shape particles are densely distributed in the matrix.

1 Introduction

In the last decades the big attention was paid to the developing of new γ-TiAl based intermetallic alloys due to their very promising properties for high temperature applications in automotive, aircraft and power industry [1]. Several processing techniques and various alloying elements have been investigated to improve their room temperature low ductility and fracture toughness, oxidation resistance at high temperatures, creep properties and raise their current operational temperature limit over 750°C [2]. Beside the additions of elements such as Mo, Nb, Ta, W, Zr or Cr, which affect the properties of these alloys through the solution hardening and changes of the microstructure, the addition of the appropriate amounts of carbon can lead to the formation of reinforcing carbide particles which can have the positive effect on mechanical properties of the final alloy at both ambient and elevated temperatures [3, 4].

The aim of the present work is to characterize chemical composition, microstructure and Vickers hardnesss of the TiAl based alloys with nominal compositions Ti-(37.5-41)Al-(3.5-4)Nb-1Mo-0.1B-xC (at.%) prepared by the induction melting in graphite crucibles. The carbon content x is varied from 3 to 7 at.%. Because of low solubility of carbon [5] in main coexisting phases present in the TiAl based alloy such as α2(Ti3Al), γ(TiAl) and β(Ti), the selected alloying with C is expected to affect solidification path and favour formation of primary carbide particles in the studied alloys during solidification [5, 6].

2 Experimental procedure

The experimental TiAl based alloys with nominal compositions Ti-(37.5-41)Al-(3.5-4)Nb-1Mo-0.1B-(3-7)C (at.%) were prepared by induction melting in graphite crucibles followed by solidification under argon atmosphere. Cylindrical ingots with a diameter of 42 mm and length of 15 mm were cut longitudinally using wire electrical discharge machining. Microstructural investigations were performed by optical microscopy (OM), scanning electron microscopy (SEM) and X-ray diffraction analysis (XRD). Chemical composition of the alloys was analyzed by energy-dispersive spectrometry (EDS) using JSM-6510 scanning electron microscope equipped with EDS detector. For measurements of oxygen and nitrogen content, LECO ONH836 Elemental Analyzer was used. The analysis of carbon content was performed by LECO CS844 Elemental Analyzer. OM, SEM, XRD and EDS samples were prepared using standard grinding and polishing metallographic techniques. After mechanical polishing the samples for OM were chemically etched in a reagent consisting of 100 ml H_2O, 10 ml HNO_3 and 3 ml HF. Vickers hardness and Vickers microhardness measurements were performed on polished and slightly etched surfaces at an applied load of 294 N (HV30) and 0.098 N (HV_m), respectively. Grain size, volume fraction of the coexisting phases and interlamellar α_2-α_2 spacing were determined from the digitalized micrographs using computer image analyzer.

3 Results and discussion

The average chemical composition of the investigated alloys measured by EDS is shown in the table 1. Because of the limitation of EDS method in measuring of light elements, the content of boron is taken from independent analysis to be 0.1 at.% of the master alloy. The average content of O, N and C measured by the elemental analysers LECO in the central part of the ingots is shown in the table 2. Because of the expected inhomogeneity in the carbon content due to the effect of the graphite crucible, the carbon content was analysed also in the vicinity of the surface connected with the crucible during melting. As seen in table 1, only slight increase of carbon content in the surface regions in comparison with the central part is detected in the ingots TiAl-3C and TiAl-5C alloyed with 3 and 5 at.% of C. On the contrary, the ingot TiAl-7C shows the difference in carbon content between the central part and surface regions 3.3 at.%. This inhomogeneous distribution of carbon in the relationship with the inhomogeneity of microstructure of the ingots is discussed later in the paper.

Table 1. Chemical composition measured by EDS (at.%).

Ingot	Nominal composition	Al	Ti	Nb	Mo
TiAl-3C	Ti-41Al-4Nb-1Mo-0.1B-3C	40.82±0.10	52.69±0.14	4.36±0.05	1.08±0.12
TiAl-5C	Ti-39.5Al-3.5Nb-1Mo-0.1B-5C	39.25±0.29	53.67±0.30	4.28±0.07	1.13±0.10
TiAl-7C	Ti-37.5Al-3.5Nb-1Mo-0.1B-7C	37.92±1.02	54.16±1.18	4.07±0.18	0.96±0.12

Table 2. Content of O, N and C measured by the analysers of the light elements LECO.

Ingot	O (wt.%)	N (wt.%)	C (wt.% / at.%) centre	C (wt.% / at.%) surface region
TiAl-3C	0.0861	0.0284	0.87 / 2.60	0.93 / 2.80
TiAl-5C	0.0886	0.0490	1.50 / 4.50	1.60 / 4.80
TiAl-7C	0.0982	0.0719	1.90 / 5.70	3.00 / 9.00

3.1 Microstructure of the ingots

The microstructure of the studied alloys is composed of fine-grained lamellar $\alpha_2(Ti_3Al)+\gamma(TiAl)$ matrix, β-phase (Ti-based solid solution with cubic crystal structure), thin layer of γ-phase formed on the grain boundaries, boride particles and carbide particles dispersed in the whole volume of the ingots, as shown in figure 1a. XRD patterns in figure 1b indicate that the crystal structure of the carbides belongs to Ti_2AlC phase. The Ti_2AlC peaks in the XRD patterns increase with the increase of

IOP Conf. Series: Materials Science and Engineering **179** (2017) 012038 doi:10.1088/1757-899X/179/1/012038

carbon content, what is in a very good agreement with the results of Song et al. [7]. Formation of TiB and TiB_2 particles is also proved very well by the corresponding peaks in the XRD patterns in figure 1b.

The shape of the carbide particles is elongated. Their average aspect ratio (length/width) is measured to be 5.0 and 4.4 in the central part of the TiAl-3C and TiAl-7C ingots, respectively. In the vicinity of the surface connected with the crucible, the thin needle shaped particles with a length over 100 μm are formed, especially in the ingot TiAl-3C (figure 2a). Large clusters of small rod shaped carbide particles with smooth surface and aspect ratio of about 2.5 are observed close to the surface of the TiAl-5C and TiAl-7C ingots (figure 2b).

The volume fraction of carbide particles increases from 8 vol.% in the TiAl-3C to 18 vol.% in the ingot TiAl-7C. The volume fraction of the β-phase varies between 7 and 10 vol.% in all three alloys.

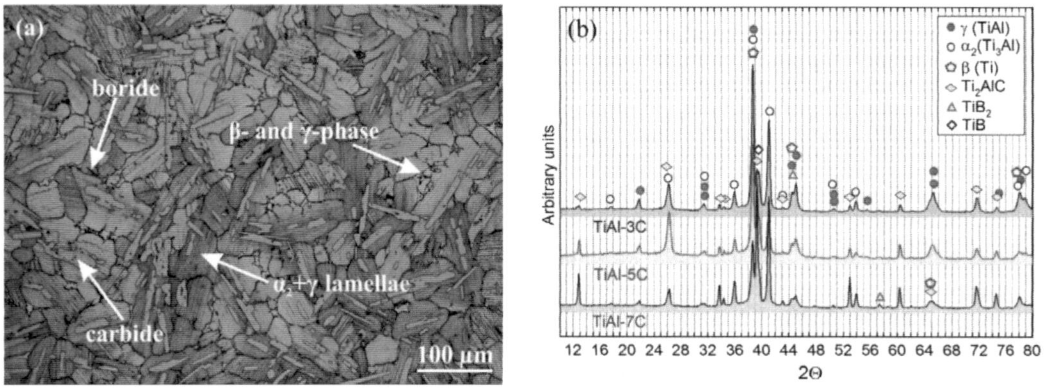

Figure 1. (a) Microstructure of the ingot TiAl-7C (OM); (b) XRD patterns of the ingots.

The lamellar grains in the central part of the ingots maintain the elongated shape of the carbide particles with the average length of around 80 μm and aspect ratio of 2. Only in the surface regions with the extensive clusters of small rod shaped carbide particles, the morphology of grains is more equiaxed with the average size of (40 ± 10) μm. The orientation of α_2+γ lamellae is parallel to the longitudinal axis of the carbide particles in the majority of grains, what indicates a defined crystallographic orientation relationship between the carbide particles and matrix formed during the solidification. The average interlamellar α_2-α_2 spacing is measured to be (300 ± 20) nm in all studied ingots.

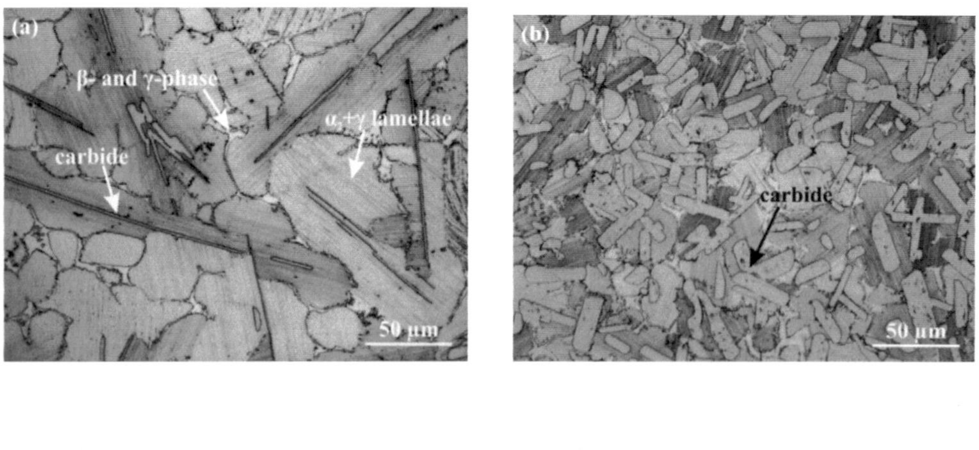

Figure 2. Microstructure of the ingots in the vicinity of the surface (OM) (a) TiAl-3C; (b) TiAl-7C.

3.2 Distribution of alloying elements

The chemical compositions of α_2+γ lamellar grains, β-phase formed at the grain boundaries and carbide particles (figure 3) are shown in tables 3 and 4.

Table 3. Chemical composition of the matrix measured by EDS (at.%).

Position (figure 3)	Al	Ti	Nb	Mo
1	41.21±0.40	52.49±0.47	4.21±0.15	0.98±0.08
2	35.69±0.13	55.08±0.43	5.55±0.24	3.68±0.12

Table 4. Chemical composition of the carbide particles measured by EDS (at.%).

Position (figure 3)	C	Al	Ti	Nb	Mo
3	38.41±1.00	0.86±0.50	58.49±2.14	2.23±0.71	0.03±0.04
4	24.72±0.93	24.49±1.44	47.51±1.05	3.25±0.20	0.04±0.04

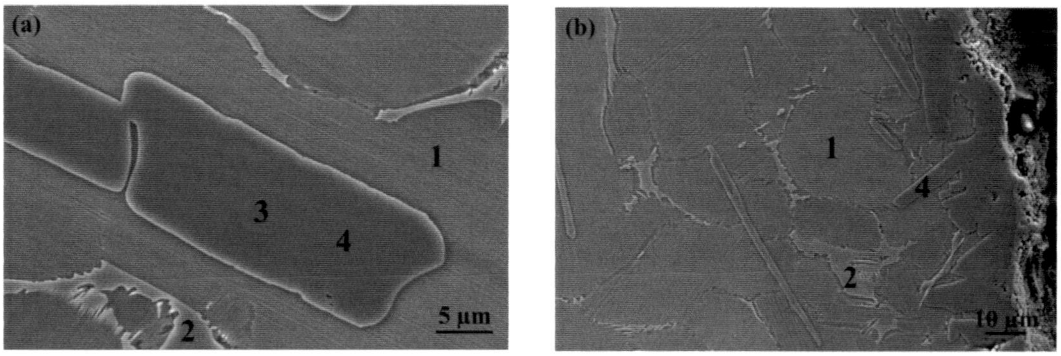

Figure 3. $\alpha_{2+}\gamma$ lamellar grains (1), β-phase (2) and carbide particles (3, 4) (SEM); (a) in the center of the ingots; (b) in the vicinity of the surface connected with the crucible during the melting.

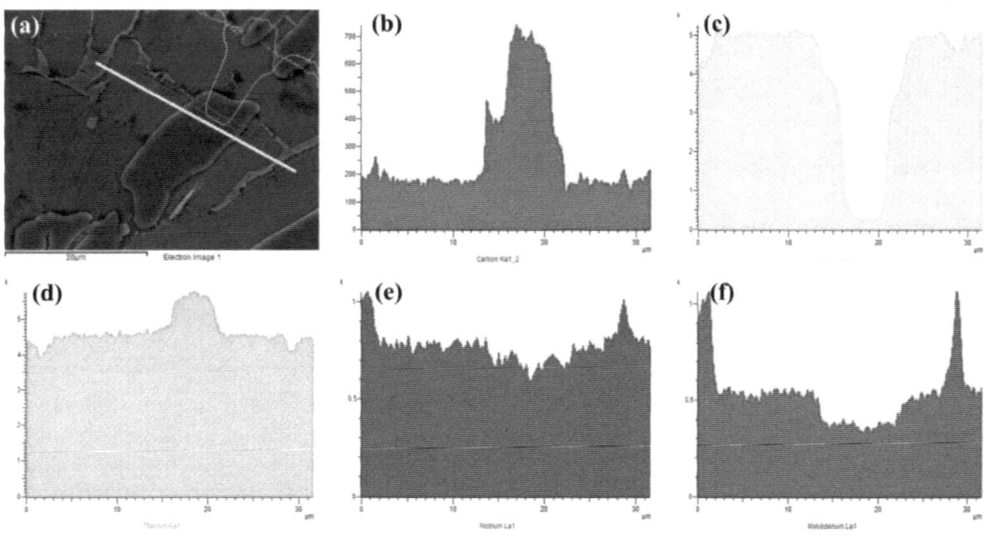

Figure 4. EDS line scan elemental profiles along the yellow line: (a) SEM image; (b) carbon, (c) aluminium, (d) titanium, (e) niobium, (f) molybdenum.

According to the atom probe tomography measurements of the local chemical composition of Ti-Al-Nb alloys with 0.5 - 0.75 at.% of C, reported in [8, 9], solubility limits of C in the α_2, γ and β phases reach the following 1.2 at.%, 0.25 - 0.3 at.% and 0.0 at.%, respectively. Because of the limitation of the EDS method applied in our study, low carbon content in the matrix of the studied alloys couldn't be measured precisely. However, as seen in figure 4b, the line scan across the all coexisting phases shows the presence of carbon not only in the carbide particles, but also its segregation on the γ/β interfaces and grain boundaries. The evidence of the accumulation of C in the vicinity of γ/β interface was reported also by Klein et al. [8].

For the measurements of the higher amounts of carbon in carbide particles, the standardization of the EDS analyzer on certified Ti_2AlC standard was used. The carbide particles are identified by EDS measurements and XRD analysis to belong to Ti_2AlC phase, as seen in table 4 and figure 1b. However, figure 3a and 4a indicates that the large carbide particles contain small central area enriched by carbon and depleted by Al. The similar core depleted by Al inside the particles was also observed in TiAl matrix composites prepared by arc-melting methods [7, 10, 11]. The surface layer shown in Figure 3b is formed by Ti_2AlC type of carbide, what was proved by both XRD patterns (figure 1b) and EDS measurements (table 4).

Nb and Mo as strong β-phase stabilizers [2] are concentrated on the grain boundaries, where the β phase is formed (figure 4e and 4f). Small increase of Nb in the β phase in comparison with the lamellar grains and three times higher content of Mo in the β phase than in the lamellar matrix (table 3) indicate the stronger stabilization effect of Mo than of Nb [7]. Carbide particles are free of Mo, but contain of about 3 at.% of Nb.

3.3 *Hardness and microhardness*

The average hardness HV30 of the ingots is not affected by the carbon content. All average values fall within the standard deviation interval of the measurements (table 5). Higher standard deviation of HV30 values in the TiAl-7 ingot is caused by the inhomogeneous distribution of the carbide particles and different size and morphology of the grains in the central part (figure 1a) and in the vicinity of the surface (figure 2b). While in the central part of the ingot TiAl-7C the average hardness is about 370 HV30, the hardness reaches the values over 400 HV30 in the regions with high density of small rod shape particles in the matrix with nearly equiaxed grains (figure 2b).

Table 5. Vickers hardness HV30 of the ingots and Vickers microhardness HV_m of the lamellar grains.

	TiAl-3C	TiAl-5C	TiAl-7C
HV30 of the ingots	386±9	384±6	385±16
HV_m of the lamellar grains	562±37	576±42	588±22

The Vickers microhardness HV_m of the lamellar matrix is slightly increasing with the increase of the carbon content, but this increase is within the standard deviations of the measurements (table 5). The average microhardness of the β-phase in all three ingots reaches the same value of about 580±48 HV_m as the lamellar matrix, and the microhardness of the carbide particles is determined to be around 800 HV_m.

5 Conclusions

The achieved results of the study of TiAl based alloys with nominal compositions Ti-(37.5-41)Al-(3.5-4)Nb-1Mo-0.1B-(3-7)C (at.%) can be summarized as follows:

1. The microstructure of the studied alloys is composed of fine-grained $\alpha_2(Ti_3Al)+\gamma(TiAl)$ lamellar matrix with the β-phase and thin layer of γ-phase formed on the grain boundaries and Ti_2AlC, TiB and TiB_2 particles dispersed in the whole volume of the ingots.
2. The shape of Ti_2AlC particles in the central part of the ingots is elongated. The volume fraction of the carbide particles increases with the amount of carbon from 8 to 18 vol.%. In the vicinity of the surface of ingots, long thin needle shaped carbide particles are formed mainly in the ingots with lower amounts of carbon (3-5 at.%). Large clusters of small rod shaped carbide particles are observed in the alloys with 5 and 7 at.% of C.
3. The lamellar grains in the central part of the ingots maintain the elongated shape of the carbide particles. The orientation of $\alpha_2+\gamma$ lamellae is parallel to the longitudinal axis of the carbide particles in the majority of grains.
4. Carbon is distributed inhomogenously in the carbide particles and in the matrix. The core of the large Ti_2AlC particles is enriched by carbon and depleted by Al in all studied ingots. The increase of C content in the vicinity of γ/β interface and along grain boundaries is detected.
5. The average Vickers hardness of the studied alloys and Vickers microhardness of lamellar grains, β-phase and carbide particles don't show the significant variations with the carbon content. The hardness considerably exceeds the average values in the surface regions with small rod shape particles densely distributed in the matrix with nearly equiaxed grains.

Acknowledgements

This work is the result of the projects of the Slovak Research and Development Agency under the contract APVV-15-0660 and APVV-0434-10. The financial support of the Slovak Grant Agency for Science under the contract VEGA 2/0125/16 is acknowledged.

References

[1] Kothari K, Radhakrishnan R and Wereley N M 2012 *Prog. Aerosp. Sci.* **55** 1
[2] Clemens H and Mayer S 2013 *Adv. Eng. Mater.* **15** 191
[3] Appel F, Paul J D H and Oehring M 2011 Gamma Titanium Aluminide Alloys (Willey-VCH Verlag &Co. KGaA) 277
[4] Kevorkijan V and Škapin S D 2009 *Metall. Mater. Eng.* **15** 75
[5] Schwaighofer E, Rashkova B, Clemens H, Stark A and Mayer S 2014 *Intermetallics* **46** 173
[6] Kamyshnykova K and Lapin J 2016 Proc. 6th Int. Conf. MTSM 2016 (Split) 67
[7] Song X J, Cui H Z, Hou N, Wei N, Han Y, Tian J and Song Q 2016 *Intermetallics* **42** 13586
[8] Klein T, Schachermayer M, Mendez-Martin F, Schöberl T, Rashkova B, Clemens H and Mayer S 2015 *Acta Mater.* **94** 205
[9] Scheu C, Stergar E, Schober M, Cha L, Clemens H, Bartels A, Schimansky F P and Cerezo A 2009 *Acta Mater.* **57** 1504
[10] Cao L, Wang H W, Zou C M and Wei Z J 2009 *J. Alloys Compd.* **484** 816
[11] Wei Z, Wang H, Jin Y, Zhang H and Zeng S 2002 *J. Mater. Sci.* **27** 1809

Correlation between standard Charpy and sub-size Charpy test results of selected steels in upper shelf region

P Konopík[1], J Džugan[1], T Bucki[2], S Rzepa[3], M Rund[1] and R Procházka[1]

[1]COMTES FHT a.s., Dobřany, Czech Republic, EU
[2]Kielce University of Technology, Kielce, Poland, EU
[3]AGH University of Science and Technology, Kraków, Poland, EU

E-mail: pavel.konopik@comtesfht.cz

Abstract. Absorbed energy obtained from impact Charpy tests is one of the most important values in many applications, for example in residual lifetime assessment of components in service. Minimal absorbed energy is often the value crucial for extending components service life, e.g. turbines, boilers and steam lines. Using a portable electric discharge sampling equipment (EDSE), it is possible to sample experimental material non-destructively and subsequently produce mini-Charpy specimens. This paper presents a new approach in correlation from sub-size to standard Charpy test results.

1 Introduction

Degradation of the properties of metal materials can lead to loss of reliability and safety of machines and metal structures in power plants during their operation. However, a number of these machines or metal structures could be used for longer time than has been predicted by their designed lifetime.

A component's residual life can be evaluated by standard mechanical test techniques, such as the tensile test, the fatigue test, the Charpy or the fracture toughness test. However, for these tests there is usually insufficient material to sample non-invasively from the component. Therefore, non-destructive techniques are being developed as well as testing methods using mini-samples.

One of the widely used methods of mini-samples testing is the Small Punch Test (SPT) [1]. The SPT is usually based on conversion of the obtained results into conventional mechanical characteristics [1].

However, it requires known correlation parameters determined for the specific material and these correlation parameters must be verified for each new material. On the other hand, there exist small specimen testing techniques respecting the same loading mode as in the case of standard samples. Namely, it is the micro-tensile test [4]-7], the miniature fracture toughness test [8-9], the miniature fatigue test [10-[10] and the mini-Charpy test for DBTT determination [12[15].

Furthermore, these methods maintain minimal material requirements without requiring previously established correlations or at least they use a much more reliable type of correlation (without necessity to measure wide range material database).

Above mentioned miniaturized testing techniques have already been verified. This paper intends to extend these testing methods to correlation from mini-Charpy results to standard Charpy results in the upper shelf region.

2 Experimental material

Five structural steels used mostly in automotive or power industries were used as experimental materials.

The materials were delivered as semi-products in the form of a rod or a tube. Three of them were experimental steels (m06, m08, m11) and the others were 34CrNiMo6 (m15) and AK1TD steels. Chemical composition of each material is stated in table 1 and their mechanical properties in table 2.

Table 1. Chemical composition (wt. %).

Material	C	Si	Mn	P	S	Cr	Mo	Ni	Cu	W
m06	0.305	0.275	0.750	0.008	0.002	1.025	0.200	0.150	0.100	-
m08	0.180	0.280	1.400	0.004	0.001	0.200	0.027	0.067	0.067	-
m11	0.170	0.330	1.450	0.008	0.002	0.250	0.053	0.133	0.133	-
m15	0.340	0.200	0.650	0.013	0.018	1.500	0.225	1.500	-	-
AK1TD	0.048	0.260	0.480	0.009	0.001	10.68	0.370	1.670	0.028	1.376

Table 2. Tensile test results.

Material	YS (MPa)	UTS (MPa)	Uniform Elongation (%)	Elongation 5D (%)	Area Reduction (%)
m6	969,1	1039,4	5,4	17,9	65,9
m8	1145,5	1252,1	2,5	14,3	70,2
m11	981,1	1022,1	4,7	17,5	70,9
m15	932,0	1034,2	5,4	15,6	55,8
AK1TD	837.1	941,9	6,3	18,1	58,4

3 Experiment

In order to describe the fracture behaviour of the material depending on the size factor in the upper shelf region, only materials showing full ductile damage at room temperature were chosen. In all cases, full size Charpy specimen as well as mini-Charpy specimens were tested. Moreover, in order to better understand the fracture behaviour, standard Charpy specimens with reduced thickness were tested in the case of three materials. Furthermore, most of the materials were tested in longitudinal and transversal direction.

Charpy-V specimens with a height of 10 mm and various widths (10 mm, 7.5 mm, 5 mm, 2.5 mm) and mini-Charpy specimens with a cross-section of 3x4 mm [16-17] are shown in figure 1.

Impact tests were performed using a 300 J pendulum with striking edge radius of 2 mm for standard specimens and a 15 J pendulum with striking edge radius of 2 mm for mini-Charpy specimens. For each testing condition, at least three repetitions were performed. After the test, absorbed energy KV was measured and instrumented record was captured. An example of instrumented records is depicted in figure 2. All records show full ductile damage which is consistent with fracture appearance of broken specimens. Absorbed energy was transformed to the notch toughness KCV according to equation (1). Results are summarized in table 3 and the graphical representation of measured values is depicted in figure 3.

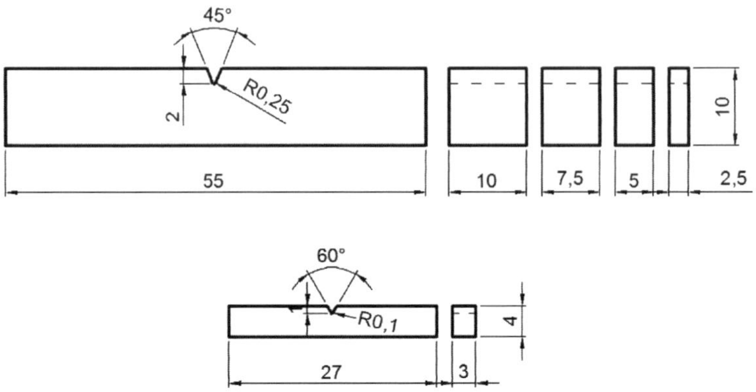

Figure 1. Dimensions of used specimens.[16-17].

$$KCV = \frac{KV}{area\ under\ notch} \tag{1}$$

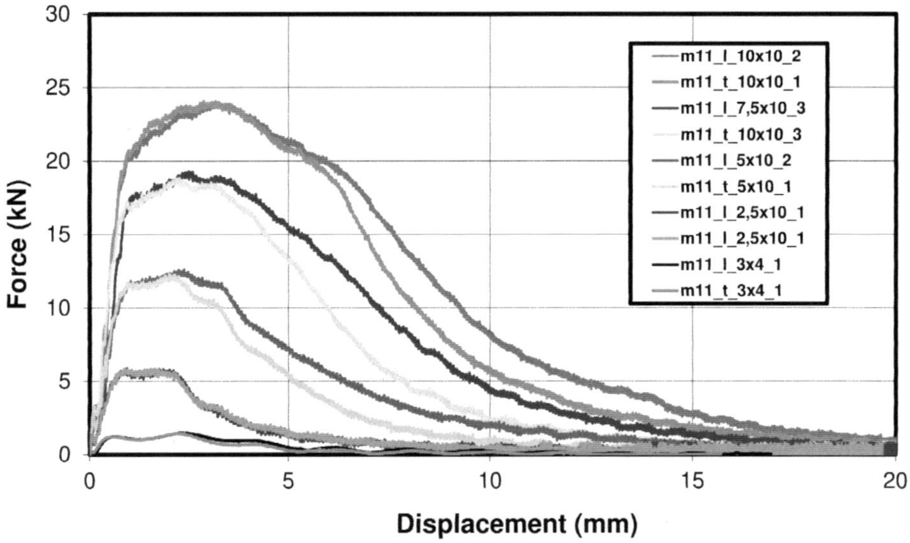

Figure 2. Example of instrumented Charpy records, material m11.

Table 3. Charpy test results.

Specimen Size (mm)	AUN (mm²)	m06 Long. (J/cm²)	m06 Trans. (J/cm²)	m08 Long. (J/cm²)	m11 Long. (J/cm²)	m11 Trans. (J/cm²)	m15 Long. (J/cm²)	AK1TD Long. (J/cm²)
10x10	80	147.1	166.0	203.0	277.8	247.2	111.7	147.6
10x7,5	60				238.6	193.9		
10x5	40			157.1	187.9	148.8		125.9
10x2,5	20			121.7	123.7	114.3		85.3
3x4x27	9	77.4	81.8	99.1	113.9	96.9	62.9	73.4

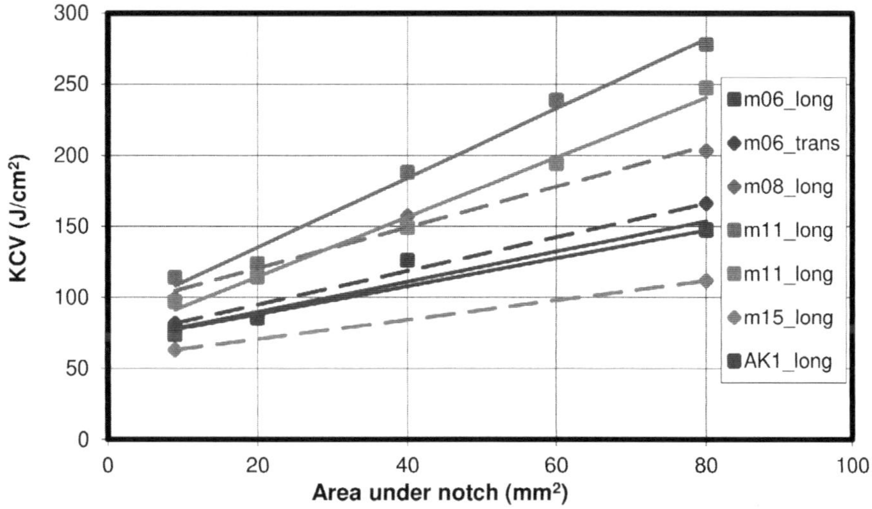

Figure 3. Graphical representation of measured values.

4 Results discussion

Measured data show that notch toughness KCV decreases with decreasing area under the notch (AUN) and therefore with decreasing specimen thickness. As each material has different absorbed energy, KCV value can be normalized using the following equation (2):

$$KCV_{NS} = \frac{KCV_S}{KCV_{10x10}} \cdot 100 \quad (2)$$

Where:

KCV_{NS} is normalized notch toughness of specific specimen size (%),
KCV_S is notch toughness of specific specimens size (J/cm2) and
KCV_{10x10} is notch toughness of full size specimen (J/cm2).

Table 4 summarizes Charpy test results converted to normalized notch toughness. Normalized notch toughness depending on the area under the notch is depicted in figure 4. All KCV_{NS} values are inside the envelope formed by the upper (UL) and lower line (LL) which can be described using equations (3) and (4). Further work will be focused on the verification of these borders for other materials.

Table 4. Charpy test results converted to normalized notch toughness.

Specimen Size [mm]	AUN [mm²]	m06 Long. %	m06 Trans. %	m08 Long. %	m11 Long. %	m11 Trans. %	m15 Long. %	AK1TD Long. %
10x10	80	100.0	100.0	100.0	100.0	100.0	100.0	100.0
10x7.5	60				85.9	78.5		
10x5	40			77.4	67.6	60.2		85.3
10x2.5	20			59.9	44.5	46.2		57.8
3x4x27	9	52.6	49.2	48.8	41.0	39.2	56.3	49.7

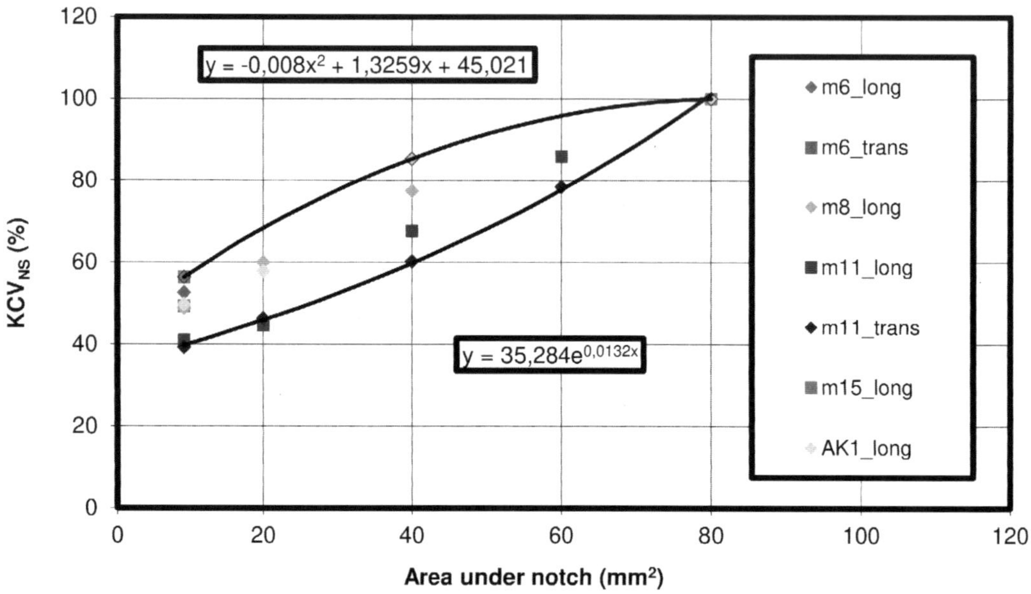

Figure 4. Normalized notch toughness depending on the area under the notch.

$$KCV_{NS/UL} = -0.008 \cdot AUN^2 + 1.3259 \cdot AUN + 45.021 \quad (3)$$

$$KCV_{NS/LL} = 35.284 \cdot e^{0.0132} \quad (4)$$

5 Conclusion

A new approach in correlation from sub-size to standard Charpy test results was proposed. In total, 24 different types of tests were performed with at least three repetitions and therefore more than 72 instrumented Charpy impact tests were performed. Furthermore, for all materials, chemical analyses were performed as well as tensile properties investigation. It can be assumed that the proposed correlation is valid for similar types of material investigated in this paper. However, other materials should be investigated to confirm the general validity of the suggested correlation or, alternatively, to specify its limits.

Acknowledgements

This paper was created by project Development of West-Bohemian Centre of Materials and Metallurgy No.: LO1412, financed by the MEYES of the Czech Republic.

References

[1] Small punch test method for metallic materials. 2007 Part B: A code of practice for small punch testing for tensile and fracture behaviour. CWA 15627

[2] Konopík P and Džugan J 2012 *2nd Int. Conf. SSTT on Det. of Mech. Prop. of Mat. by Small Punch and other Miniature Testing Techn.* Determination of Tensile Properties of Low Carbon Steel and Alloyed Steel 34CrNiMo6 by Small Punch Test and Micro-Tensile Test (Ostrava: Ocelot sro) pp 319-328.

[3] WANG. Z.-X. et al 2008 Small punch testing for assessing the fracture properties of the reactor

vessel steel with different thicknesses. *Nuclear Engineering and Design* Volume 238. Issue 12. Pages 3186-3193

[4] Konopik P, Dzugan J and Rund M 2014 *METAL 2014 - 23rd Int. Conf. on Metallurgy and Mat. (Brno)* Dynamic tensile and micro-tensile testing using DIC method (Ostrava: TANGER Ltd.) pp 498-503

[5] Prochazka R Dzugan J and Kover M 2015 *Archives of Materials Science and Engineering* Miniature specimen tensile testing of AZ31 alloy processed by ECAP **Volume 76 Issue 2** December ISSN: 18972764 (Gliwice: International OCSCO World Press) Pages 134-139

[6] Džugan J, Procházka R and Konopík P 2015 *Small Specimen Test Techniques* Micro-tensile test technique development and application to mechanical property determination **6th Volume, ASTM STP 1576** Sokolov M A and Lucon E (West Conshohocken: ASTM International) pp 12-29

[7] Rund M. Procházka R. Konopik P Dzugan J and Folgar H 2015 *Procedia Engineering* Investigation of Sample-size Influence on Tensile Test Results at Different Strain Rates, **Vol. 114** (Amsterdam: Elsevier science bv) pp 410-15

[8] Konopík P, Dzugan J and Prochazka R 2013 *METAL 2013 – 22nd Int. Conf. on Metallurgy and Mat. (Brno)* Determination of fracture toughness and tensile properties of structural steels by small punch test and micro-tensile test (Ostrava: TANGER Ltd.) pp 722 – 27

[9] Konopik P, Dzugan J and Rund M 2015 *METAL 2015 – 24th Int. Conf. on Metallurgy and Mat. (Brno)* Determination of fracture toughness in the upper shelf region using small sample test techniques (Ostrava: TANGER Ltd.) pp 710-15

[10] Dzugan J, Konopik P., Rund M and Prochazka R 2015 *ASME Pressure vessels and piping conf. - 2015* Determination of local tensile and fatigue properties with the use of sub-sized specimens, **Volume 1A: Codes and Standards** (New York: Amer. Soc. Mech. Eng.)

[11] Dzugan J Novy Z Konopik and P Motycka P 2010 *METAL 2010 - 19th Int. Conf. on Metallurgy and Mat. (Brno)* Improvement of fatigue properties of 34crnimo6 steel by controlled thermomechanical treatment (Ostrava: TANGER Ltd.) pp 421-426

[12] Konopík P, Dzugan J and Prochazka R 2013 *Mat. Sci. and Tech. Conf. and Exh.* Evaluation of local mechanical properties of steel weld by miniature testing technique, pp 2404-11

[13] Dzugan J, Konopik P and Prochazka 2016 *Materials Science Forum* SPD processed materials mechanical properties determination with the use of miniature specimens **Volume 879** (Zurich: Trans tech publications ltd) pp 471-476

[14] Towers O L 1986 Testing of sub-size Charpy specimens: Part 1 - the influence of thickness on the ductile/brittle transition *Metal Construction* **18(3)** pp 171-176

[15] Wallin K 1994 Methodology for selecting Charpy toughness criteria for thin high strength steels: Part 1 - determining the fracture toughness *Jernkontorets Forskning. Report from Working Group* **4013(89)** p 28

[16] ASTM E23:12c *Standard Test Methods for Notched Bar Impact Testing of Metallic Materials* p 17

[17] ISO 14556:2000 Steel - Charpy V-notch pendulum impact test - Instrumented test method *European Committee for Standardization*

Comparison of irradiated 15Kh2MFA material mechanical properties using conventional testing methods and innovative approach of small punch testing (SPT) and automated ball indentation (ABIT)

R Kopriva[1], P Petelova[1], I Eliasova[1], M Kytka[1] and M Culek[1]

[1]UJV Rez, a. s., Integrity and Technical Engineering Division, Hlavni 130, Rez, 250 68 Husinec, Czech Republic

radim.kopriva@ujv.cz

Abstract. Article describes two innovative testing methods – Small Punch Testing (SPT) and Automated Ball Indentation Test (ABIT) – which are based on the determination and evaluation of material properties from miniaturized testing specimens. These methods are very promising due to minimum material needed for testing and also in case of testing highly irradiated materials of components that are not included in standard surveillance programs. The test results were obtained for reactor pressure vessel (RPV) base material 15Ch2MFA in both states - initial unirradiated and irradiated. Subsequently results were compared with standard tensile tests to prove applicability of these testing methods for the evaluation of degradation of irradiated structural materials of nuclear power plants.

1 Introduction

Current trend of the extension of the nuclear power plant (NPP) components operational lifetime requires precise information of the structural materials degradation for the assurance of safe and reliable long-term operation. Special attention should be paid to the verification of integrity and strength of materials of hardly replaceable components of the nuclear power plant's primary circuit such as reactor pressure vessel and internals.

Evaluation of the mechanical properties in the present conventional way is usually based on the relatively high consumption of the testing materials, whose availability is often very limited and in several cases archive materials are not available completely. Several years ago, UJV Rez, Integrity and Technical Engineering Division started to pay considerable attention to the employment of perspective semi-destructive testing techniques in the process of irradiated NPP structural materials degradation evaluation – small punch tests and automated ball indentation testing. These methods are valuable for the determination of mechanical properties without affecting the overall integrity of evaluated component and their semi-destructive approach in connection with the small volume of material necessary for the testing. This approach allows focusing on the locations of the operated component which are subject to higher stresses, neutron flux or thermal ageing.

2 Testing methods – fundamentals

Small punch test method is based on the testing of thin clamped specimens using hemispherical punch or ball with the direct measurement of the specimen deflection on the opposite side of the specimen. Example of test data and the scheme of the test setup are shown on the Fig. 1. Obtained results can be correlated with standard tensile test results and this technique can be also used for the determination of fracture toughness properties and ductile to brittle transition temperature (DBTT).

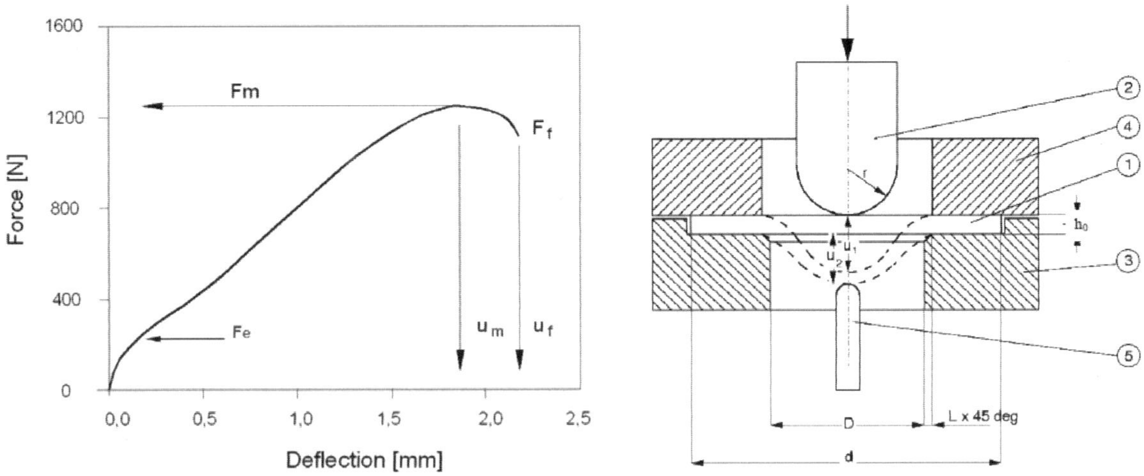

Figure 1. Example of test data (left), scheme of the test setup (right).

Automated ball indentation testing is based on the principle of strain controlled multiple indentations at a single penetration location on a polished surface by a small hemispherical indenter (various diameters from 0.508 to 2.5 mm) [1]. Instrumented test method is fully automatic and does not require the diameter of the indentation to be measured after testing. Scheme of the indent [2] and example of the test result is shown on the Fig. 2.

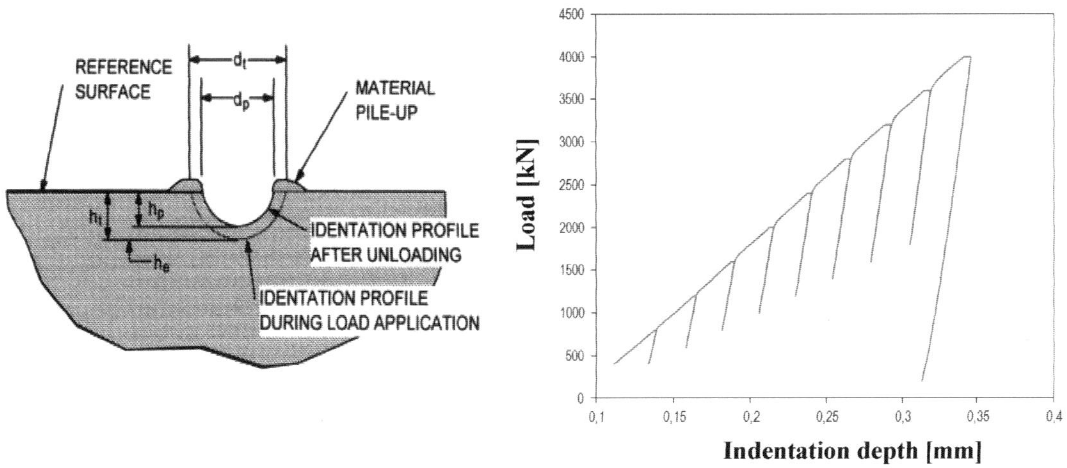

Figure 2. Scheme of the indent during the test (left), example of test data (right).

To determine the applicability of chosen test methods to the irradiated materials it was necessary to perform preparatory and supporting activities, e.g. finite element modelling and metallography of chosen material. FEM analyses, employed into process of ABI testing, were essential to determine the minimal necessary thickness of testing material in the terms of the plastic deformation affected zone size. Acquired results were also beneficial to determine the minimal distance of indents from the edge of the testing specimens due to the fact that most of the experimental work was performed on the standard Charpy type specimens. Metallography analyses in initial and irradiated state of material were performed mainly to determine that testing material has sufficiently fine grain structure to fulfill the recommendation to have at least 6 grains per diameter of the indenter.

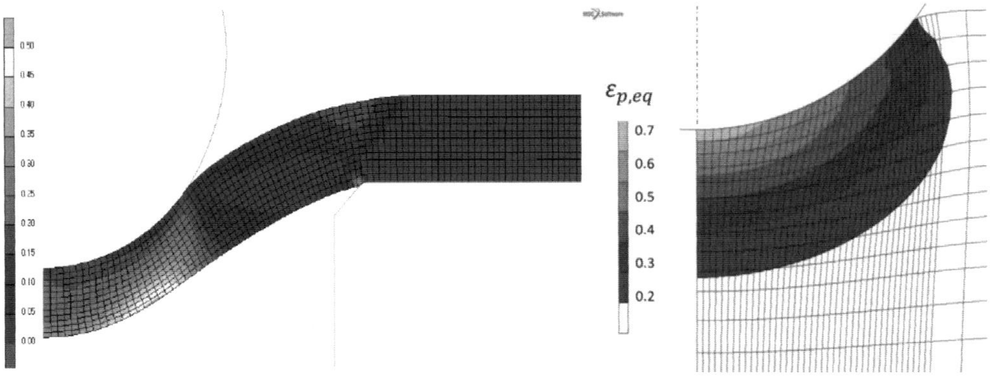

Figure 3. Examples of FEM analyses results for SPT (left) and ABI testing (right).

3 Testing results

3.1 Static tensile tests results

One of the first steps within the testing was to prepare data base of standard static tensile tests results for subsequent preparation of correlations with perspective methods of semi-destructive testing, SPT and ABIT.

For the realization of tests were machined tensile testing specimens with 4 mm diameter and 22 mm gauge length in the transverse orientation in original block of material. Tests were carried out at room temperature and +265 °C in accordance with the ISO 6892 [3, 4] using video extensometer for precise measurements of elongation. Results are summarized in the table 1.

Table 1. Tensile test results for the correlation with SPT and ABIT.

Number of specimens	Test Temperature [°C]	Neutron fluence (E > 0.5 MeV) [10^{18} cm^{-2}]	Average $R_{p0,2}$ [MPa]	Average R_m [MPa]
3	24	unirradiated	525	646
3	265	unirradiated	469	549
3	24	65.7	663	747
3	265	65.7	571	668

3.2 Small punch tests results

Small punch tests were carried out on the electro mechanical testing machine INSTRON in accordance with CWA15627 [5] document on the round specimens with the diameter of 8 mm and thickness of 0.5 mm (prepared with electric discharge machining and polishing).

Evaluated yield strength data were obtained from the SPT, by empirical correlations between SP and standardized test results based on the UJV Rez large database of ferritic base materials. It was observed that the nature of the load – deflection curve (Fig. 1) varies significantly with the specimen thickness h_0. The effect of test specimen thickness can be eliminated by correlating yield stress with parameter Fe/h_0^2. Results of tests are summarized in table. 2

Table 2. Results of Small Punch Testing (SPT).

Number of specimens	Test Temperature [°C]	Neutron fluence (E > 0.5 MeV) [10^{18} cm^{-2}]	Average $F_e/(h_0)^2$ [MPa]	Average $R_{p0,2}$ [MPa]	Average $F_m/(u_m . h_o)$ [MPa]	Average R_m [MPa]
15	24	unirradiated	843	510	2311	655
15	265	unirradiated	1029	495	1871	563
3	24	65.7	1999	660	3112	755
3	265	65.7	1577	588	2268	624

3.3 Automated ball indentation tests results

Experimental data were evaluated in several steps. For the yield strength assessment, for each loading cycle was determined force and the depth of indent h_t during load application (Fig. 2). From the depth of indent parameter d_t is calculated (idealized diameter of contact area) and is defined by the formula (1):

$$d_t = 2 \cdot \sqrt{h_t \cdot D - h_t^2} \qquad (1)$$

All data points of the dependence are fitted by a curve defined by the Mayer's law (2) -parameter A with the Mayer's index m is determined from the linear regression:

$$\frac{P}{d_t^2} = A \cdot \left(\frac{d_t}{D}\right)^{m-2} \qquad (2)$$

Parameter A is used for the calculation of yield strength approximation (3), where β_m is the parameter for evaluated material:

$$R_{p0,2}(APR) = \beta_m \cdot A \qquad (3)$$

Final value of the yield strength is calculated from the formula (4), where parameters a and b are determined by the calibration for the evaluated type of material.

$$R_{p0,2} = a + b \cdot R_{p0,2}(APR) \qquad (4)$$

For the ultimate strength evaluation, for each load cycle is determined a pair of values εp and σt where εp is logarithmic plastic deformation and σt is proportional to the actual stress in one-dimensional homogeneous deformation. If the resulting stress-strain curve is expressed by the exponential equation (5) it is possible to use obtained strain hardening coefficient K and strain hardening exponent *n* to estimate the ultimate strength approximation (6).

$$\sigma_t = K \cdot \varepsilon_p^n \qquad (5)$$

$$R_m(APR) = K \cdot \left(\frac{n}{e}\right)^n \qquad (6)$$

Calculation of the ultimate tensile strength is carried out in an analogous manner as for the yield strength (7), where parameters c and d are determined by the calibration for the evaluated type of material.

$$R_m = c + d \cdot R_m(APR) \quad (7)$$

Evaluated results of ABI testing for both unirradiated and irradiated specimens were correlated with the results from standard tensile testing. Summarized results for the various types of indenter diameters are summarized in the Table 3.

Table 3. Results of Automated Ball Indentation (ABIT).

Test Temperature [°C]	Indenter Diameter [mm]	Neutron fluence (E > 0.5 MeV) [10^{18} cm^{-2}]	Average $R_{p0,2}$ [MPa]	Average R_m [MPa]
24	2.500	unirradiated	488	612
	1.575	unirradiated	484	638
	0.762	unirradiated	489	632
	0.508	unirradiated	512	633
265	1.575	unirradiated	434	559
	0.762	unirradiated	444	552
24	2.500	65.7	554	693
	0.762	65.7	572	694
265	2.500	65.7	461	572
	0.762	65.7	477	591

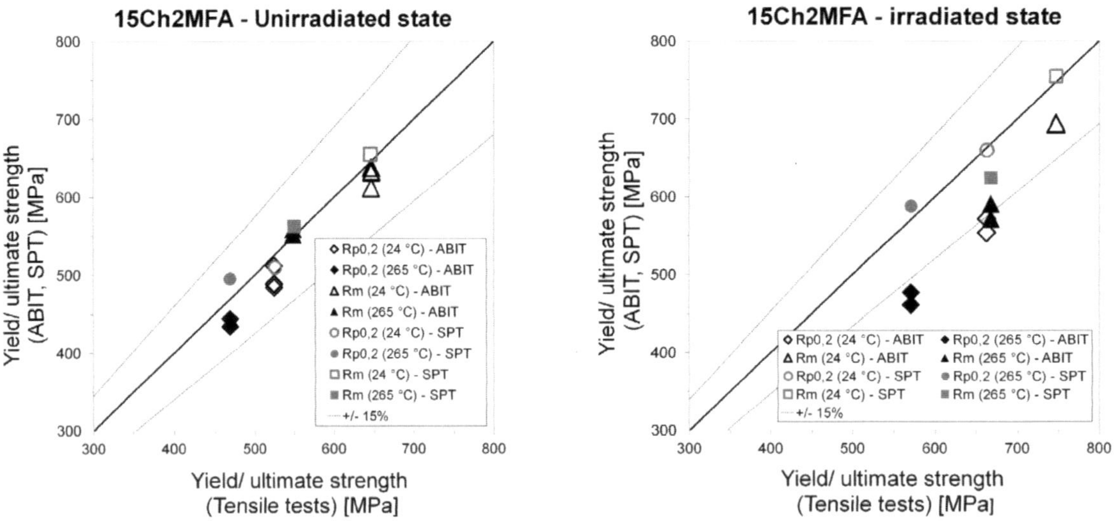

Figure 4. Correlation of standard tensile test results with the results of SPT and ABIT in initial unirradiated state (left) and irradiated state (right).

4 Conclusions

Based on obtained results on the 15Ch2MFA base material, correlations between standard tensile test and ABI tests was prepared as well as correlation with Small Punch test results (Fig. 4).

Both semi-destructive testing methods showed to be very promising for assessment of structuralirradiated NPP materials degradation. Evaluated results of ABIT and SPT show very good correlation with results from standard tensile testing.

At present, for the employment of the method into standard portfolio of laboratory test techniques is necessary to enlarge the experimental data volume for different types of used indenters, materials and to perform testing in temperature wider range from -190 °C to +350 °C.

Acknowledgement

This paper includes results created within the project TA03011266 "Development of innovative semi-destructive method of high active material evaluation for nuclear reactor components lifetime assessment" and the project TA02020811 "Development of methodology for evaluation of properties degradation of irradiated materials hardly replaceable components of nuclear power plants using penetration tests", both supported by Technology Agency of the Czech Republic.

References

[1] ASTM Draft, Tests Methods for Automated Ball Indentation (ABI) Testing of Metallic Samples and Structures to Determine Tensile Properties and Stress – Strain Curves, ISO/TC 164/SC 3 N1109

[2] Haggag F M , Wang J A , Sokolov M A , Murty K L.: Use of portable/in situ stress-strain microprobe system to measure stress-strain behaviour and damage in metallic materials and structures, in Non-traditional methods of sensing stress, strain and damage in materials and structures, ASTM STP 1318, Lucas and Stubbs eds. ASTM, Philadelphia, 1997

[3] ISO 6892-1-Metallic materials - Tensile testing - Part 1: Method of test at room temperature

[4] ISO 6892-2-Metallic materials - Tensile testing - Part 2: Method of test at elevated temperature

[5] CEN Workshop Agreement, CWA15627:2006, "Small Punch Test Method for Metallic Materials"

Impact of quenching temperature and isothermal holding time during austempering on bainite content in high-silicon steel

V Kotěšovec, I Vorel, Š Jeníček, J Káňa, K Ibrahim

University of West Bohemia, Regional Technological Institute, Univerzitní 22, Plzeň, Czech Republic

E-mail: vkotesov@rti.zcu.cz

Abstract. Silicon plays an important role in the manufacture and processing of steel. It is involved in metallurgical processes in the melt, improves castability, and, together with aluminium, belongs to the elements which suppress cementite formation during heat treatment of steels. The last-named aspect offers a great potential for developing high-strength steels with excellent ductility. The amount of bainite in the microstructure has a substantial impact in this respect. Bainite fraction depends mainly on temperature and on the isothermal holding time in the bainitic transformation region. In steels with low silicon levels, isothermal bainitic transformation continues until all austenite has transformed to bainite. At higher silicon levels, approximately 2 weight percent, the bainitic transformation is expected to be incomplete. The resulting bainite fraction would then be dictated by the thermodynamic equilibrium in the austenite-bainite system. As a result, one could control microstructural evolution, and thus the mechanical properties in high-strength bainitic-martensitic steels

1 Introduction

Isothermal bainitic transformation involves austenite decomposition into bainitic ferrite and carbides. In the course of this transformation, only interstitial atoms, i.e. mainly carbon, can diffuse. Hence, the formation and growth of carbides is conditional on the migration of carbon. Simply put, carbides absorb the carbon atoms that are rejected from the newly-forming bainite laths. Carbide precipitation can be prevented by adding a certain amount of silicon, at a level that is sufficient from the perspective of thermodynamic stability of the carbide-silicon system [1].

When steel contains enough silicon, carbon cannot form carbides during bainitic reaction and, eventually, diffuses into austenite which surrounds the emerging bainite laths [2, 3] (Fig. 1). Austenite thus continues to absorb carbon in the course of transformation until metastable equilibrium between bainitic ferrite and carbon-enriched austenite is established. At this point, the bainitic transformation stops, and therefore some austenite remains untransformed [4, 5].

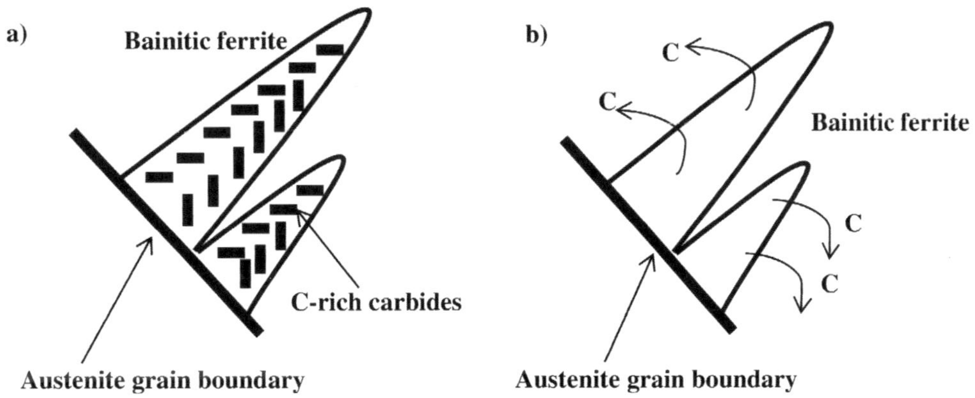

Figure 1. Effect of silicon content on the mechanism of austenite-bainite transformation a) low-silicon steel, b) high-silicon steel.

2 Experimental programme

The purpose of this experimental programme was to verify that bainitic transformation remains incomplete in a low-alloy steel with 2 % silicon and 0.43 % carbon (Table 1).

Table 1. Chemical composition of experimental steel.

C	Si	Mn	Cr	Mo	Nb	P	S	M_s [°C]	M_f [°C]
0.43	2.03	0.56	1.33	0.16	0.03	0.005	0.003	298	178

Austempering sequences were proposed with various temperatures, at which quenching was interrupted to enable bainitic transformation (Fig. 2). The choice of these temperatures was based on calculations of thermodynamic quantities with respect to the maximum attainable bainite fraction at various quenching temperatures in the bainitic transformation region [6]. (Figure 3)

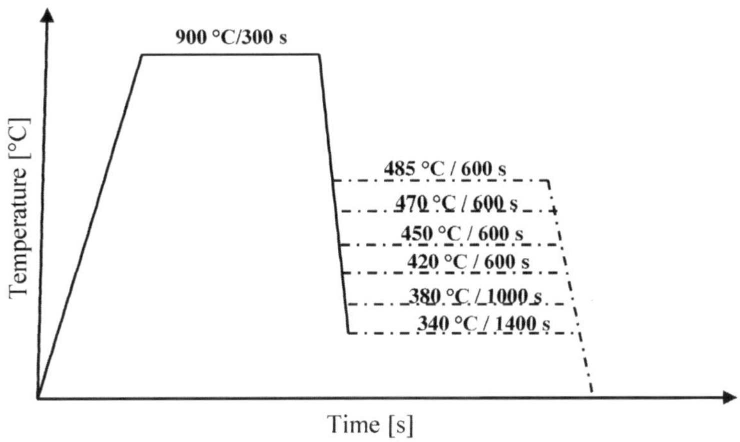

Figure 2. Experimental heat treatment sequences with quenching temperatures in bainitic transformation range.

The holding periods at bainitic transformation temperatures were based on the TTT diagram calculated using JMatPro software. They were intended to provide enough time for the bainitic reaction to complete (Figure 4). The results of the theoretical calculations were compared with experimental data on phase fractions obtained by image analysis. The experimental specimens were threaded round bars with a diameter of 8 mm. Prior to this experiment, they had been annealed at 1000 °C for 20 minutes. Heating and cooling was carried out in a thermomechanical simulator with an induction-resistance heating capability.

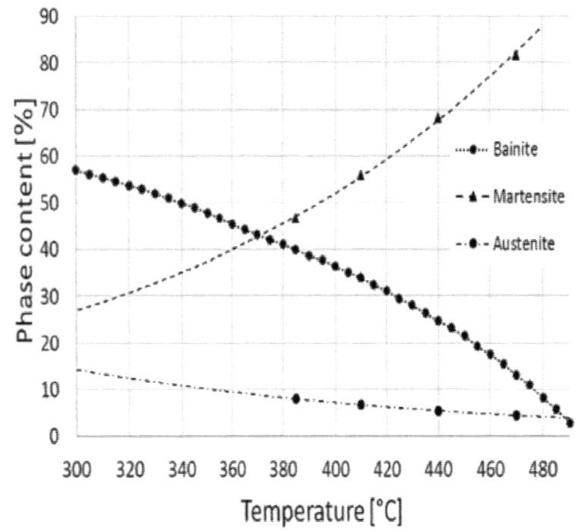

Figure 3. Theoretical fractions of bainite, martensite, and austenite at various quenching temperatures in the bainitic transformation region.

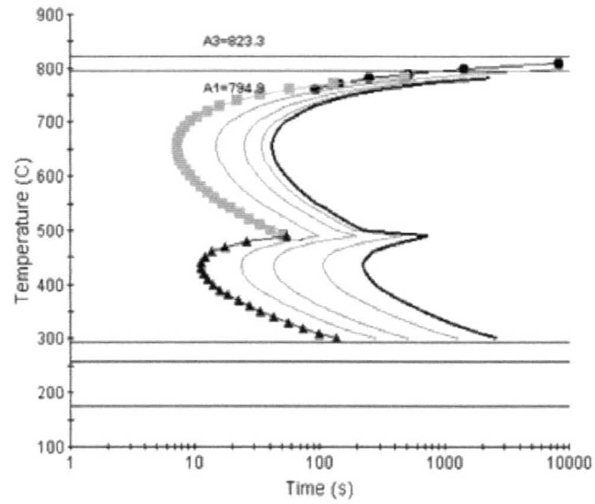

Figure 4. TTT diagram of experimental steel constructed on the basis of calculations using JMatPro program.

3 Results and discussion

Microstructures of the specimens were examined using scanning electron microscopy. They were revealed by etching with 4 % picric acid, 1 % hydrochloric acid, and 3 % nital. Phase compositions were determined using image analysis. Phase fractions were measured on 6 micrographs from each specimen using the software provided with a TESCAN microscope. The results were compared with the outcomes of thermodynamic calculations.

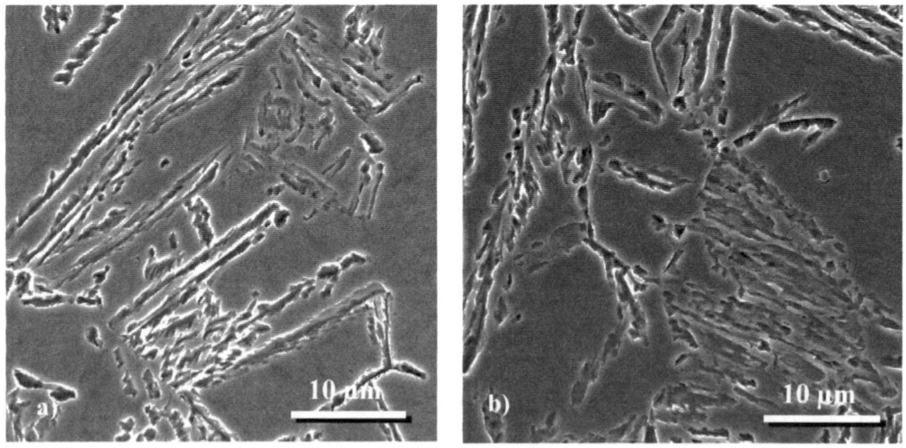

Figure 5. Micrographs of specimens upon interrupted quenching in bainitic transformation region.

a) specimen 485 °C/600 s – 28 % bainite, b) specimen 470 °C/600 s – 34 % bainite

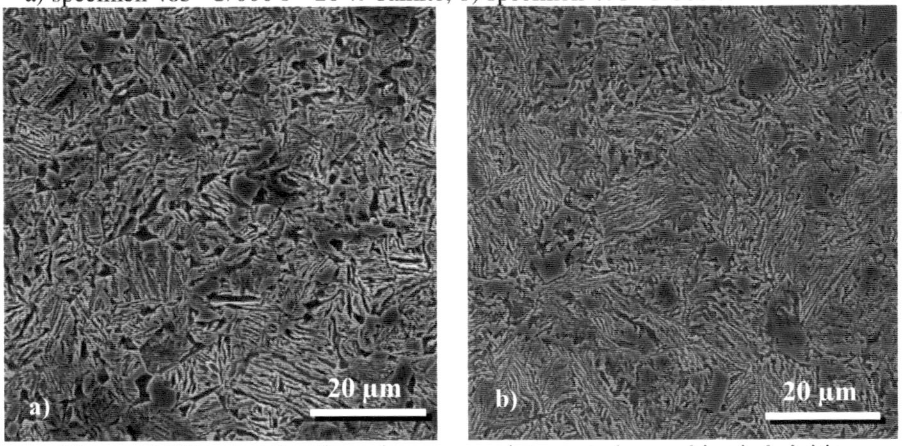

Figure 6. Micrographs of specimens upon interrupted quenching in bainitic transformation region
a) specimen 450 °C/600 s – 51 % bainite, b) specimen 420 °C/600 s – 63 % bainite.

Using scanning electron microscopy, martensite, the M-A constituent, and bainite were identified in individual specimens (Figs. 5, 6, and 7). The amount of bainite increased with decreasing quenching temperature in the bainitic transformation region. The lowest bainite fraction was found in specimen 485 °C/600 s: the amount was 28 % (Figure 5). The discrepancy between the theoretical value and the actual bainite fraction in this case was 23 %.

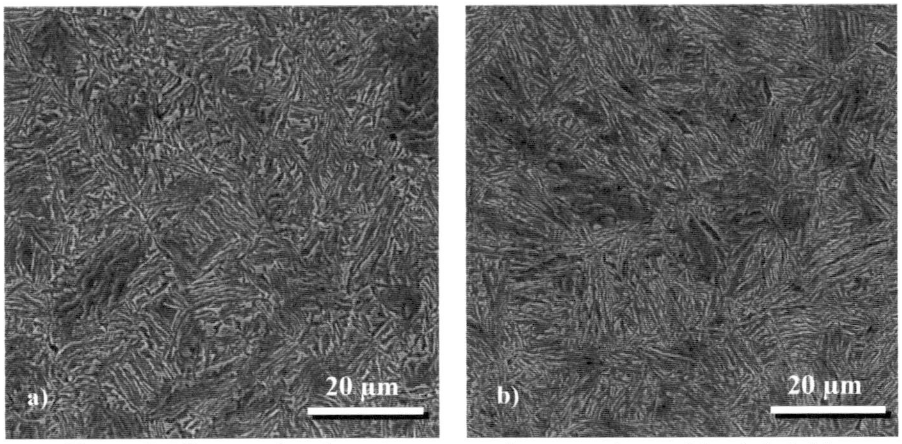

Figure 7. Micrographs of specimens upon interrupted quenching in bainitic transformation region
a) specimen 380 °C/1000 s – 74 % bainite, b) specimen 340 °C/1400 s – 86 % bainite.

The highest bainite fraction, 86 %, was found in specimen 340 °C/1200 s (Figure 7). The difference to the theoretical value was 36 % (Table 2). No experimental specimen showed complete transformation of austenite to bainite.

Table 2. Bainite fractions in experimental specimens – comparison between measured and predicted values.

Specimen designation	Bainite fraction [%] **Measured**/predicted value	Martensite and retained austenite fraction [%] **Measured**/predicted value
485/600	**28**/5	**72**/95
470/600	**34**/10	**66**/90
450/600	**51**/20	**49**/80
420/600	**63**/31	**37**/69
380/1000	**74**/41	**26**/59
340/1400	**86**/50	**14**/50

4 Conclusion

A study of incompleteness of bainitic transformation by interrupting quenching at bainitic transformation temperature was conducted on a low-alloy steel that contained 0.43 % carbon, 2 % silicon, 0.56 % manganese, and 1.33 % chromium. Using thermodynamic calculations, heat treatment sequences were proposed, and the resulting bainite fractions were predicted. The holding times at individual quenching temperatures were determined from the TTT diagram calculated using JMatPro software, and were intended to enable complete austenite-bainite transformation. Microstructure observation showed that the bainitic transformation was not complete. The transformation stopped after a certain amount of bainite had formed. The microstructures of all the experimental specimens consisted of bainite and martensite or the M-A constituent. The amount of bainite increased with the quenching temperature. The largest fraction – 86 % – was obtained when the quench was interrupted at 340 °C for 1400 seconds. The results of this experiment highlight the indisputable effect of silicon on phase transformation or more precisely on the thermodynamic equilibrium of the austenite-bainite system during isothermal treatment.

Acknowledgements
This paper presents results obtained under the project LO1502 Development of Regional Technological Institute. The project is subsidised by the Ministry of Education of the Czech Republic from specific resources of the state budget for research and development.

References

[1] Bhadeshia H K D H and Edmonds D V 1983 Bainite in silicon steels: new composition-property approach Part 1. Met. Sci. 17:411- 419
[2] Bhadeshia H K D H 1982 Bainite: Overall transformation kinetics. Journal de Physique Colloq. C4 12:443–448
[3] Ko T and Cottrell S A 1952 The formation of bainite J. Iron Steel Inst. 172:307–313
[4] Peet M J and Babu S S and Miller M K and Bhadeshia H K D H 2004 Three–dimensional atom probe analysis of carbon distribution in low–temperature bainite. Scripta Materialia 50
[5] Speer J G and Edmonds D V and Rizzo F C and Matlock D K 2004 Partioning of carbon from supersaturated plates of ferrite, with application to steel processing and fundamentals of the bainite transformation Current Opinion in Solid State and Materials Science 8:219–237
[6] Caballero F G and Bhadeshia H K D H 2002 Very Strong Low Temperature Bainite Materials Science and Technology, 18 279-284

Influence of the input parameters on the efficiency of plaster sanding with alundum abrasive discs

D Krajcarz[1, 2], S Spadło[1] and P Młynarczyk[1]

[1] Kielce University of Technology, Division of Materials Science and Munitions Technology, Al. Tysiąclecia Państwa Polskiego 7, 25-314 Kielce, Poland

[2]E-mail: d.krajcarz@wp.pl

Abstract. The paper presents test results concerning the relationship between selected input parameters and the process efficiency for the sanding of plaster surfaces with alundum abrasive discs. The input parameters under study were the size of the abrasive grains, the force exerted by the plaster sample pressing against the abrasive disc and the no-load rotational speed of the abrasive disc. The experimental data illustrating the relationship between the process efficiency and the particular input parameters were used to select the optimum plaster sanding conditions.

1 Introduction

Sanding is used to remove excess material from wrought or cast cylindrical objects and to finish metallic and non-metallic surfaces, including gypsum plaster. In the engineering nomenclature, the terms grinding, honing, superfinish or lapping are applied to describe similar processes [1-3]. The tools used for this purpose require flexible mesh discs with resin-bonded abrasive grains ranging from 60 to 240 in size [4, 5]. The aim of the study was to determine the effect of the process conditions on the quality of plaster sanding [6, 7].

2 Experiment

The tests were conducted using Flexa abrasive discs with resin-bonded alundum (Al_2O_3) [8] grains differing in size (60, 120 and 180). The aggregates on the abrasive meshes were irregular in shape. The tests were performed using a specially developed test stand comprising a drywall sander with interchangeable abrasive discs and a system to feed a rod-like plaster sample [9] 50 mm in diameter. The test variables were the force exerted by the plaster sample pressing against the abrasive disc, the no-load rotational speed of the abrasive wheel, expressed by revolutions per minute and the size of the abrasive grains.

2.1 Test conditions

The experiments were carried out according to the Box Behnken 3x3 factorial design of experiments [10-12] (with three factors and three levels). In the experiments, the variables were: the size of the abrasive grains (Z), the force exerted by the plaster sample pressing against the abrasive disc (F) and the no-load rotational speed of the abrasive disc (n) expressed by revolutions per minute. One-factor experiments were performed for the input parameters in the predetermined ranges of variability. The design of experiments (DOE) was then generated using the DOE module of Statistica 10. Table 1 provides actual values of the input parameters and the values of the process efficiency (E_f) calculated on the basis of the experimental data.

Table 1. Design of experiments and the measurement results.

Test No.	Input parameters			Process efficiency
	Z, mesh	*F*, N	*n*, rpm	E_f, cm³/min
1.	60	2	2250	136.92
2.	180	2	2250	75.48
3.	60	6	2250	367.97
4.	180	6	2250	210.27
5.	60	4	1500	67.67
6.	180	4	1500	41.76
7.	60	4	3000	327.08
8.	180	4	3000	203.02
9.	120	2	1500	56.07
10.	120	6	3000	105.13
11.	120	2	3000	107.05
12.	120	6	3000	367.97
13.	120	4	2250	173.16
14.	120	4	2250	178.41
15.	120	4	2250	168.21

3 Results and discussion

Table 2 shows the input parameters and the corresponding coefficients of partial correlation. The strongest linear correlation (*L*) was reported for the parameter *n(L)*, where $k = 0.64$. From the values of the coefficients of partial correlation it is clear that there is another significant positive correlation for the parameter *F*, where $k = 0.59$. The values of the quadratic nonlinear regression (*Q*) did not reach the level of significance $k > |0.5|$ in any of the cases.

Table 2. Coefficients of partial correlation.

Input parameter	Coefficient of correlation
Z(L)	-0.32
Z(Q)	-0.07
F(L)	0. 59
F(Q)	-0.06
n(L)	0.64
n(Q)	0.13

Figure 1 shows the relationship between the process efficiency and two parameters, i.e. the force *F* and the grain size *Z* (at $n = 2250$ rpm). From the diagram it is evident that in the analysed range of variability, the maximum efficiency is achieved for the abrasive discs with a mesh size of 60. An increase in the force *F* in the whole range of variability causes an increase in the sanding efficiency.

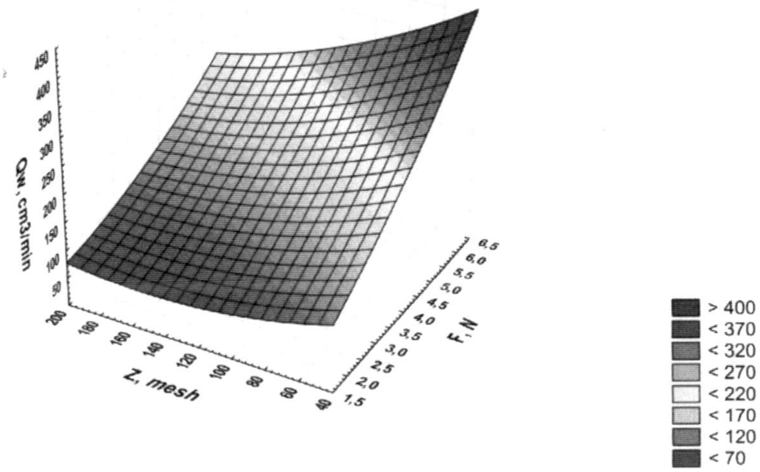

Figure 1. Process efficiency versus the force exerted F by the plaster sample pressing against the abrasive disc versus the size of the abrasive grains Z (at n = 2250 rpm).

Figure 2 shows the relationship between the plaster sanding efficiency and two input parameters: n and Z (at $F = 4$ N). As can be seen from the diagram, the process efficiency increases with increasing rotational speed n in relation to a change in the grain size in the test range. In the analysed range of variability, the maximum process efficiency is reported for the abrasive discs with a mesh size of 60.

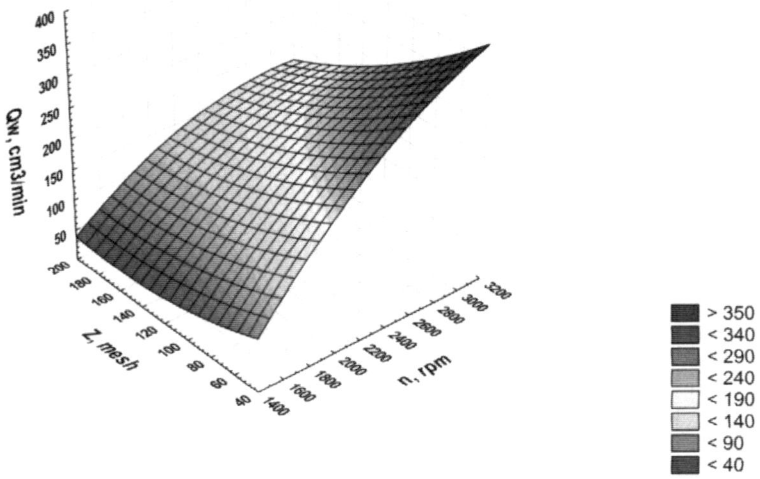

Figure 2. Process efficiency versus the rotational speed of the abrasive disc n versus the size of the abrasive grains Z (at F = 4 N).

Figure 3 shows the relationship between the efficiency of the sanding process and two input parameters: the rotational speed n and the force exerted by the plaster sample F (for $Z = 120$ mesh). From the diagram it is evident that an increase in either of the parameters leads to an increase in the process efficiency. The maximum efficiency is achieved when both parameters of the experimental design are high.

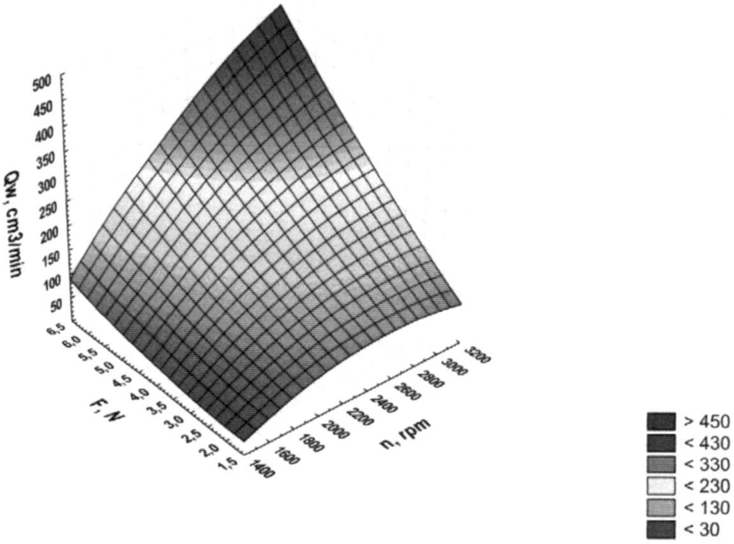

Figure 3. Process efficiency versus the force exerted by the plaster sample pressing against the abrasive disc N versus the rotational speed of the disc n (for $Z = 120$ mesh).

4 Conclusions

The values of the coefficients of partial correlation obtained for the linear regression confirm a considerable influence of the rotational speed of the abrasive disc *n(L)* and the force exerted by the plaster sample *F(L)* on the process efficiency (positive correlation, $k = 0.64$ and $k = 0.59$, respectively).

The coefficients of partial correlation obtained for linear regression show that the size of the abrasive grains *Z(L)* had the least influence on the efficiency of the plaster sanding process (negative correlation, $k = -0.32$).

The values of the quadratic nonlinear regression did not reach the level of significance in any of the cases considered. For the parameters affecting the efficiency of the sanding process, i.e. the force exerted by the plaster sample *F(Q)* and the size of the abrasive grains *Z(Q)*, the coefficient of correlation fluctuated around 0, which indicates that the relationship is practically linear.

References

[1] Chłądzyński S and Wieczorek M 2007 Porównanie właściwości gipsów budowlanych badanych krajowymi i europejskimi metodami, *Izolacje*, no. 2, pp 38-42

[2] Filipowski R and Marciniak M 2000 *Techniki obróbki mechanicznej i erozyjnej*, Warszawa: Oficyna wydawnicza Politechniki Warszawskiej.

[3] Pszczołowski W and Rosienkiewicz P 1995 Obróbka ścierna narzędziami nasypowymi, WNT, pp 448

[4] Czerniki S 2012 Gładzie gipsowe – teoria i praktyka, *Izolacje*, no. 6 pp 46-49

[5] Idzikowski P 2006 Gładzie i tynki gipsowe – wskazówki wykonawcze. *Izolacje*, no. 2 pp 60-62

[6] Szponar R 1996 Tynki gipsowe, *Materiały Budowlane*, no. 10, pp 13

[7] Walawko W 1998 Tynki gipsowe – nowa jakość ścian i sufitów, *Materiały Budowlane*, no. 10, pp 21-24
[8] Patnaik P 2002 Handbook of Inorganic Chemicals. McGraw-Hill
[9] EN 13279-1:2004 Gypsum binders and gypsum plasters. Part 1: Definitions and requirements
[10] Spadło S and Krajcarz D and Dudek D 2015 Wpływ wybranych parametrów procesu przecinania strugą wodno-ścierną na dokładność geometryczną i jakość powierzchni otworów cylindrycznych, *Mechanik*, no. 8-9, pp 308-312
[11] Tekindal M and Bayrak H and Ozkaya B and Genc Y 2012 Box- behnken experimental design in factorial experiments: the importance of bread for nutrition and health. *Turkish Journal of Field Crops*, vol. 17(2), pp 115-123
[12] Ferreira S L C nd Bruns R E and Ferreira H S and Matos G D and David J M and Brandão G C and Silva E G P and Portugal L A and Reis P S and Souza A S and Santos W N L 2007 Box-Behnken design: Analternative for the optimization of analyticalmethods, *AnalyticaChimica Acta*, vol. 597(2), pp 179–186

Aluminium alloys with transition metals prepared by powder metallurgy

V Kucera[1], **F Prusa**[1], **D Vojtech**[1]

[1]Department of Metals and Corrosion Engineering, University of chemistry and technology Prague, Technická 5, 16628 Prague 6, Czech Republic

E-mail: kucerao@vscht.cz

Abstract. Powder metallurgy represented by mechanical alloying and spark plasma sintering was used for preparation of the AlFe16 and the AlSi20Fe16 alloys. Microstructure of the both alloys consisted of very fine intermetallic phases homogenously dispersed in the matrix of α-Al solid solution. Fine nature of microstructure led to promising results of compressive stress-strain tests performed at laboratory and elevated temperature of 400 °C. The compressive strengths of the AlSi20Fe16 and the AlFe16 alloys at laboratory temperature were 780 MPa and 508 MPa, respectively. Elevated temperature resulted in drop of the compressive strengths to 480 MPa and 211 MPa, respectively. However, the results of investigated alloys outperformed the thermally stable AlSi12Cu1Mg1Ni1 (wt. %) used as reference material.

1 Introduction

Aluminium and its alloys belong to the most widespread used metallic materials, especially in automotive or aircraft industry, due to its low density and good strength to weight ratio [1, 2]. The primary production of pure aluminium from ore consumes a huge amount of energy and it is potentially harmful for the environment. Hence, recycling of aluminum alloys becomes more and more important, since it consumes much less energy. The main disadvantage of recycling is the accumulation of the impurities and other alloying elements in the melt [3-5]. It is the iron, which is the common impurity in aluminium alloys and also other transition metals that negatively influences the mechanical properties. In the alloys produced by conventional casting technologies, transition metals form relatively large intermetallic phases due to their low solubility in α-Al solid solution (max. solubility for Fe is 0.052 wt.% at 655 °C) [6]. In the Al-Si-Fe based alloys these intermetallic phases are α-Al_8Fe_2Si with hexagonal structure and monoclinic or orthorhombic β-Al_5FeSi phase, while a monoclinic $Al_{13}Fe_4$, also described in literature as Al_3Fe, is formed in the Al-Fe alloys. The morphology of the phases is needle-like or platelet-like and its dimension depends on the content of the iron as well as on the cooling rate during solidification. The length of β-phase, commonly considered as the most harmful phase in the Al-Si-Fe alloys, increases with iron content. The detrimental effect on mechanical properties, particularly on ductility, is caused by the brittle nature of the phase, which acts as potential site for microcrack propagation and growth resulting in subsequent decohesion failure [6-10]. To prevent the formation of brittle and relatively large intermetallic phases, powder metallurgy is used as an alternative to process aluminium alloys with high content of transition metals [6, 11, 12]. On the other hand, transition metals are characterized with low diffusion coefficients in α-Al solid solution. Therefore, aluminium alloys with high content of these elements are considered as materials for elevated temperatures applications [12, 13]. Thus, the refinement of the potentially harmful phases down to sub-micrometer sizes represents the way how to utilize such materials, while the mechanical properties can outperform currently used alloys [14].

Mechanical alloying (MA) belongs partially among the severe plastic deformation (SPD) methods. It was originally developed to manufacture oxide-dispersion strengthened (ODS) alloys. However its use is increasing and materials with fully amorphous or nanocrystalline microstructure, disordered solid solutions and supersaturated solid solution are produced by this method. During the MA, particles of the powders are trapped between milling balls and wall of the mill and are repeatedly flattened, cold welded, fractured and re-welded again. Therefore, particles of the powders are heavily deformed, which leads to increase of crystal defects such as vacancies, dislocations, stacking faults and increased number of grain boundaries [15-17]. To preserve fine character of the microstructure obtained by the MA, prepared powders are consolidated via spark plasma sintering (SPS). The SPS is loosely described as control flow of current (up to 20 000 A) throughout a conductive die (mostly fabricated of graphite or WC) filled with powder, which is at the same time compressed by a moving pistons. A high thermal efficiency and heating rate enables consolidation at relatively lower temperatures and reduce the total time of compaction [12, 18].

In this work, two aluminium alloys with high content of iron were prepared by short-term MA and subsequently compacted via SPS. The prepared MA+SPS alloys were investigated to determine the phase composition, microstructure as well as mechanical properties and thermal stability.

2 Materials and methods

The AlFe16 and AlSi20Fe16 (in wt. %) alloys were prepared by short-term MA and compacted via SPS. Pure powders of Al (purity 99.7 %), Fe (purity 99.9 %) and Si (purity 99.9 %) were blended together at the proper ratio and placed into a steel mould together with milling balls both made from AISI 420 stainless steel, sealed and flushed with argon gas to prevent oxidation during alloying. The ball-to-powder mass ratio was 30:1. Mechanical alloying was done by planetary milling device Retsch PM 100 with rotational speed of 400 rpm while the direction of rotation was reversed every 30 min. The total time of MA was 8 hours.

Prepared powder alloys were pre-compacted by the pressure of 360 MPa for 5 minutes at the LabTest 5.250SP1-VM universal testing machine forming semi-compact samples further used for the compaction via SPS. For this purpose, the FCT Systeme HP-D 10 was used. The heating rate was 200 °C·min^{-1} till it reached final 500 °C. After the temperature was reached, the sample was compressed by a pressure of 48 MPa. The total time the sample remained at the 500 °C was 10 minutes followed by a spontaneous cooling.

Simultaneously, both alloys were prepared by common casting metallurgy simulating slow solidification. The AlFe16 alloy was prepared by melting the master alloy AlFe25 (in wt. %) with pure aluminium (purity of 99.62 %) in electric resistance furnace. The melt was casted into a brass mould forming a slowly solidified ingot with 50 mm in diameter and 150 mm in height. The AlSi20Fe16 alloy was prepared by melting pure aluminium and silicon (purity of 99.9%) together with AlFe25 (wt. %) master alloy by induction under a protective argon atmosphere and then poured into a brass mould forming an ingot measuring 20 mm in diameter and 150 mm in height.

Chemical and phase composition was studied by XRF and XRD (ARL 9400 XP and PANalytical X'Pert Pro, Cu Kα1 λ=1.54059·10^{-10} m) analyses.

Compact samples were cut using a cutting machine (Leco Vari/Cut VC-50) equipped with a diamond blade into specimens that were used for further microstructural and mechanical investigations. Samples for microstructure observation were prepared by standard metallography procedure including grinding, polishing and etching the samples in Keller reagent (190 ml distilled water, 5 ml HNO_3, 3 ml HCl, 2 ml HF). Microstructure was observed by optical microscope (Olympus PME-3) and scanning electron microscope (Tescan Vega 3 LMU, 20 kV, SE + BSE detectors) equipped with energy dispersive spectroscopy detector (Oxford Instruments INCA 350, 20 mm^2). The compressive tests were performed by the LabTest universal testing machine (already mentioned) with strain speed of 1 mm/min at laboratory temperature and at elevated temperature of 400 °C. The height to base ratio of the samples for compressive tests was 1.5:1.

As a reference material, the commercial thermally stable casting AlSi12Cu1Mg1Ni1 (wt. %) alloy, obtained in form of an ingot from our external supplier, was used. The alloy was heat treated by the T6 regime combining solution annealing (510 °C/5 h), water quenching and artificial aging (230 °C/6 h).

3 Results and discussion

3.1 X-ray diffraction analysis of the prepared alloys

The X-ray diffraction patterns of as-mixed powders and both alloys prepared by MA+SPS are shown in **Figure 1**. The phase composition of the AlFe16 as-cast alloy corresponds to the α-Al solid solution and to the most thermally stable intermetallic phase $Al_{13}Fe_4$ (not shown). This concurs with the binary Al-Fe phase diagram, where intermetallic $Al_{13}Fe_4$ phase crystallizes from the melt with 16 wt. % of Fe followed by eutectic reaction L → α-Al + $Al_{13}Fe_4$ at eutectic temperature. [6]. The XRD pattern of as-mixed powders (**Figure 1**a) shows only peaks corresponding to pure Al and Fe. After the MA+SPS of the both materials, slight peak broadening corresponding to grain refinement and to the increase of the internal stresses as well as presence of intermetallic phases is clearly visible. Krasnowski and Kulik [15] observed similar behavior in the MA AlFe20 (at. %) alloy, where supersaturated solid solution α-Al was formed during the first hours of the process. More importantly, the formation of new metastable Al_5Fe_2 phase was observed after 25 hours of MA. In comparison, the alloy containing 15 to 17 at. % Fe developed fully amorphous microstructure after 40 hours of MA.

In case of AlFe16 binary alloy none of new phases were observed after 8 hours of MA. However, after consolidation via SPS intermetallic phase $Al_{13}Fe_4$ was formed and the phase composition of consolidated sample composed of the α-Al solid solution and of intermetallic $Al_{13}Fe_4$ phase. The XRD diffraction pattern of the AlSi20Fe16 alloy is depicted in figure 1b. Phase composition of the as-cast alloys (not shown) consisted of α-Al, Si and intermetallic phases identified as $Al_9Fe_2Si_2$ and Al_3FeSi_2, commonly designated in the Al-Fe-Si ternary phase diagram as τ_6 and τ_4 [19]. As in previous case, XRD diffraction pattern of as-mixed powders contained only peaks of pure Al, Si and Fe which presence partially diminished during the MA+SPS allowing to form the $Al_9Fe_2Si_2$ intermetallic phase.

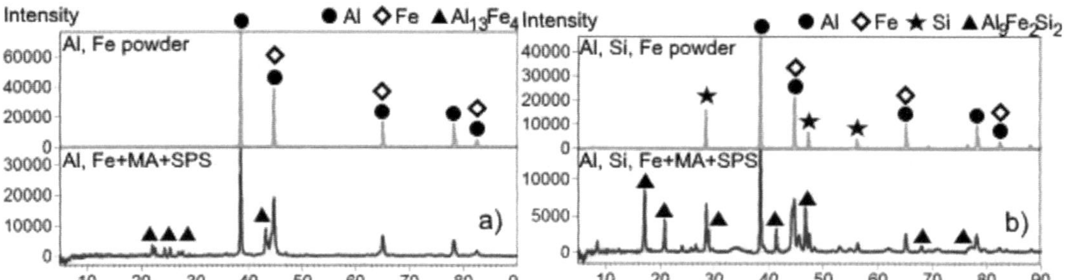

Figure 1. XRD patterns of the as-mixed powders and MA+SPS alloys: a) AlFe16, b) AlSi20Fe16.

3.2 Study of microstructure

The SEM micrographs of the as-cast and MA+SPS alloys are shown in **Figure 2** and **Figure 3**. The microstructure of the both as-cast AlFe16 (**Figure 2**a) and AlSi20Fe16 (**Figure 3**a) alloys was characterized by presence of relatively large intermetallic phases with the longitudinal size reaching several hundreds of micrometres identified as $Al_{13}Fe_4$ (Figure 2a) and $Al_9Fe_2Si_2$, Al_3FeSi_2 (Figure 3a), respectively. Further, the microstructures of both the AlFe16 and AlSi20Fe16 alloys contained eutectic mixture of α-Al + $Al_{13}Fe_4$ and α-Al + Si, respectively. Furthermore particles of primary silicon are seen in the microstructure of AlSi20Fe16, which is predictable due to the high content of Si. Microstructure of both the MA+SPS alloys consists of very fine intermetallic $Al_{13}Fe_4$ (**Figure 2**b) and $Al_9Fe_2Si_2$ (**Figure 3**b) phases and of very fine particles of Si, which were homogenously dispersed in the α-Al solid solution.

Figure 2. SEM micrographs of AlFe16 alloy: a) as-cast, b) MA+SPS.

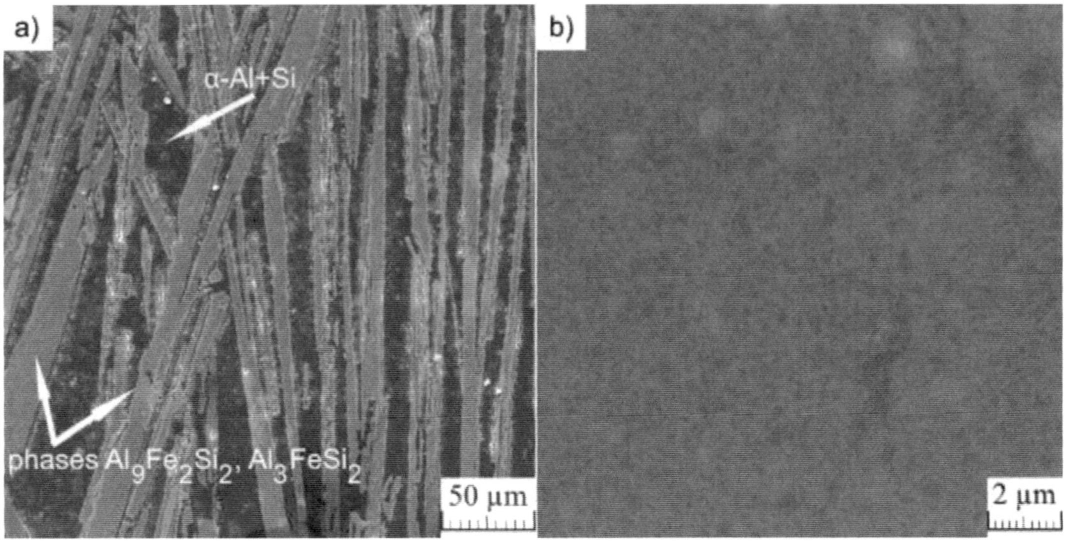

Figure 3. SEM micrographs of AlSi20Fe16 alloy: a) as-cast, b) MA+SPS.

3.3 Compressive stress-strain tests

The compressive stress-strain tests of the MA+SPS alloys were measured at laboratory and elevated temperature of 400 °C and compared to the commercial thermally stable casting AlSi12Cu1Mg1Ni1 alloy. The results are shown in **Figure 4** and in **Table 1**, respectively. The highest compressive strength

(CS) measured at laboratory temperature Figure 4a) was observed by the AlSi20Fe16 alloy reaching 758 MPa. The CS of the MA+SPS AlFe16 alloy was similar to the reference AlSi12Cu1Mg1Ni1 alloy. However, both alloys prepared by MA+SPS exhibited lower plasticity compared to the reference alloy, which can be explained by large internal stresses and strains of the crystallographic lattice induced during the MA and also by a relatively high volume fraction of intermetallic phases as well as by very fine character of present α-Al grains. Figure 4b depicts compressive stress-strain curves at 400 °C. The values of CS could not be determined due to the continuous character of the deformation with absence of fracture. Both the MA+SPS alloys indicated that high value of iron plays important role in thermal stability due to its low diffusion coefficient in α-Al. Compressive yield strength (CYS) of the MA+SPS AlFe16 and AlSi20Fe16 alloys were 211 MPa and 480 MPa, respectively. Such a high value of CYS in the later mentioned alloy was caused by high content of Si particles as well as by high volume fraction of intermetallic phases containing Fe and Si. Thus, both MA+SPS prepared alloys evinced better thermal stability compared to reference AlSi12Cu1Mg1Ni1 alloy that reached CYS of only 100 MPa. Another fact that should be mentioned is the enormous increase in plastic deformation reaching several tens of percent.

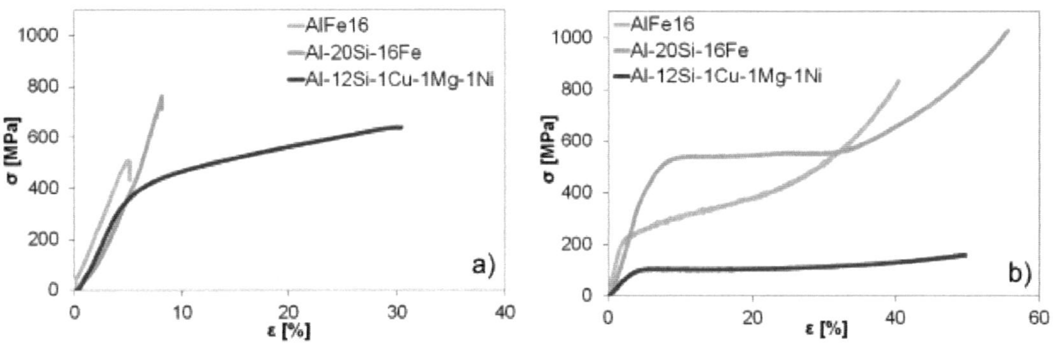

Figure 4. The compression stress-strain curves of tested alloys at laboratory temperature a) and elevated temperature of 400 °C b).

Table 1. Results from compression tests measured at laboratory and elevated temperature of 400 °C (CYS-compressive yield strength; CS-compressive strength).

Sample	Lab. temperature		400 °C	
	CYS [MPa]	CS[MPa]	CYS [MPa]	CS [MPa]
AlFe16	-	508	211	-
AlSi20Fe16	-	758	480	-
AlSi12Cu1Mg1Ni1	430	680	100	-

4 Conclusion

Two aluminium alloys AlFe16 and AlSi20Fe16 were prepared by MA followed by consolidation via SPS. Microstructure of both the MA+SPS alloys consisted of very fine intermetallic phases identified as $Al_{13}Fe_4$ and $Al_9Fe_2Si_2$, respectively, homogenously dispersed in α-Al solid solution. The fine character of the microstructure led to very promising mechanical properties. The highest value of CS (758 MPa) at laboratory temperature was achieved by the Al20Si16Fe alloy, however it showed absence of plastic deformation. Both alloys showed improved thermal stability compared to the reference casting AlSi12Cu1Mg1Ni1 alloy. In fact, the CYS of the MA+SPS AlFe16 alloy was almost two times higher compared to the reference alloy while the AlSi20Fe16 alloy showed almost five times higher CYS.

Acknowledgment
The authors wish to thank the Czech Science Foundation (project no. P108/12/G043) for its financial support of this research.

References

[1] Park W-W, B-S You and N J Kim 1996 *Materials & Design* **17**(5): p. 255-259.
[2] Choi Y S, et al. 1999 *Journal of Materials Science* **34**(9): p. 2163-2168.
[3] Schmitz C, J Domagala and P Haag, *Handbook of Aluminium Recycling*, ed. C. Schmitz. 2007: Vulkan-Verlag GmbH.
[4] Osawa Y, et al. 2007 *MATERIALS TRANSACTIONS* **48**(9): p. 2467-2475.
[5] Malakhov D V, D Panahi and M Gallerneault 2010 *Calphad* **34**(2): p. 159-166.
[6] Belov N A, A A Aksenov and D G Eskin, *Advances in Metallic Alloys*. Iron in Aluminium Alloys: Impurity and Alloying Element, ed. J.N. Fridlyander and D.G. Eskin. Vol. 2. 2002: Taylor & Francis. 360.
[7] Fang X, et al. 2007 *Materials Science and Engineering: A* **445–446**: p. 65-72.
[8] Eidhed W 2008 *Journal of Materials Science & Technology* **24**(1): p. 45 - 47.
[9] Taylor J A 2012 *Procedia Materials Science* **1**: p. 19-33.
[10] Khalifa W, et al. 2005 *Metallurgical and Materials Transactions A* **36**(4): p. 1017-1032.
[11] Dobromyslov A V, et al. 2015 *The Physics of Metals and Metallography* **116**(9): p. 942-950.
[12] Sasaki T T, T Ohkubo and K Hono 2009 *Acta Materialia* **57**(12): p. 3529-3538.
[13] Průša F and D Vojtěch 2013 *Materials Science and Engineering: A* **565**: p. 13-20.
[14] Průša F, et al. 2014 *Materials Science and Engineering: A* **603**: p. 141-149.
[15] Krasnowski M and T Kulik 2009 *Materials Chemistry and Physics* **116**(2–3): p. 631-637.
[16] Minamino Y, et al. 2004 *Science and Technology of Advanced Materials* **5**(1–2): p. 133-143.
[17] Suryanarayana C 2001 *Progress in Materials Science* **46**(1–2): p. 1-184.
[18] Suárez M, et al., *Challenges and Opportunities for Spark Plasma Sintering: A Key Technology for a New Generation of Materials*, in *Sintering Applications*, B. Ertuğ, Editor. 2013, InTech.
[19] Marker M C J, et al. 2011 *Intermetallics* **19**(12): p. 1919-1929.

IOP Conf. Series: Materials Science and Engineering **179** (2017) 012044 doi:10.1088/1757-899X/179/1/012044

High ductility of bainite-based microstructure of middle carbon steel 42SiMn

L Kučerová, M Bystrianský, Š Jeníček

RTI, UWB in Pilsen, Universitni 8, 30614 Pilsen, Czech Republic

E-mail: skal@rti.zcu.cz

Abstract. Heat and thermo-mechanical treatments with various processing parameters were applied to middle carbon low alloyed 42SiMn steel. The aim of the treatment was to obtain multiphase microstructure typical for TRIP (Transformation induced plasticity) steel and to achieve the best combination of ultimate tensile strength and ductility. TRIP steels typically possess about 5-15% of metastable retained austenite, which can transform to martensite during plastic deformation. The gradual phase transformation during loading postpones the onset of necking, thus increasing ultimate tensile strength and ductility at the same time. Manganese and silicon, used as the main alloying elements of the experimental steel, are employed to increase austenite stability and to hinder cementite precipitation during the treatment. All proposed methods of heat and thermo-mechanical treatment contain bainitic hold at 400 °C or 425 °C. The final microstructures were very complex, consisting of bainite, ferrite, very small areas of extremely fine perlite lamellas, about 10% of retained austenite and M-A constituent (austenitic islands partially transformed to martensite). Even though pearlite and martensite are undesirable microstructure in TRIP steel, the tensile strength ranged from 850 to 1065 MPa and ductility A_{5mm} from 26 to 47 %.

1 Introduction

Various grades of advanced high strength steels have been proposed in the last two decades [1]. Many of them require particular chemical compositions, based either on heavy alloying or on special micro alloying. Each group of AHSS has slightly different combination of mechanical properties, starting from relatively low strength and high ductility of dual phase steels [2] to very high strengths with lower ductility of martensitic steels after Q-P (quenching and partitioning) processing [3-7]. Somewhere in the middle are TRIP (transformation induced plasticity) steels which can possess interesting combinations of relatively high strengths and very good ductility [8, 9]. Another advantage of TRIP steels is that they can reach these properties even without special alloying. The most typical chemical composition of TRIP steels consist of 0.2-0.4% C, 1-2%Mn and 1-2%Si [10, 11] and the first conventional TRIP steels kept their carbon contents at the lower boundary of the interval, around 0.2%. Many other alloying concepts have been also investigated to improve mechanical properties or workability of the steel.

The main principle which is responsible for good strength to ductility balance is mechanically induced transformation of retained austenite to martensite, which can occur during cold plastic deformation. The gradual transformation postpones the necking of the steel, which increases homogeneous plastic deformation. It has been shown that about 10% of properly stabilized retained austenite is ideal to ensure the utilization of TRIP effect. To retain sufficient amount of austenite to room temperature in low alloyed steel, special heat or thermo-mechanical treatment has to be applied. The aim of the treatment is not only to produce desired semi-product but mainly to produce suitable multiphase microstructure and also to increase carbon content in remaining austenite during the cooling and thus decrease its martensite start temperature below room temperature. Higher carbon content in the steel would naturally help to achieve this goal and it would also increase the hardness and strength of the steel. Conventional steel with middle carbon contents usually pay for increase in strength by deteriorating ductility. This does not have to be the case for TRIP steels, as this article is going to suggest.

2 Experimental program

2.1. Characterization of used steel

Middle carbon, low alloyed steel with very simple alloying concept was used in this work. Besides 0.4% C, it was further alloyed only by 0.6% Mn and 2% Si. Manganese was added to support austenite stabilization, as an austenite-forming element. Silicon can hinder cementite precipitation during the cooling, which in the end effect also promotes austenite stabilization, as more carbon is than available to diffuse into remaining austenite. TTT (time temperature transformation) diagram of the steel was calculated in JMatPro software (Figure 1).

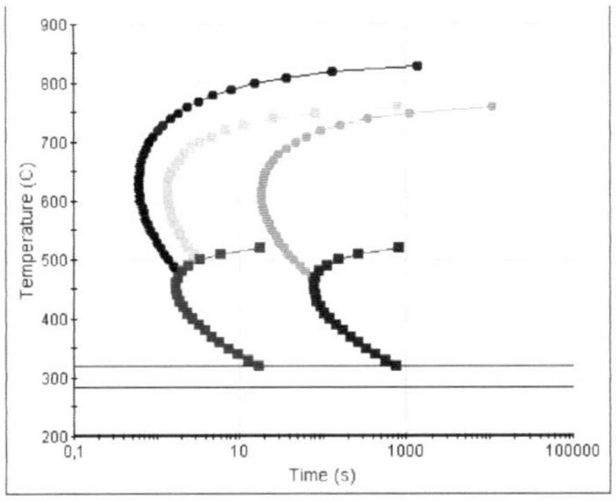

Figure 1 TTT diagram of 0.4C-0.6Mn-2Si steel.

Mechanical properties after the processing were obtained by tensile test on small flat samples with the length of active part being 5 mm and the cross section 2 x 1.2 mm. Volume fraction of retained austenite was determined by X-ray diffraction phase analysis using automatic powder diffractometer AXS Bruker D8 Discover with HI-STAR detector and Co lamp ($\lambda K\alpha = 0.1790307$ nm). Focusing polycapillary lens was used to achieve X-ray spot with 0.5 mm diameter. Resulting microstructures were analyzed by scanning electron microscope Zeiss EVO.

2.2. Heat and thermo-mechanical treatment

Three methods of thermo-mechanical treatment and six methods of heat treatment were tested on experimental steel (Table 1). Two soaking temperatures were chosen, the higher temperature of 1000 °C ensures full austenisation of the steel, while the lower temperature of 850 °C lies in the intercritical austenite - ferrite region. Soaking hold of 100 s was performed for all processing methods used in this work. Two compressive deformations were carried out during the cooling of the samples from soaking temperature. The size of each deformation corresponded to 10% of actual size of the sample (i.e. the second deformation was slightly smaller). Deformation temperatures of 900 °C and 800 °C or 720 °C were chosen on the base of previous experiments [5, 6]. Two cooling rates were tested, 15 °C/s and 30 °C/s ,to cool the samples down to bainitic hold. Bainitic hold lasted 600 s for all of the methods and was applied either at 425 °C or 400 °C.

IOP Conf. Series: Materials Science and Engineering **179** (2017) 012044 doi:10.1088/1757-899X/179/1/012044

Processing of all samples was carried out at thermo-mechanical simulator, which allows precise control of thermal and deformation parameters and good repeatability of processing parameters. The active part of processed samples is cylindrical with 8 mm diameter and the length of 16 mm. The geometry of the sample was optimised to achieve homogeneous thermal field distribution in active part of the specimen.

Table 1. Processing parameters (T_S – soaking temperature, $T_{deformation}$– temperature of compressive deformations, T_B – bainitic hold temperature), mechanical properties and retained austenite (RA) volume fraction.

T_S [°C]	$T_{deformation}$ [°C]	Cooling rate [°C/s]	T_B [°C]	Rm [MPa]	A_{5mm} [%]	RA [%]
1000	900, 800	30	425	992	37	13
1000	900, 720	30	425	925	42	12
1000	900, 720	30	400	963	34	14
1000	-	30	425	1013	29	14
1000	-	30	400	1068	26	13
850	-	15	425	891	37	11
850	-	15	400	869	34	5
850	-	30	425	932	45	9
850	-	30	400	934	47	11

3 Results

Thermo-mechanical treatment with higher temperature of the second deformation 800 °C resulted in very coarse bainitic microstructure with small amount of fine grains of proeutectoid ferrite and small pearlitic areas (Figure 2). The microstructure had high strength of 992 MPa with very good ductility A_{5mm} of 37%. There was 13% of retained austenite, which formed the laths of upper bainite.

Lowering of the temperature of the second deformation to 720 °C slightly decreased tensile strength to 925 MPa and improved ductility A_{5mm} to 42%. There was higher amount of free ferrite in the microstructure, bainitic blocks were finer even though individual bainitic laths were generally thicker than in the previous case (Figure 3). Very fine pearlitic areas were also observed in the microstructure.

Lower bainitic hold temperature of 400 °C further prolonged bainitic laths of M-A constituent and increased the amount of bulk M-A constituent islands (Figure 4). Despite of 14% of retained austenite in the microstructure, the ductility A_{5mm} of this sample reached only 34%.

Heat treatment was first repeated with the same thermal schedule, which was used in thermo-mechanical processing. Rather coarse microstructures made of mainly upper bainite were found in the samples soaked at 1000 °C and cooled by 30 °C/s to either 420 °C or 400 °C (Figure 5). In both cases, distinctively lamellar bainitic microstructures possessed high strengths above 1000 MPa. Lower temperature of bainitic hold increased the strength to 1068 MPa at the cost of relatively low ductility A_{5mm} 26%. No traced of pearlite were observed in either microstructure suggesting deformations applied during the cooling accelerated pearlite transformation in all the samples during thermo-mechanical treatment.

To increase ferrite fraction in the final microstructure without risking pearlite formation at lower cooling rates, soaking temperature of the next four treatments was kept at 850 °C. The combinations of two cooling rates, 15 °C/s and 30 °C/s, and two bainitic holds, 425 °C and 400 °C, were tested.

Lower soaking temperature resulted in all the cases in visible refinement of the final microstructure (Figure 6, Figure 7).

Slower cooling rate 15 °C/s caused pearlite formation in the sample held at 425 °C. Small pearlitic colonies with typical size below two micrometers were quite frequent in the microstructure (Figure 6). It should be however noted that inter-lamellar distance was very small. Tensile strength was just below 900 MPa with ductility 37%. Mechanical properties deteriorate further to 869 MPa tensile strength and 34% ductility A_{5mm} with the drop of bainitic hold to 400 °C. These relatively small changes in mechanical properties are accompanied by more significant changes in the final microstructure, which is now predominantly ferritic-pearlitic with a few bulky islands of M-A constituent (Figure 7). The pearlite blocks are coarser than in the previous case and the laths in them are sometimes thicker. There is only 5% of retained austenite in the microstructure, mostly of the bulky type.

On the other hand, lower soaking temperature combined with higher cooling rate of 30 °C/s resulted into the best combinations of mechanical properties and austenite volume fractions around 10%. The microstructure obtained after bainitic hold at 425 °C had very homogeneous distribution of fine equiaxed grains of free ferrite and bainitic laths with occasional occurrence of M-A constituent islands (Figure 8). Few laths of very fine pearlite were also sporadically found in the microstructure. The combination of high ultimate tensile strength of 932 MPa with ductility A_{5mm} of 45% prove that the presence of pearlite in this form and amount does not necessarily have a negative effect on mechanical properties.

Decrease of bainitic hold temperature to 400 °C further increased the ductility A_{5mm} to 47% without the change in tensile strength. The microstructure consisted of higher amount of free ferrite, most of the bainitic blocks are smaller than in previous case, however there are also heterogeneously distributed blocks with the length of bainitic laths around 10 micrometers (Figure 9).

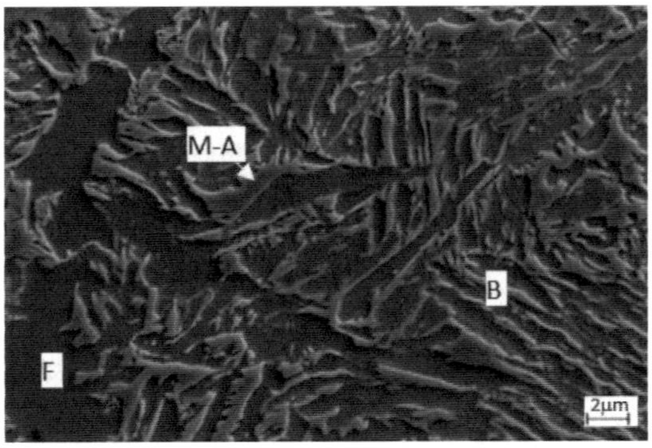

Figure 2 Thermo-mechanical treatment: 1000°C/100s, deformations at 900°C and 800°C, 30°C/s, hold at 425°C/600s.

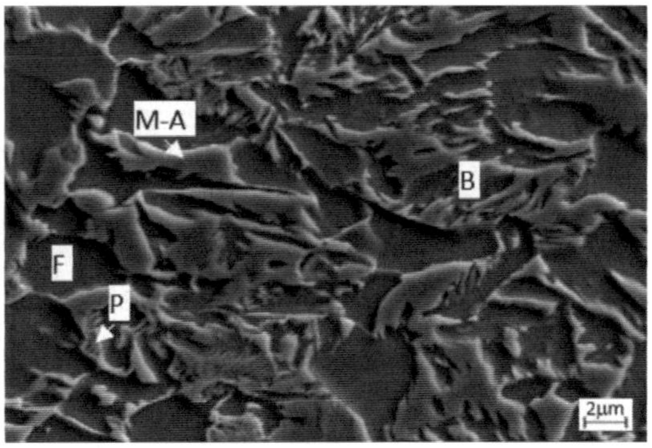

Figure 3 Thermo-mechanical treatment: 1000°C/100s, deformations at 900°C and 720°C, 30°C/s, hold at 425°C/600s.

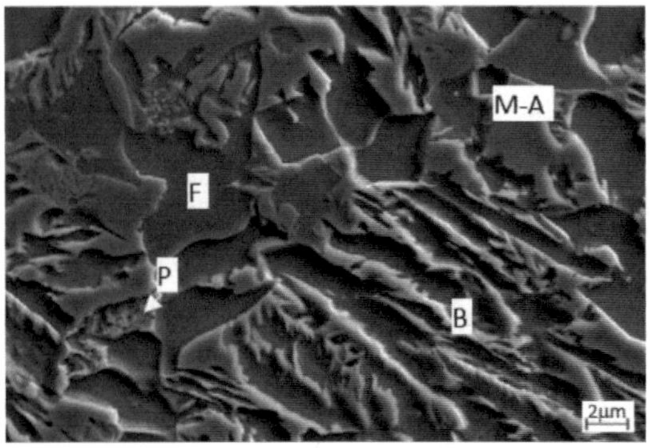

Figure 4. Thermo-mechanical treatment: 1000°C/100s, deformations at 900°C and 720°C, 30°C/s, hold at 400°C/600s.

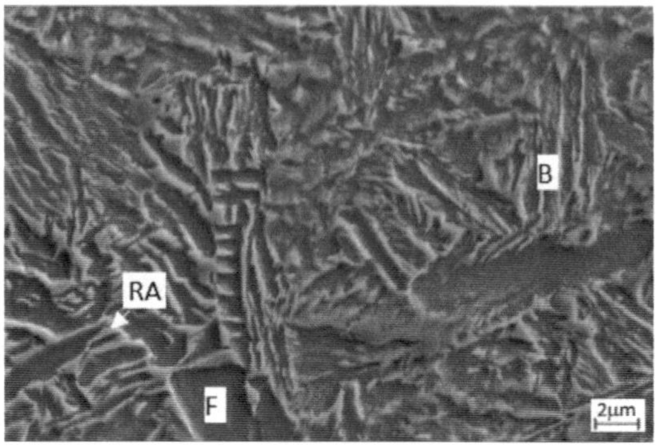

Figure 5. Heat treatment: 1000°C/100s, 30°C/s, hold at 400°C/600s

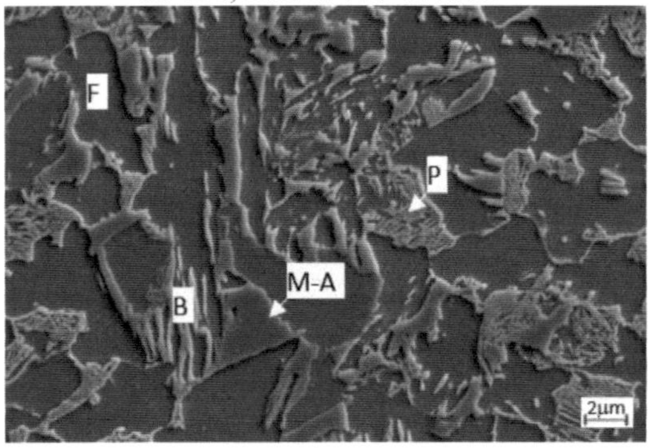

Figure 6. Heat treatment: 850°C/100s, 15°C/s, hold at 425°C/600s.

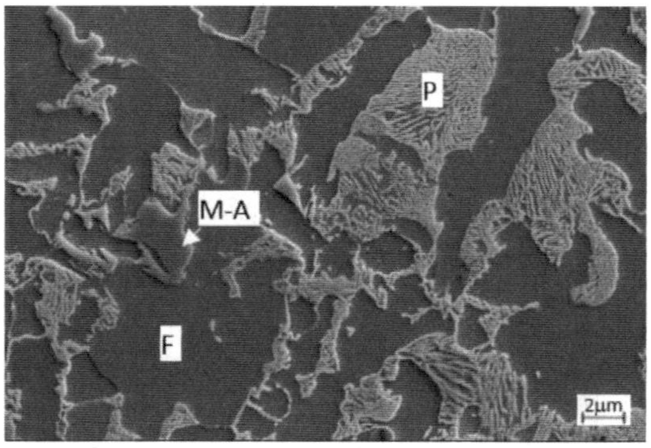

Figure 7. Heat treatment: 850°C/100s, 15°C/s, hold at 400°C/600s.

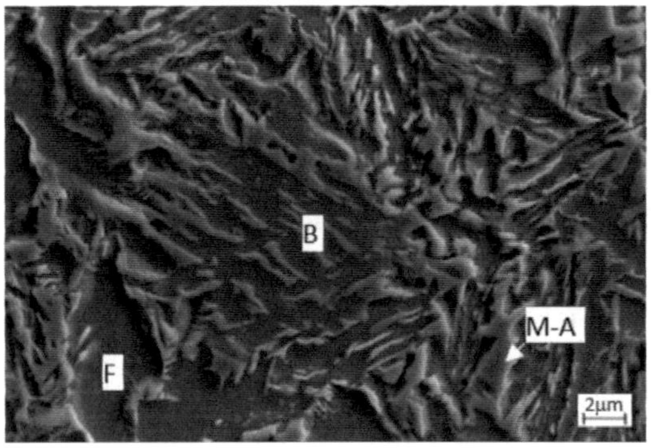

Figure 8. Heat treatment: 850°C/100s, 30°C/s, hold at 425°C/600s.

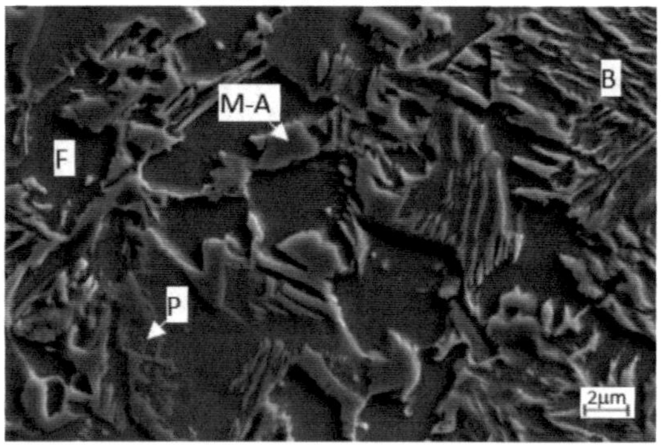

Figure 9. Heat treatment: 850°C/100s, 30°C/s, hold at 400°C/600s.

4 Conclusions

Nine methods of heat and thermo-mechanical treatments were applied to low alloyed middle carbon steel 0.4C-0.6Mn-2Si. In most of the cases, the resulting complex multiphase microstructures possessed interesting combination of relatively high ultimate tensile strength above 900 MPa with ductility A_{5mm} above 35%.

The very best ductility A_{5mm} of 47% combined with very good strength of 934 MPa were achieved after simple heat treatment consisting of heating to 850 °C with 100 s hold at the temperature and subsequent cooling by 30 °C/s to 400 °C. High ductility A_{5mm} of 45% obtained after similar heat treatment with the hold at 425 °C suggest that the first result is no exception, but rather a rule for this kind of the treatments and resulting microstructures. The final microstructures with high ductility consisted of bainite-ferrite mixture, with about 10% of retained austenite and small amount of extremely fine pearlitic areas.

Acknowledgement
The present contribution has been prepared within project LO1502 ‘Development of the Regional Technological Institute‘ under the auspices of the National Sustainability Programme I of the Ministry of Education, Youth and Sport of the Czech Republic aimed to support research, experimental development and innovation.

References

[1] Adamczyk J, Grajcar A 2005 *J. Mater. Process. Technol.* 162–163 267-74.
[2] Świątnicki W A 2015 *Science and Technology* 31 7 1 755-7.
[3] Hauserova D et al. 2011 *Procedia Eng.* 10 2961-66.
[4] Kucerova L et al. 2009 Proc Annals of DAAAM and Proceedings of the International DAAAM Symposium Ed. B. Katalinic DAAAM International Vienna 1035-6.
[5] Kucerova L, Jirkova H, Masek B 2014 *Arch. Metall. Mater* 59 3 1189-92.
[6] Jirkova H et al. 2015 *Mater Today* 2S 627-30.
[7] Kučerová L et al 2011 *J. Iron Steel Res. Int.* 18 427-31.
[8] Pereloma E V, Timokhina I B, Hodgson P D 1999 *Mater. Sci. Eng., A* 273-275, 448-52.
[9] Li Q, Huang X, Huang W 2016 *Mater. Sci. Eng., A* 662 129–35.
[10] Gao G 2014 *Acta Mater.* 76 425–33.
[11] Wiewiórowska S, Muskalski 2015 *Procedia Manufacturing* 2 181–5.

IOP Conf. Series: Materials Science and Engineering **179** (2017) 012045 doi:10.1088/1757-899X/179/1/012045

Modeling plasticity of materials with nanostructure

N R Kudinova

Department of Physics of Materials, Charles University, Prague, Czech Republic

Abstract. A new approach to modeling of the plasticity of materials with the particle size in the range from 3 to 20 nm (nanostructure) has been proposed. It is based on classical thermodynamic approach employing the surface tension of nanoparticles. Its main advantage is the minimum number of physical parameters in use. In the context of the proposed model, we calculated the dependence of the melting temperature on the nanoparticle size which is consistent with experimental data. The volume density of the surface energy of nanoparticles was also determined. This energy is assumed to be a significant part of the internal energy during deformation Yield point was interpreted as the result of changes of grains surface energy during the deformation. The obtained yield point dependence on the grain size was related to the Hall–Petch law, and this resulted in confirmation of the hypothesis on the crucial role of surface tension forces in the initial stage of plastic deformation of nanomaterials.

1 Introduction

Nanocrystalline materials exhibit potential for applications in a wide range of areas, particularly because of their high strength when compared with their coarse-grained counterparts. Classically, the Hall–Petch [*Petch*, 1953] relation predicts that the yield strength increases with decreasing grain size:

$$\tau = \tau_0 + Kd^{-\frac{1}{2}}, \tag{1}$$

where τ_0 is the friction stress in the absence of boundaries, K is a constant and d is the grain size. This relation was experimentally found to be consistent with experimental data for materials with grain sizes ranging from 1 millimeter down to 1 micrometer. Consequently, it was believed that if the average grain size decreased even further to the nanometer length scale the yield strength would continue to increase. However, experiments on many nanocrystalline materials demonstrated that if the grains reached a small enough size, typically around 10 nm, the yield strength would either remain constant or decrease with decreasing grains size [*Conrad et al.,* 2000]. This phenomenon has been termed the reverse or inverse Hall–Petch relation. A number of different mechanisms have been proposed to explain this phenomenon. As suggested by Carlton [*Carlton et al*, 2007] et al., they fall into four categories: (1) dislocation-based [*Gutkin et al,* 2003] (2) diffusion-based [*Kocks* ,1970], (3) grain-boundary shearing-based [*Kim*, 1998; *Ovid'ko et al,* 2011], (4) two-phase-based. [*Carlton et al*, 2007].

Figure 1 displays schematically the well-known break-down of the Hall-Petch behavior, i.e. the rise of the stress with decreasing grain size in polycrystalline metals, when the grain size drops into the nanometre range resulting in a peak plastic resistance at a grain size of about 10 nm.

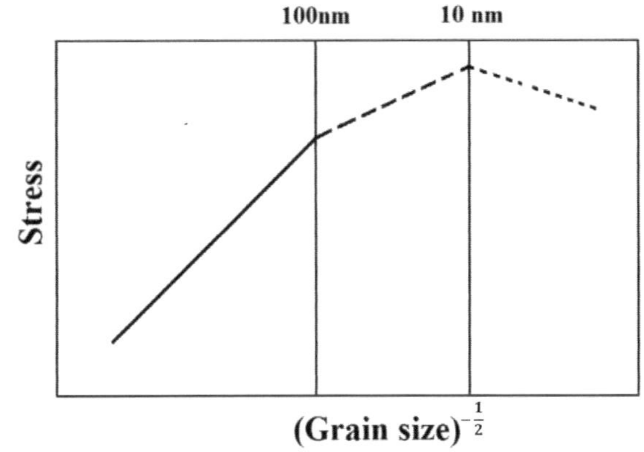

Figure 1. Stress of a material as a function of the grain size [*Kumar et al,* 2003].

2 Thermodynamical approach

In this chapter we will introduce the different approach for explaining the inverse Hall-Petch relation. It is based on the surface (or interfacial) tension - a fundamental thermodynamic parameter - related to the work spent on reversible isothermal formation of the interface [*Frank*, 1955]. Surface tension is considered as a measure of the excess free energy in the surface layer compared to the free energy in the volumes of coexisting phases [*Dadashev*, 2008].

The surfaces tension appears in many equations of physics. It determines such parameters and phenomena as the capillary pressure, wetting (contact) angle, saturated vapor pressure, adsorption, capacitance of the double electrical layer, work spent on the formation of critical nuclei, length of capillary waves, etc. Due to recent advances in the physics and chemistry of nanomaterials, interest in the phenomenon of surface tension has noticeably quickened [*Andrievskii et al*, 2005]. A central point here is the deducing of the size dependence of the surface tension. The Gibbs–Tolman–Koenig–Buff differential equation, which determines this dependence for the spherical surface, is well known [*Gibbs*, 1948; *Buff*, 1955; *Koenig*, 1950; *Tolman*, 1949; *Ono et al*, 1960].

3 Modeling of the surface tension

Gibbs–Tolman–Koenig–Buff [*Gibbs*, 1948; *Buff*, 1955; *Koenig*, 1950; *Tolman*, 1949; *Ono et al*, 1960] equation for a spherical particle is the following:

$$\frac{\partial \ln \sigma}{\partial \ln r} = \frac{\frac{2\delta}{r}\left(1+\frac{\delta}{r}+\frac{1\delta^2}{3r^2}\right)}{1+\frac{2\delta}{r}\left(1+\frac{\delta}{r}+\frac{1\delta^2}{3r^2}\right)}, \tag{2}$$

where σ, r and δ are the surface tension, the radius of the particle and the Tolmen constant, respectively. The latter being equal to the surface layer thickness of the particle.

It has been believed until recently that Eq. (2) cannot be solved analytically, i.e., expressed through elementary fractions. Such a belief seems to rely upon Tolman's original work [R. C. Tolman, 1949], where this equation was solved asymptotically and the Gibbs–Tolman–Koenig–Buff equation (in its initial form) was integrated numerically. In order to solve the Eq. (2) approximately, the theory assumes that $r \gg \delta$ [*Ono et al*, 1960], so that parenthesized expressions in the numerator and denominator on the right of (2) can be omitted. In this case, the straightforward integration yields the Tolman [1949] formula:

$$\sigma = \frac{\sigma^{(\infty)}}{1+\frac{2\delta}{r}}, \tag{3}$$

where $\sigma^{(\infty)}$ is the surface tension for a plane surface. When using the same condition ($r \gg \delta$), one obtains another analytical expression for the surface tension [*Yurov et al*, 2011]:

$$\sigma(r) = \sigma_\infty exp\left(-\frac{2\delta}{r}\right). \tag{4}$$

However, it is possible to obtain the general solution of (2), provided that δ does not depend on r. It can be expressed in an analytical form, with $s = \frac{r}{\delta}$ being introduced as the dimensionless variable. After separation of the variables in the equation (2) one obtains:

$$\ln\frac{\sigma}{\sigma^{(\infty)}} = -2\int_{r/\delta}^{\infty}\left(\frac{3s^2+3s+1}{3s^3+6s^2+6s+2}\right)\frac{ds}{s}. \tag{5}$$

The integral in (5) is calculated by expanding the integrand into elementary fractions. The final result is written in the form

$$\sigma = \frac{\sigma(\infty)r}{\delta}e^{\left(-\sum_{k=1}^{3}\frac{s_k^2\ln(\frac{r}{\delta}-s_k)}{3s_k^2+4s_k+2}\right)}, \tag{6}$$

where $s_k = \{-0.558; -0.721 + i0.822; -0.721 - i0.822\}$ - are the roots of the cubic equation $3s^3 + 6s^2 + 6s + 2 = 0$

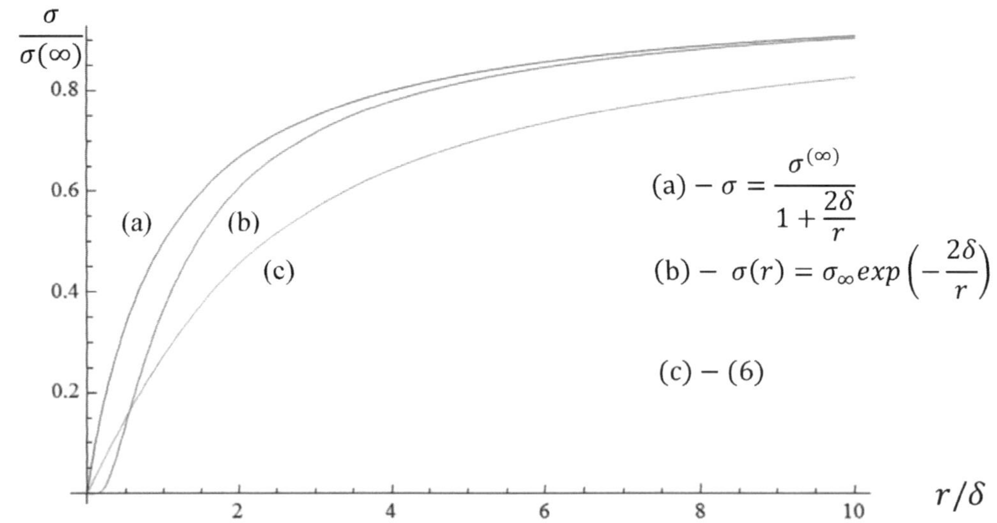

Figure 2. Scaled functions of surface tension for the exact solution (6) and the approximations (a), (b).

4 Dependence between melting temperatures and surface tension

The melting temperature of a bulk material is not dependent on its size. However, as the dimensions of a material decrease towards the atomic scale, the melting temperature scales with the material dimensions. The decrease in melting temperature can be of the order of tens to hundreds of degrees for metals with nanometer dimensions, e.g aggregated and powdered nanomaterials, metallic nanopowders of pure metals etc. This phenomenon is called melting-point depression and may be very strong in nanomaterials, which melt at temperatures hundreds of degrees lower than bulk materials [*Buffat et al*, 1976]. In Figure 3 the melting temperature dependence on particle size for Au is shown.

The variation of melting temperature is directly connected with the energy of surface tension since the probability of dislodging of individual atoms and molecules from a crystallite increases as a result of thermal motion. The effect of the surface tension of crystallites on the melting temperature has not been studied yet.

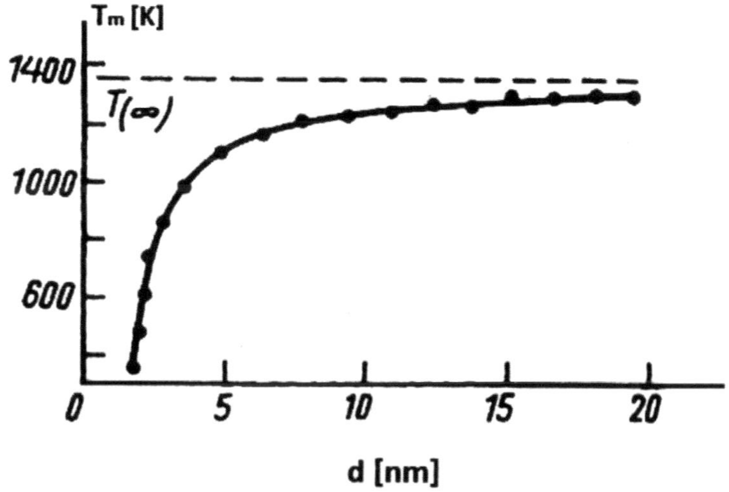

Figure 3. Melting curve for gold as a function of nanoparticle diameter, dots are the experimental data [*Buffat*, 1976].

Let us use the well-known linear dependence of physical properties on the size of the particles [*Yurov*, 2011]:

$$A(r) = A_0(1 - \frac{d}{r}) \tag{7}$$

where A_0 – is a physical property bulk sample, $A(r)$– physical property of small particle, d – the radius corresponding to the particle size at which the size effects begin to play a role.
The critical radius d may be expressed by the following formula:

$$d = \frac{2\sigma\vartheta}{RT}, \tag{8}$$

where ϑ is the – molar volume, and R is a gas constant.

From (8) one can see that the surface tension is linearly dependent on temperature:

$$\sigma = \alpha T \tag{9}$$

where $\alpha = \frac{dR}{2\vartheta} = 10^{-3} Jm^{-2}K^{-1}$ for all metals [*Yurov*, 2011].
Moreover, there is an experimental confirmation of the linear dependence of melting temperature on the surface tension [*Rehviashvili et al*, 2006].

Using the exact solution of GTKF (6), which was obtained above, and the relation (9) and assuming temperature being the physical property in eq. (7) we can derive the exact dependence of the melting temperature on the particle size:

$$T_m = \frac{T_0 r}{\delta} e^{\left(-\sum_{k=1}^{3} \frac{s_k^2 \ln(\frac{r}{\delta} - s_k)}{3s_k^2 + 4s_k + 2}\right)}, \tag{10}$$

where $s_k = \{-0.558; -0.721 + i0.822; -0.721 - i0.822\}$ - are the roots of the cubic equation $3s^3 + 6s^2 + 6s + 2 = 0$

Figure 4 shows such dependence for gold (experimental points) and calculated functions.

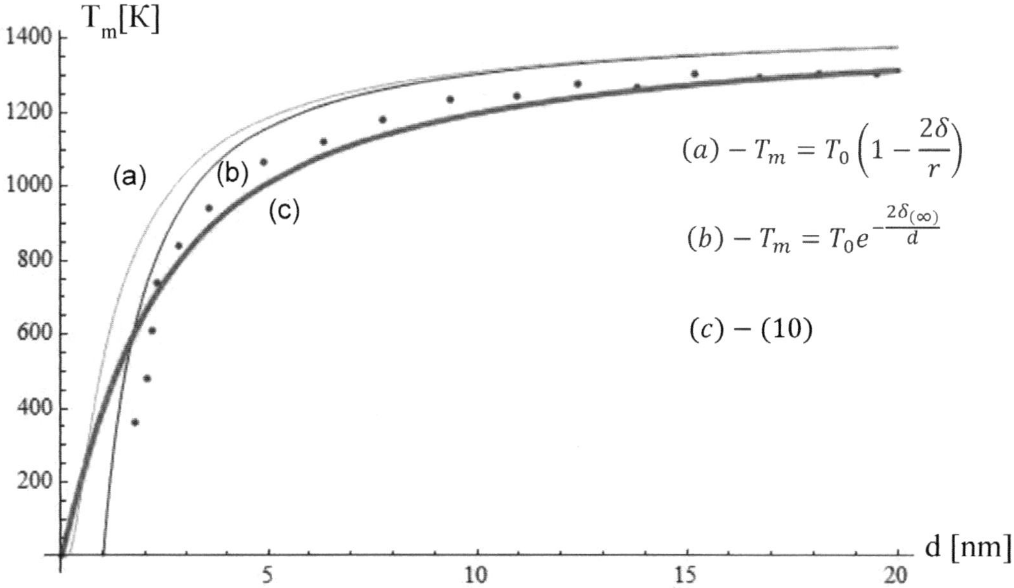

Figure 4. Dependence of the melting temperature for gold on the particle diameter d=2r. for the exact solution (10) and the approximations (a), (b).

Comparison of the plots shows that our exact solution describes well the experimental data in the range of the particle sizes from 3 to 20 nm. At the same time, the approximate functions suffer from very large (up to 100%) uncertainty. Thus the exact solution (10) provides an adequate description for variation of the melting temperature with particle size. This justifies its use in further calculations of surface energy and its density.

5 Surface tension and yield point

Nondislocation models of plasticity usually consider the relation between surface energy and the yield point. This assumption is based on the apparent similarity of the influence of the surface tension on the particle size with the Hall-Petch relation. Modeling of surface tension dependence on the particle size based on thermodynamic GTKB approach allows us to make some predictions of the influence of surface tension forces in crystallites on the yield point of the material. This is particularly true in materials whose structure elements were refined to nanometric scale.

The volume density of the surface energy of a spherical particle may be expressed by the form:

$$\rho_v = \sigma.\frac{S}{V} = \frac{\sigma}{\frac{4}{3}\pi r^3}, \tag{11}$$

where S and V is the surface and volume of the spherical particle, respectively.
Using the exact solution (6) of the Gibbs–Tolman–Koenig–Buff equation we can obtain the dependence of the volume density of surface energy on the particle size:

$$\rho_v = \sigma S = \frac{3}{r}\sigma = 3\frac{\sigma(\infty)}{\delta} e^{\left(-\sum_{k=1}^{3}\frac{s_k^2 \ln(\frac{r}{\delta}-s_k)}{3s_k^2+4s_k+2}\right)} \tag{12}$$

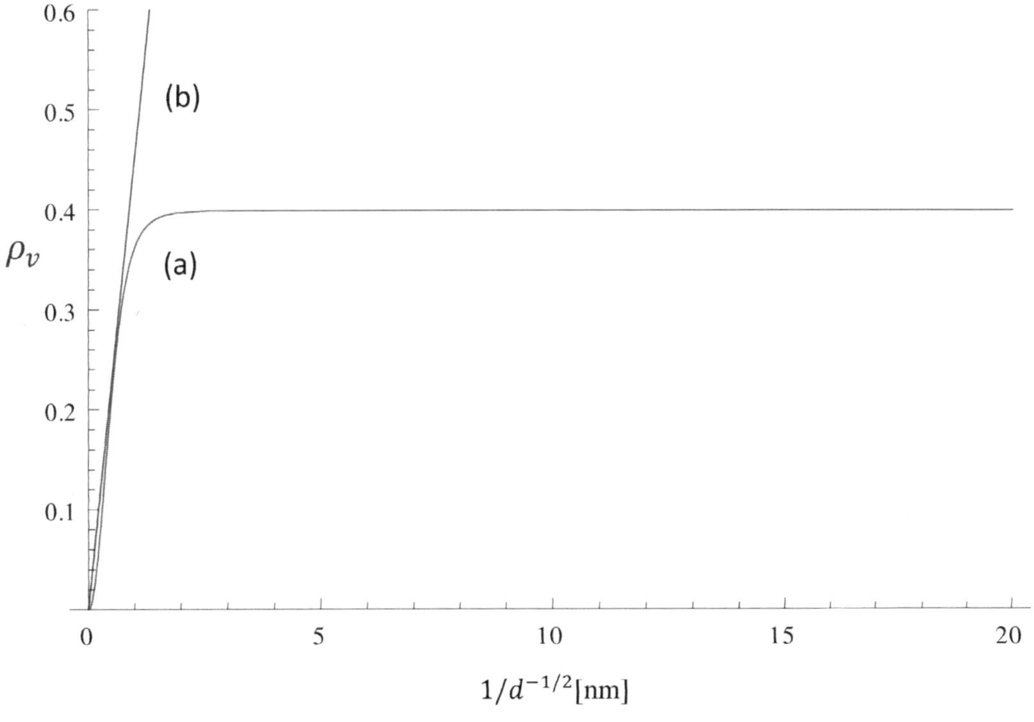

Figure 5. Scaled dependence of density of surface energy per the unit of volume for gold. a) eq. (12); b) approximation by the H-P law.

The following two regions can be distinguished in Figure 5. For grain sizes larger than 1 nm, the obtained dependence can be approximated using the H-P law, while at the sizes below 1 nm we obtain a finite value for the volume density of the surface energy. This is consistent with the well-known breakdown of the H-P relation for grains smaller than 10 nm, cf. Figure 1. Detail analysis of eq. (12) indicates that there is a limit of grain sizes ($r \approx 2\delta$) for which the volume density of surface energy reaches its limit value. This fact gives a reason to interpret plasticity as the variation of the surface energy. It means that the allowed value of the surface energy (shown in Figure 5 for gold) was exceeded in the material if it begins to flow. However, further considerations of relations between stress and the release of elastic and plastic energy at the beginning of plastic deformation as well of the relation between surface energy of free particles and grains in a solid matter are needed to improve nondislocation models of plasticity, in particular to predict yield stress values from specific surface energies without involving dislocation interactions.

6 Conclusion

The energy of surface tension can play an important role in plastic deformation of micro and nano materials. This mechanism opens the possibility for constructing new models of plastic deformation without using dislocation theory.

Exact solution of the equation Gibbs–Tolman–Koenig–Buff allows more adequate description of the dependence of the melting temperature on the particle size in the range from 3 to 20 nm, that supports the hypothesis of the crucial role of surface tension forces in plastic yielding of micro and nano disperse materials.

Acknowledgments

This work was financially supported by the Grant Agency of Charles University under the grant number 1110816.

References

[1] Andrievskii R. A.and A. V. Ragulya, Nanostructured Materials ,*Akademiya*, Moscow, 2005. [in Russian]
[2] Buffat P.H., J.P. Borel, Size effect on the melting temperature of gold particles, *Phys. Rev.* A, 13, 2287, 1976.
[3] Buff F.P., Spherical interface II. Molecular theory, *Journal Chem. Phys.*,3, , 419, 1955.
[4] Carlton, C; Ferreira, P. J. What is Behind the Inverse Hall–Petch Behavior in Nanocrystalline Materials, *Mater. Res. Soc. Symp. Proc.*,976, 2007.
[5] Conrad, H; Narayan, J ,On the grain size softening in nanocrystalline materials, *Scripta Mater.* 42 (11), 1025, 2000.
[6] Dadashev R. Kh., Thermodynamics of Surface Phenomena, *Fizmatlit*, Moscow, 2007; *Cambridge International Science*, Cambridge, 2008.
[7] Frank P, Spherical Interface. II. Molecular Theory, *J. Chem. Phys.* 23, 419, 1955.
[8] Gibbs J.W., Collected Works, Vol.1, *New Haven*, 1948.
[9] Gutkin M.Yu. and I.A. Ovid'ko, Physical Mechanics of Deformed Nanostructures. Volume 1: Nanocrystalline Materials, *Yanus*, St.Petersburg, 2003.
[10] Hall E.O, Deformation and ageing of mild steel, *Proc. Phys. Soc. London, ser. B*, 64, (1), 747, 1951.
[11] Kim H.S. A composite model for mechanical properties of nanocrystalline materails, *Scripta Materialia.*, 39 (8), 1057. 1998.
[12] Kocks U.F., The relation between polycrystal deformation and single crystal deformation, *Metal. Trans.*, (5), 1121, 1970.
[13] Koenig F.O., On the Thermodynamic Relation between Surface Tension and Curvature, *The Journal of Chemical Physics* 18, 449, 1950.
[14] Kumar K.S., H.Van Swygenhoven, S. Suresh Mechanical behavior of nanosrystalline metals and alloys, *Acta Materialia.* 51 5743, 2003.
[15] Ono S., S. Kondo. Molecular Theory of Surface Tension. Berlin: *Springer-Verlag* 1960.
[16] Ovid'ko I., M. Gutkin, N. Skiba, Strengthening and softening mechanisms in nanocrystalline materials under superplastic deformation, *Acta Materialia.*, 51, (14), 4059, 2011.
[17] Petch N.J., The cleavage strength of polycrystals. *J. Iron and Steel Inst.*, 174,1953.
[18] Rehviashvili S. Sh., E.V. Kishtikova: Model of a nanoparticle in the theory of inhomogeneous medium, *Technical Physics*, 56, 143, 2011.
[19] Tolman R. C., The effect of droplet size on surface tension, *J. Chem. Phys.* 17, 333, 1949.
[20] Yurov V.M., V.S. Portnov, N.H. Ibraev, S.A. Gurchenko, The surface tension of solid bodies, small particles and thin membrans, *Phys.-math.* №11, 2011.
[21] Yurov V.M. Superfecial tension of pure metals: *Eurasian Physical Technical journal* 8 (1), 10, 2011.

Tribological and corrosion properties of plasma nitrided and nitrocarburized 42CrMo4 steel

D Kusmic[1] and D Van Thanh [2]

[1, 2] Department of Mechanical Engineering, Faculty of Military Technology, University of Defence, Kounicova 65, 662 10 Brno, Czech Republic, EU

E-mail: [1]david.kusmic@unob.cz, [2]thanhvan.doan@unob.cz

Abstract. This article deals with tribological and corrosion resistance comparison of plasma nitrided and nitrocarburized 42CrMo4 steel used for breech mechanism in the armament production. Increasing of materials demands (like wear resistance, surface hardness, running-in properties and corrosion resistance) used for armament production and in other industrial application leads in the field of surface treatment. Experimental steel samples were plasma nitrided under different nitriding gas ratio at 500 °C for 15h and nitrocarburized for 45 min at temperature 590°C and consequently post-oxidized for 10 min at 430°C. Individual 42CrMo4 steel samples were subsequently metallographically evaluated and characterized by hardness and microhardness measuring. The wear test "ball on disc'' was realized for measuring of adhesive wear and coefficient of friction during unlubricated sliding. NSS corrosion tests were realized for corrosion resistance evaluation and expressed by corroded area and calculated corrosion rate. The corrosion resistance evaluation is by the surface corrosion-free surfaces evaluation supplemented using the laser confocal microscopy. Due to different surface treatment and plasma nitriding conditions, there are wear resistance and corrosion resistance differences evident between the plasma nitrided steel samples as well.

1 Introduction

Plasma nitriding is a chemical-heat treatment process that is widely used for increase of surface hardness, fatigue strength, corrosion resistance and to improve tribological behavior mostly of steels [1]. The treatment temperature of plasma nitriding process is usually set below the eutectoid temperature in Fe-N diagram (590 °C). It is suitable to set the nitriding temeperature below the last heat-treatment procedure temperature, to ensure unchanged microstructure of the core of prior heat-treatment (e.g. tempered structure). Thanks to diffusion of nascent nitrogen (during the nitriding process generated) a compound and diffusion layer is created. Compound (white) layer mostly consists of the ε-$Fe_{2-3}N$ and γ'-Fe_4N iron nitrides, as well as of mixture of alloying elements nitrides like Al, Cr, Mo, V [2-4]. This surface compound layer is characterized by increased hardness and good corrosion resistance. Increased corrosion resistance can be additionally increased by post-oxidation process, applied directly after plasma nitriding process in the device [5, 6]. The formed oxide layer is able to cover the micropores and improves the corrosion resistance of nitrided steel [7-10].

This paper is focused on evaluation of wear and corrosion resistance of plasma nitrided and nitrocarburized 42CrMo4 (AISI 4137/4140) steel and compared to tempered one. Plasma nitriding process was applied under different nitriding gas ratio (H_2:N_2) for 15 h. The wear tests "pin on disc'' were employed to find out the adhesive wear and coefficient of friction during unlubricated sliding (dry sliding). The corrosion resistance was tested using the NSS corrosion test according with ISO

9227 standard, visually and gravimetric evaluated. After removing the corrosion products, the surfaces were evaluated using the laser confocal microscopy. Results of wear and corrosion tests were further supplemented by metallographic documentation and measuring of tie-layer thickness. Thickness and microhardness of created layers were measured by Vickers microhardness method in accordance with DIN 50190 standard.

2 Experimental

For this study was used 42CrMo4 (AISI 4137/4140) steel with the following chemical composition (in wt.%): 0.40 C, 1.08 Cr, 0.63 Mn, 0.27 Si, 0.15 Mo, 0.10 Ni, 0.0019 S, 0.0012 P. Chemical composition was verified using the Q4 Tasman device, calibrated by the Fe 130 and Fe 140 standards. Experimental samples were heated at 850 °C for 20 min, oil quenched, tempered at 550°C for 40 min for micro-structure homogenization and to achieve standard mechanical properties. All the surfaces of samples were grinded and cleaned in ethanol prior plasma nitriding and nitrocarburizing (known as Tenifer®).

Plasma nitriding was performed under two nitriding gas ratio in the RUBIG PN 60/60 device (marked as PN1 and PN2), for plasma nitriding conditions see table 1. Tenifer® (marked as NC) essentially is a technology including nitrocarburizing process in salt bath of SURSULF solution for 45 min under temperature of 590 °C followed by oxidation process in salt bath of OXINIT solution for 10 min at 430 °C for corrosion-resistant oxides Fe_3O_4 (magnetite) and Fe_2O_3 (hematite) creation (marked as NC).

Table 1. Plasma nitriding process conditions.

Process	Temperature *(°C)*	Duration *(h)*	Pressure *(Pa)*	Bias *(V)*	Gas flow *(l.h^{-1})* H_2	N_2
Plasma cleaning	480	0.5	80	800	20	2
PN1- $3H_2$:$1N_2$	500	15	280	530	24	8
PN2- $1H_2$:$3N_2$	500	15	280	530	8	24

Thickness of diffusion layer was evaluated by microhardness testing in accordance with DIN 50190 standard using the automatic microhardness tester LECO LM 247 AT equipped with the AMH43 software allowed to investigate diffusion layer depth in direction from surface to the core at 50 g load and 10 s dwell. The final microhardness profile and measuring of diffusion layer depth was determined by 18 indentations with variable indentation spacing. The compound and diffusion layer characteristics are summarized in table 2.

Table 2. The thickness of compound layer and layer case depth.

	Case depth *(μm)*	Thickness of compound layer *(μm)*	Thickness of oxide layer *(μm)*	Microhardness *(HV0.05)*
Tempered	-	-	-	366
PN 1 - $3H_2$:$1N_2$	200	3.6	-	681
PN 2 - $1H_2$:$3N_2$	240	7.4	-	778
NC	120	16.5	1	772

For metallographic testing, all samples were crosswise cutted, wet grinded using silicon carbide paper with grit size from 80 to 2000, subsequently polished and finally etched by Nital. The cross-structure documentation and compound layer thickness measuring was realized using the opto-digital microscope OLYMPUS DSX 500. The bainitic–sorbitic microstructure of tempered sample and structure of nitrided and nitrocarburized layers can be seen in figure 1.

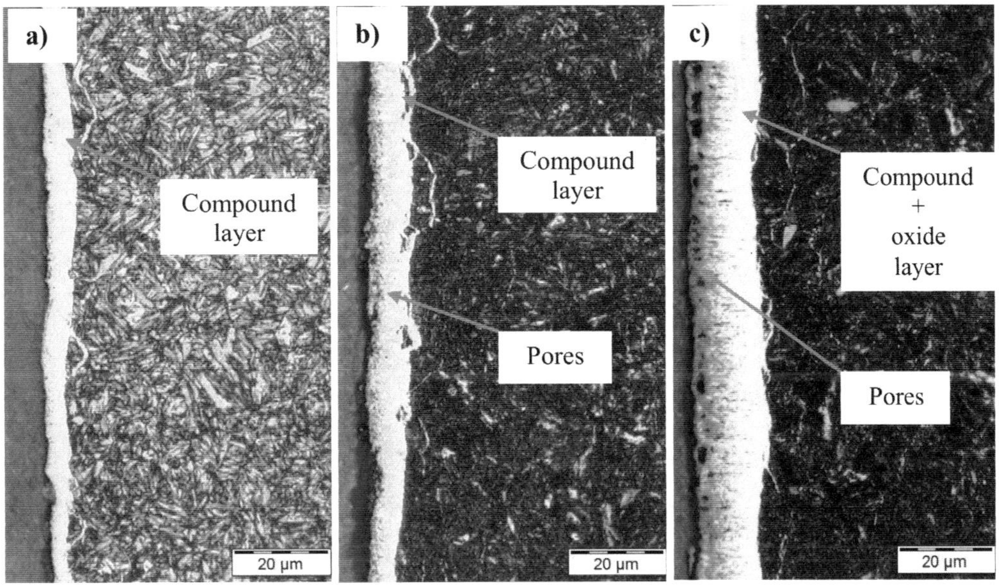

Figure 1. Cross-sectional microstructures (a) PN1 ($3H_2$:$1N_2$, 15h), (b) PN2 ($1H_2$:$3N_2$, 15h) and (c) NC (nitrocarburized).

It can be clearly seen that the case depth (see table 2.) and thickness of compound layer increases with N_2 ratio in nitriding gas mixture from 5.4 µm as seen in Figure 1(a) for PN1 sample to 8.6 µm in Figure 1(b) for PN2 sample gas mixture. Increased N_2 ratio in nitriding gas mixture has a significant influence to compound layer porosity, which increases with compound layer thickness, can be seen in figure 1(b). Increased porosity of compound layer (reached 16.5 µm), can be seen in figure 1(c) of nitrocarburized steel. Generally better corrosion resistance of plasma nitrided steel under $1H_2$:$3N_2$ nitriding gas mixture (formed predominantly by ε-$Fe_{2-3}N$ nitride) can be affected. Increased porosity of compound layer can cause reduction of corrosion resistance and delamination of stressed compound layer. Contrariwise significant improving of corrosion resistance of nitrocarburized steel can be expected, due to post-oxidation process. Thanks to post-oxidation process covering of pores by 1 µm thick oxide layer (dominantly formed by Fe_3O_4 oxides) can heal the surface, see figure 1. (c) [10].

3 Results and discussions

3.1 Wear tests

The wear tests "ball on disc'' in accordance with ASTM G99-95 standard were performed using the CETR UMT-3 tribometer, for coefficient of friction and in combination with the Talysurf CLI 1000 profilometer to evaluate the wear track for wear rate calculation:

$$W = \frac{V}{F.s}\left(m^3.N^{-1}.m^{-1}\right) \tag{1}$$

where W is the wear per unit load per sliding distance, V (m3) is the worn volume, F (N) is the applied load and s (m) is the sliding distance. The measurement of wear trace profile was realized using the Talysurf CLI 1000 profilometer. Four measurements of canopy profiles were used for calculating the worn volume (see figure 2):

$$V = S.\Pi.2r\left(m^3\right) \tag{2}$$

where the S is the average canopy area (m^2) and r the radius of wear test measurement (m).

For wear tests was used the carbide wolfram ball of diameter 6.3 mm with hardness of 92 HRA. The tests parameters were set up as following: applied load of 20 N, rotary velocity of 500 rpm and 13500 cycles.
The friction coefficient is a dimensionless quantity, defined as quotient of normal and friction force. Hence, this value changes continually with surface state, duration and environment [11].
The dependence of friction coefficient vs. time of treated steel samples – carbide wolfram contact is plotted in figure 3.

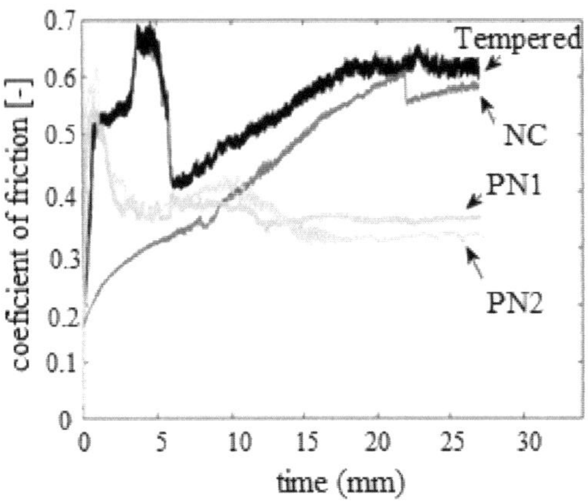

Figure 3. Coefficient of friction.

The tempered steel sample shows the highest value of friction coefficient (μ = 0.62). The nitrocarburized (NC) sample indicated the second greatest coefficient of friction ($\mu = 0.58$).
The nitrided samples showed the lowest value of friction coefficient and it is very stable. The sample PN2 ($1H_2{:}3N_2$) provides the lowest coefficient ($\mu = 0.32$) and sample PN1 ($\mu = 0.35$).
The coefficient of friction of nitrocarburized steel samples gradient is not as steep as others at to beginning of measurement, but after 20 min of measuring reached high stable value of coefficient of friction. The coefficient of friction μ was considered as an average value of stable section. In this article experiment, the wear rate is not proportional to the values of friction coefficient.

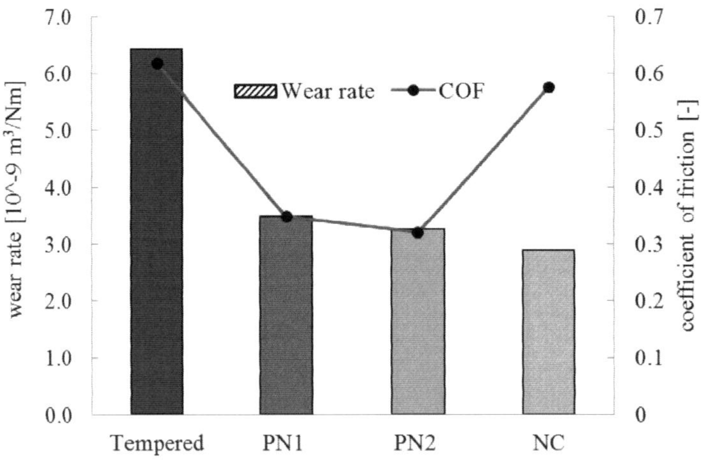

Figure 4. Tribological characters.

The interrelationship between coefficient of friction and calculated wear rate, as in aqution (1), are summarized in figure 4. The tempered steel sample shows the highest wear rate and friction coefficient. Although nitrocarburizing enhances the value of $\mu = 0.58$ but wear rate of this sample is the lowest. Sample PN2 exhibits very positive results, that both two values of coefficient of friction and wear rate are very low (see figure 4).

3.2 *Corrosion behaviour*

The tempered, plasma nitrided and nitrocarburized samples were exposed to a 5 % neutral sodium chloride solution (NSS) fog in accordance with ISO 9227 standard in the VLM GmbH SAL 400-FL corrosion chamber under following conditions: the temperature of 35 ± 2 °C, 5 % neutral sodium chloride solution, the amount of vapor condensation was set for 1÷2 ml.h^{-1} on square of 80 cm^2, pH 6.5 ÷ 7.2, the suspension angle of 20° from the vertical line, after every exposition period (2, 4, 8, 24, 48, 72, 96, 144 and 196 hours) were the exposed samples documented (see figure 5). The samples were degreased before the corrosion testing by the ethylalcohol.

The corrosion resistance was during the evaluation periods after drying documented and visually evaluated using the QuickPHOTO Industrial 2.3 software with Phase analyses application, as seen in figure 5.

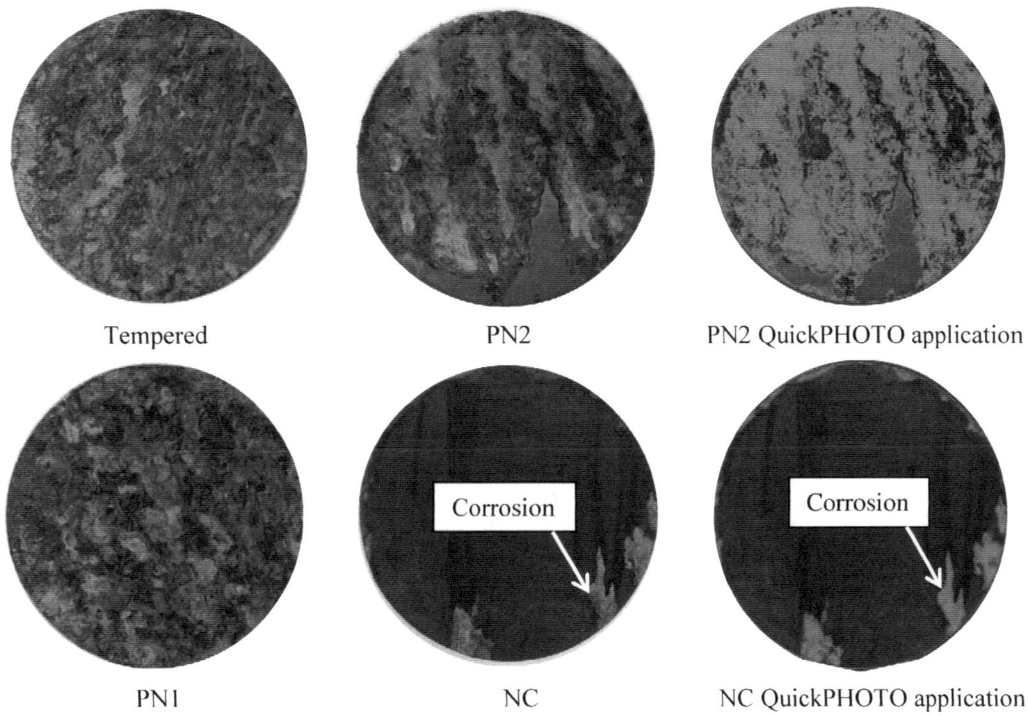

Figure 5. Visual evaluation after 72h in NSS.

Uniform type of corrosion attack of tempered steel sample was expected and combined corrosion attack or pitting of plasma nitrided and nitrocarburized and subsequently post-oxidized steel samples. The best results achieved the nitrocarburized and subsequently post-oxidized steel samples. Good corrosion resistance can be attributed to Fe_3O_4 and Fe_2O_3 oxides. Oxide layer formed during the post-oxidizing process, covers the micropores in the compound layer and ensure passivity [12]. Increased corrosion resistance comparing to tempered sample was confirmed for PN1 and PN2 samples, it is evident that the thickness and compactness of compound layer plays a role (see figure 1, table 2). The corrosion progress of corroded surface area vs. exposure period is plotted in figure 6.

During the corrosion tests the weight gain was measured and the corrosion rate was calculated:

$$K_{corr} = \frac{m_t}{S.t} \left(mg.cm^{-2}.h^{-1} \right) \tag{3}$$

IOP Conf. Series: Materials Science and Engineering **179** (2017) 012046 doi:10.1088/1757-899X/179/1/012046

where the m_t (mg) is weight gain in evaluated time period defined by $m_t = m_h - m_{int}$ (mg), the m_h (mg) is the measured weight in evaluated time period and m_{int} (mg) is the initial weight before corrosion test, S (cm^2) is the total surface area and t (h) is the evaluated time period.

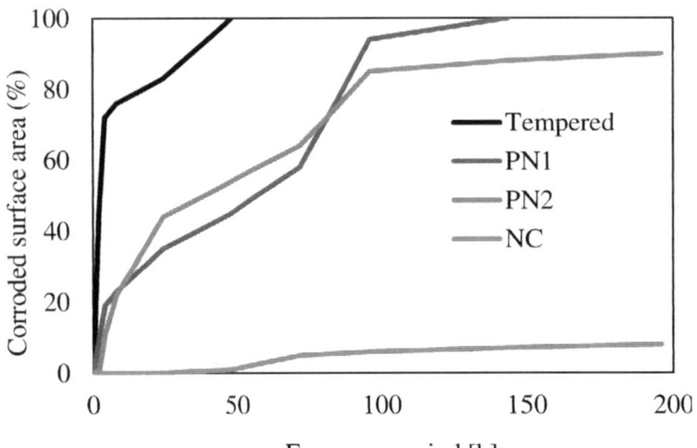

Figure 6. Corroded surface area vs. exposure (NSS).

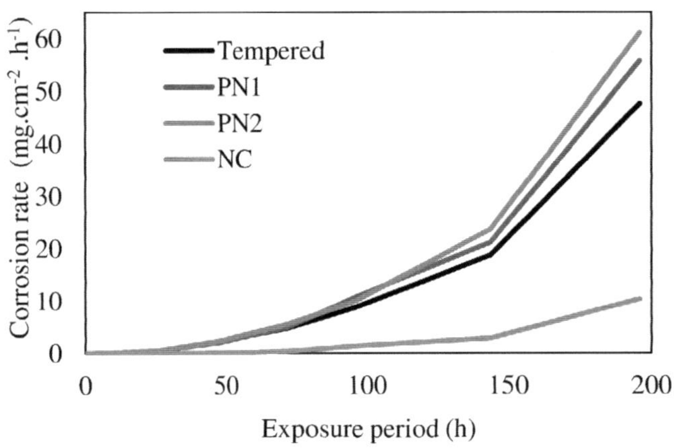

Figure 7. Corrosion rate - weight gain (NSS).

Corrosion rates of all tested samples are plotted in figure 7. It is evident that the trend of corroded surface area progress (figure 6) in not same as the corrosion rate progress (figure 7). It was necessary to investigate the corrosion mechanism of plasma nitrided samples PN1 and PN2, why is the corrosion rate of sample PN1 and PN2 higher than of tempered sample.

For corrosion mechanism investigation were all corroded surfaces in accordance with ISO 8407 standard chemically and mechanically cleaned. For this purpose was used the solution of 500 ml of HCl, 3.5 ml of urotropin and 496.5 ml of H_2O and afterwards the topography was by the laser confocal microscopy (Olympus OLS 3000) studied (see figure 8) and the pits could be measured. Thanks to laser confocal microscopy was the type of corrosion attack clear. Tempered sample was corroded by uniform corrosion sample PN1 and PN2 by mixed corrosion attack (uniform and pitting corrosion) and NC sample dominantly by pitting. The depth of pitting could be by pitting corrosion factor expressed in accordance with CSN ISO 11463: 1997 (identical to ISO 11463: 1995 standard):

$$PF = \frac{deepest\ penetration}{average\ penetration} \ (\text{-}) \qquad (4)$$

if the pitting factor (PF) is equal to value 1, then represents uniform corrosion. Higher value of pitting factor represents deeper pits penetration. Values of PF were set as following: 1 (tempered sample), 1.34 (PN1), 1.22 (PN2) and 1.24 (NC). Corrosion attack of plasma nitrided 42CrMo4 steel is a combined one, a combination of general and pitting corrosion with increased pitting penetration. Increased pitting penetration is in this case on compound layer compactness and porosity dependent (see figure 1 b).

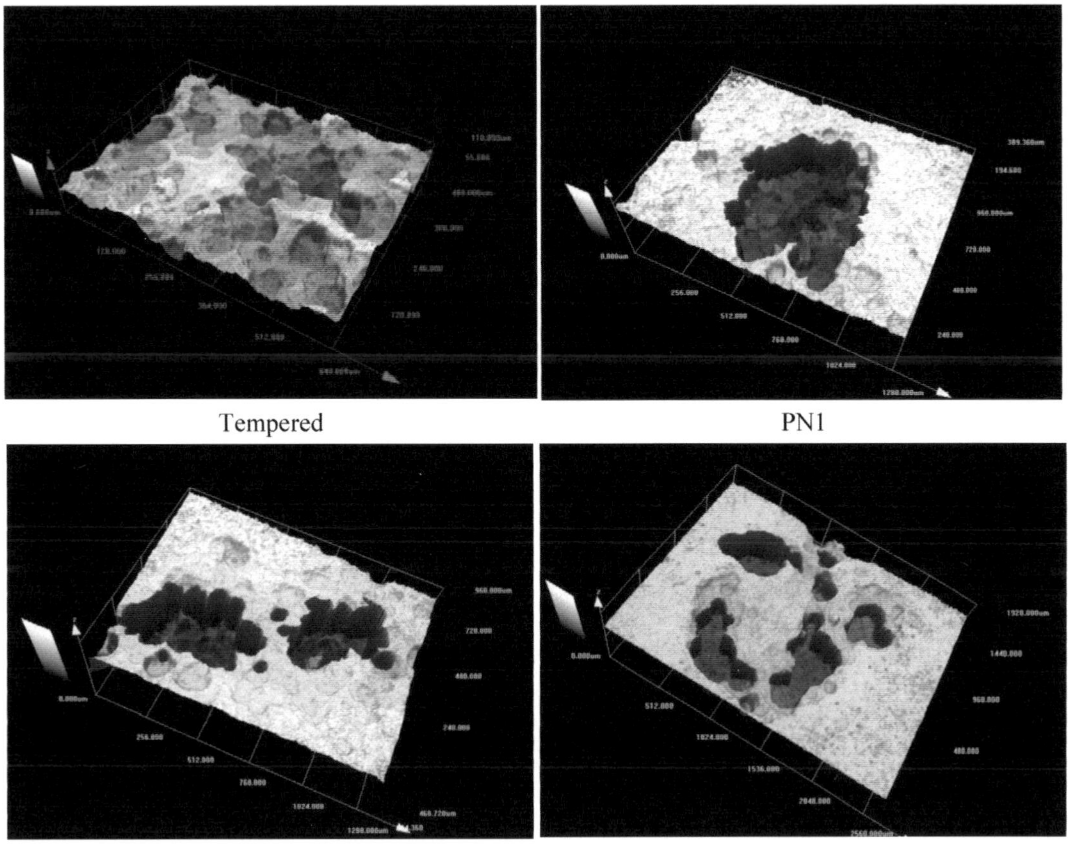

Tempered PN1

PN2 NC

Figure 8. 3D surface evaluation (Olympus OLS 3000).

4 Conclusions

42CrMo4 steel samples were prepared for wear and corrosion resistance evaluation. Tempered steel samples were compared to plasma nitrided for 15 h in different nitriding gas ratio ($3H_2$:$1N_2$ for PN1 and $1H_2$:$3N_2$ for PN2) and to Tenifer® technology (NC - nitrocarburizing followed by post-oxidation process).

Using the reverse ratio nitriding atmosphere has a positive influence to wear resistance and to value of coefficient of friction (see figure 4). Increased wear resistance of PN2 sample could be attributed to increased hardness, thickness and phase composition of compound layer (increased ratio of $Fe_{2-3}N$). Increased porosity of compound layer was found of the PN2 ($1H_2$:$3N_2$) and NC samples. This porosity negatively influences the corrosion resistance of plasma nitrided steel. Increased pitting and general corrosion was evident. This combination of corrosion attack negatively influences the corrosion rate (weight gain). The visual evaluation of corrosion resistance was found as biased and therefore a combination of visual and gravimetric evaluation was used.

Best corrosion resistance reached Tenifer® thanks to 1μm thin compact Fe_3O_4 oxide layer covering the surface pores. This surface layer is especially resistant to abrasion, corrosion and galling.

Acknowledgements

The paper has been prepared thanks to the support of the project *The Development of Technologies, Design of Firearms, Ammunition, Instrumentation, Engineering of Materials and Military Infrastructure "VÝZBROJ (DZRO K201)."* and *"Surface technology in applications special techniques SV16-216."*

References

[1] Maniee A. Mahboubi F and Soleimani R 2014 The study of tribological and corrosion behavior of plasma nitride 34CrNiMo6 steel under hot and cold wall conditions *Journal Materials and Design.* vol. 60 2014

[2] Pokorny Z. Hruby V and Studeny Z 2016 Effect of nitrogen on surface of layers *Journal Metallic Materials.* vol. 54 no. 2 2016

[3] Pye D 2003 Practical nitriding and ferritic nitrocarburizing ed ASM International (Materials Park: Ohio) chapter 4 pp 31– 35

[4] Nikolussi M. Leinweber A et al. 2007 Examination of phase transformations in the system Fe-N-C by means of nitrocarburising reactions and secondary annealing experiments, the α+ε two-phase equilibrium *Journal Material Research.* vol. 98 no.11 pp 1086-92

[5] Dong-Cherng W 2009 Plasma nitriding of plastic mold steel to increase wear and corrosion properties *Journal Surface&Coatings Technology.* vol. 204 pp 511-19

[6] Yang L. Liang W. Dandan Z and Lie Shen 2010 Improvement of corrosion resistance of nitrided low alloy steel by plasma post-oxidation *Journal Applied Surface Science.* vol. 256 Issue 13 pp 4149-52

[7] Basu A. MAJUMDAR J D et al. 2008 Corrosion resistance improvement of high carbon low alloy steel by plasma nitriding *Journal Materials Letters.* vol. 62 pp 3117–20

[8] Kusmic D and Dobrocky D 2015 Corrosion Resistance of Plasma Nitrided Structural Steels *Journal Manufacturing Technology.* vol. 15 no. 1 pp 64-69

[9] Ebrahimi M. Heydarzadeh Sohi M et al. 2010 Effect of plasma nitriding temperature on the corrosion behavior of AISI 4140 steel before and after oxidation *Journal Surface & Coatings Technology.* vol. 205 pp 261-66

[10] Marusic K. Otmacic H et al. 2006 Modification of carbon steel surface by the Tenifer® process of nitrocarburizing and post-oxidation *Journal Surface & Coatings Technology.* vol. 201 pp 3415-21

[11] Wu J. Liu H et al. 2015 Enhancement of corrosion resistance for plasma nitride AISI 4110 steel by plain air plasma post-oxidizing *Journal of Alloys and Compounds.* vol. 632 pp. 397-01

Evaluation of the effectiveness of the use of horizontal and vertical rolls in the "Rolling-pressing" process on the basis of the stress-strain state studying

S Lezhnev[1], A Naizabekov[1], E Panin[2,3], T Koinov[3], I Mazur[4], A Arbuz [2]

[1] Rudny industrial institute, Rudny, Republic of Kazakhstan; [2] Karaganda state industrial university, Temirtau, Republic of Kazakhstan; [3] University of chemical technology and metallurgy, Sofia, Bulgaria; [4] Lipetsk state technical university, Lipetsk, Russia

E-mail: sergey_legnev@mail.ru

Abstract. In this work was performed a comparative analysis of the efficiency of horizontal and vertical rolls in the "rolling-pressing" process. The comparison was performed based on the study of stress-strain state of both variants with the aid of computer simulation in the program DEFORM-3D based on the finite element method. For analysis of the stress state was used the Lode-Nadai coefficient, allowing you to determine which type of deformation is realized at a specific point – tension, compression or shear. For analysis of the strain state was used equivalent strain, which allows to estimate the common level of accumulated strain. In the comparative analysis of parameters of SSS was revealed that the use of vertical rolls at the exit from the matrix during "rolling-pressing" process, allows you to achieve a more favorable distribution of deformation along the entire length of the deformable workpiece.

1 Introduction

In the current shortage of energy and raw materials there is very actual problem of energy-saving technologies, including the obtaining of materials with high strength and ductility.

This unique combination of properties is typial for subultra - and ultrafine-grained (SUFG and UFG) form of structural state of materials with a predominance of large-angular grain boundaries. The results of the application of sub-ultrafine-grained materials in the engineering and medical fields, show them a great future and the demand for products made of such materials.

However, demand growth is greatly restricted by the high cost of production of such materials due to high energy and labour intensity of their production. The most common and studied method of obtaining SUFG and UFG is equal channel angular pressing (ECAP) [1]. However, the disadvantage of this and many other known processes lies in their discrete nature, i.e. the impossibility of treating articles of relatively large length and the necessity of a large number of treatment cycles.

To solve the problem on introduction in manufacture of technology for sub-ultrafine-grained materials by severe plastic deformation (SPD), we developed a combined technology of deformation "rolling - equal channel angular pressing", allowing to obtain high quality bars with square or rectangular cross-section. In works [2-5] have presented a comprehensive studies of this process, that show advantages over the classic ECA-pressing. However, in several works [1, 6-7] it is known that to achieve grain size of the order of $10^{-7}\div10^{-9}$ m necessary for the formation of sub-ultrafine-grained structure, is required to provide a deformable metal, the equivalent strain $\varepsilon > 3$. To achieve this level of development is required a multi-cycle deformation. In particular, the implementation of ECA-pressing need for at least 9 cycles [8], the implementation of combined process "rolling - equal channel angular pressing" reduces desired number of cycles to 6-7 [4, 5], which is still high. Therefore, further development of this combined process is the search of possible schemes of implementation of the process, allowing to reduce the number of cycles of deformation required to obtain sub-ultra-fine-grained structure.

One of the possible variants of this process is the use of universal stand instead of two horizontal (figure 1). In this case, the workpiece when pushing in the matrix will be deformed for height and at the exit from the matrix, once in the second rolls will be pulled out of the matrix, obtaining a compression in width. In this case, the change of direction of deformation has a certain analogy with the BC route [9], known from a number of works on ECA-pressing. It in multi-pass pressing the direction of the workpiece deformation changes along the longitudinal axis at 90° to that in the proposed scheme corresponds to rolling in the vertical rolls.

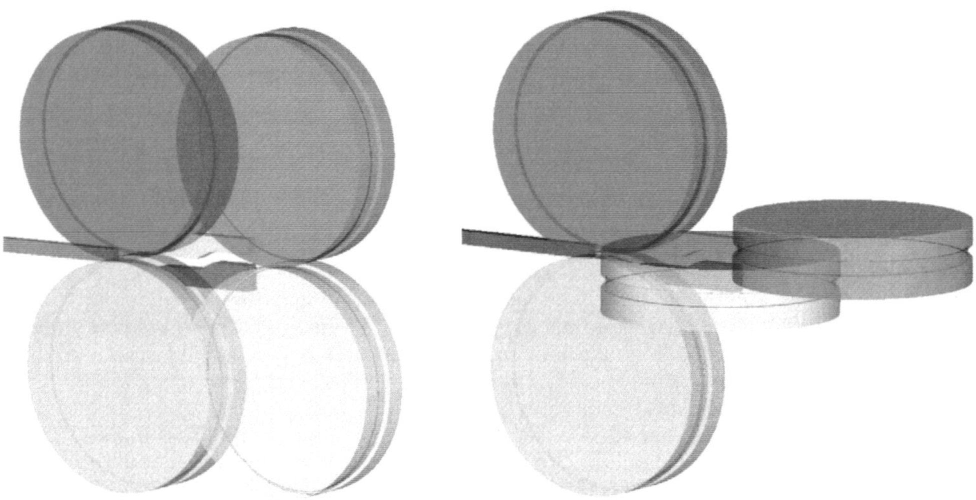

a) b)

a – with horizontal rolls at the exit of the matrix; b – with vertical rollers at the exit of the matrix.

Figure 1. Possible variants of rolls location in the combined process "rolling-equal channel angular pressing".

2 Initial model

To confirm the hypothesis about the effectiveness of using vertical rolls instead of horizontal rolls was performed a simulation of the combined process "rolling – equal channel angular pressing" in the program DEFORM-3D. The dimensions of the original workpiece were 20x15x350 mm. Conditions and assumptions in modelling:

– workpiece material in the initial state (before deformation) was isotropic and there was no initial stress and strain;

– workpiece was meshed from 180 000 finite elements with an average edge length of an element was equal to 0.5 mm;

– initial temperature of the billet was assumed to 1150 °C, also was attended deformation heating and heat transfer between the workpiece, tool and environment;

– tool was made absolutely rigid; the tool geometry (3D model) was created using the program KOMPAS 3D V15 and saved with the extension STL;

– model of the billet material was assumed to elastic-plastic;

– material adopted in the simulation is AISI 1015 steel corresponding to steel 15, the hardening curves were taken from the DEFORM library;
– coefficients of friction between the tool and the workpiece were selected based on recommendations from [2] equal to 0.5 at the contact of the workpiece with the rolls, and 0.1 on the contact of the workpiece with the matrix;
– roll rotation speed was equal to 60 rpm.
After the calculation of both models, in the workpiece were established 70 points along the longitudinal axis in increments of 5 mm, which was carried out measurements of the parameters of the stress-strain state (SSS) (figure 2). For the analysis, we selected stage of the process, when the workpiece is simultaneously rolled at both pairs of rolls.

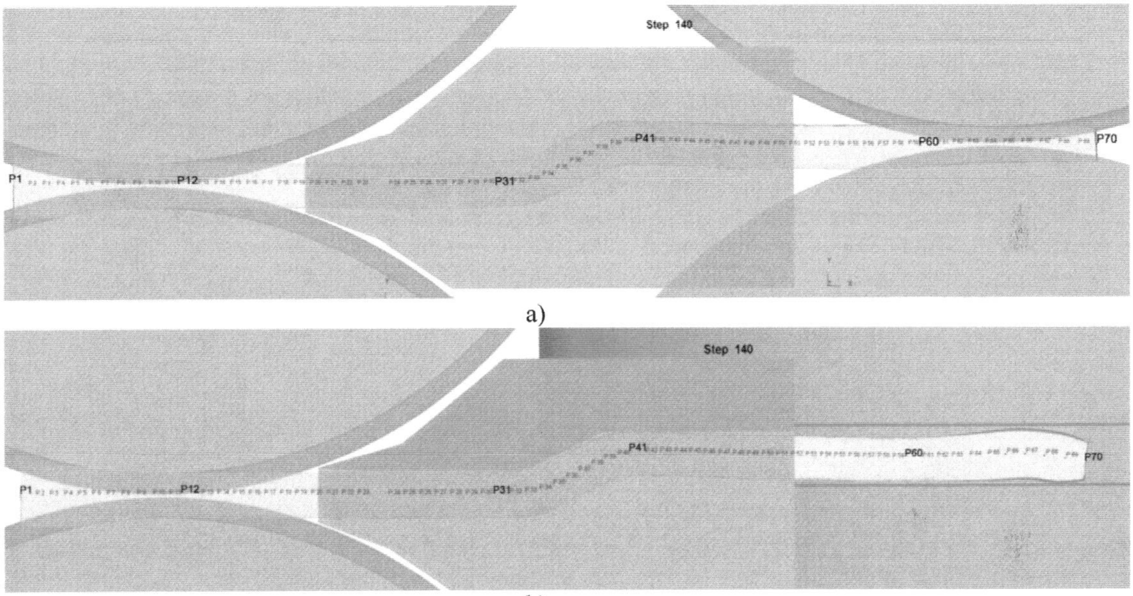

a)

b)

a – with horizontal rolls at the exit of the matrix; b – with vertical rollers at the exit of the matrix.

Figure 2. The location of the points to analyze the SSS parameters.

Conventionally the workpiece can be divided into the following areas:
1) area of rolling in the first pair of rolls (points 1÷12);
2) distance between rolling areas in the 1st pair of rolls and junction of channels in the matrix (points 13÷31);
3) area of junction of the channels in the matrix (points 32÷41);
4) distance between area of junction of channels in the matrix and area of rolling in the 2nd pair of rolls (points 42÷60);
5) area of rolling in the second pair of rolls (points 61÷70).

Since the studied models differ only in the location in space of the second pair of rolls, it is logical to assume that the values of the parameters will have differences only in the last area, in the rest areas they are identical.

3 Stress state

For analysis of the stress state, it was decided to use the coefficient of Lode-Nadai [10]. This coefficient allows to assess the nature of deformation occurring in the workpiece, i.e., to determine which type of deformation is realized at a specific point – tension, compression or shear.

Calculation of Lode-Nadai coefficient is calculated by the formula:

$$\mu = 2 \cdot \frac{\sigma_2 - \sigma_3}{\sigma_1 - \sigma_3} - 1, \quad (1)$$

where $\sigma_1, \sigma_2, \sigma_3$ – main stresses, MPa.

The value of the coefficient varies from -1 to 1. A value from 0 to 1 corresponds to compression; from 0 to -1 corresponds to the tension; the value of the coefficient tends to 0, corresponds to the shear [10].

During calculation of Lode-Nadai coefficient were obtained the following results (figure 3). As expected, the results in the first three zones were completely identical. In the first zone during rolling present as a region of compression (μ = 0,6÷0,9) and the area of tension (μ= -0,6÷-0,8), which is the result of the peculiarities of the formation of the deformation zone during rolling, which includes areas of lag and advance. Similar results were obtained in the last area, when rolling in the second pair of rolls. In the second zone, in its entirety develops a compressive strain (μ = 0,95÷1), which is a consequence of backpressure from the inclined channel matrix. In the third area when the passage of the workpiece through the channels of the matrix, in the deformable metal develop two types of deformation: compression (μ = 0,6÷0,85) and shear (μ = 0÷0,2), which is the most favorable scheme for the grinding of the original grain.

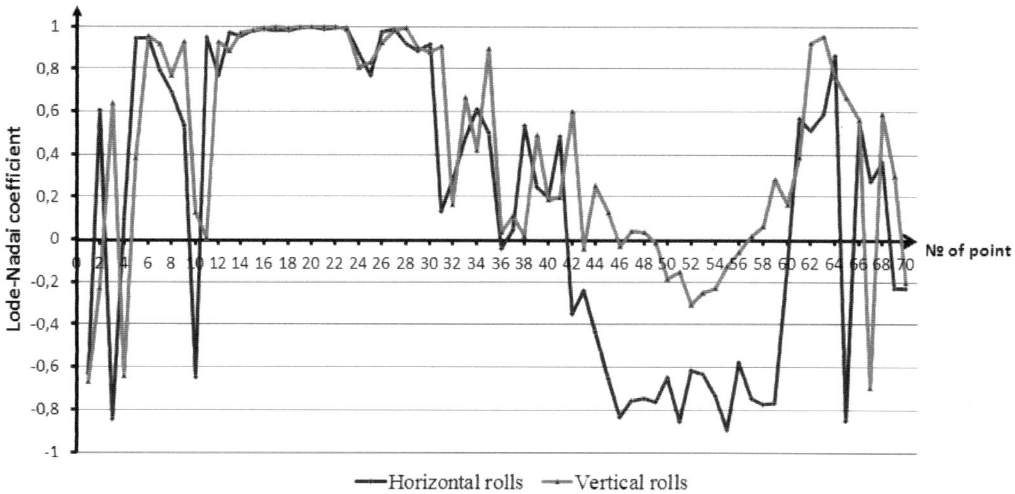

Figure 3. Lode-Nadai coefficient.

The most interesting are the values of Lode-Nadai coefficient obtained in the 4th zone, which characterizes the distance between the areas of the joints of channels in the matrix and rolling in the 2nd pair of rolls. Here the values of Lode-Nadai coefficient with horizontal and vertical rolls have significant differences. This is due to the kinetics and the forming of the workpiece in contact with the second pair of rolls. If you have a second pair of horizontal rolls, the billet gets re-compression in height, while the required amount of compression for a stable process flow is much less than in the first pair of rolls. As a result, the workpiece receives a significantly lower amount of widening. As a result, in the fourth area develop mainly tensile stresses (μ= -0,6 ÷ -0,8), which is a negative factor.

In the case of vertical rolls at the exit of equal channel step matrix the billet gets a second compression in width, the value of which is commensurate with the compression in the first pair of rolls. As a result, in the fourth area occurs backpressure from the rolls, thereby, tensile stresses are significantly reduced, until the appearance of zones of compression (μ= -0,3÷0,3).

4 Strain state

For analysis of the strain state was used indicator such as equivalent strain, which in some works is referred as intensity of deformation. This parameter allows to assess the common level of accumulated strain at any stage of the process.

Equivalent strain is determined by the formula

$$\varepsilon_{EQV} = \frac{\sqrt{2}}{3}\sqrt{(\varepsilon_1 - \varepsilon_2)^2 + (\varepsilon_2 - \varepsilon_3)^2 + (\varepsilon_3 - \varepsilon_1)^2}, \quad (2)$$

where ε_1 , ε_2 , ε_3 - main strains.

In the study of this parameter were obtained results presented in figure 4. When comparing models with horizontal and vertical rolls, it was established that in the first four zones the values of equivalent strain are identical. This difference in values appears only in the last area, where due to the change in the direction of compression, in the vertical rolls significantly increases the level of processing, thus the value of the equivalent strain increased from 0,55÷0.8 in the horizontal rolls to 0.9÷1.15 in the vertical rolls.

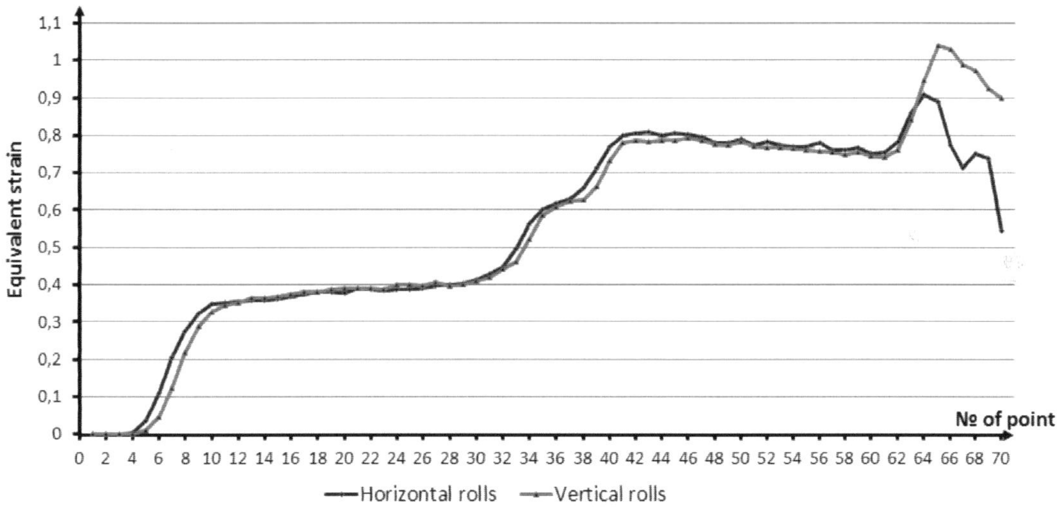

Figure 4. Equivalent strain.

5 Conclusions

In this work was performed a comparative analysis of the efficiency of horizontal and vertical rolls in the "rolling-pressing" process. The comparison was performed based on the study of stress-strain state of both variants with the aid of computer simulation in the program DEFORM-3D based on the finite element method. For analysis of the stress state was used the Lode-Nadai coefficient, allowing you to determine which type of deformation is realized at a specific point – tension, compression or shear. For analysis of the strain state was used equivalent strain, which allows to estimate the common level of accumulated strain. In the comparative analysis of parameters of SSS was revealed that the use of vertical rolls at the exit from the matrix during "rolling-pressing" process, allows you to achieve a more favorable distribution of deformation along the entire length of the deformable workpiece.

References

[1] Valiev R and Alexandrov I 2007 Bulk nanostructured metallic materials (Moscow: Akademkniga) p 398

[2] Panin E, Lezhnev S, Naizabekov A and Koinov T 2016 Theoretical bases of combined process "rolling – equal-channel step pressing" *J. Chem. Technol. Metall.* JCTM51(5) pp 594-602

[3] Lezhnev S, Naizabekov A and Panin E 2011 Theoretical studies of the joint "extrusion-rolling" process aimed at making sub-ultra fine - grained structure metal, Proc. Int. Conf. on Metallurgy and Materials (Brno, Czech Republic) pp 272-7

[4] Lezhnev S, Panin E and Volokitina I 2013 Research of combined process "rolling-pressing" influence on the microstructure and mechanical properties of aluminium *Adv. Mat. Res.* AMR814 pp 68-75

[5] Naizabekov A, Lezhnev S, Panin E and Volokitina I 2014 Influence of Combined Process "Rolling-pressing" on Microstructure and Mechanical Properties of Copper *Proc. Eng.* PE81 pp 1499–1504

[6] Valiev R and Langdon T 2006 Principles of equal-channel angular pressing as a processing tool for grain refinement *Prog. Mater. Sci.* PMS51 pp 881-981

[7] Raab G 2005 Plastic flow at equal channel angular processing in parallel channels *Mat. Sci. Eng. A* MSEA410–411 pp 230–233

[8] Horita Z, Smith D, Furukawa M, Nemoto M, Valiev R and Langdon T 1996 An Investigation of Grain Boundaries in Submicrometer-Grained Al-Mg Solid Solution Alloys Using High-Resolution Electron Microscopy *J Mater Res* JMR11(8) pp 1880-90

[9] Valiev R, Islamgaliev R and Alexandrov I 2000 Principles of equal-channel angular pressing as a processing tool for grain refinement *Prog. Mater. Sci.* PMS45(2) pp 103-89

[10] Segal V, Reznikov V, Kopylov V, Pavlik D and Malyshev V 1994 Processes of plastic structure formation of metals (Minsk: Science and technology) p 232

Hydrogen effect on compression mechanical properties of TiNb alloys at elevated temperatures

M Losertová, O Štefek, I Schindler, L Szkandera and P Kudela

VŠB - Technical university of Ostrava, Faculty of Metallurgy and Materials Engineering, Ostrava - Poruba, Czech Republic

E-mail: mlosertova@vsb.cz

Abstract. The study of this work was focused on the hydrogen effect on hot deformation behavior of hydrogen charged TiNb based alloys at three temperatures in comparison with non-charged specimens. The Ti24Nb and Ti26Nb (at.%) alloys were heat treated by three step regime in argon or hydrogen atmospheres. The hot compression tests were performed on a Gleeble 3800 machine at 800, 750 and 700°C with compression strain of $5x10^{-3}$ s^{-1} and deformation degree of 50 %. The microstructure resulting from heat treatment as well as from isothermal compression test was analyzed using optical and scanning electron microscopies. Measurement of microhardness revealed that higher microhardness values for Ti24Nb after thermo-hydrogen treatment corresponded to fine grained microstructure and strengthening by grain boundaries. Hydrogen contents determined using LECO RH600 analyzer showed that Ti26Nb contained higher amount of hydrogen due to the stabilization of beta phase by higher Nb concentration. Based on the hot compression test results, the plasticity at elevated temperatures was evaluated. The evolution of uniaxial compressive test curves showed the stabilization of flow stress due to the hydrogen for high deformations at lower temperatures. The differences in hot compression behavior for both niobium contents on the one hand and for hydrogen charged and non-charged specimens on the other hand were observed.

1 Introduction

The mechanical properties of most engineering metallic materials and especially of titanium alloys are generally deteriorated by hydrogen presence. The brittleness of alloys induced by the hydrogen is observed at room temperature as well as at elevated temperatures. Nevertheless, in some cases the hydrogen can act as beneficial element in improving particular properties. Titanium alloys show high strength-to-weight ratio, good corrosion resistance and biocompatibility, so they are of particular interest for use as biocompatible implants in orthopedics and traumatology [1].

Generally, workability of titanium alloys at room temperature is difficult and forming must be performed at elevated temperature. As it was discussed in [2-4], it is possible to improve hot workability of titanium and titanium alpha-beta alloys by using of hydrogen as a temporary alloying element in special heat treatment technology, so-called the thermo-hydrogen treatment (THT). The temporary hydrogen alloying allows hot working of alpha-beta titanium alloys at lowers temperatures with lower flow stresses. The microstructure refinement resulting from the THT technology allows improving fatigue and strength properties.

The study of this work was focused on the hydrogen effect on hot deformation behavior of hydrogen charged TiNb based alloys at three temperatures in comparison with non-charged specimens. The Ti24Nb and Ti26Nb (at.%) alloys that have predominantly beta phase structure were heat treated by three step regime in argon or hydrogen atmospheres.

2 Experimental

Experimental specimens of Ti-24Nb and Ti26Nb (at.%) alloys that nominal compositions are plotted in figure 1 were prepared of forged bars. Cylindrical samples of 10 mm in diameter and 12 mm in height were machined and heat treated in high temperature hydrogen charging furnace. Three steps thermo-hydrogen treatment according to [5, 6] and described in figure 2 was realized on four specimens in flowing hydrogen (5N).

Considering higher solubility of hydrogen [7] in predominantly beta phase TiNb, the time of hydrogenation steps was modified:

(1) heating to 600 °C and holding for 30 minutes,
(2) increasing temperature to 850 °C and holding for 18 minutes,
(3) cooling to 590 °C and holding for 60 minutes,
(4) hydrogen-cooling to 300 °C and then argon-cooling to room temperature.

In order to compare the high temperature deformation behavior with and without hydrogen in the alloy structure the same procedure of heat treatment was performed on four samples in flowing argon (5N).

The hydrogen charged and non-charged specimens were submitted to isothermal compression at constant strain rate of $5x10^{-3}$ s^{-1} on a Gleeble-3800 machine. The isothermal compression tests were performed at three selected temperatures: 800, 750 and 700 °C. Prior to compression reaching 50.0 % of maximum height reduction, the specimens were held for 3 min at the nominal deformation temperature and after the test, the samples were air-cooled to the room temperature. During isothermal compression, the flow stress was recorded as a function of strain.

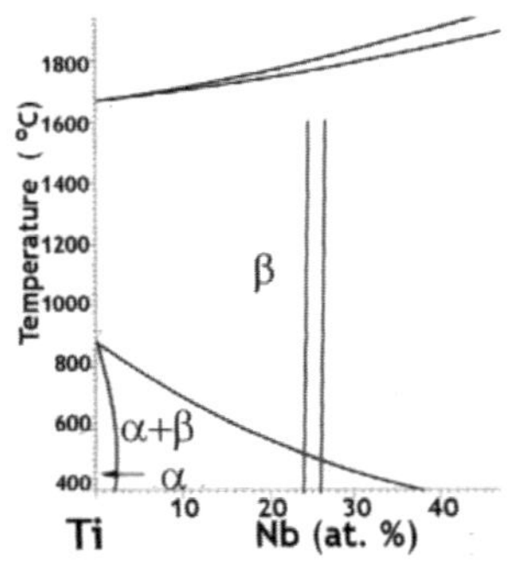

Figure 1. Binary Ti-Nb diagram for Ti rich alloys: ocurrence of alpha and beta phases with increasing Nb content .

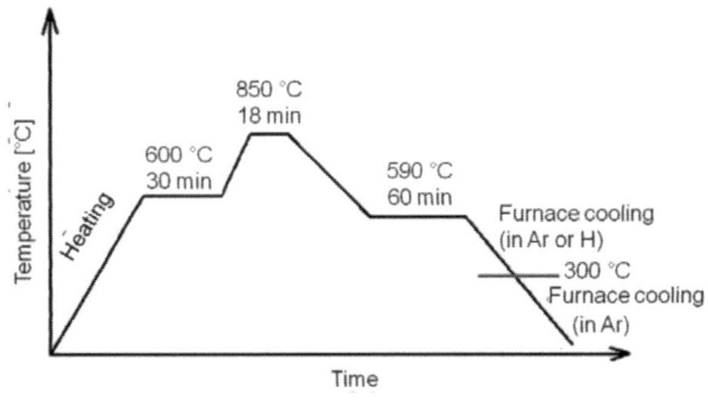

Figure 2. Three steps regime of thermo-hydrogen treatment (THT) before subsequent isothermal compression test.

The metallographic study was realized on the specimens ground and polished by conventional methods and etched using Kroll's reagent (2 HF : 4 HNO_3 : 80 H_2O). The amount of hydrogen in the specimens was determined before and after the compression test by hydrogen determinator LECO RH 600.

3 Results and discussion

The macroscopic view of specimens after the compression tests with an amount of reduction in height of 50 % is illustrated in figure 3. The surfaces of hydrogen charged specimens were cracked (red boxes in figure 3 b) while the non-charged specimens were cracked rather rarely on the surface.

The flow stresses recorded for both alloys are depending on the temperature as well as on the niobium content. It is clear that niobium is beta stabilizer, so for the Ti26Nb the flow stresses are lower then for Ti24Nb due to the higher content of more ductile beta phase. Comparing the deformation behavior at three different temperatures of hydrogen non-charged and charged specimens, it is clear that with increasing temperature the flow stress is decreasing and by reason of this the hot forming is improving for both composition and both heat treated stages.

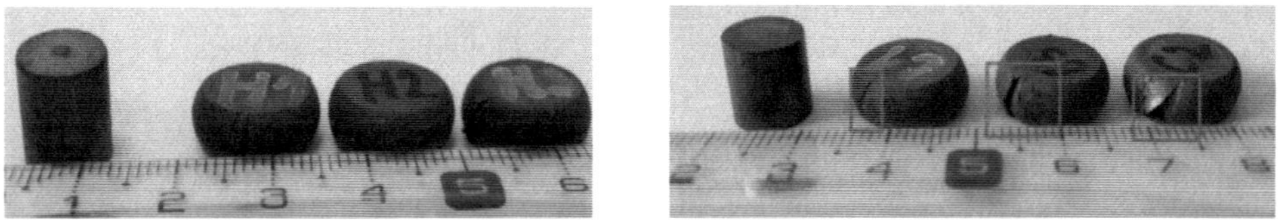

Figure 3. Specimens of TiNb alloys after the isothermal compression test: a) hydrogen non-charged, b) hydrogen charged (cracking in red boxes).

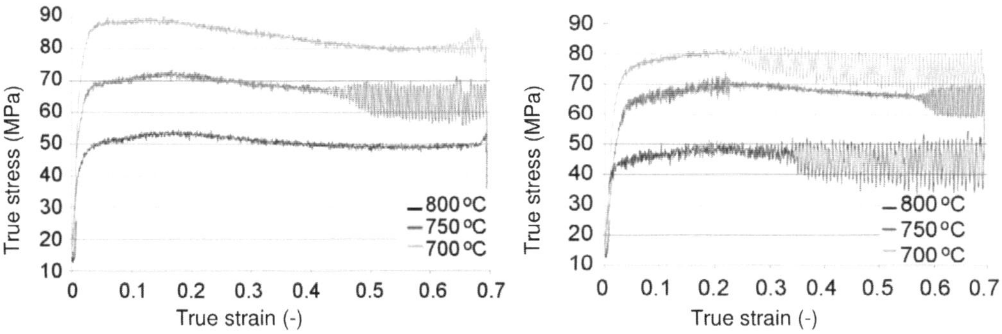

Figure 4. True stress - true strain curves for hydrogen non-charged: a) Ti24Nb; b) Ti26Nb.

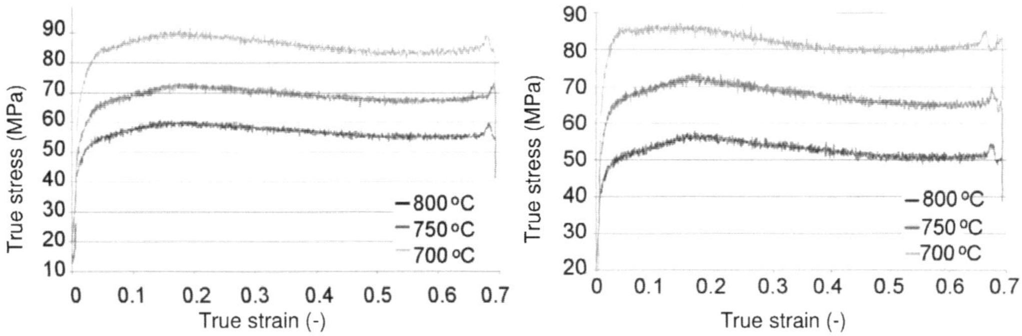

Figure 5. True stress - true strain curves for hydrogen charged: a) Ti24Nb; b) Ti26Nb.

In the case of the hydrogen non-charged specimens unusual behavior was registred during compression tests (figure 4). Large instabilities of flow stress are more evident for the specimens with higher niobium content and could be related with lower rate of dynamic recrystallization. The hydrogen effect on development of compression test was inconsistent for both alloy compositions.

Unlike the Ti24Nb alloy the hydrogen content in the Ti26Nb specimens increased the flow stress, but on the other side the flow stress instabilities were suppressed for both alloys. Indeed, the evolution of uniaxial compressive test curves (figure 5) showed the stabilization of flow stress due to the hydrogen for high deformations at all three temperatures and for both compositions of TiNb alloy. The differences in hot compression behavior for both niobium contents on the one hand and for hydrogen charged and non-charged stages on the other hand are related with increasing niobium and hydrogen amounts effecting strengthening by solid solution [8].
The hydrogen amounts measured after the THT and after the isothermal compression test (CT) are represented in figure 6. The highest amount (2572 wt. ppm) was determined in Ti26Nb after the THT that corresponds with higher hydrogen solubility in beta phase. Nevertheless, the decrease of hydrogen content after the CT at 800 °C was more important just for the alloys with more niobium. Unfortunately, the hydrogen promoted cracking on the surfaces of compression tested specimens.

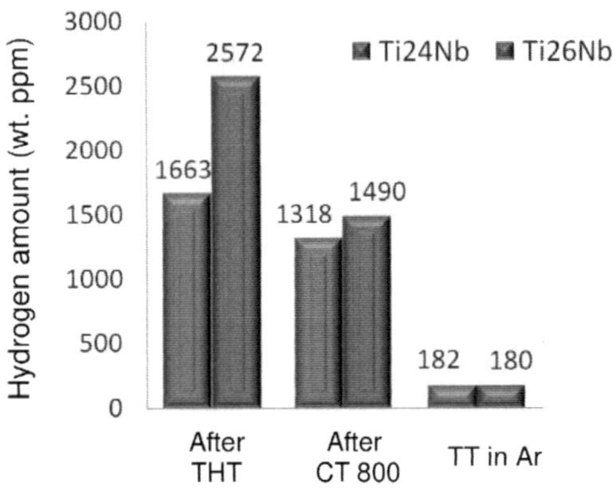

Figure 6. Amount of hydrogen in tested specimens .

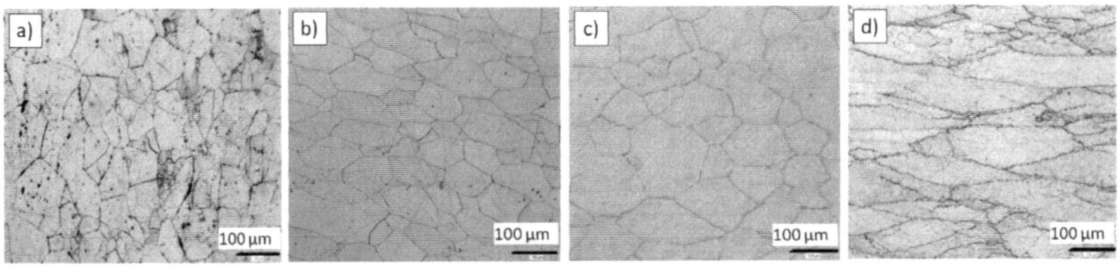

Figure 7. Optical micrograph of Ti24Nb alloy treated by three steps heat treatment in Ar (a) and deformed by compression test at 700 °C (b), 750 °C (c) and 800 °C (d).

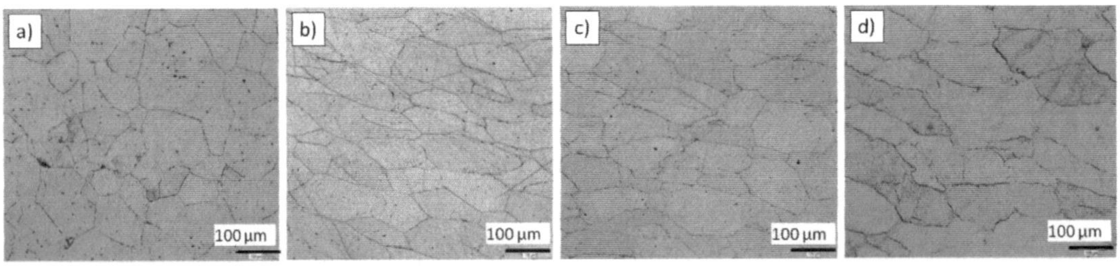

Figure 8. Optical micrograph of Ti26Nb alloy treated by three steps heat treatment in Ar (a) and deformed by compression test at 700 °C (b), 750 °C (c) and 800 °C (d).

IOP Conf. Series: Materials Science and Engineering **179** (2017) 012048 doi:10.1088/1757-899X/179/1/012048

The microstructure of hydrogen non-charged samples was formed of beta grains with fine alpha precipitates (figures 8 a and 9 a). After the hot compression tests, the microstructures of both alloys were dynamically recrystallized to large grains with very fine subgrains (figures 7 d and 8 d). The hydrogen charging led to the important refining of microstructure after the THT as well as after the CT, as seen in figures 9 and 10.

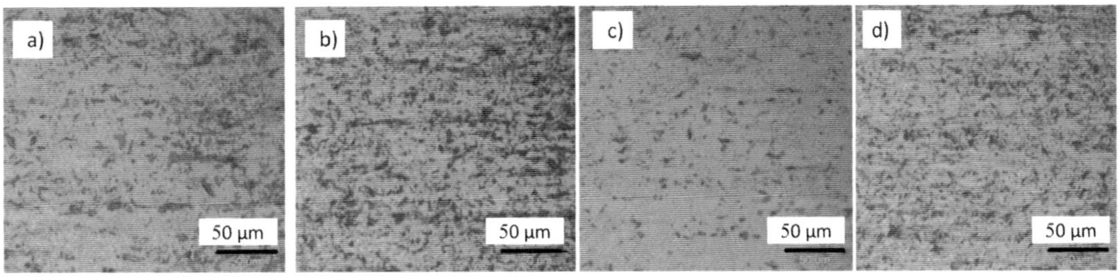

Figure 9. Optical micrograph of Ti24Nb alloy treated by three steps heat treatment in hydrogen (a) and deformed by compression test at 700 °C (b), 750 °C (c) and 800 °C (d).

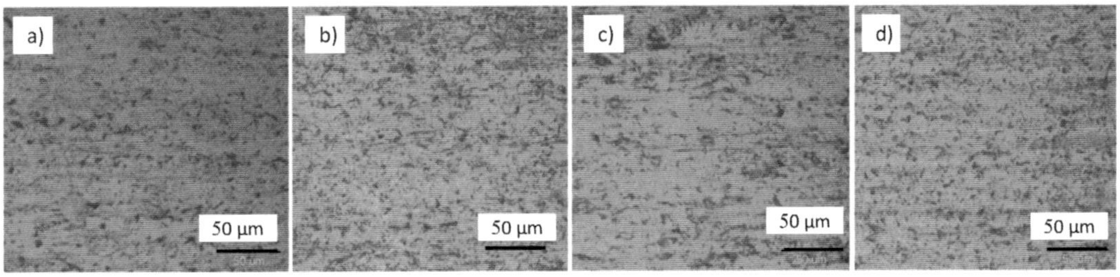

Figure 10. Optical micrograph of Ti26Nb alloy treated by three steps heat treatment in hydrogen (a) and deformed by compression test at 700 °C (b), 750 °C (c) and 800 °C (d).

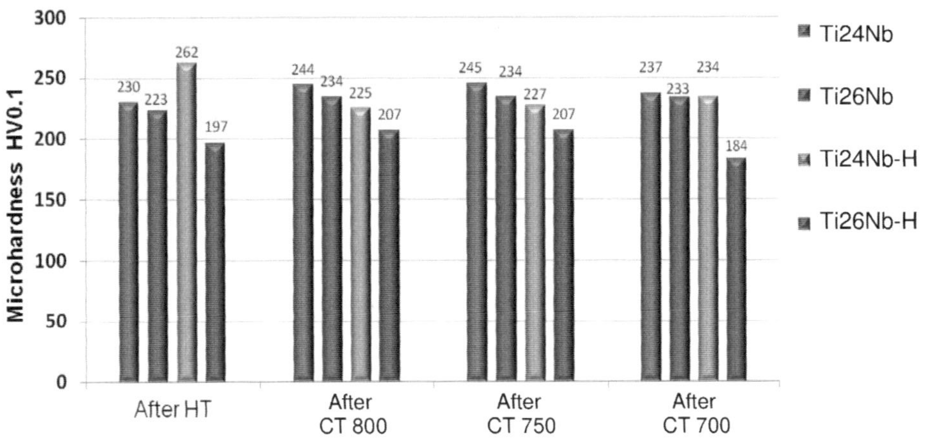

Figure 11. Microhardness of TiNb samples before and after compression test (CT) at 800, 750 a 700 °C.

The values of microhardness measured for hydrogen non-charged Ti24Nb alloy are higher in all treated stages (figure 11) and are related with precipitation of fine particles of alpha phase in beta grains and on grain boundaries. The finer microstructure of Ti24Nb after the THT caused the higher microhardness unlike the specimens with higher niobium content in which the hydrogen decreased the microhardness in all treated stages. As for the compression test curves, understanding the inconsistency of hydrogen influence on microstructure together with mechanical properties is presently limited and must be completed by other phase analysis. The hydrogen on the one side allows stabilizing more ductile beta phase but on the other side provides strengthening by solid solution. Nevertheless, it is clear that thermo-hydrogen treatment in the case of TiNb alloys is not as favorable as it is for Ti6AlV, at least for the niobium contents studied in this work.

4 Conclusions

The hydrogen effect on the microstructure and flow stress was studied using isothermal compression test at three different temperatures for two TiNb alloy compositions. Comparing the results obtained for hydrogen non-charged and hydrogen charged specimens it was found that the hydrogen content as high as 1663 and 2572 wt. ppm for Ti24Nb and Ti26Nb, respectively, has an obvious benefit effect on the supression of instabilities in true stress - true strain curves.

However, the higher amount of hydrogen induced increasing flow stress and resulted in cracking of specimens. The benefit effect of hydrogen on increasing formability of TiNb based alloys at lower temperatures was not evident as it was observed in the case of Ti6Al4V alloy.

Acknowledgement

This article has been elaborated in the framework of the projects No. TA03010804 and No. TH01020487 financed by Technology Agency of the Czech Republic and No. LO1203 "Regional Materials Science and Technology Centre - Feasibility Program" funded by Ministry of Education, Youth and Sports of the Czech Republic.

References

[1] Niinomi M 2008 Mechanical biocompatibilities of titanium alloys for biomedical applications *J. of the Mech. Behav.Biomed.Mat.* 1/1 (2008) 30-42

[2] Senkov O N and Froes F H 1999 Thermohydrogen processing of titanium alloys *Int.J. Hydrogen Energy* **24** (1999) 565-576

[3] Murzinova M A, Mazurski M I, Salishev G A and Afonichev D D 1997 Application of reversible hydrogen alloying for formation of submicrocrystalline structure in (alpha + beta) titanium alloys *Proceedings of the Selected Papers of the 1st International Conference on Hydrogen Treatment of Materials* International Association for Hydrogen Energy 22/2-3 (1997) 201-204

[4] Li M Q, Luo J and Niu Y 2010 Effect of the hydrogen content on the deformation behavior in the isothermal compression of Ti600 alloy *Mat. Sci. Eng.* A **527** (2010) 6626-6632

[5] Fang T Y and Wang W H 1998 Microstructural features of thermo mechanical processing in a Ti-6Al-4V alloy *Materials Chem. Physics* **56** (1998) 35, PII S 0254-0584(98)00137-0.

[6] Losertová M, Štamborská M and Schindler I 2014 Influence of thermo hydrogen treatment on hot deformation behavior and microstructural evolution of Ti-6Al-4V alloy Archives of Materials Science and Engineering **66** (2014) 2 62-66

[7] *Phase Diagrams of Binary Hydrogen Alloys*, Ed. by F.D. Manchester, ASM International (2000) 238-258

[8] Zong Y Y, Huang S, Wang Y, Shan D B Strenghening versus softening mechanisms by hydrogen addition in beta-Ti40 alloy *Int. J. of Hydrogen Energy* **39** (2014) 3498-3504

Influence of rotary swaging and subsequent age hardening on properties of EN AW 6082 aluminium alloy

L Maleček[1], J Palán[1], J Nacházel[1], J Dlouhý[1],

[1] COMTES FHT a.s., Dobřany, The Czech Republic, EU

E-mail: ladislav.malecek@comtesfht.cz, jan.palan@comtesfht.cz, jan.nachazel@comtesfht.cz, jaromir.dlouhy@comtesfht.cz

Abstract. Mechanical properties and microstructure of EN AW 6082 were investigated. The aluminium alloy was processed by combining the solution annealing, plastic deformation and artificial age hardening, respectively. The initial state of the investigated material was provided in the form of extruded rods with the diameter of 12 mm. For the solution annealing the temperature 530 °C was chosen. The plastic deformation was realized by rotary swaging at ambient temperature and the investigated material was rotary swaged from 12 to 10 mm in diameter. The effect of the age hardening temperature and time was studied at temperatures of 120 and 160 °C and times 1 - 12 hours. The impact of processing parameters on mechanical properties was assessed by tensile testing and hardness measurement. Metallographic examination was carried out by light optical microscopy (LOM) and scanning electron microscopy using electron backscatter diffraction (SEM-EBSD).

1 Introduction

The rising interest in ultrafine-grained (UFG) metallic materials has led to an intensive development of forming methods based on SPD processes [1, 2]. SPD (Severe Plastic Deformation) is a generic term that refers to processes which employ large plastic deformation to achieve notable grain refinement down to the nanometre scale. Grain refinement brings higher strength properties and ductility which surpass those found in today's conventional materials.

SPD processes have found increased use in the forming of aluminium alloys. This also applies to 6xxx-series aluminium alloys (Al-Mg-Si alloys). They can undergo precipitation strengthening, thanks to varying solubility of the Mg_2Si intermetallic phase in aluminium [3]. One of these alloys is the EN AW Al 6082 which is the subject of this investigation.

A number of studies of SPD forming and subsequent artificial age hardening of aluminium alloys have been published in recent years, such as [4-15]. Most of them [4–14] focused on equal channel angular pressing (ECAP) or its modified variants [15]. Only a handful of studies have been devoted to rotary swaging or to combinations of rotary swaging and artificial age hardening.

For instance, the effects of rotary swaging (RS) on mechanical properties and microstructural evolution in commercial purity aluminium are reported in [16]. The authors of [17] dealt with RS and artificial age hardening in EN AW 2024 (Al-Cu-Mg alloy). The present study was carried out to gather information about the effects of RS and subsequent artificial age hardening on the properties of the EN AW Al 6082 alloy.

Plastic deformation was applied by rotary swaging (RS) (Figure 1). RS is an incremental metal forming process for the reduction of cross-sections of bars, tubes, wires or rods. Sets of four dies perform small, high-frequency, simultaneous radial movements (oscillations). With every stroke of the die, a part of the workpiece is formed. Die movement is generated by means of cams. The kinematics are equivalent to those of a planetary gear. This assembly is completely accessible, which makes it easy to carry out any maintenance work [18].

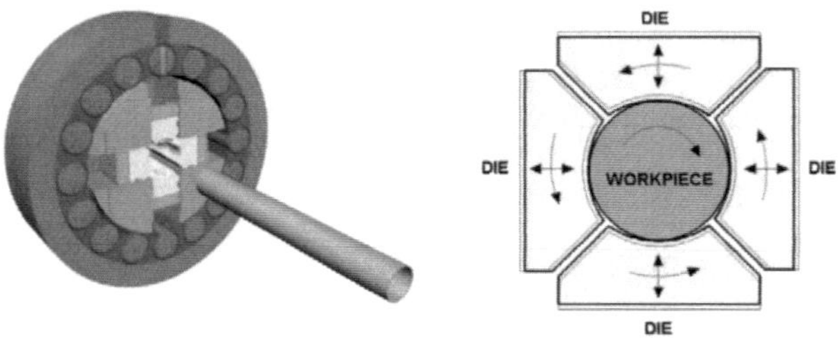

Figure 1. Rotary swaging configuration.

2 Experimental material and procedure

Testing was performed on extruded round bars whose length and initial diameter were 0.5 m and 12 mm, respectively. Table 1 gives the chemical composition of the EN AW 6082 alloy that was used for this testing. The temper of the as-received bars was T6511.

Table 1. Chemical composition of EN AW 6082 (wt. %).

Si	Fe	Cu	Mn	Mg	Cr	Zn	Al
1.00	0.41	0.11	0.58	0.65	0.13	0.048	Balance

The treatment route used for the bars was as follows:

1. Solution annealing at 530 °C for 1.5 hours, and subsequent water quenching
2. Forging in the rotary swaging machine at ambient temperature; reduction of diameter: ∅ 12 mm → ∅ 10 mm; feed rate of 100 mm/s (6000 mm/min); rotational speed of dies of 24.2 rps (1450 rpm).
3. Artificial age hardening at 120 °C and 160 °C for defined periods:
   - at 120 °C for 1, 3, 6, and 12 hrs.
   - at 160 °C for 1, 3, 6, and 12 hrs.

The schematic profile of the entire experimental route is plotted in Figure 2.

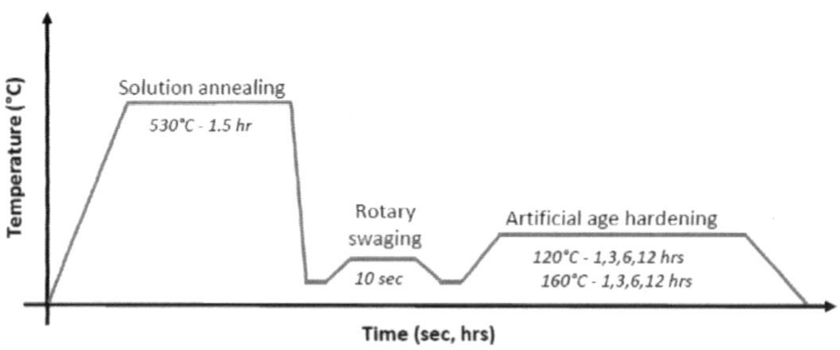

Figure 2. Experimental route.

3 Results and discussion

3.1 *Mechanical properties*

Anisotropy of mechanical properties was evaluated by measuring hardness of the material. HV1 scale was employed. The values were measured using the STRUERS DuranScan tester on transverse sections through the bar, from one edge to the other. The resulting hardness profiles are plotted in Figure 3.

Mechanical properties were analysed with the aid of tensile testing. Specimens of 6 mm diameter and 30 mm gauge length for quasi-static tensile testing were made from samples taken in longitudinal direction of the parts. The samples were tested on a ZWICK Z250 testing machine. The length change was measured with strain gauge clip-on extensometer. The data was evaluated in accordance with the EN ISO 6892-1 standard. The properties (offset yield stress YS, ultimate tensile strength UTS, elongation A_g, elongation A_5, and reduction of area RA) are plotted in Figure 4. The values of mechanical properties are listed in Table 2.

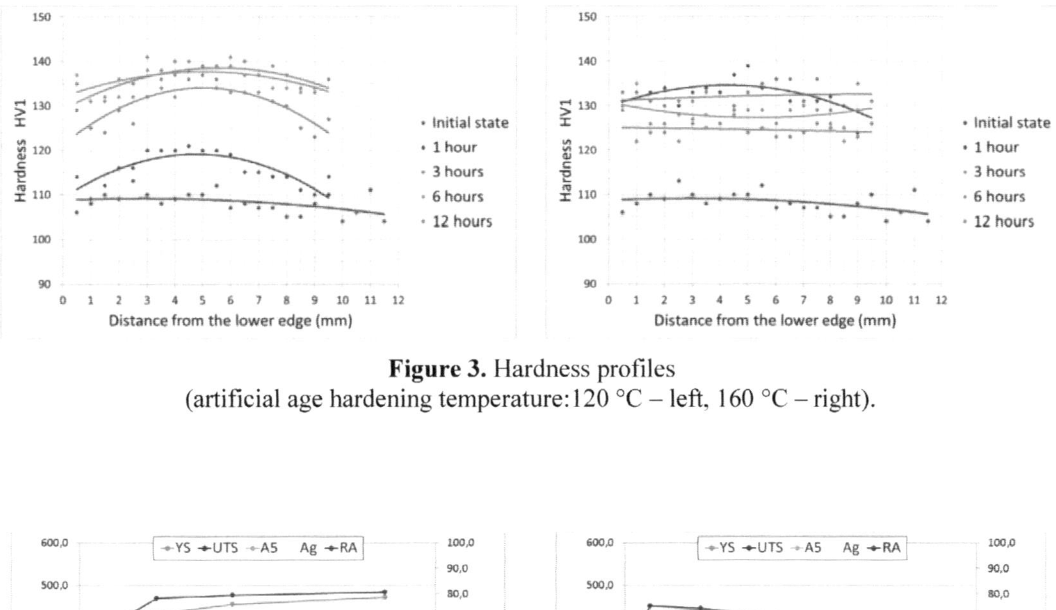

Figure 3. Hardness profiles
(artificial age hardening temperature:120 °C – left, 160 °C – right).

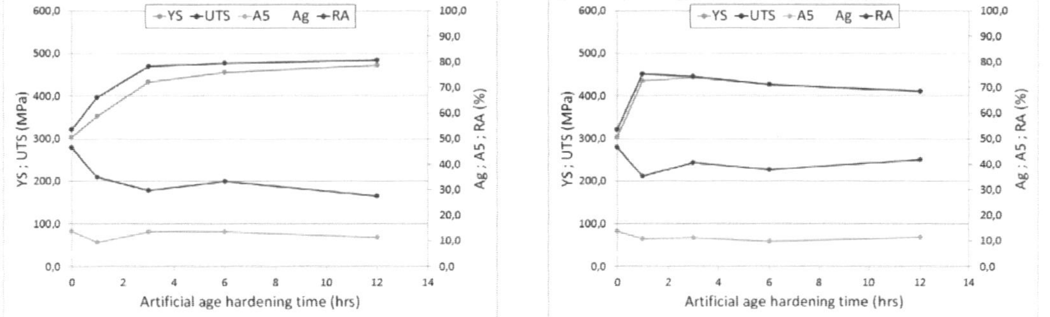

Figure 4. Representative tensile test results
(artificial age hardening temperature:120 °C – left, 160 °C – right).
Note: Artifical age hardening time 0 hrs represents the initial state

Table 2. Mechanical properties (YS, UTS, A_g, A_5, and RA) before (i.e. as-received condition) and after the experimental treatment, t = artificial age hardening temperature, τ = artificial age hardening time.

t - τ (°C - hrs)	YS (MPa)	UTS (MPa)	A_g (%)	A_5 (%)	RA (%)
Initial state	303.4	321.9	4.8	13.8	46.7
120 °C – 1 hr	352.3	396.2	5.2	9.3	34.9
120 °C – 3 hrs	432.1	469.4	7.7	13.5	29.6
120 °C – 6 hrs	455.1	476.5	7.0	13.6	33.3
120 °C – 12 hrs	472.5	483.4	6.3	11.4	27.6
160 °C – 1 hr	435.1	451.2	4.9	10.7	35.2
160 °C – 3 hrs	443.8	445.6	4.2	11.1	40.6
160 °C – 6 hrs	426.7	427.1	4.0	9.7	37.9
160 °C – 12 hrs	410.8	410.9	3.5	11.4	41.8

3.2 *Microstructure*

The microstructure analysis was first performed under the Carl Zeiss – Observer.Z1m light microscope. The specimens were ground, polished, and etched on longitudinal and transverse cross-sections. For this purpose, all samples were cold-mounted. After polishing, anodization (J = 0.2 A/cm2; 60 s) in Barker´s reagent (200 ml H_2O + 5 ml HBF_4) was performed. Polarized light mode was used for observation. Figure 5 shows the relevant optical micrographs.

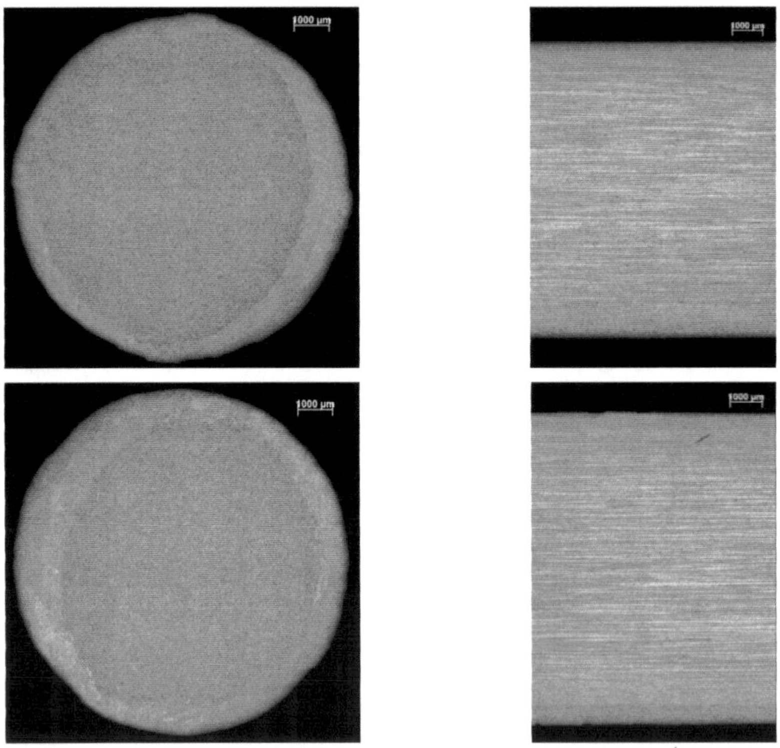

Figure 5. Micrographs of etched specimens after artificial age hardening taken under polarized light illumination; transverse section (left) and longitudinal section (right). Artifical age hardening conditions: 120 °C – 12 hrs (top), 160 °C – 12 hrs (bottom).

The micrographs of the etched sections show grains elongated in the longitudinal direction. The surface layer contains coarse grains. This was caused by the as-received condition of the material (T6511). During solution annealing, recrystallization took place in this region. Using optical microscopic observation, no difference was found between specimens age-hardened at different temperatures and/or for different times.

Subsequently, an analysis by electron backscattered diffraction (EBSD) was performed to evaluate grain size, and detect boundaries and misorientation on selected longitudinal sections. The JEOL JSM-7400F scanning electron microscope equipped with the Oxford Nordlys EBSD camera was employed. Specimens were prepared by ion polishing to avoid the surface deformation. EBSD maps were acquired across 500×500-point grids with a step size of 0.1 μm.

Three specimens were chosen for EBSD analysis: the as-received material, the material treated at 120 °C – 12 hrs, and that treated at 160 °C – 12 hrs (artificial age hardening conditions). Neither this analysis revealed any significant differences in the grain shape or size. The reduction of cross section was reflected in the traces of slip bands within grains inclined 45° to the direction of deformation.

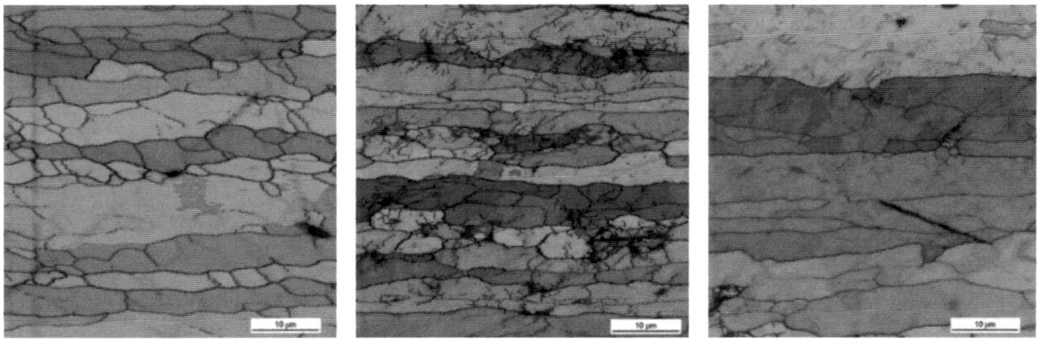

Initial state

120 °C / 12 hrs (artifical age hardening conditions)

160 °C / 12 hrs (artificial age hardening conditions)

Figure 6. EBSD maps of longitudinal sections through selected specimens.

5 Conclusion

The objective of this study was to identify the effect of rotary swaging and subsequent artificial age hardening on mechanical properties and microstructure of the EN AW 6082 aluminium alloy. The experimental treatment route led to an increase in strength properties (YS, UTS, HV1), and to reduced ductility properties (A_5 and RA).

Artificial age hardening at 120 °C resulted in decreasing hardness on the transverse section from the core towards the surface. This is probably due to recrystallized grains in the surface region of the bar. On the other hand, this decrease was not seen in the material treated at 160 °C, which is why this effect will be studied with respect to this temperature.

Optical and electron microscopy were insufficient for revealing changes in the microstructure. The morphologies of grains were similar in all specimens. Hence, one can assume that the key strengthening mechanism which led to differences in the mechanical properties of specimens upon artificial age hardening was precipitation strengthening.

Transmission electron microscopy should be able to reveal the associated changes. It will be part of the follow-up study on this material.

Acknowledgement
This paper is one of the outcomes of the project entitled Development of West-Bohemian Centre of Materials and Metallurgy No.: LO1412, which is funded by the Ministry of Education of the Czech Republic.

References

[1] Azusjima A, Kopp R 2008 Severe plastic deformation (SPD) processes for metals. *CIRP Annal - manufacturing Technology.* **57** 716-735.
[2] Valiev R, Langdon T 2011 Achieving exceptional grain refinement through severe plastic deformation: New approaches for improving the processing technology. *Metallurgical and Materials Transactions A.* **42** 2942-2951.
[3] Michna Š, Lukáč I et al. 2005 Encyklopedie hliníku (Prešov: Adin, s.r.o.) p 722
[4] Fujda M, Kvačkaj T et al. 2008 Improvement of mechanical properties for EN AW 6082 aluminium alloy using equal-channel angular pressing (ECAP) and post-ECAP aging. *Journal of Metals, Materials and Minerals.* **18** 81-87.
[5] Lacková P, Buršák M et al. 2015 Mechanical and corrosion properties of aluminium alloy EN AW 6082 after severe plastic deformation (SPD). *Metalurgija.* **54** 181-183.
[6] Kvačkaj T, Fujda M et al. 2015 Ultra fine microstructure and properties formation of EN AW 6082 alloy. *JSASS Space Tech. Japan.* **7** 85-91.
[7] Roven H J et al. 2005 Mechanical properties of aluminium alloys processed by SPD: comparison of different alloy systems and possible product areas. *Materials Science and Engineering: A.* **410-411** 426-429.
[8] Cabibbo M et al. 2005 Influence of severe plastic deformations on secondary phase precipitation in a 6082 Al-Mg-Si alloy. *Metallurgical and Materials Transactions A.* **36** 1353-1364.
[9] Baig M et al. 2005 Thermo-mechanical responses of an aluminum alloy processed by equal channel angular pressing. *Materials & Design.* **57** 510-519.
[10] Angella G et al. 2004 Aging behaviour and mechanical properties of a solution treated and ECAP processed 6082 alloy. *Materials Transactions.* **45** 2282-2287.
[11] Angella G et al. 2004 Intermetallic particle evolution during ECAP processing of a 6082 alloy. *Materials Transactions.* **45** 2182-2186.
[12] Dadbakhsh S et al. 2010 Strengthening study on 6082 Al alloy after combination of aging treatment and ECAP process. *Materials Science and Engineering A.* **527** 4758-4766.
[13] Fujda M, Sláma P et al. 2013 Thermal stability of the ultrafine grained EN AW 6082 aluminium alloy. *Kovove Materialy (Metallic materials).* **51** 117-122.
[14] Kovařík T, Zrník J, Cieslar M 2010 Proceedings of the conference METAL 2015 (Roznov pod Radhostem) vol 19 (Ostrava: TANGER Ltd.) p 517
[15] Kubina T, Köver M et al. 2015 Link between technological parameters and material properties of EN 6082 processed by ECAP-CONFORM and precipitation hardening. *International Journal of Material and Mechanical Engineering.* **4** 36-43.
[16] Abdulstaar M A et al. 2015 Severe plastic deformation of commercial purity aluminum by rotary swaging: Microstructure evolution and mechanical properties. *Materials Science and Engineering A.* **565** 351-358.
[17] Köver M, Sláma P et al. 2015 Impact of rotary swaging and age hardening on mechanical properties of EN AW 2024. *Journal of Achievements in Materials and Manufacturing Engineering.* **72** 61-66.
[18] Rotary swaging principle [online]. Available on: http://www.hmp.de/en/products/rotary-swaging/index.html, [cit. 2016-08-15].

IOP Conf. Series: Materials Science and Engineering **179** (2017) 012050 doi:10.1088/1757-899X/179/1/012050

Processing of bulk Al7075 alloy by spark plasma sintering

P Málek[1], O Molnárová[1], J Cinert[2,3], F Lukáč[2], T Chráska[2]

[1]Charles University, Faculty of Mathematics and Physics, Department of Physics of Materials, Prague, Czech Republic, EU
[2]Institute of Plasma Physics of the CAS, Prague, Czech republic, EU
[3]Czech Technical University in Prague, Faculty of Electrical Engineering, Department of Electrotechnology, Czech Republic, EU

E-mail: malek@met.mff.cuni.cz

Abstract. The main advantages of powder metallurgy processing route are the possibility to produce near-net-shape compacts and to minimize the finish machining and material loss. The main problem in particle consolidation process is to suppress porosity, to remove oxide layers, and to retain the microstructure of powder materials. Spark plasma sintering (SPS) combines concurrent uniaxial pressure and direct heating by a pulsed DC current. Sintering occurs at relatively low temperatures for a short time and does not influence significantly the microstructure in the interiors of original powder particles. The efficiency of SPS in producing compacts with low porosity might be dependent on the distribution of particle size in original powder material. The gas atomized Al7075 powder was sieved to several charges and then sintered by SPS. Microstructure of sintered compacts was studied by light and scanning electron microscopy. The phase composition was investigated using X-ray diffraction. The mechanical behaviour was tested by bending tests.

1 Introduction

Powder metallurgy (PM) represents a very effective tool for processing metallic materials that can achieve properties superior to their ingot metallurgical counterparts. Atomization techniques make possible to produce powder particles at cooling rates up to 10^5 Ks^{-1} which ensures the formation of materials far from thermodynamic equilibrium. The content of dissolved alloying elements might be enhanced significantly above the equilibrium solid solubility limit [1] and the microstructure of atomized powder particles is frequently fine-grained. The above mentioned aspects can improve especially the strength of PM materials.

Powder consolidation represents the main problem in the PM processing route. The current consolidation methods (e. g. high isostatic pressing) require generally a high temperature and a long time of sintering. During this period, the material tends to reach its thermodynamic equilibrium and many gains from atomization can be lost. The spark plasma sintering/field assisted sintering technology (SPS/FAST) [2] combines applied uniaxial pressure with heating by low voltage pulsed DC current flowing through the sample. High current density and large Joule heat can be evolved at contact points of powder particles where the temperature can highly exceed the set one. Sintering occurs at these points and the interiors of powder particles are nearly unaffected and retain the microstructure formed during atomization. This short time of exposition to elevated temperatures helps to avoid undesirable processes like recrystallization or grain growth. It is well known that the density and microstructure of sintered compacts can be influenced by many parameters of SPS (applied stress, heating rate, holding time, sintering temperature) [3]. On the contrary, there are scarce data on the influence of powder particle size distribution which can play an important role especially at the beginning of sintering during re-arrangement of powder particles.

The commercial Al7075 alloy was chosen for our experiments. This alloy is a typical precipitation strengthened material which derives its high strength from the precipitation sequence occurring either at room temperature (natural ageing) or at elevated temperatures (artificial ageing) after solution treatment. Our recent research has confirmed that fully dense compacts can be prepared from gas atomized powders using spark plasma sintering [4]. A bimodal microstructure with numerous small grains located along boundaries of original powder particles was observed after sintering [5]. There are two possible explanations how such bimodal microstructure can be formed:

- The small powder particles present in the gas atomized powder are re-arranged during the initial stage of sintering
- New small grains are formed during sintering at contact points between original powder particles where a high current density can result in a local melting and following solidification of the material

To solve this problem, the original gas atomized powder with a broad distribution of particle sizes was sieved to several charges and these charges were sintered separately. The microstructure, phase composition, porosity level, and mechanical properties were studied on these samples.

2 Experimentals

The nitrogen atomized Al7075 powder was supplied by Nanoval GmbH & Co. KG, Berlin. The powder was sieved into 5 charges (< 20 μm, 20 – 25 μm, 25 – 32 μm, 32 – 40 μm, and 40 – 50 μm) in an aqueous environment. The powders were sintered using pulsed electric current sintering system – model SPS 10-4 (Thermal technology, USA) at 425 °C under the pressure of 100 MPa for 4 minutes. The phase composition was verified by X-ray diffraction (XRD) using a diffractometer D8 Discover (Bruker AXS). Quantitative Rietveld analysis was performed to estimate the weight fraction of identified phases.

The microstructure of sintered compacts was studied by scanning electron microscopy (SEM) using microscopes EVO MA 15 (Carl Zeiss SMT) and FEI Quanta 200F. Image analysis was performed for porosity estimation. The chemical composition was determined by energy dispersive spectroscopy (EDS) and is given in table 1. The arrangement of grains in sintered compacts was verified by light microscopy (LM) using microscope Olympus IX70. The three-point bending tests were performed at room temperature using Instron 1362 machine.

Table 1. The chemical composition determined by from energy dispersive spectroscopy.

Element	Zn	Mg	Cu	Al
Wt. %	6.2	2.7	1.8	Balance

3 Experimental results and discussion

Figure 1 shows the SEM micrograph of powders of the finest (< 20 μm) and coarsest (40 – 50 μm) charges, respectively. The typical feature of the finest charge is a tendency to agglomeration of powder particles to objects of the size approaching nearly 1 mm. The coarsest charge contains not only powder particles exceeding 40 μm but also numerous very fine particles. The distribution of particle sizes is given in figure 2.

The phase composition of all compacts is given in figure 3. The compact prepared from the finest charge exhibits only Al matrix diffraction peaks. The absence of diffraction peaks from intermetallic phases might reflect the segregation free microstructure of the finest powder particles due to very rapid solidification. Some undulations of the diffraction curve can be explained by the presence of aluminium oxide hydroxide. This phase can be formed during sieving process occurring in water. All other compacts contain the $MgZn_2$ phase (~ 3 vol. %) and also the Al_2CuMg phase (~ 1 vol. %).

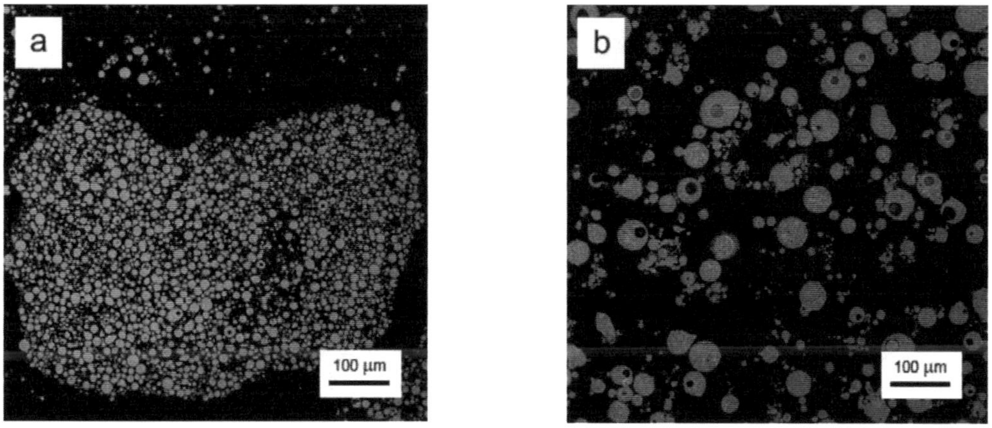

Figure 1. SEM micrographs of sieved powders, charge < 20 µm (a), charge 40 - 50 µm (b).

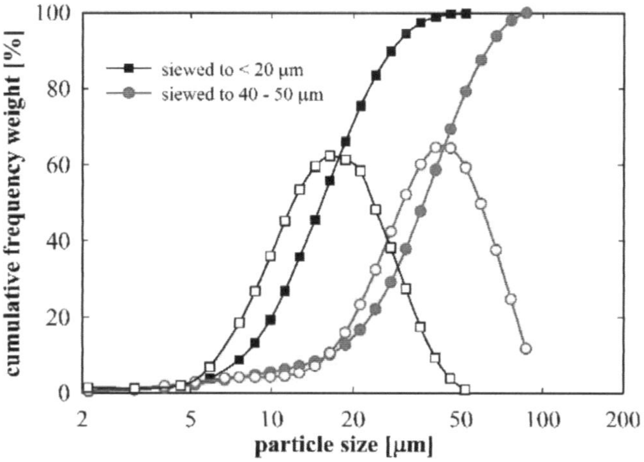

Figure 2. The distribution of particle size.

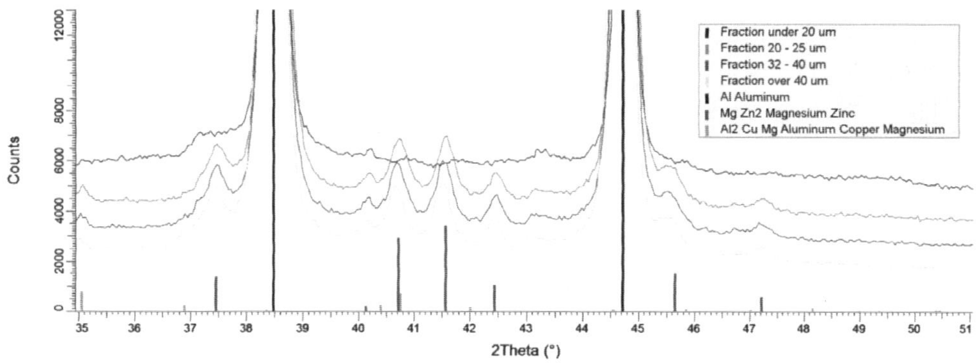

Figure 3. A part of the XRD pattern for sintered compacts.

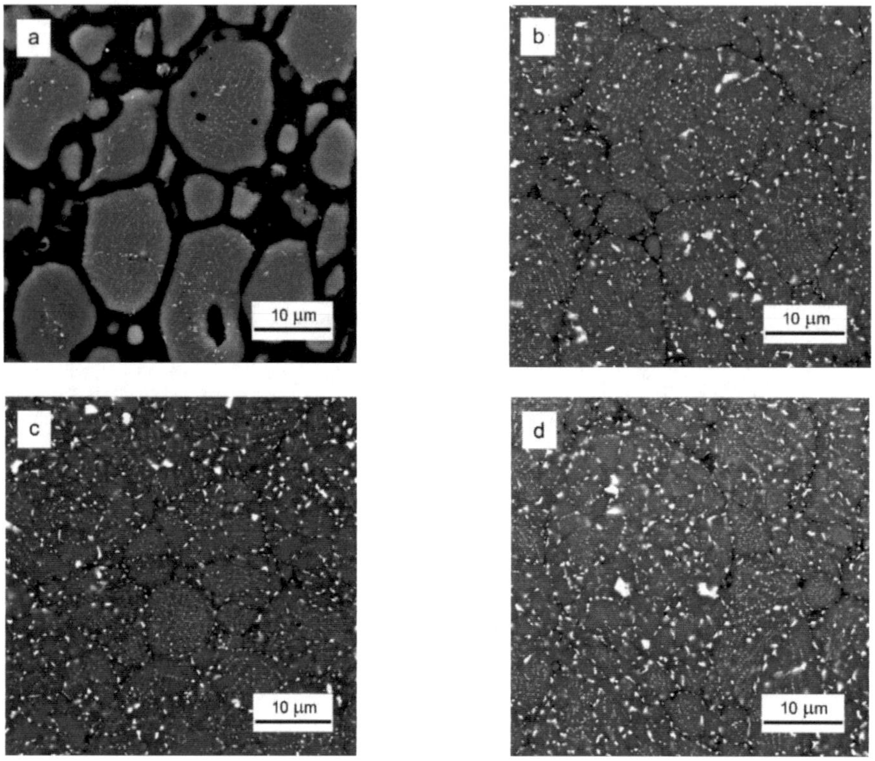

Figure 4. The microstructure of sintered compacts: charge < 20 µm (a), charge 20- 25 µm (b), charge 25 – 32 µm (c), charge 40 – 50 µm (d), back scattered electron contrast.

The microstructure of the compact prepared from the finest charge is completely different from all other compacts (figure 4). The corresponding micrograph reveals that nearly no necks between original powder particles were formed during SPS. The measurement of porosity revealed the volume fraction of pores exceeding 10 %. No coarse precipitates were observed in the interior of powder particles and the smallest ones seem to be even free of precipitates. This observation agrees well with the result of XRD analysis. The extent of porosity decreases with increasing size of powder particles. The compact prepared from the coarsest charge exhibits the volume fraction of pores not exceeding 1 %. Two types of precipitates can be distinguished - coarser precipitates are located especially at the boundaries between original powder particles and finer precipitates are located in the interior of original powder particles, either along cell boundaries or homogeneously in the cell interiors.

The EDS analysis was performed to find the distribution of individual elements within the sintered compacts. Figure 5 documents an abnormally enhanced amount of oxygen at places between original powder particles in the finest compact (exceeding even 10 wt. %). These places are also rich in Mg. The distribution of other alloying elements (Zn, Cu) is homogeneous within the measurement accuracy. Much lower oxygen content was found in the coarsest compact (up to 0.5 wt. %). For comparison, the oxygen content below 0.2 wt. % was found in the compact prepared from non-sieved powder. The increased oxygen content in sieved materials can result from sieving process performed in aqueous environment. An enhanced Mg content was observed both at boundaries and in precipitates. Cu was found especially in the coarsest precipitates (see arrows), Zn mostly in finer precipitates. It can be concluded on the base of XRD analysis that coarse precipitates are probably formed by the Al_2CuMg phase and the finer, more homogeneously distributed, precipitates are formed preferentially by the $MgZn_2$ phase.

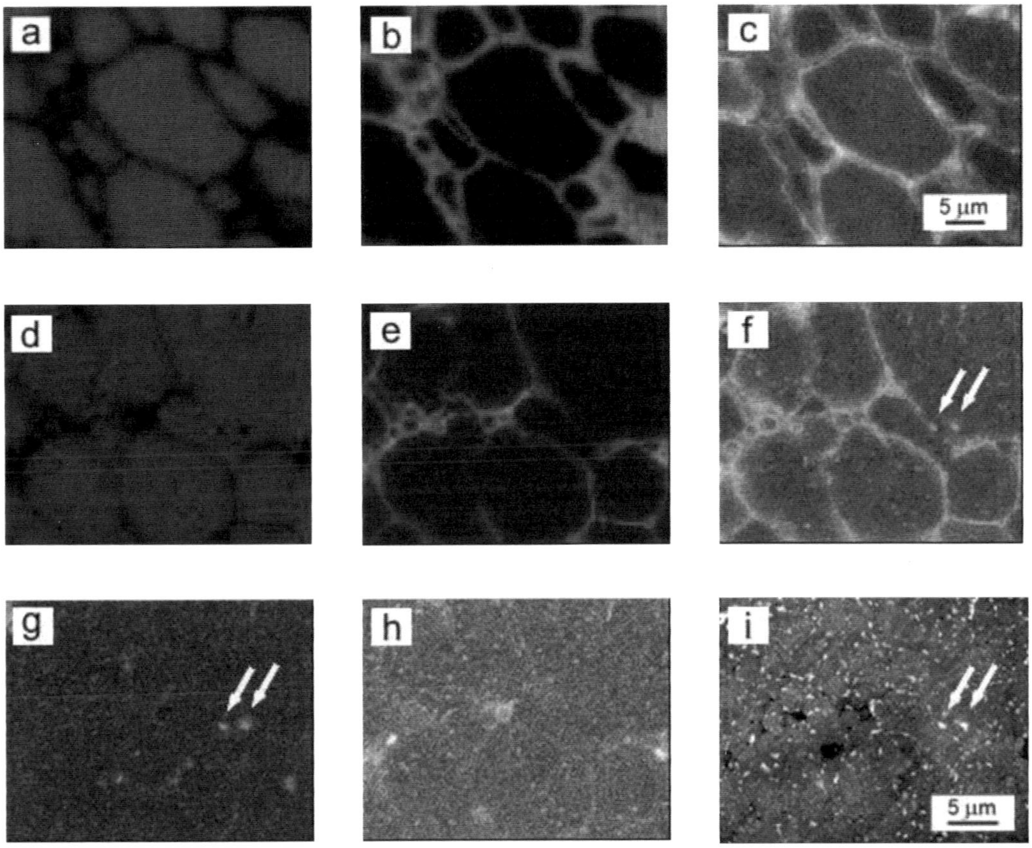

Figure 5. Distribution of Al (a), O (b), and Mg (c) in the finest charge, distribution of Al (d), O (e), Mg (f), Cu (g), and Zn (h) in the coarsest charge, corresponding secondary electron image in (i).

Figure 6 shows the typical microstructure of the compact from the coarsest charge. At some places a perfect contact was formed immediately between large powder particles, at other places colonies of smaller grains were observed. SEM micrograph in figure 4b documented presence of imperfect contacts at these places. It can be therefore concluded that these small grains are small powder particles re-arranged during the initial stage of sintering [6] and not new grains formed from the melt during sintering.

Three-point bending tests at room temperature were used to estimate the ability of plastic deformation of sintered compacts. Whereas the compact prepared from non-sieved powder (containing a broad distribution of particle sizes) revealed plastic deformation, all compacts prepared from sieved powders exhibited fracture already within the region of elastic deformation. A tendency to increasing fracture stress with increasing powder particle size was observed. The explanation can be found in the increased oxide hydroxide phases in sieved materials. Simultaneously, the presence of powder particles with different sizes in non-sieved material improved the re-arrangement of powder particles during sintering and, therefore, less volume fraction of pores.

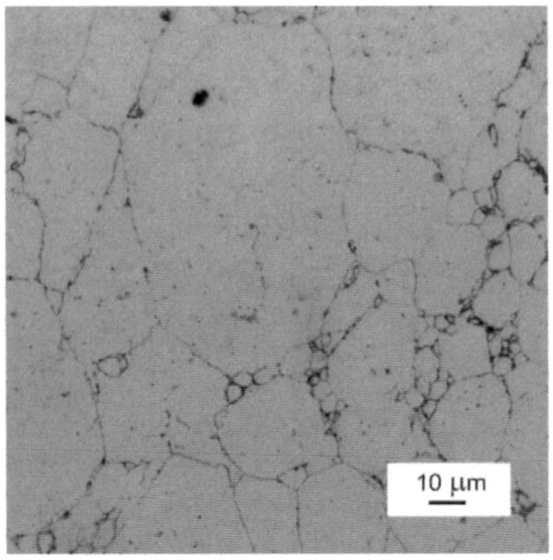

Figure 6. Microstructure of the compact prepared from the coarsest charge, LM.

4 Conclusions

Powder particle size influences the phase composition of sintered Al7075 compacts. Higher solidification rate in small particles results in a segregation free microstructure, coarser particles contain numerous precipitates of the $MgZn_2$ and Al_2CuMg phases. Sieving of the powder material to different charges in aqueous environment supports formation of oxide hydroxide layers along boundaries of original powder particles, especially in the finest material. The compact prepared from the finest charge contains the highest volume fraction of this phase and exhibits the worst ability to form necks and also the worst deformation behaviour in bending tests. On the contrary, broader distribution of powder particle sizes supports sintering and leads to the best plasticity of sintered compacts. Re-arrangement of powder particles of different sizes plays an important role in the sintering process.

Acknowledgements

This work was financially supported by the grant GACR 15-15609S. One author O. M. is grateful for financial support from the grant SVV-2016-260213.

References

[1] Jones H 1982 Rapid solidification of metals and alloys, Monograph 8 The institution of Metallurgists

[2] Orru R, Licheri R, Locci A M, Cincotti A and Cao G 2009 Consolidation/synthesis of materials by electric current activated/assisted sintering *Mater. Sci. Eng. R* **63** 127-287

[3] Sweet G A, Brochu M, Hexemer Jr R L, Donaldson I W and Bishop D P 2014 Microstructure and mechanical properties of air atomized aluminium powder consolidated via spark plasma sintering *Mater. Sci. Eng. A* **608** 273-282

[4] Becker H, Dopita M, Stráská J, Málek P, Vilémová M and Rafaja D 2015 Microstructure and properties of spark plasma sintered Al-Zn-Mg-Cu alloy *Acta Physica Polonica A* **128** 602-605

[5] Molnárová O, Málek P and Becker H 2015 The investigation of the Al7075 + 1 wt. % Zr alloy prepared using spark plasma sintering technology, Proc. Conf. Metal 2015 1221-1226

[6] Devaraj S, Sankaran S and Kumar R 2013 Influence of spark plasma sintering temperature on the densification, microstructure and mechanical properties of Al-4.5 wt.%_Cu alloy *Acta Metall Sin.* **26** 761-771

IOP Conf. Series: Materials Science and Engineering **179** (2017) 012051 doi:10.1088/1757-899X/179/1/012051

FEM model and experimental measurement of clinched joint

P Malý, F Lopot and J Sojka

Czech Technical University in Prague, Faculty of Mechanical Engineering, Department of Designing and Machine Components, Technická 4, 166 07 Prague 6, Czech Republic

E-mail: pavel.maly@fs.cvut.cz

Abstract. This work describes creation of FE model of clinched joint and brings information about its verification with help of experiment. Also the sensitivity of the simulation model to the selected parameters (size of the model and coefficient of friction) and boundary conditions is studied so that the model is close to the reality. Deformation and uniaxial stress near the joint obtained with help of strain gauge is the output from the unidirectional, nondestructive test. Comparison of two values of stress in the chosen place obtained by experiment and FE simulation is the principle for verification of FE model and its adjustment. Based on the results it can be said that the FE model corresponds with the reality with the desired accuracy.

1 Introduction

Clinched joining of sheet-metal plates is a technology permanently applied in many branches of production thanks to many advantages that are e.g. high productivity, outstanding quality and its repeatability or possibility to connect sheet-metal plates with surface treatment. It is the reason why this type of joining material is studied in many articles where the process or mechanical properties are described. More information can be found e.g. in [1], simulation or experimental methods are mentioned in [2], [3], [4] and [5]. The other general example of application of numerical and experimental analysis is presented in [6]. This study concerns with the round clinched joints (figure 1) and describes the process of creation and study of FE model and its properties with usage of an experiment. This text brings information that is important for understanding of clinched joint properties from the simulation approach point of view.

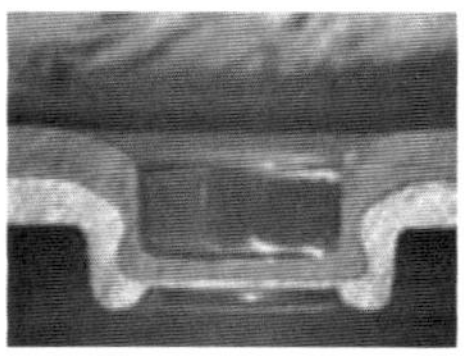

Figure 1. Round clinched joint, [1].

2 Method

The work is solved in two related approaches: FE model of the clinched joint as a first approach is designed to correspond with the second approach, with the experiment.

2.1 Experiment
The set of 5 specimens was subjected to the experimental measurement to get the information about behavior of the joint under the real loading in the perpendicular direction to the axis of the joint.

2.1.1 Measurement protocol and instrumentation. Each specimen was subjected to the uniaxial, unidirectional, nondestructive test. The controlled variable was tension force acting on the specimen and smoothly increasing from 0 N up to 3000 N. The measurement was performed on the special single purpose testing stand that was set up for the experiment, the specimens were clamped in jaws, loading was applied with a hydraulic system and force was measured directly on the jaws with the force sensor HBM S9M/5kN.

2.1.2 Investigated set of specimens. The specimen consists of two sheet-metal plates from low carbon steel DX51D with cross-section 40×3 mm and length 280 mm connected in the middle by a round joint with characteristic diameter of 10 mm. The deformation near the joint on the upper sheet-metal plate in the direction of loading was measured using wire strain gauge from HBM – see figure 2.

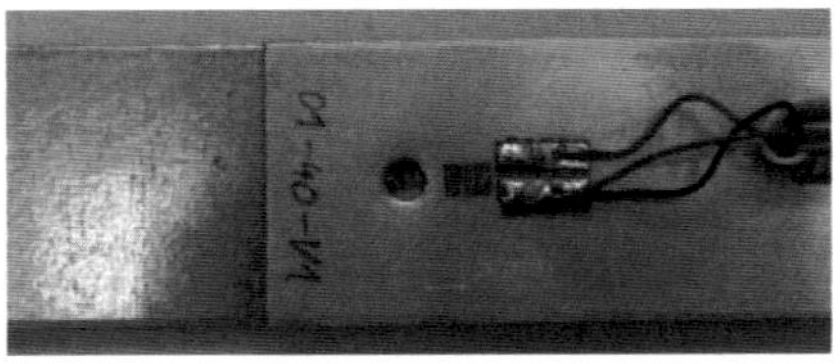

Figure 2. Strain gauge placement near the joint.

2.1.3 Data acquisition and processing. The test was controlled and the data were measured by National Instrument apparatus with LabView software. Data were processed by Matlab software.

2.1.4 Experiment outcomes. The resultant stress near the clinched joint in the place of the strain gauge for the loading force of 3000 N was obtained. It was compressive stress with the value of -29.5±2.6 N/mm^2 (average ± standard deviation).

2.2 Finite element modelling
Simulation model of the clinched joint is created as a deformable 3D model that consist of two plates connected by the joint. The clinched joint, i.e. interlock of upper and bottom sheet-metal plates, is created with help of curves with different radius. The shape is measured from the macrosection of the clinched joint with characteristic diameter of 10 mm (figure 3).

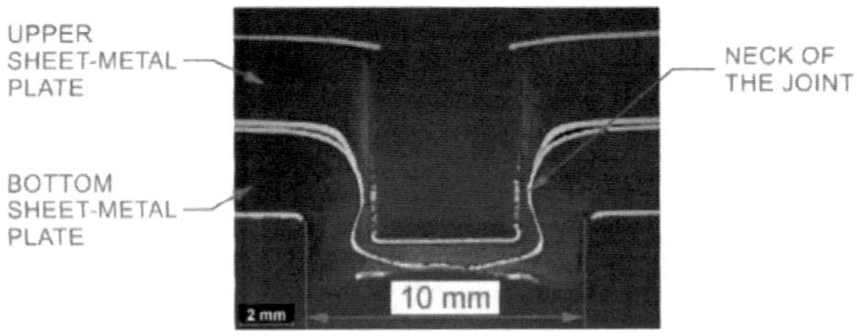

Figure 3. Macrosections of the clinched joints.

2.2.1 Model setup. The whole simulation model can be seen in figure 4. The symmetry of the model in the plane XY, i.e. the half of the model, is applied in the simulations. The dimensions of each plate are: length L = 75 mm or 280 mm, width of the half of the model W = 40 mm and the thickness T = 3 mm. The boundary conditions were applied in the reference points (RP-1 and RP-2) that are placed at the end of each plate in the middle of the thickness. The reference points are connected to the adjacent surfaces width length of 20 mm using continuum distributing couplings. The used material is steel with isotropic elastic behavior (Young's modulus 210 GPa, Poisson's ratio 0.3), the deformations are expected to be small, therefore nonlinear effects are not assumed in the calculation. Parts are meshed using 8-node linear bricks with reduced integration (C3D8R).

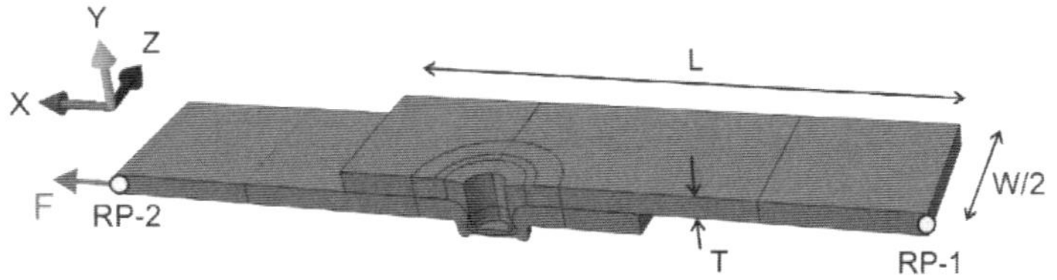

Figure 4. Whole model for simulation.

2.2.2 Loading. The applied force F is in X direction and has the value of 3000 N which corresponds to the load and conditions of experimental testing. The force is much lower than the limited force (shear strength) of the joint which is approximately 4000 N for the tested type and size of clinched joint, therefore no plastic deformation is assumed.

2.2.3 Joint model. The model of the clinched joint, i.e. the interlock between the sheet-metal plates, is created using tie connection in the bottom part of the joint. Standard hard surface-to-surface contact with tangential behavior (frictionless, or penalty friction with friction coefficient *f*) is used for the contact of other surfaces, see figure 5.

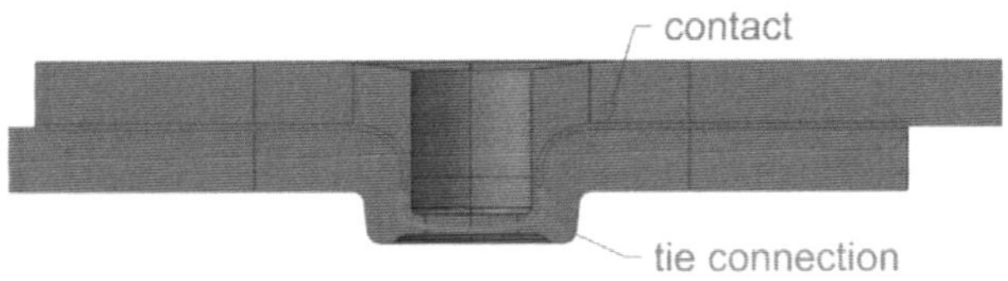

Figure 5. Model of the clinched joint.

2.2.4 Simulations. Set of ten simulations of the connection with different boundary conditions and parameters (length of the plate and friction) were done so that the model is analyzed and its robustness is verified. The simulations and the parameters are summarized in the table 1. The notation is used from Abaqus/CAE software.

Table 1. Simulation models and parameters

Simulations	Boundary condition RP-1	Boundary condition RP-2	Length L [mm]	Friction f [-]
C1	encastre	U2, U3, UR1, UR2, UR3	75	frictionless
C2	encastre	U2, U3, UR1, UR2, UR3	75	0.1
C3	encastre	U2, U3, UR1, UR2, UR3	75	0.2
C4	encastre	U2, U3, UR1, UR2	75	frictionless
C5	U1, U2, U3, UR1, UR2	U2, U3, UR1, UR2	75	frictionless
C6	encastre	U2, U3, UR1, UR2, UR3	280	frictionless
C7	encastre	U2, U3, UR1, UR2	280	frictionless
C8	U1, U2, U3, UR1, UR2	U2, U3, UR1, UR2	280	frictionless
C9	encastre	U2, U3, UR1, UR2, UR3	280	0.1
C10	encastre	U2, U3, UR1, UR2, UR3	280	0.2

Boundary condition in the reference point RP-1 is either encastre for the first case, or the displacement is restricted in all directions (X, Y, Z) and rotation is restricted only in the axes X and Y for the second case. The force F is applied in the reference point RP-2 hence the displacement in X direction is enabled for all simulations.

Other displacements and rotations in the point RP-2 are restricted for different simulation according to the table 1. Two sizes of the plates were used in the model, length L = 75 mm and L = 280 mm, the last one corresponds to the size of real testing specimens. The influence of the friction was verified by simulations for three cases, the frictionless contacts and penalty contact with constant friction coefficient of value 0.1 and 0.2.

3 Results

The resultant stress in X direction (S11) on the surface of the upper plate is chosen for analyzes and comparison of the joint model simulations. This direction corresponds to the direction of loading and the direction of the real stress measured by the strain gauge. The course of this stress S11 along the curve near the joint can be plotted and its value in the place of interest, i.e. the position of the strain gauge, is obtained. This value is used for evaluation and comparison of the simulations and verification with experimental measurement. The set of nodes and elements along the curve where the values of stress S11 are obtained can be seen in figure 6.

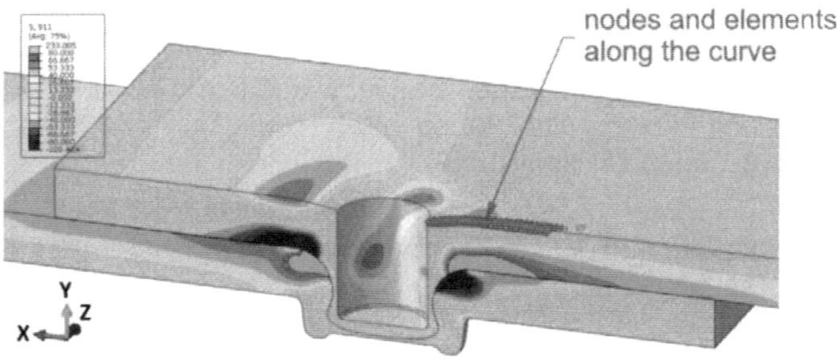

Figure 6. Nodes and elements along the curve for the stress S11

The example of the S11 development along the curve can be seen in figure 7. The position is measured from the edge of the hole of the joint on the upper plate as is shown for strain gauge position (figure 2).

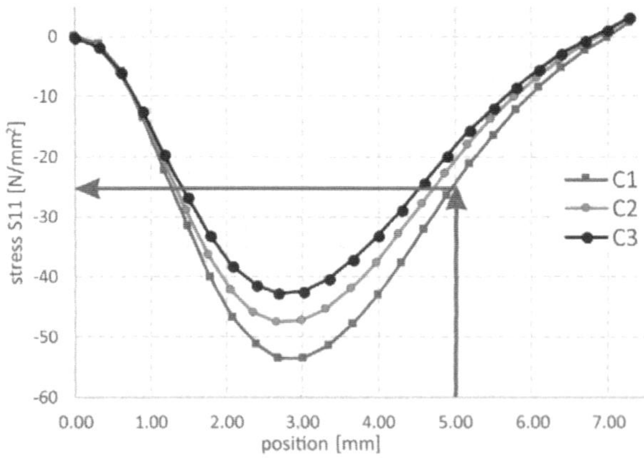

Figure 7. Development of S11 stress along the curve near the joint.

The calculated values of S11 for the all simulations in the place of interest (in the distance of 5 mm from the joint) are summarized in table 2.

Table 2. Result stress S11 [N/mm^2] in the position of the interest.

Simulation	C1	C2	C3	C4	C5	C6	C7	C8	C9	C10
S11 [N/mm^2]	-25	-21	-18	-22	-28	-32	-28	-34	-28	-24

The results of simulations using different plate size are summarized in table 3 where each pair of simulations (C1 vs. C6, C4 vs. C7 and C5 vs. C8) has the same boundary condition. It is obvious that the difference of the result stress is between 6 and 7 N/mm^2 which is approximately 20 %.

Table 3. Result stress S11 [N/mm^2] for different size of the plates.

	C1	C4	C5	C6	C7	C8	difference
Parameter: size, (C1 vs. C6)	-25			-32			7
Parameter: size, (C4 vs. C7)		-22			-28		6
Parameter: size, (C5 vs. C8)			-28			-34	6

Results of simulations with different friction are in table 4. All these simulations were performed with the same boundary conditions, the difference is assumed between adjacent values of friction (frictionless, f = 0.1 and f = 0.2) and has the value 3 and 4 N/mm^2 which is approximately 15 %.

Table 4. Result stress S11 [N/mm^2] for different friction.

	C1	C2	C3	C6	C9	C10	difference	
Parameter: friction, (C1 vs. C2 vs. C3)	-25	-21	-18				4	3
Parameter: friction, (C6 vs. C9 vs. C10)				-32	-28	-24	4	4

The influence of different boundary conditions for two lengths of the plates can be described by the results that are summarized in table 5. Comparison of results are focused on the size of plates with the length L = 280 mm because the simulation model is the same as the testing specimens. The difference between simulations C6 and C7 is 4 N/mm^2 (approx. 13 %), when the rotation in the place of acting force (RP-2) in axis Z is enabled for simulation C7. When the rotation in axis Z is enabled in the place of encastre (RP-1) in simulation C8, the difference between simulation C7 and C8 is 6 N/mm^2 (approx. 18 %). Thus, the change of boundary condition in RP-1 has greater influence than change of boundary condition in the place of acting force.

Table 5. Result stress S11 for different boundary conditions.

	C1	C4	C5	C6	C7	C8
Parameter: boundary conditions (C1vs. C4 vs. C5)	-25	-22	-28			
Parameter: boundary condition (C6 vs. C7 vs. C8)				-32	-28	-34

4 Conclusion

The FE model of clinched joint was created and analyzed using set of ten simulations. With respect to the application of the model, each simulation differs in chosen input parameters, i.e. size of the plate, friction and boundary conditions. Thus, the robustness and the sensitivity of the simulation model to the range of the input parameters were verified.

The change of the plate size influences the simulation quite significantly, so it is important to have the same dimensions of the testing specimen and simulation model. It can be said that the friction coefficient between surfaces in contacts is also important especially when the range of f is large. However, range of friction coefficient is often smaller than the one that was simulated. The results imply the necessity to choose the boundary conditions that correspond to the real conditions which can be often very difficult.
The verification of the FE model with the reality was done with help of experiment. The set of 5 specimens was subjected to the unidirectional, nondestructive test and the stress was obtained with help of strain gauge. Comparing of measured stress (-29.5±2.6 N/mm^2) and stress from simulations (-32 N/mm^2 for simulation C6 and -28 N/mm^2 for simulation C9) showed that the selected FE models correspond with the reality.

References

[1] TOX®-Product Range [online]. [cit. 2016-09-20]. Available on http://www.tox-de.com/assets/countries/EN/pdf/TOX_Product_Range_00_en.pdf
[2] Coppieters S 2012 Experimental and numerical study of clinched connections. PhD Thesis, Katholieke Universiteit Leuven – Faculty of Engineering (Leuven, Belgium)
[3] Hamel V, Roelandt J M, Gacel J N and Schmit F 2000 Finite element modeling of clinch forming with automatic remeshing *Computers and Structures* vol 77 issue 2 pp 185-200 http://dx.doi.org/10.1016/S0045-7949(99)00207-2
[4] Oudjene M and Ben-Ayed L 2008 On the parametrical study of clinch joining of metallic sheets using the Taguchi method *Engineering Structures* vol 30 issue 6 pp 1782–1788 http://dx.doi.org/10.1016/j.engstruct.2007.10.017
[5] Xiaocong H 2010 Recent development in finite element analysis of clinched joints *Int J Adv Manuf Technol* vol 48 issue 5 pp 607-612 http://dx.doi.org/10.1007/s00170-009-2306-2
[6] Semrád K, Draganová K 2016 Numerical-experimental analysis of the tensile properties of the composite wing hinge connection of the ultra-light sport aircraft Int. Conf. on Měření, diagnostika, spolehlivost palubních soustav letadel (Brno: Univerzita obrany) pp 101-8 ISBN 978-80-7231-377-8

A method for assessing the wear of sanding abrasive discs differing in the type of grains (SiC, Al2O3(95A))

P Młynarczyk[1, 2], S Spadło[1] and D Krajcarz[1]

[1] Kielce University of Technology, Division of Materials Science and Munitions Technology, Al. Tysiąclecia Państwa Polskiego 7, 25-314 Kielce, Poland

[2]piotrm@tu.kielce.pl

Abstract. This paper describes a method for studying the wear of abrasive mesh discs differing in the type of grains employed to sand plaster. The tests were conducted at a specially developed and constructed setup. The investigations involved analysing the influence of the basic input parameters (the size of the abrasive grains covering the disc surface, the force exerted by the plaster sample pressing against the abrasive disc and the no-load speed of the abrasive disc) on the efficiency of the sanding process. The paper also describes the structural details of the experimental setup.

1 Introduction

Sanding process is included in the "conventional" group of machining methods together with the turning, milling, drilling, etc. opposite to "non-traditional" or "non-conventional" group processes such as EDM[1,2], water jet cutting (AWJ)[3], laser cutting[4,5], microwelding[6,7] etc. The wear of abrasive discs used for plaster sanding [8,9] can be assessed by applying different forces to a rod-like plaster sample pushed through the feed sleeve. The force is dependent on the mass of the weights attached. Weights differing in mass can be attached interchangeably. A change in the force exerted by the plaster sample pressing against the abrasive disc causes a change in the level of the process intensity. The sanding process is also affected by the rotational speed of the abrasive wheel selected on the tool. The aim of this paper is to present the test method and analyse the results concerning the abrasive wear behaviour of mesh discs.

2 Description of the test stand

Figure 1 shows a diagram of the stand used in the study. The stand designed to test the wear of abrasive discs consisted of a sleeve attached firmly to the sander flange, which acted as a guard protecting the operator. The random orbit sander used for the testing was mounted tightly to the stand. Plaster samples [10-12] in the form of rods were placed in the sleeve and pushed by a pusher to touch the surface of the abrasive disc. The force exerted by the plaster sample was dependent on the mass of the weights attached to the cord (2, 4 or 6 N).

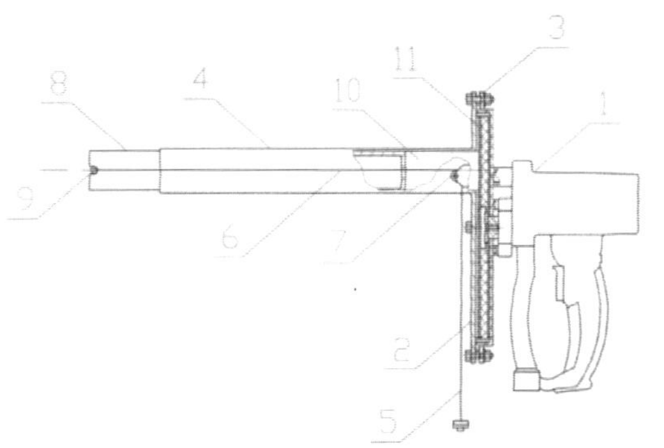

Figure 1. Stand to test the wear of abrasive discs: 1-random orbit sander, 2-feed sleeve flange,

3-mounting ring, 4-feed sleeve, 5-weight, 6-cord, 7-feed rolls, 8-pusher, 9-cord fitting system, 10-rod-like plaster sample, 11-abrasive disc.

3 Materials

The study was performed to compare two types of abrasive mesh discs 180 mm in diameter. The resin-bonded abrasive grains were silicon carbide (SiC)[13] and alundum (Al_2O_3(95A)). Figure 2 shows SEM images of the abrasive aggregates under study.

a)  b)

Figure 2. Abrasive grains a) silicon carbide, b) alundum.

4 Input parameters selected to study the wear of abrasive discs

The experiments were conducted using a specially designed and constructed test stand to assess the wear of abrasive discs. The input parameters were predetermined on the basis of the authors' earlier experiments, the literature data [9,14-16] and the information provided by the abrasive mesh producer.

The tests were carried out under the following conditions:

- no-load speed - 2250 rpm (average set value in the range of 1500-3000 rpm) measured for slow speeds,
- force exerted by the plaster sample, size of the abrasive grains – 120.

5 Analysis of the wear of abrasive discs

The experimental results were used to determine the abrasive wear behaviour of the mesh discs. The study involved assessing the efficiency of the sanding process as a function of time. It was necessary to determine the limit at which the process became less efficient. The measurement results are presented in Table 1.

Table 1. Measurement data.

Layer No.	Sanding time per layer [s]	
	SiC	*Al_2O_3(95A)*
1.	36	56
2.	75	127
3.	114	197
4.	179	225
5.	180	500
6.	301	-
7.	475	-

Each layer of gypsum plaster [17] removed from the sample had the same volume (19.6 cm^3). The sanding was continued until the volumetric efficiency of the process (Q_w) decreased to less than 0.05 cm^3/s.

The comparative analysis of the abrasive meshes showed that higher efficiency of the sanding process was achieved for the discs with SiC grains. It was less than 0.05 cm^3/s after the removal of seven layers, with a total volume of plaster removed being 137.2 cm^3. For the discs with Al_2O_3 (95A) grains, the sanding efficiency decreased below the assumed value after five layers (98 cm^3) were removed. The tests revealed that the efficiency of plaster sanding was higher for the discs with SiC aggregates than for the discs with Al_2O_3 (95A) aggregates.

Figure 3 is a graphical representation of the experimental data concerning the relationship between the efficiency of the sanding process and the sanding time per layer for four different abrasive discs.

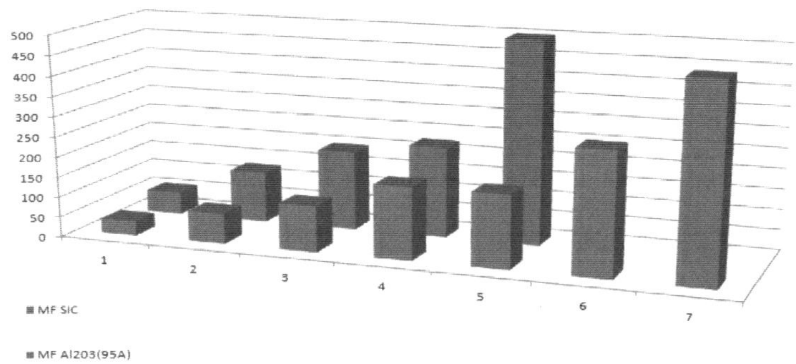

Figure 3. Efficiency of the sanding process expressed as the time of removal of a predetermined volume of plaster as a function of the sanding time per layer.

6 Influence of the input parameters on the process efficiency

The tests to determine the influence of the input parameters on the efficiency of the sanding process were conducted according to a 3x3 factorial design (with three factors and three levels), as proposed by Box Behnken [18-19]. In the tests, the three factors were:

- the size of the abrasive grains *(Z)*,
- the force exerted by the plaster sample pressing against the abrasive disc *(F)* expressed by newtons (*N*),
- the no-load rotational speed of the abrasive disc (*n*) expressed by revolutions per minute (*rpm*).

The design of experiments was generated using Statistica 10 for the ranges of variability of the particular input parameters determined in one-factor experiments. Table 2 summarises the actual values of the input parameters and the values of the sanding efficiency (Q_w) calculated from the experimental results.

Table 2. Design of experiments and the measurement data.

Test No.	Input parameters			Sanding efficiency SiC	Al_2O_3(95A)
	Z, grit size	*F*, N	*n*, rpm	Q_w, cm^3/min	Q_w, cm^3/min
1.	60	2	2250	117.75	111.08
2.	180	2	2250	90.58	79.56
3.	60	6	2250	255.98	163.54
4.	180	6	2250	133.81	125.27
5.	60	4	1500	103.29	99.79
6.	180	4	1500	79.56	71.80
7.	60	4	3000	327.08	452.88
8.	180	4	3000	203.02	346.32
9.	120	2	1500	60.08	62.63
10.	120	6	3000	85.33	122.66
11.	120	2	3000	122.66	255.98
12.	120	6	3000	235.50	235.50
13.	120	4	2250	154.93	136.92
14.	120	4	2250	159.12	140.18
15.	120	4	2250	147.19	127.99

7 Results and discussion

Table 3 compares the input parameters for the two types of abrasive grains. The strongest linear correlation was obtained for the parameter *n(L)*, which was $k_{SiC} = 0.71$ and $k_{Al2O3(95A)} = 0.80$ for SiC and Al_2O_3 (95A), respectively. In no other case did the value of the linear or quadratic regression reach the level of significance $k > |0.5|$.

Table 3. Coefficients of partial correlation.

Input parameter	Coefficient of correlation SiC	Al_2O_3 (95A)
Z(L)	-0.38	-0.18
Z(Q)	-0.18	-0.13
F(L)	0.41	0.12
F(Q)	0.21	0.24
n(L)	0.71	0.80
n(Q)	-0.01	-0.37

8 Conclusions

From the experimental results it was possible to determine the efficiency of the gypsum plaster sanding process performed by means of abrasive wheel tools, which was expressed as a function of time. The results showed that the process efficiency and the disc durability were higher for the discs with SiC grains.

The values of the coefficients of partial correlation obtained for the linear regression confirmed a significant influence of the rotational speed of the abrasive disc *n(L)* on the efficiency of the sanding process(positive correlation, $k = 0.71$ for the discs with SiC grains and $k = 0.80$ for the discs with Al2O3 (95A) grains).

The analysis of the values of the coefficients of partial correlation obtained for the linear regression showed that the efficiency of the sanding process was least dependent on the size of the abrasive grains *Z(L)* (negative correlation, $k = -0.18$) and the press load *F(L)* (positive correlation, $k = 0.12$) for the disc with Al2O3 (95A) abrasive aggregates.

References

[1] Spadło S., Młynarczyk P., Depczyński W. "Investigation of the Selected Properties of the Superficial Layer Alloying with the Tungsten Electrodes" Proceedings of 24th International Conference on Metallurgy and Materials, METAL 2015, pp: 863-867
[2] Spadło S., Młynarczyk P., Łakomiec K. "Influence of the of electrical discharge alloying methods on the surface quality of carbon steel", The International Journal of Advanced Manufacturing Technology, 2016, DOI 10.1007/s00170-016-9168-1
[3] Hlavac L. M., Spadło S., Krajcarz D., Hlavacova I. M. "Influence Traverse Speed on Surface Quality after Water-jet Cutting for Hardox Steel" Proceedings of 24th International Conference on Metallurgy and Materials, METAL 2015, pp: 723-728
[4] Grabas B, Tofil S, Napadłek W. "An Analysis Of Laser-Welded Nicr-Ir And Nicr-Pt Micro Joints On Spark Plug Electrodes In Biogas-Fuelled Engines" Archives of Metallurgy and Materials Volume: 61, Issue 2, 2016, pp: 267-287
[5] Kurp P., Mucha Z., Tofil S., Mulczyk K. "The Influence of Treatment Parameters on the Microstructure, Properties and Bend Angle of Laser Formed Construction Bars" Archives of Metallurgy and Materials Volume 61, Issue 2A, 2016, pp: 1151–1156
[6] Depczyński W., Spadło S., Młynarczyk P., Ziach E., Hepner P. "The Selected Properties of Porous Layers formed by Pulse Microwelding Technique" Proceedings of 24th International Conference on Metallurgy and Materials, METAL 2015, pp: 1087-1092
[7] Młynarczyk P., Spadło S., Depczyński W. "The Selected Properties of the Connection Superalloy Haynes H 230® using Microwelding" Proceedings of 24th International Conference on Metallurgy and Materials, METAL 2015, pp: 792-797
[8] Chłądzyński S., Pichniarczyk P., (2008) Tynki gipsowe – rodzaje, właściwości i ich trwałość, *Izolacje*. R.13, No. 2 pp 59-64.
[9] Idzikowski P., (2006) Gładzie i tynki gipsowe – wskazówki wykonawcze. *Izolacje*, No. 2 pp 60-62.
[10] Czerniki S., (2012) Gładzie gipsowe – teoria i praktyka, *Izolacje*, No. 6 pp 46-49.
[11] EN 13279-1:2004 Gypsum binders and gypsum plasters. Part 1: Definitions and requirements.
[12] Walawko W. (1998) Tynki gipsowe – nowa jakość ścian i sufitów, *Materiały Budowlane*, No. 10, pp 21-24.
[13] Weimer, A. W. (1997)Carbide, nitride, and boride materials synthesis and processing. *Springer*. p 115.
[14] Chłądzyński S. (2008), Spoiwa gipsowe w budownictwie,*Medium Dom Wydawniczy,* Warszawa
[15] Filipowski R., Marciniak M., (2000) *Techniki obróbki mechanicznej i erozyjnej*, Warszawa: Oficyna wydawnicza Politechniki Warszawskiej,.
[16] Pszczołowski W., Rosienkiewicz P. (1995) Obróbka ścierna narzędziami nasypowymi, WNT, p 448.
[17] EwaGlos, (2015), European Illustrated Glossary Of Conservation Terms For Wall Paintings And Architectural Surfaces.
[18] Kocabaş, Z (2001) An Application and interpretation of second order response surface model. Ankara Üniversitesi Tarım Bilimleri Dergisi. *Journal of Agricultural Sciences 7*, pp 121-128.
[19] Ferreira S L C,BrunsR E, Ferreira H S, MatosG D, DavidJ M, BrandãoG C, Silva E G P, PortugalL A, ReisP S, SouzaA S, Santos W N L (2007) Box-Behnken design: Analternative for the optimization of analyticalmethods,*AnalyticaChimica Acta*, 597(2), pp 179–186

Effects of the pouring temperature on the formation of the bonding zone between AZ91 and AlSi17 in the compound casting process

R Mola[1], T Bucki[1] and A Dziadoń[1]

[1]Kielce University of Technology, Kielce, Poland, EU

E-mail: rmola@tu.kielce.pl

Abstract. The compound casting process was used to join AZ91 magnesium alloy to AlSi17 aluminium alloy. Liquid AZ91 was poured onto a solid AlSi17 insert placed in a steel mould heated to 370 °C. The experimental results showed that the temperature of the AZ91 melt affected the formation of the bonding zone between the two alloys. A continuous bonding zone was formed by applying a pouring temperature of 650 °C. The use of higher temperatures, i.e. 680 °C and 700 °C, did not lead to the formation of a continuous metallurgical transition zone at the AZ91/AlSi17 interface. The bonding zone was analysed using an optical microscope and a scanning electron microscope equipped with an energy dispersive X-ray (EDS) detector. The structural constituents of the bonding zone near the AlSi17 alloy were: an Al_3Mg_2 intermetallic phase, primary Si particles surrounded by a rim of an Mg_2Si intermetallic phase and fine Mg_2Si particles. The area of the bonding zone that was adjacent to the AZ91 alloy had a eutectic structure composed of an $Mg_{17}Al_{12}$ intermetallic phase and a solid solution of Al and Si in Mg.

1 Introduction

Compound casting is an economical manufacturing process employed to produce bimetal parts. It involves casting one type of metal or alloy, in a molten state, directly onto another metal or alloy, which is solid. The diffusion process that takes part during casting contributes to the formation of a continuous metallurgical bonding zone between the two metal materials. In recent years, much research has been conducted to use compound casting to join dissimilar metal materials, for instance, cast steel to chromium cast iron [1], steel to cast iron [2-4], steel to Al [5], Al to Cu [6], Al alloy to Al [7], Mg alloy to Mg alloy [8] and Al to Mg [9,10].

Because of the increasing needs of the automotive industry for lightweight metal alloys and composites, the fabrication of Mg/Al bimetals seems to be a promising solution. These materials have unique properties, which make them suitable for specific applications. Mg/Al compound materials can be produced by twin-roll casting [11], hot rolling [12, 13], extrusion [14], explosive cladding [15], diffusion bonding [16] and, as mentioned above, by compound casting. The compound casting technique has proved to be well-suited for fabricating Mg/Al items with intricate shapes.

This paper analyses the application of the compound casting method to join AZ91 magnesium cast alloy to AlSi17 aluminium cast alloy. It focuses on the influence of the pouring temperature on the formation of the boding zone between the two alloys. It also discusses the microstructure of the AZ91/ AlSi17 interface.

2 Experiment procedure

Molten AZ91 magnesium alloy and solid AlSi17 aluminium alloy were used in the compound casting process. The AlSi17 was in the form of cylindrical inserts with a diameter of 30 mm and a thickness of 5 mm cut from an ingot. The inserts were ground using up to 800 grit SiC paper and then cleaned with ethanol. The casting process began by placing an AlSi17 insert at the bottom of a steel mould and both were heated up to 370 °C. In the meantime, AZ91 was melted in pure argon up to 650 °C, 680 °C or 700 °C and then cast onto the AlSi17 insert under normal atmospheric conditions.

After the casting process was completed, the AZ91/AlSi17 bimetal specimens were cut and prepared for metallographic observations using a STRUERS automatic polishing machine. Polishing was performed with colloidal silica. The microstructure of the bonding zone was observed by means of a Nikon ECLIPSE MA 200 optical microscope and a JEOL JSM-5400 scanning electron microscope. The chemical composition of the interfacial area was studied using an X-ray energy dispersive spectrometer (EDS). The phase composition of the bonding zone was determined on the basis of the EDS quantitative results and the phase diagrams for Mg-Al [17] and Mg-Al-Si [18].

3 Results and discussion

The analysis was conducted for AZ91/AlSi17 specimens produced at three different temperatures. From the results it is clear that a pouring temperature of 650 °C led to the formation of a continuous, uniform and defect-free bonding zone at the AZ91/AlSi17 interface (see figure 1).

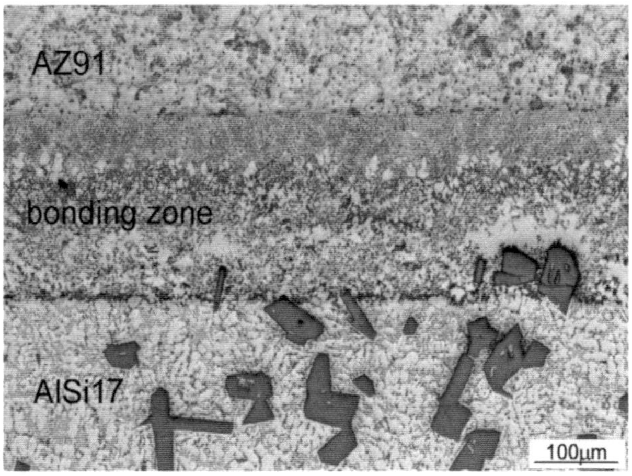

Figure 1. Optical micrograph of the bonding zone for the AZ91/AlSi17 bimetal specimens fabricated by compound casting at a pouring temperature of 650 °C and a mould temperature of 370 °C.

Figure 2 presents an SEM image of the bonding zone with the corresponding EDS line scan results. The elemental distribution indicates that the zone is rich in Mg, Al and Si. Higher concentrations of Al and Si and a lower concentration of Mg were observed on the AlSi17 side. The interfacial reaction between the molten AZ91 and the solid AlSi17 insert occurring during the compound casting process led to the formation of a transition layer between the two materials, which had a multiphase structure.

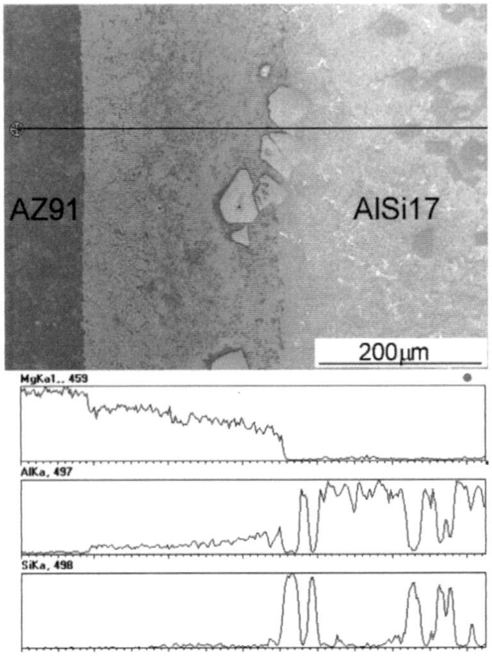

Figure 2. SEM image of the bonding zone and the corresponding EDS line scan results.

Figure 3 shows details of the microstructure of the bonding zone on the AZ91 side. In the area adjacent to the AZ91 substrate, there is a two-phase mixture with a lighter phase (marked 1) and a darker phase (marked 2). Below, we can see large light dendrites (marked 3). The results of the EDS quantitative analysis conducted at three points (see figure 3) are provided in Table 1. With reference to the Mg-Al phase diagram and the results of the quantitative analysis it can be concluded that the two-phase mixture is a eutectic composed of an $Mg_{17}Al_{12}$ intermetallic phase and a solid solution of Al and Si in Mg. The chemical composition of the large light dendrites correspond to that of an $Mg_{17}Al_{12}$ intermetallic phase.

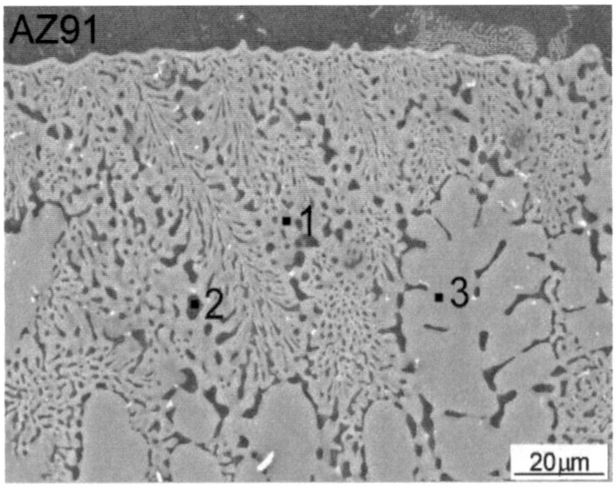

Figure 3. SEM image of the bonding zone on the AZ91 side.

Table 1. Results of the EDS analysis corresponding to the points shown in figure 3.

Point	Mg (at. %)	Al (at. %)	Si (at. %)
1	64.47	35.27	0.26
2	84.46	15.13	0.41
3	61.39	38.61	-

Figure 4 shows the microstructure of the central area of the bonding zone. On the AZ91 side in the upper part of the image, there are dendrites of an $Mg_{17}Al_{12}$ phase and a two-phase eutectic. Below the eutectic, we can see grey fine-grained particles, distributed irregularly throughout the light matrix and a large light particle surrounded by the grey phase. The EDS analysis at point 1 (99.91 at. % Si and 0.09 at. % Al) indicates that the large light particle is a primary silicon crystal. The composition of the grey area near the primary silicon (analysis at point 2: 64.54 at. % Mg, 33.44 at. % Si and 2.02 at. % Al) and the grey particles irregularly distributed throughout the light matrix (analysis at a grey particle: 63.52 at. % Mg, 34.30 at. % Si, 2.18 at. % Al) correspond to that composition of the of an Mg_2Si intermetallic phase. It is evident that silicon reacted with magnesium, which gave rise to the formation of the Mg_2Si intermetallic phase. The composition of the light matrix (analysis at point 3: 60.34 at. % Al, 39.66 at. % Mg) is similar to that of an Al_3Mg_2 intermetallic phase.

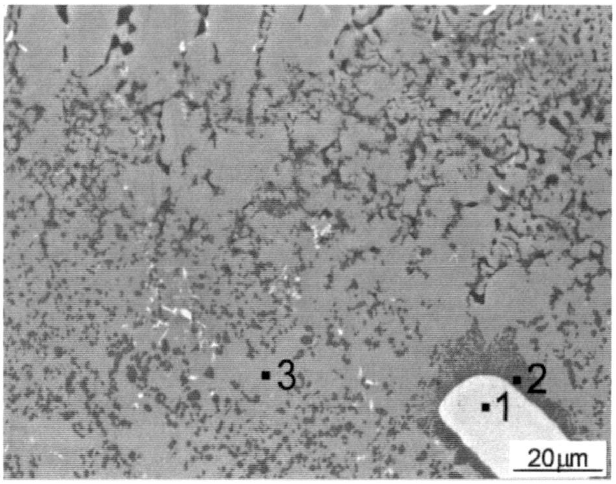

Figure 4. SEM image of the central area of the bonding zone.

Figure 5 shows an SEM image of the bonding zone on the AlSi17 side. A light primary Si particle can be seen in the area adjacent to the AlSi17 alloy. The Si particle is surrounded by a rim of an Mg_2Si intermetallic phase, like the Si particles found in the central area of the bonding zone. Fine particles of the Mg_2Si phase in the Al_3Mg_2 phase matrix were also observed in the area adjacent to the AlSi17 alloy. The analysis of the microstructure revealed that there were small white particles distributed irregularly over the entire AZ91/AlSi17 interface (see figures 3, 4 and 5). The EDS results indicated that the particles were rich in Al, Ni, Mg and Fe (analysis at a light particle: 64.95 at. % Al, 17.36 at. % Ni, 10.18 at. % Mg and 7.5 at. % Fe).

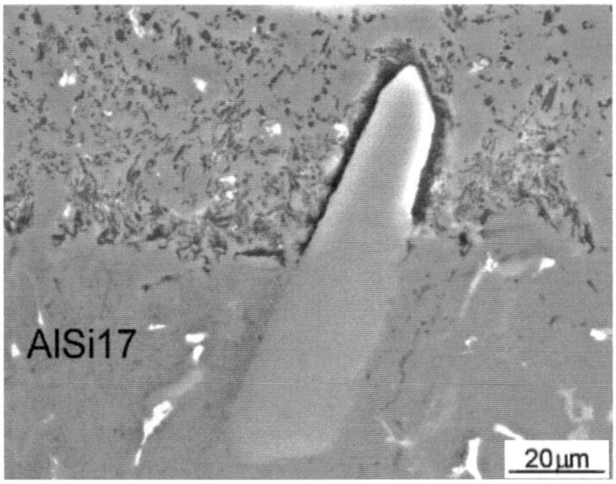

Figure 5. SEM image of the bonding zone on the AlSi17 side.

When higher pouring temperatures were used, i.e. 680 °C and 700 °C, the bonding zone that formed between the AZ91 and the AlSi17 was not continuous. Local discontinuities at the AZ91/AlSi17 interface were reported, as shown in figure 6.

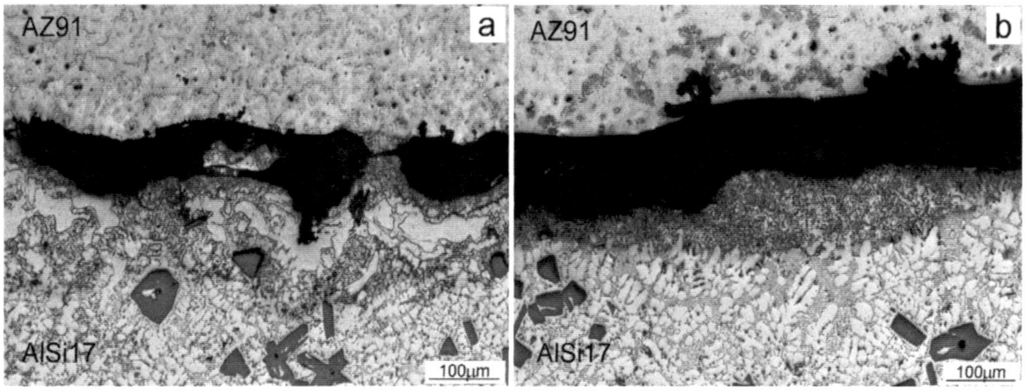

Figure 6. Optical micrographs of the AZ91/AlSi17 interface in bimetal specimens fabricated by compound casting. Casting conditions: (a) pouring temperature of 680 °C and a mould temperature of 370 °C, (b) pouring temperature of 700 °C and a mould temperature of 370 °C.

Below the discontinuities we can observe a change in the microstructure of the AlSi17 alloy. This suggests that at the beginning of the casting process there was a direct contact between the AZ91 and AlSi17 alloys, which enabled a reaction between them and the formation of the bonding zone, but then the continuity at the reactive interface was lost.

4 Conclusions

The results show that the AZ91 alloy was successfully joined to the AlSi17 alloy by compound casting. When the pouring temperature was 650 °C, a continuous bonding zone with a thickness of 200 µm formed at the AZ91/AlSi17 interface. At higher pouring temperatures, i.e. 680 °C and 700 °C, local discontinuities in the bonding zone were observed. The bonding zone adjacent to the AlSi17 alloy was composed of Si primary particles surrounded by a rim of Mg_2Si, Mg_2Si and Al_3Mg_2 intermetallic phases. The area of the bonding zone adjacent to the AZ91 alloy had a eutectic structure composed of an $Mg_{17}Al_{12}$ intermetallic phase and a solid solution of Al and Si in Mg.

References

[1] Kilarski J and Suchoń J 2006 Odlewy warstwowe staliwo - żeliwo *Archives of Foundry* **6(22)** pp 266-271
[2] Avci A, Ilkaya N, Simsir M M and Akdemir A 2009 Mechanical and microstructural properties of low-carbon steel-plate-reinforced gray cast iron *J. Mater. Process. Technol.* **209** pp 1410-1416
[3] Tenerowicz S, Suchoń J and Cholewa M 2010 Studies of the transition zone in steel - chromium cast iron bimetallic casting *Archives of Foundry Engineering* **10(1)** pp 199-204
[4] Wróbel T and Szajnar J 2015 Bimetallic casting: ferritic stainless steel – grey cast iron *Arch. of Metall. and Mater.* **60** pp 2361-2365
[5] Choe K H, Park K S, Kang B H, Cho G S, Kim K Y, Lee K W, Kim M H, Ikenaga A and Koroyasu S 2008 Study of the Interface between Steel Insert and Aluminum Casting in EPC *J. Mater. Sci. Technol.* **24** pp 60-64
[6] Zare G R, Divandari M and Arabi H 2013 Investigation on interface of Al/Cu couples in compound casting *Mater. Sci. Technol.* **29** pp 190-196
[7] Papis K J M, Hallstedt B, Löffler J F and Uggowitzer P J 2008 Interface formation in aluminium - aluminium compound casting *Acta Mater.* **56** pp 3036-3043
[8] Papis K J M, Löffler J F and Uggowitzer P J 2010 Interface formation between liquid and solid Mg alloys - an approach to continuosly metallurgic joining of magnesium parts *Mater. Sci. Eng.* **A 527** pp 2274-2279
[9] Hajjari E, Divandari M, Razavi S H, Emami S M, Homma T and Kamado S 2011 Dissimilar joining of Al/Mg light metals by compound casting process *J. Mater. Sci.* **46** pp 6491-6499
[10] Mola R, Bucki T and Dziadoń A 2016 Formation of Al - alloyed Layer on Magnesium by Casting *Archives of Foundry Engineering* **16(1)** pp 112-128
[11] Bae J H, Prasada Rao A K, Kim K H and Kim N J 2011 Cladding of Mg alloy with Al by twin - roll casting *Scripta Mater.* **64** pp 836-839
[12] Zhang X P, Yang T H, Castagne S and Wang J T 2011 Microstructure, bonding strength and thickness ratio of Al/Mg/Al alloy laminated composites prepared by hot rolling *Mater. Sci. Eng.* **A 528** pp 1954-1960
[13] Wierzba A, Mróz S, Szota P, Stefanik A and Mola R 2015 The influence of the asymetric ARB process on the properties of Al-Mg-Al multi-layer sheets *Arch. of Metall. and Mater.* **60** pp 2821-2825
[14] Tokunaga T, Matsuura K and Ohno M 2012 Aluminum coating on magnesium-based alloy by hot extrusion and its characteristics *Mat. Trans.* **53** pp 1034-1041
[15] Mróz S, Stradomski G, Dyja H and Galka A 2015 Using the explosive cladding method for production of Mg-Al bimetallic bars *Arch. Civ. Mech. Eng.* **15** pp 317-323
[16] Li Y, Liu P, Wang J and Ma H 2008 XRD and SEM analysis near the diffusion bonding interface of Mg/Al dissimilar materials *Vacuum* **82** pp 15-19
[17] Okamoto H 1998 Al-Mg (Aluminum-Magnesium) *J. Phase Equilibria* **19** p 598
[18] Feufel H, Gödecke T and Lukas H L 1997 Investigation of the Al-Mg-Si system by experiments and thermodynamic calculations *J. Alloys Compd.* **247** pp 31-42

IOP Conf. Series: Materials Science and Engineering **179** (2017) 012054 doi:10.1088/1757-899X/179/1/012054

Experimental and computational evaluation of rolling bearing steel durability

L Nohál and M Vaculka

Jedovnická 8, 628 00 Brno, Czech Republic, EU

E-mail: libor.nohal@zkl.cz

Abstract. Rolling element bearings are widely-used machine components and their failure can result in damage to the whole machine. A bearing failure can be caused by many factors. In most cases it is damage on the raceway surface as a result of rolling contact fatigue (RCF). This article describes the fatigue analysis consists of determining service-life of a roller bearing using a stress-strain analysis with finite element method and subsequent numeric calculation using software fe-safe with application of multiaxial fatigue criterion. These theoretical results are compared to the experiments carried out on AXMAT test-rig with more accurate defect detection using acoustic emission method. Numerical service-life calculation can be applied as tool for fatigue life prediction of full scale bearing with sufficient correlation with experimental results.

1 Introduction

Nowadays it is known several approaches for prediction of service life of machine parts. One of approaches on the highest level is computational modelling using multiaxial fatigue criteria (MFC). There are a lot of different MFC which are divided to several categories. Usually they differ to criteria used for low-cycle fatigue (strain-based criteria, energy-based criteria) and high-cycle fatigue (stress-based criteria). Machine parts like roller bearings, gears or train wheels are exposed to limit state which is called roller contact fatigue (RCF). If there are repeating contacts of objects, RCF is reached. The crack initiates in a subsurface layer and propagates to the surface. After that a small piece of material is rubbed and this defect is called pitting. In the article [1] is mentioned a possibility of using Brown-Miller MFC for RCF description. This is the strain-based criterion with the assumption of initiation in the critical plane (plane of maximal shear strain range). This criterion is primarily designed for low-cycle fatigue. Software fe-safe allows calculation of service-life from stress-strain analysis using finite element method applying Brown-Miller criterion. Article [2] proposes other criteria for description RCF. Stress-based Dang Van criterion is one of them. Second appropriate criterion is Liu-Mahadevan [3] which is based on critical plane of maximal shear stress range and it is used for high-cycle fatigue. Simultaneously it includes its mean stress correction. In article [4] is mentioned good match of life results according to Liu-Mahadevan with the experiment of bearing steel.

This paper compares approaches for possible prediction of service-life at multiaxial stress state and its contribution is possible verification with experimental approach. The experiment is based on tests series executed on testing machine AXMAT. It is the system of roller bearing ring, rolling elements and a sample. This system is axial loaded and causes the highest values of stress in the sample. RCF is reached primarily on the surface of the sample.

2 Experimental procedure

2.1 Experimental test-rig

The experimental RCF apparatus employed in this study is a flat washer-type RCF test-rig with AE (acoustic emission) and vibration monitoring systems. This special test-rig, as shown in Figure 1, is designed for life tests of thrust bearings (smaller size) and an evaluation of the rolling contact fatigue resistance of the material. It consists of a mechanical loading lever, an electrical motor, a specimen holder, a catch driver, a supporting frame and the monitoring system. The speed of the electrical motor can be adjusted by a frequency converter to the required level. This allows standard RCF tests to be performed, including tests at low speed. [5]

The upper ring of the test bearing is clamped in the holder and the lower ring is fastened in the catch driver. In the case of testing material specimens, the specimen is fastened in the holder in place of the upper bearing ring. The holder is stationary and the catch driver is driven by a shaft attached to an electrical motor. For a standard RCF test, the speed is set at 1380 min^{-1}. The holder is equipped with a polyamide safety element to offload in case of specimen overheating. The Hertzian pressure for rolling contact tests can be set using the combination of weights in the range from 2000 to 6000 MPa. [5]

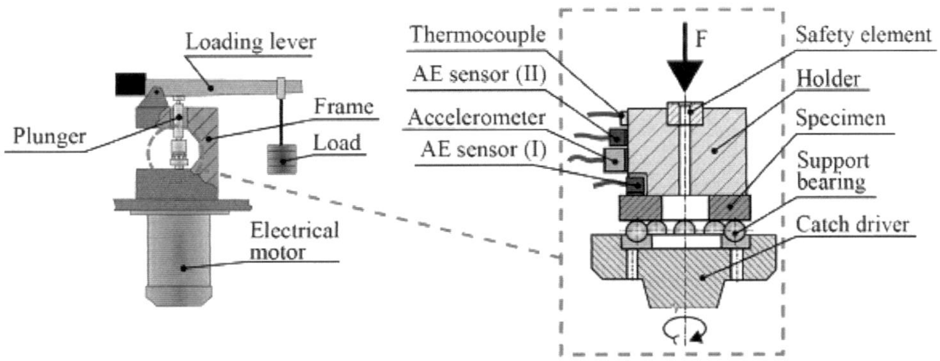

Figure 1. Experimental test-rig layout [5].

2.2 Specimens

The disc shape specimens with the dimensions 28×10×5 mm (outer diameter, inner diameter, thickness) were made from the steel 100Cr6 ISO 683-17. The specimens were heat-treated and ground at roughness Ra 0.1. The supporting bearings ZKL 51102A P6 had 21 balls with 3.175 mm in diameter and were lubricated by lithium complex soap based grease MOGUL LV 2-3.

Table 1. Chemical composition of the 100Cr6 steel.

Element	C	Si	Mn	P	S	Cr	Mo	Al	Cu
Wt. %	0.97	0.24	0.35	0.014	0.004	1.51	0.017	0.029	0.03

3 Results

3.1 Rolling contact fatigue tests

Test was carried out on 20 samples of steel 100Cr6 and durability results were processed by the two-parameter Weibull distribution and determined 10% quantile durability L_{10}, see table 2.

The life of a sample is the time from test start to pitting initiation. The vibration level and AE (acoustic emission) parameters are used for determination of the sample life. These parameters (vibration level, temperature, RMS AE and AE mean frequency) monitored during the test of one material sample are shown in figure 2.The process of experimental evaluation of steel durability using the acoustic emission method is described in more detail in articles [5] and [6].

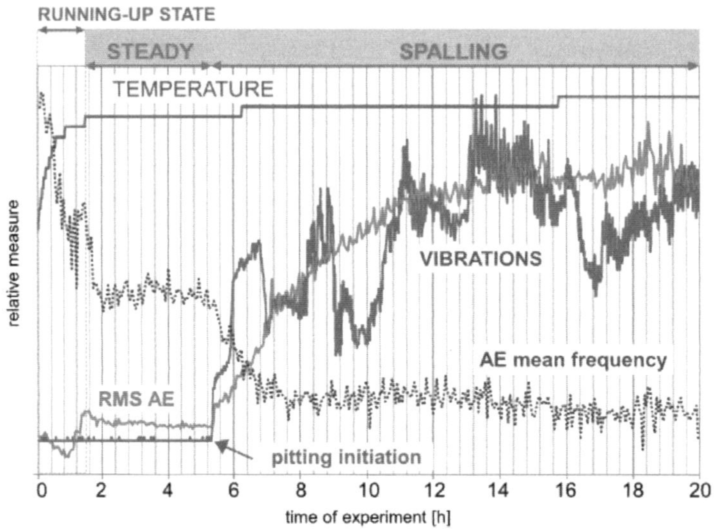

Figure 2. Observation of test: AE, vibration, temperature [6].

3.2 Comparison of theoretical and experimental results

The results based on numerical modelling of multiaxial fatigue criteria and tests are compared in table 2.Given the results presented in [7] was chosen for this criterion "SCALE" in fe-safe software to 0.93. With this setting the lifetime best matches the standard ISO 281.Higher value of experimentally obtained sample lifetimeis caused by the ability to detect only more developed defects. Numerical solution determines the time to fatigue crack initiation, not a pitting formation.

Table 2. Comparison of sample lifetime.

Method	Brown-Miller + Morrow	Experimental (L_{10})
Sample lifetime (min)	261	335

4 Conclusion

The lifetime of steel samples based on numerical modelling and experimental results was compared. A rolling contact fatigue test using the acoustic emission method was undertaken on a flat washer test-rig and the temperature, vibration level and AE parameters were processed by the two-parameter Weibull distribution. The calculation of service-life from stress-strain analysis using finite element method applying Brown-Miller criterion in software Fe-safe were done. The results shows that for more accurate calculation will be appropriate use other criteria suitable for high-cycle fatigue than Brown-Miller.

It can be concluded that calculation of service-life from stress-strain analysis using finite element method applying proper multiaxial fatigue criteria can be used for bearing or material sample lifetime prediction.

Acknowledgements
The research leading to these results has received **funding from the MEYS under the National Sustainability Programme I (Project NETME CENTRE PLUS LO1202).**

References

[1] Dambaugh, G. F. (2006). Fatigue Considerations of High Strength Rolling Bearing Steels. Retrieved October 2, 2016, from http://www.simulia.com/download/scc-papers/Industrial/fatigue-considerations-high-strength-rolling-bear-2006-F.pdf

[2] Ciavarella, M., & Monno, F. (2010). A comparison of multiaxial fatigue criteria as applied to rolling contact fatigue. Tribology International, 43(11), 2139-2144. doi:10.1016/j.triboint.2010.06.003

[3] Liu, Y., & Mahadevan, S. (2005). Multiaxial high-cycle fatigue criterion and life prediction for metals. International Journal of Fatigue, 27(7), 790-800. doi:10.1016/j.ijfatigue.2005.01.003

[4] Foletti, S., Beretta, S., & Tarantino, M. (2014). Multiaxial fatigue criteria versus experiments for small crack under rolling contact fatigue. International Journal of Fatigue, 58, 181-192. doi:10.1016/j.ijfatigue.2013.05.006

[5] Nohál, L.; Hort, F.; Dvořáček, J. and Mazal, P. An Experimental Investigation of Rolling Contact Fatigue of Steels Using Acoustic Emission Method. *INSIGHT*, 2013, roč. 55, č. 12, s. 665-669. ISSN: 1354- 2575.

[6] Chlada M, Nohál L, Převorovský Z, ConditionMonitoringofThrustBallBeatingsusing continousAE. In Proceedings 32nd European Conference on Acoustic Emission Testing. Praha, 2016.

[7] VACULKA, M. *Návrh metodiky výpočtu životnosti valivého ložiska.* Brno: Vysoké učení technické v Brně, Fakulta strojního inženýrství, 2016. 84 s. Vedoucí diplomové práce prof. RNDr. Ing. Jan Vrbka, DrSc., dr. h. c.

The effect of annealing on mechanical and structural properties of UFG titanium grade 2

J Palán[1], T Kubina[1], P Motyčka[1],

[1] COMTES FHT a.s., Dobřany, Czech Republic, EU

E-mail: jan.palan@comtesfht.cz, tomas.kubina@comtesfht.cz, petr.motycka@comtesfht.cz

Abstract. The present paper explores the effect of annealing on mechanical and structural properties of ultrafine-grained (UFG) Ti Grade 2 after it was mechanically worked in either CONFORM SPD or rotary swaging (RS) machine. The main objective of this study was to optimize the annealing schedule in order to achieve recovery in the severely deformed structure and at the same time retain the mechanical properties of the worked material. The temperatures of the recovery processes were determined by means of dilatometry. It was found that in an ultrafine-grained material the temperatures of the recovery processes are shifted towards lower values. Based on dilatometric analysis, annealing schedules were developed and mechanical and structural properties were evaluated.

1 Introduction

Technologies which make use of severe plastic deformation (SPD) have been receiving major attention. SPD leads to considerable grain refinement and therefore to improved mechanical properties. The relationship between mechanical properties and grain size is described by the Hall-Petch equation. The best-known SPD technique is equal channel angular pressing (ECAP) where material is forced through two channels which connect at an angle. High hydrostatic pressure, shear deformation and sufficiently low temperature lead to intensive grain refinement [1]. Research into processing of pure titanium by SPD techniques has been pursued for quite some time [2]. Today, ultrafine-grained materials can even be produced using the CONFORM ECAP (CONFORM SPD) method [3]. The main advantage of this method is its continuous process [4]. As proven in the early study [2,4], processing of Ti by CONFORM ECAP provides higher strength without any substantial decrease in elongation. The strengthening effect is however limited by a certain number of passes, or more accurately, by certain grain refinement achieved [5]. After this number of passes, mechanical properties cease to rise. This can be explained by the fact that beyond a certain limit of grain refinement, equiaxed grains merely rotate without becoming finer [6]. Nevertheless, ultimate strength and yield stress can still be improved through classical work hardening. Process routes that involved working in the CONFORM ECAP machine and subsequent drawing in a wire-drawing machine have already been tested [5]. It should be noted that subsequent working causes changes in the microstructural characteristics of the material, and reduces its ductility. What is meant by the changes in microstructural characteristics is substantially higher dislocation density. The material that had been worked in this manner then exhibits reduced toughness and high residual stresses due to non-uniform deformation. These negatively affect its behaviour, for instance during machining. Workpiece fracture may occur due to superposition of machining-induced stresses and the residual stresses caused by previous working (work hardening). Figure 1 shows a fractured implant. It was manufactured from a material that had been worked by CONFORM SPD and rotary swaging. Residual stresses may also compromise the shape stability of the product and cause the Bauschinger effect. In addition, they may have impact on the fatigue strength of the material. Nevertheless, one can improve all these properties by appropriate heat treatment which relieves residual stresses and provides recovery of the structure. Its temperature must be optimal to ensure that these desirable effects are not accompanied by a decrease in strength. An important factor which plays a role in choosing the annealing temperature is the accumulated strain. Generally, accumulated strain energy introduced by severe deformation causes recovery processes to start at lower temperatures. The present work studied this shift of recovery temperatures, and the effect of temperature on mechanical properties of commercial-purity Grade 2 titanium in various heat-treated conditions.

Figure 1. Fractured implant. The implant material was treated as follows: CONFORM SPD (3 passes) + rotary swaging (cross section area reduction=80 %). A detail of the failure location is shown on the right.

2 Material and experiment description

The material used was CP Grade 2 titanium with the chemical composition given in Table 1. This composition was measured by means of a Bruker Q4 Tasman optical emission spectrometer, and a Bruker G8 Galileo gas analyser. The diameter of the Ti rods was 10 mm.

Table 1. Chemical composition of feedstock in weight percent (wt. %)

Fe	O	C	H	N	Ti
0.046	0.12	0.023	0.0026	0.0076	99.822

The processing was carried out in CONFORM SPD machine (Figure 2 – left), type 315i, with a modified die chamber, and in the HMP R4-4 rotary swaging machine. In the CONFORM SPD process, the temperature was 220 °C, the wheel speed was 0.5 rpm, and the angle of the die chamber was 90°. Three passes through the CONFORM SPD machine were completed. The product cross-section was identical to that of the feedstock. Rotary swaging (Figure 2 – right), the subsequent process, was carried out at ambient temperature. In this operation, the cross-section area was reduced by 20 % in each pass. The total area reduction was 80 %.

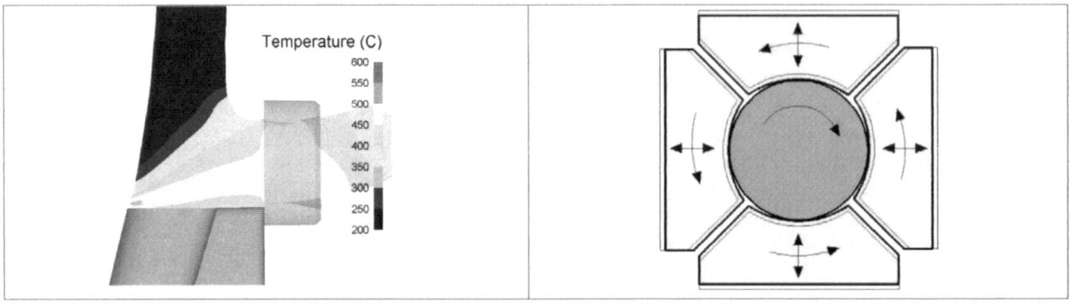

Figure 2. 3D model of the CONFORM SPD process (left) and the detail of the shear deformation zone inside the die chamber (middle). On the right picture, the principle of rotary swaging is shown.

For the purposes of observation in the transmission electron microscope (TEM), thin foils were prepared with final electrolytic thinning in a Tenupol 5 device, using a solution of 300 ml CH3OH + 175 ml 2-butanol + 30 ml HClO4 at -10 °C and a voltage of 40 V. The TEM analysis was performed in a JEOL 200CX instrument with an acceleration voltage of 200 kV. Selective electron diffraction was used for the determination of the phases. Grain size was measured using the linear intercept method.

The effect of deformation on the thermal expansion of titanium, and the temperature range for recovery were explored using a Linseis L75 Platinum horizontal dilatometer with an Al_2O_3 specimen chamber and a pull bar. The temperature changes were monitored with a thermocouple on specimens of 5 mm diameter and with lengths of approximately 20 mm. Nitrogen (N2) was used as the protective gas. After heating to 950 °C at a rate of 3 K/min, the specimen was cooled at 20 K/min to 600 °C. Then the specimen was left to cool in air to the ambient temperature.

The specimens were annealed at various temperatures in a range between 150–500 °C for various times in a box furnace. The specimens were placed into a heated furnace. Cooling took place in air. Prior to annealing, the material was subjected to three passes through the CONFORM SPD machine and to rotary swaging with an overall cross section area reduction of 80 %. Hardness was measured on specimens annealed according to the initially proposed annealing sequences. Based on their values, new annealing sequences were developed. Specimens annealed using these sequences were then tensile-tested.

3 Microstructure

CONFORM SPD passes led to considerable refinement of the initial microstructure. The initial mean grain size was 5390 nm (Figure 3 left). After three passes, it dropped to 390 nm (Figure 4 right). After three passes, the resulting grains were equiaxed and the dislocation density in them was non-uniform. Figure 4 shows the microstructure upon 3 passes through CONFORM SPD and after rotary swaging with a cross section area reduction of 35 %. Microstructural changes were visible: higher dislocation density and grains elongated in the direction of material flow.

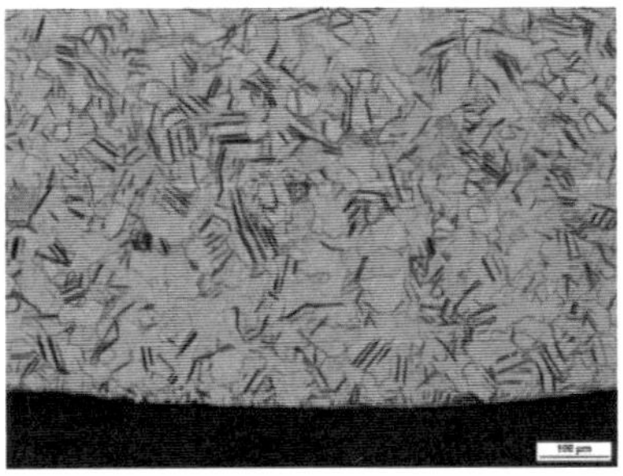

Figure 3. The micrograph on the left shows the initial structure with the mean grain size of 5390 nm. The micrograph on the right shows the structure upon three passes through CONFORM SPD; the mean grain size is 390 nm. Both micrographs are of the transverse section through the wire.

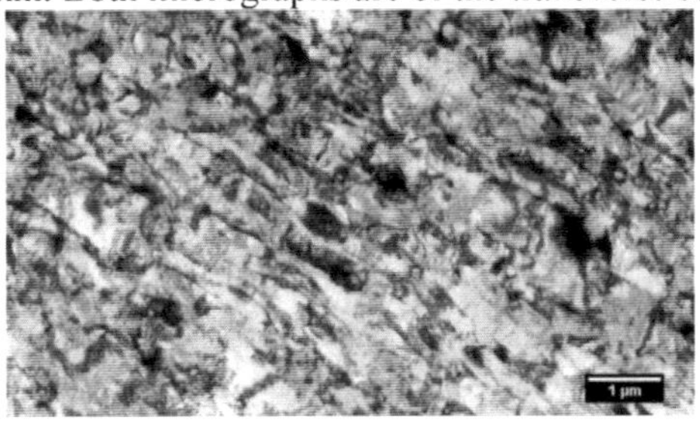

Figure 4. Microstructure upon 3 passes through CONFORM SPD and after rotary swaging (cross section area reduction of 35 %). Detail of microstructure on a longitudinal section through the wire.

4 Dilatometric measurement

The changes in length during heating is shown in Figure 5. The length change of the undeformed specimen (as-received condition) does not change significantly, and is in a good agreement with literature data [17]. The length of the deformed specimen increases more rapidly and then somewhat more slowly again twice – within two temperature intervals. The increase is assumed to be linked with stress relief but the shape of the peak suggests that several processes might operate concurrently. The minor decrease in the expansion rated is assumed to be linked either with the recovery and recrystallization or rearrangement of the atoms of alloying elements into lattice defects. The temperature ranges of these phenomena are listed in Table 2.

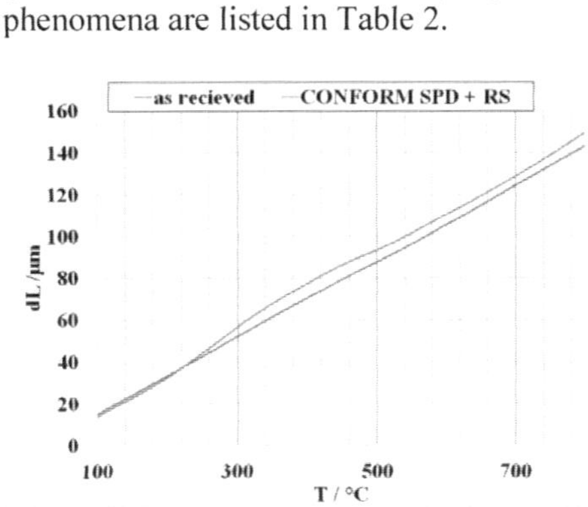

Figure 5. Length change during the heating for as received state and for state after three passes through the CONFORM SPD device and after rotary swaging with the area reduction of 80%.

Table 2. Peak start and end temperatures for changes in length during heating.

Condition	Peak 1 start /°C	Peak 1 end /°C	Peak 2 start /°C	Peak 2 end /°C
CONFORM SPD + RS	175	350	430	535

5 Mechanical properties

Hardness measurement

The hardness measurement was provided for the sample after CONFORM SPD and rotary swaging for different annealing temperatures and times. Figure 6 indicates that annealing in the 150–300 °C interval caused no substantial decrease in hardness. Where hardness became lower, the decrease was only slight. One can assume that the structure becomes rearranged in order to reduce the total energy stored, and that residual stresses are relieved, which is evidenced by dilatometric measurement. At temperatures of 350–500 °C, the decrease in hardness is substantial because recrystallization sets in.

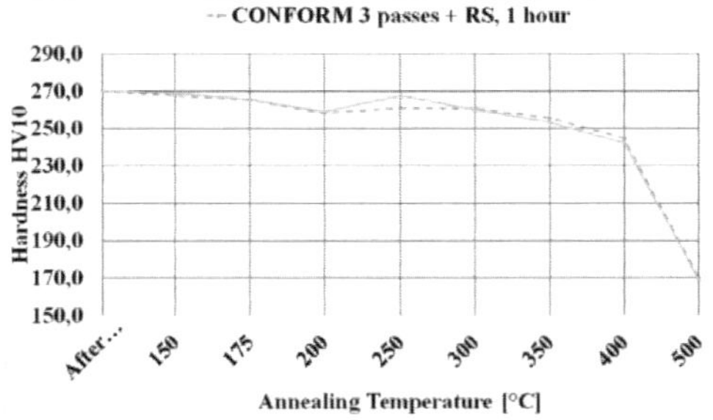

Figure 6. Hardness vs. annealing temperature and time for sample after CONFORM SPD (3 passes)+RS (Area reduction of 80%).

Tensile testing

Tensile properties found by testing are given in Table 3Table 2. Table 3 indicates that the largest increase in mechanical properties was obtained by the first pass through the CONFORM SPD machine. The subsequent passes led to smaller increments. It is important to note that the increase in ultimate strength and yield stress upon CONFORM SPD is not offset by a decrease in ductility. The improvement in mechanical properties is mainly due to grain refinement. A further increase in mechanical properties was obtained by rotary swaging applied after the three passes through the CONFORM SPD machine.

Table 3. Mechanical properties after CONFORM SPD processing and after CONFORM SPD + rotary swaging. Ultimate tensile strength (UTS); offset yield (OYS); reduction of area (RA); Elongation (A_5).

Condition	0.2 OYS [MPa]	UTS [MPa]	A_5 [%]	RA [%]
as received	370	480	25	52
CONFORM SPD – 1 pass	540	580	23	62
CONFORM SPD – 2 passes	560	600	23	62
CONFORM SPD – 3 passes	570	623	20	64
CONFORM SPD – 3 passes + RS (80 % area reduction)	980	1057	12	56

Tensile testing after annealing

Figure 7 and Figure 8 show the evolution of mechanical properties with annealing temperature and time for sample after three passes on the CONFORM SPD device and rotary swaging with the area reduction of 80%. Figure 7 shows the response at 250 °C. It is clear that annealing for one hour causes a slight decrease in ultimate strength but a more pronounced decrease in yield stress. At annealing times between one hour and four hours, the values of mechanical properties do not change and the structure exhibits good thermal stability. Mechanical properties decrease slightly again after annealing for five hours. Similar behaviour was found at the temperature of 300 °C (Figure 8). The only difference is that mechanical properties decrease more steeply at the higher temperature. The annealing had no effect on the A5 elongation value. Another finding was that the ratio of ultimate strength and yield stress increased after only one-hour annealing. One can therefore conclude that annealing improves the toughness of the material.

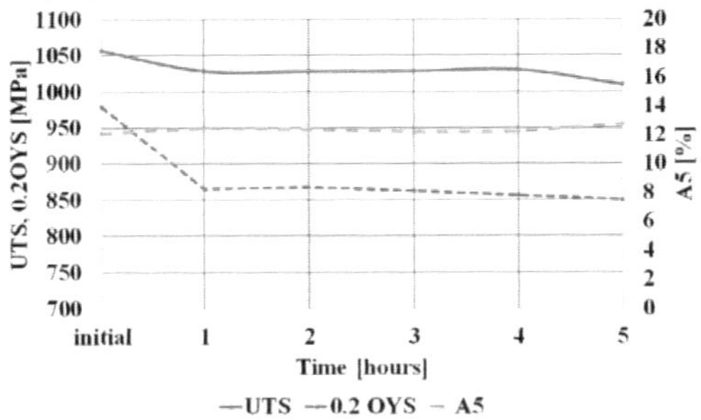

Figure 7. Evolution of mechanical properties at the annealing temperature of 250 °C. The sample CONFORM SPD (3 passes) + RS (Area reduction of 80%).

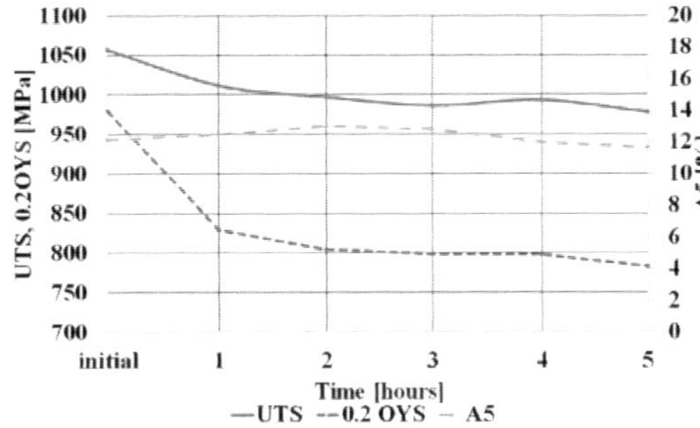

Figure 8. Evolution of mechanical properties at the annealing temperature of 300 °C. The sample CONFORM SPD (3 passes) + RS (Area reduction of 80%).

6 Conclusion

The purpose of this study was to analyze the effect of annealing on mechanical and microstructural properties of ultrafine-grained Grade 2 titanium.

Forming by CONFORM SPD led to considerable refinement of the initial microstructure. The mean grain size was 390 nm. The resulting grains were equiaxed and the dislocation density in them was non-uniform. Thanks to the grain refinement, the yield stress and ultimate strength increased without the ductility of the material being reduced. After that, the material was rotary-swaged. The grains in the microstructure produced by rotary swaging were not equiaxed any more: they were elongated in the forming direction, with higher dislocation density. Working led to a large amount of work hardening. It was reflected in the increased ultimate strength and yield stress and in the reduced A5 elongation.

Dilatometric analysis revealed that the temperatures of recovery in the deformed specimen had decreased. The reason is that the activation energy for recovery processes was reduced by the deformation applied in CONFORM SPD and rotary swaging. Dilatometry also showed that the amount of residual stresses from forming had been relieved.

The effect of annealing in the temperature interval of 150–500 °C on hardness was explored. In the 150–300 °C range, hardness did not decrease substantially. The structure showed a good thermal stability.

The materials annealed at 250 °C and 300 °C were tested in tension. As short annealing as one hour caused a slight decrease in ultimate strength and a more pronounced decrease in yield stress. The decrease was steeper after annealing at 300 °C. On one hand, annealing led to a slight decrease in ultimate strength but on the other hand, the ratio of ultimate strength and yield stress increased, which suggests better toughness. Future research will focus on microstructural evolution after annealing.

Acknowledgement
This paper was developed under the project entitled Development of West-Bohemian Centre of Materials and Metallurgy No.: LO1412, which is financed by the Ministry of Education of the Czech Republic.

References

[1] Valiev R.Z et al. 2000 Bulk nanostructured materials from severe plastic deformation. Progress in Materials Science. 45 103-189.
[2] Kubina T et al. 2015 Preparation and thermal stability of ultra-fine and nano-grained commercially pure titanium wires using CONFORM equipment. Materiali in Technologije. 49 213-217.
[3] Raab G et al. 2004 Continuous processing of ultrafine grained Al by ECAP–Conform. Materials Science and Engineering: A. 382 30-34.
[4] Duchek M et al. 2013 Development of the production of ultrafine-grained titanium with the Conform equipment. Materiali in Technologije. 47 515-518.
[5] Gunderov D.V et al. 2013 Evolution of microstructure, macrotexture and mechanical properties of commercially pure Ti during ECAP-conform processing and drawing. Materials Science and Engineering: A. 562 128-136.
[6] Mishra A et al. 2007 Microstructural evolution in copper subjected to severe plastic deformation: Experiments and analysis. Acta Materialia. 55 13-28.
[7] Wei Q, Cheng S et al. 2004 Effect of nanocrystalline and ultrafine grain sizes on the strain rate sensitivity and activation volume: fcc versus bcc metals. Materials Science and Engineering: A. 381 71-79.
[8] Ostrovska L, Vistejnova L et al. 2016 Biological evaluation of ultra-fine titanium with improved mechanical strength for dental implant engineering. Journal of Materials Science. 51 3097-3110.
[9] Zrník J. et al. 2007 Evropská strategie výrobních procesů (Ostrava: Repronis)
[10] Raab G et al. 2004 Continuous processing of ultrafine grained Al by ECAP–Conform. Materials Science and Engineering: A. 382 30-34.
[11] Zemko M et al. 2014 Technological aspects of preparation of nanostructured titanium wire using a CONFORM machine. IOP Conference Series: Materials Science and Engineering. 63 p ??.
[12] Gunderov D.V et al. 2013 Evolution of microstructure, macrotexture and mechanical properties of commercially pure Ti during ECAP-conform processing and drawing. Materials Science and Engineering: A. 562 128-136.
[13] Semenova I.P et al. 2008 Strength and fatigue properties enhancement in ultrafine-grained Ti produced by severe plastic deformation. Journal of Materials Science. 43 7354-7359.
[14] Palán J et al. 2015 Continuous extrusion of commercially pure titanium GRADE. 4. Journal of Achievements in Materials and Manufacturing Engineering. 69 33-37.
[15] Humphreys F.J et al. 2004 Recrystallization and related annealing phenomena (Boston: Elsevier) p 574.
[16] Eivani A.R. et al. 2016 Mechanism of the formation of peripheral coarse grain structure in hot extrusion of Al-4.5Zn-1Mg. Philosophical Magazine. 96 1188-1196.
[17] Hindert P 1943 Thermal expansion of titanium. Journal of Research of the nationa Bureau of Standards. 30 101-105.
[18] Reglitz G et al. 2015 Combined volumetric, energetic and microstructural defect analysis of ECAP-processed nickel. Acta Materialia. 103 396-406.

IOP Conf. Series: Materials Science and Engineering **179** (2017) 012056 doi:10.1088/1757-899X/179/1/012056

Implementation of American weld connection standards into finite element computations

M Peč, P Vosynek, F Šebek and T Návrat

Institute of Solid Mechanics, Mechatronics and Biomechanics; Faculty of Mechanical Engineering; Brno University of Technology; Technická 2896/2, 616 69 Brno, Czech Republic, EU

E-mail: sebek@fme.vutbr.cz

Abstract. Presented paper summarizes the weld assessment on the basis of national standards. The approach concerning the static loading has been still not adequately developed in standards. The standard of American Institute of Steel Construction was considered as the best candidate for implementation into the finite element method using post-processing of a particular weld. A comparative analysis of maximum allowable loadings of weld connection models loaded in-plane and out-of-plane has been performed. Results obtained by the finite element method and American and Czech standards were compared and the conclusions of applicability and reliability were made.

1 Introduction

The assessment of welded structures in national standards as American Institute of Steel Construction (AISC), British Standards (BS), Czech technical standards (ČSN) and others is related only to simple weld connections. When complicated welded structure is intended to asses, it is meant in general geometry and topology, general loading or combination of general geometry and loading, there is a problem because the weld assessment is either difficult or impossible.

There are two main categories of the weld assessment – connections with either static of cyclic loading. Those are developed on a different levels. Cyclically loaded welded structures are precisely documented and implemented into the Finite Element Method (FEM) in scope of assessment of complex configurations. However, the assessment of statically loaded welded connections is developed only on the basis of analytical formulas. Therefore, the implementation of standards regarding static loading into FEM is needed for assessment of complex welded structures.

2 Methods

Four basic approaches are the nominal stress, the structural stress, the notch stress and the fracture mechanics [1, 2]. The first three approaches differ in using various stresses (Figure 1). The latter approach is based on the completely different background and it is stated only for completeness. Thereafter, the stress-based methods might be distinguished in two main categories – the national standards (AISC, BS or ČSN) and special approaches, such as International Institute of Welding (IIW), Det Norske Veritas (DNV) or Forschungskuratorium Maschinenbau (FKM). The state-of-the-art suggests that the standard of AISC is probably the most suitable for assessment of complex welded structures [3].

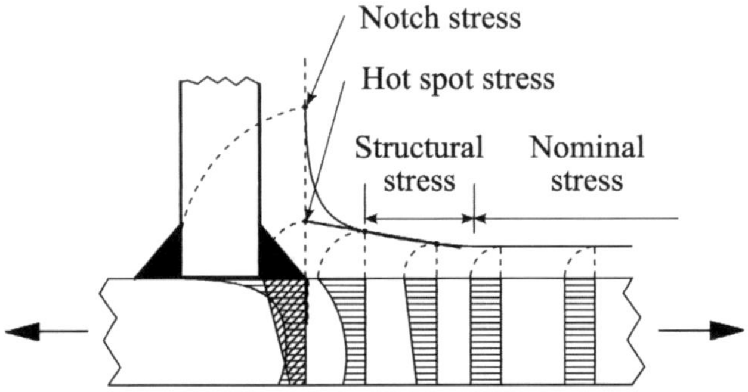

Figure 1. Approaches to the weld assessment [9].

3 Methodology

Standard configurations and loadings of welded connections are stated in the standard of AISC using the form of coefficients which are empirical. Those coefficients were obtained using Instantaneous Centers of Rotation Method (ICRM). Each part is able to withstand different loading when the weld joint is consisted of several partial welds variously oriented, and the loading can be redistributed into the whole weld group. Then, it is assumed that the whole weld group rotates around one Instantaneous Centre (IC). The principle of ICRM was described by Lesik and Kennedy [4, 11] and Miazga and Kennedy [12]. The description of the method can be summarized into following points and it can be illustratively depicted as in Figure 2. The group of fillet welds eccentrically loaded rotate around the IC which is iteratively obtained from the equations of static equilibrium. The weld is divided into the finite number of segments where the reaction forces are calculated. Maximum deformations of segments are computed in dependence of angle between the reaction forces and weld segments axes, and the critical segment is determined. The maximum external load is defined on the basis of this critical segment.

Another possible approach to assessment of welded structures might lie in deploying more sophisticated material model [5] or the damage [6], but the process would not be linear anymore.

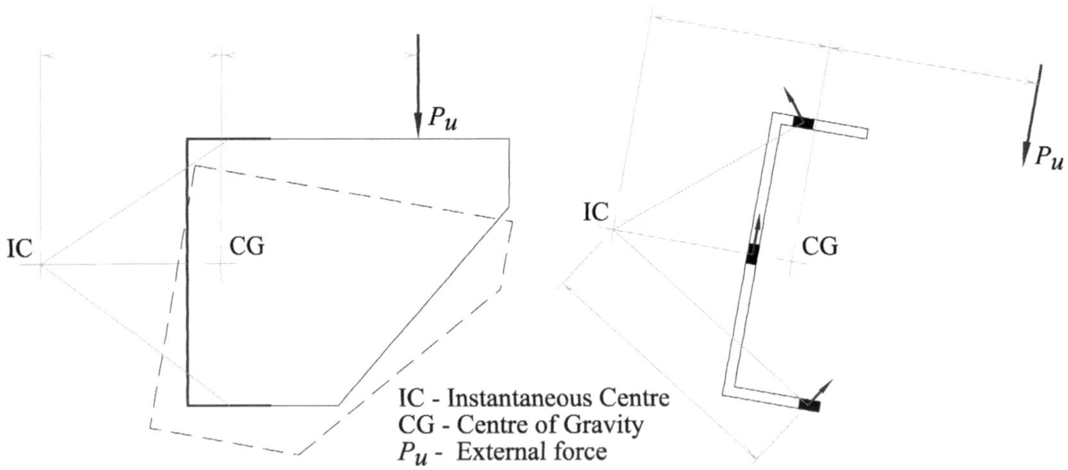

Figure 2. The principle of considered method [7].

4 Analytical solution

There were created three computational models, analytical and numerical, for comparison purposes (Figure 3). First two cases for assessment of the weld strength of joints exposed to in-plane loading and one case regarding out-of-plane loading.

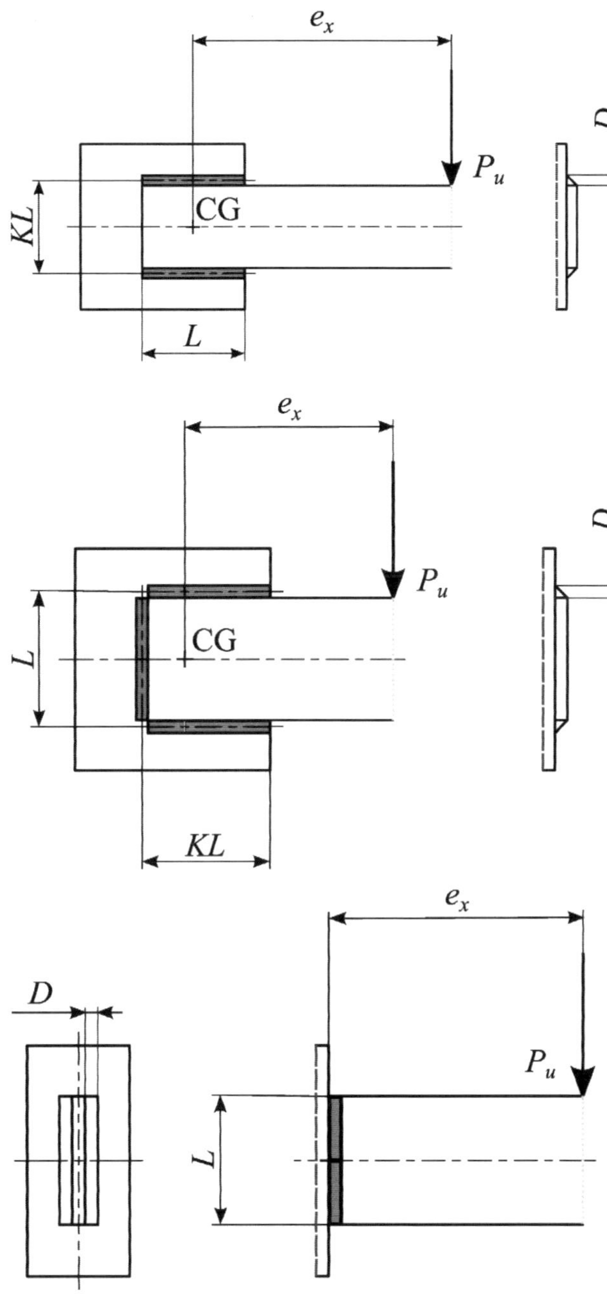

Figure 3. Basic weld configurations [3].

Table 1. Comparison of results from standards.

Case	AISC F_{stand} [N]	ČSN F_{stand} [N]	Δ [%]
1	65762	45067	–32
2	131312	86484	–34
3	52418	36513	–30

All the cases were assessed by standard of AISC [7, 10], ČSN [8] and by developed algorithm. Geometry was designed so as the weld strengths can be assessed by the standards, of course. The electrode E70 was used as a weld material with the strength F_{EXX} = 480 MPa, therefore the electrode strength coefficient was C_1 = 1.00. The base material was S355 steel with yield stress f_y = 355 MPa and ultimate tensile strength f_u = 510 MPa. Only the assessment of weld connections follow, so the base material was not assessed. Particular parameters for cases 1–3 are the weld size D = 9.525, 9.525, 9.525 mm, e_x = 203.2, 203.2, 203.2 mm, KL = 50.8, 101.6, 0 mm and characteristic length of the weld group L = 101.6, 101.6, 101.6 mm. Based on these dimensions, the coefficient of geometry and angle of loading of the weld group, which is tabulated, was determined as C = 0.616, 1.230, 0.491. Resulting maximum external loads of weld groups F_{stand} solved by AISC (Equation 1) and ČSN (Equation 2) are given in Table 1 where Δ represent percent difference of results. The expression for external force reads:

$$P_u = CC_1DL \tag{1}$$

$$\sqrt{\sigma_\perp^2 + 3\left(\tau_\perp^2 + \tau_\parallel^2\right)} \leq \frac{f_u}{\beta_w \gamma_{M2}} \tag{2}$$

In Equation 2, $\sigma_\perp$ is the normal stress perpendicular to the load-carrying cross-section of the weld, $\tau_\perp$ is the shear stress perpendicular to the weld axis and lying in the load-carrying cross-section, $\tau_\parallel$ is the shear stress parallel with the weld axis, β_w = 0.9 is the correlation coefficient according to the material class and γ_{M2} = 1.25 is the coefficient of weld connections reliability.

5 Numerical solution

The method taken over American standard is implemented into FEM with following advantages. There is a possibility to create an arbitrary weld, the determination of requested angles for the assessment is more accurate and there is no need to use the iterative algorithm of finding IC.

Particular computational models were defined for FEM as follows. All materials were assumed as homogeneous, isotropic and linearly elastic. The material of weld has E = 210000 MPa Young's modulus and μ = 0.3 Poisson's ratio. The material of base metal had E = 2100000 MPa and μ = 0.3. Boundary conditions were the same for all cases. Dashed line represents the loci where the metal was fixed and the dotted line represents the loci of applied external force (Figure 3). The discretization of model was conducted at two levels according to the standard. Each weld was divided into ten segments (Figure 4) where the reaction forces were determined. The whole model was discretized by mapped mesh with a linear 2 mm elements.

The most important part is a post-processing. Reaction forces were determined at particular weld segments on the surfaces loaded in shear – red-colored surfaces in Figure 3. All nodes belonging to particular segment were selected on this surface and components of reaction forces F_x, F_y and F_z were calculated in the Centre of Gravity (CG) of a particular surface (Figure 5). Using these forces, the angle θ between weld axis and segment force F_{dos} was computed. Based on angle θ, the maximum allowable force on particular segment $F_{pr,R}$ was computed via the Equation 3, where Φ = 0.75 is the resistance factor. The Equation 3 is extended version of Equation 1.

$$F_{pr,R} = 0.6\Phi F_{EXX}\left[1 + 0.5\sin^{1.5}(\theta)\right] \tag{3}$$

There is a criterion for estimation of maximum external force using FEM, F_{FEM}, in Equation 4. The allowable external load can be calculated for a problem with an arbitrary external load with sequential linear interpolation so as p_{dos} = 1 at a critical segment.

$$p_{dos} = \frac{F_{pr,R}}{F_{dos}} \geq 1 \tag{4}$$

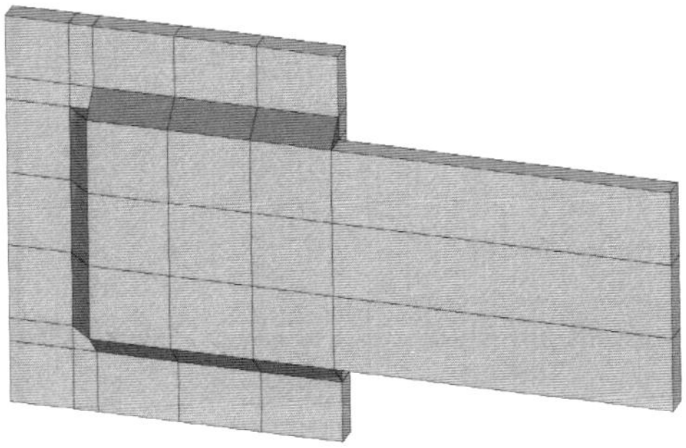

Figure 4. Division into three segments.

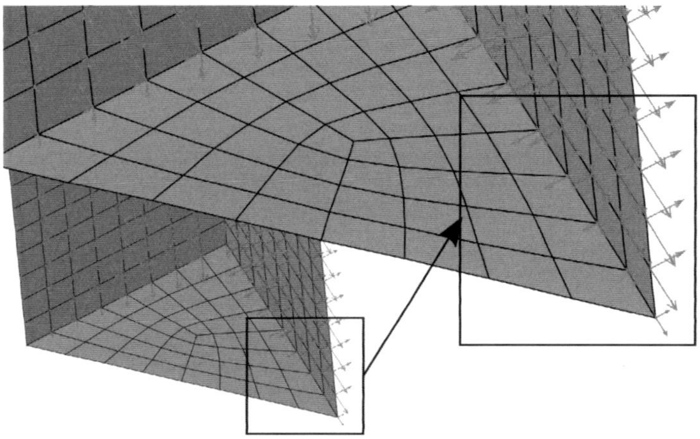

Figure 5. The estimation of forces [3].

Resulting maximum allowable external forces determined by this methodology are given in Table 2 where those are compared to values from standards. Based on these results, there was designed a weld group according to Figure 6 where the weld material is the same as in previous cases and h_1 = 76.2 mm, h_2 = 101.6 mm, h_3 = 508 mm, v_1 = 50.8 mm, v_2 = 152.4 mm, v_3 = 76.2 mm and D = 9.525 mm. It is not possible to assess this case according to the standard but it can be done by proposed algorithm. Results for particular loading are given in Table 3 depicting the linear dependency. The interpolation of forces so as p_{dos} = 1 gives maximum external force of F_{FEM} = 178624 N which should the weld safely withstand.

Table 2. Comparison of cases according to AISC and FEM.

Case	F_{FEM} [N]	F_{stand} [N]	Δ [%]
1	46379	65762	–30
2	118961	131312	–9
3	60000	52418	15

Table 3. Results of illustrative case.

F_{FEM} [N]	F_{dos} [N]	AISC $F_{pr,R}$ [N]	AISC θ [deg]	p_{dos} [–]

100000	9142			1.79
200000	18285	16330	21	0.89
300000	27427			0.60

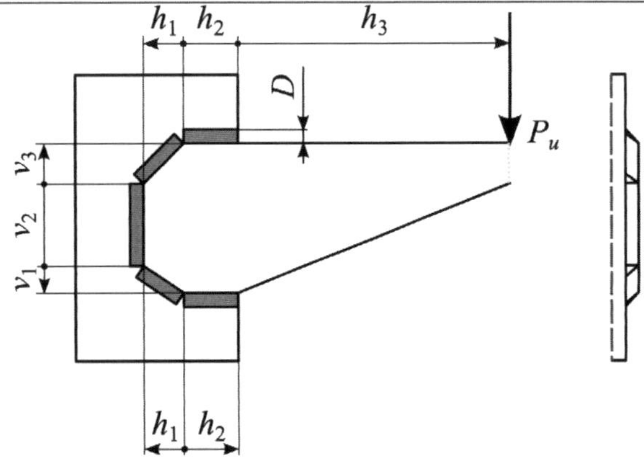

Figure 6. Non-standard weld group [3].

6 Conclusions

There has been developed the algorithm for assessment of complex welded structures loaded in-plane and out-of-plane according to American standard of AISC which is based on solution in elastic region and ICRM. The advantage is in concerning residual stress and variability of weld joint given by the term 0.6Φ in Equation 3. When certain rules are complied, the methodology can be applied to assessment of welds with results between values obtained by AISC and ČSN – which is more conservative. It is always necessary to perform experimental testing and weld inspection [12] even when the approach is derived from standard.

Acknowledgements

This work is an output of project NETME CENTRE PLUS (LO1202) created with financial support from the Ministry of Education, Youth and Sports under the „National Sustainability Programme I".

References

[1] Maddox S 1991 Fatigue strength of welded structures (England: Cambridge)

[2] Radaj D 1990 Design and analysis of fatigue resistant welded structures (England: Cambridge)

[3] Peč M 2015 Analysis of load capacity of selected welded structural joints under static and cyclic load (Brno: Brno University of Technology)

[4] Lesik D F and Kennedy D J L 1988 Ultimate strength of eccentrically loaded fillet welded connections (Edmonton: University of Alberta)

[5] Peč M, Kubík P, Šebek F, Návrat T and Petruška J 2016 Modeling of the blast load effects in explicit dynamics, Proc. of the 22nd Int. Conf. Engineering Mechanics (Svratka) eds Zolotarev I and Radolf V (Svratka: Academy of Sciences of the Czech Republic) pp 442–445

[6] Šebek F, Kubík P and Petruška J 2014 Localization problem of coupled ductile failure models compared to uncoupled ones, Proc. of the 20th Int. Conf. Engineering Mechanics (Svratka) ed Fuis V (Svratka: Brno University of Technology) pp 632–635

[7] American Institute of Steel Construction 1994 Manual of steel construction: load and resistance factor design (Chicago)

[8] ČSN EN 1993-1-8 2006 Eurocode 3: Design of steel structures - Part 1-8: Design of joints (Praha)

[9] Aygül M 2012 Fatigue analysis of welded structures using the finite element method (Gothenburg: Chalmers University of Technology)

[10] American Institute of Steel Construction 2005 Steel construction manual (Chicago)

[11] Lesik D F and Kennedy D J L 1990 Ultimate strength of fillet welded connections loaded in plane *Can. J. Civ. Eng.* **17** pp 55–67
[12] Miazga G S and Kennedy D J L 1989 Behaviour of fillet welds as a function of the angle of loading *Can. J. Civ. Eng.* **16** pp 583–599

Fe-Mn(Al, Si) TWIP steel – strengthening characteristics and weldability

P Podany[1], M Koukolikova[1], T Kubina[1], R Prochazka[1], A Franc[2]

[1] COMTES FHT a.s., Prumyslova 995, 334 41 Dobrany, Czech Republic

[2] University of West Bohemia, New Technologies - Research Centre, Univerzitni 8, 306 14 Plzen, Czech Republic

E-mail: pavel.podany@comtesfht.cz

Abstract. Twinning Induced Plasticity steel, or TWIP steel, has had increased interest in recent years from various industry sectors. This is due to it being lightweight, strong, and ductile; which are all properties that are useful in the automotive and aerospace industries. These steels potentially can offer lighter weight vehicles and parts with increased strength and other mechanical properties. This combination could offer greater fuel efficiency and performance while at the same time improving the safety features of the vehicle. This steel is characterised by being a high alloy steel, specifically having a high manganese content. It also has a fully austenitic microstructure at room temperature, which is a unique characteristic. But, for TWIP steel to be useful in various industrial sectors, it must have good weldability. This paper deals with the description of the strengthening due to the cold rolling on experimental heats of manganese steel with TRIP/TWIP effect. Impacts on microstructure, yield strength and tensile strength are described. Also, the weldability of experimental TWIP steel by studying the properties of weld joints after laser welding is described.

1 Introduction

Manganese steels with TWIP effect have attracted considerate attention recently since it has excellent tensile strength–ductility combination [1-3]. Their strengthening mechanism can be explained by the presence of alternative deformation mechanisms, such as the creation of twins (TWIP effect), phase transitions produced by strain (TRIP) and plasticity induced by shear bands [4-6]. An important parameter determining the type of deformation mechanism in these steels is their stacking fault energy (SFE).

According to Dumay et al., for SFE values greater than 18 mJ/m2, the TWIP effect tends to occur, while for lower values, the TRIP effect is predominant, with an α' martensite phase being formed for SFE smaller than 12 mJ/m^2. The SFE also depends on the chemical composition and test temperature of the steel. In alloys with a Mn content (wt. %) less than 15%, the TRIP effect dominates, while for a Mn content higher than 25% the TWIP effect is dominant. On the other hand, in alloys with a Mn content between 15% and 25%, the TRIP and TWIP effects can coexist [7] and combine final properties of both types of steels. It is recognised that TRIP steels show better mechanical strength whereas TWIP steels yield better elongation to fracture [8].

The role of aluminium as a solute in the manganese austenitic steels is twofold. It increases the stacking fault energy of the steel and hence reduces the probability of mechanical twinning during deformation. The higher stacking fault energy at the same time eliminates the formation of ε

martensite. Both of these effects lead to a greater resistance of the steel to hydrogen embrittlement. [9].

The role of the aluminium addition has therefore been investigated in this study for the cold rolled austenitic Fe–15.0Mn–0.1C-0.5÷1.5Al (wt. %) steel in order to establish the relation between mechanical properties, microstructure evolution, occurrences of phase transformation, strain-hardening mechanisms and deformation behaviour.

2 Experiment description

Experimental heats were manufactured in a vacuum induction type furnace and cast into a round ingot mould. The chemical composition of both heats is given in table 1. The only difference between the two heats is the amount of aluminium (1.40 % vs 0.38 %).

Table 1. Chemical composition of experimental heats

Heat nr.	Element [wt. %]				
	Mn	Si	**Al**	C	Fe
T15-81	15.1	1.58	**0.40**	0.12	bal.
T15-82	15.0	1.54	**1.40**	0.10	bal.

After cooling, they were reheated in a furnace to the forging temperature of 1100 °C. In a universal hydraulic press, these ingots were then forged into slabs of 280×130 mm cross-section. The slabs were, in turn, hot-rolled to strips of a final thickness of 8 mm. The rolled strips of both heats were annealed at 850 °C for 2 hours. After annealing, the strips were cold rolled in 6 passes. The final reduction of both strips was 40 %. For the study of the microstructure behaviour and its development after annealing, the samples from cold rolled strips were then annealed at 950°C for 3 hours and at 750 °C for 1 hour.

The laser welding was done on ThruDisk 5002 with welding head BEO D70. The laser wavelength was 1030 nm (Yb: YAG) and it was connected to the fibre with 400 μm diameter. The focal length was 200 mm. Welding speed was 20 millimetres per second. Argon was used as a shielding gas at a flow rate of 24 litres per minute.

The mechanical properties were measured on standard samples (on cold rolled strips) and on mini-tensile test samples (on welded specimens). This method of mechanical properties measurement on a small amount of experimental material has proved successful in earlier research [10-16]. Standard tensile tests were carried out according to EN ISO 6892-1: Metallic materials – Tensile testing – Part 1: Test method at room temperature. The chosen strain rate was 0.001 s-1. Tests were executed using electromechanical testing system Zwick. Prior to testing, dimensions of samples were measured and recorded.

The microstructures were documented using a Zeiss Axio Observer optical microscope. EBSD analysis was performed and scanning electron micrographs taken by means of JEOL 7400F microscope and an HKL Nordlys EBSD camera from Oxford Instruments.

Phase analysis by X-ray diffraction was carried out at room temperature using a Bruker D8 Discover diffractometer. The diffracted radiation was detected by means of a planar detector. A cobalt X-ray source has been used ($\lambda K\alpha$ = 0.1790307 nm). The instrument was equipped with a polycapillary lens focusing the primary X-ray beam into a circular spot with a diameter of 0,5 mm.

3 Results and discussion

3.1. Microstructure evolution of sheets after cold rolling and after annealing

3.1.1. Cold rolling

The microstructure of sheets in the initial state (annealed after hot rolling) is very complex. Etching in Klemm's reagent reveals ε martensite colourless γ (austenite) yellow to brown and α′ martensite blue to dark brown (see figure 1 a). HCP (ε) martensite forms little islands in α′ phase. The identification of all phases complies with the results EBSD analysis (see figure 1 b and caption).

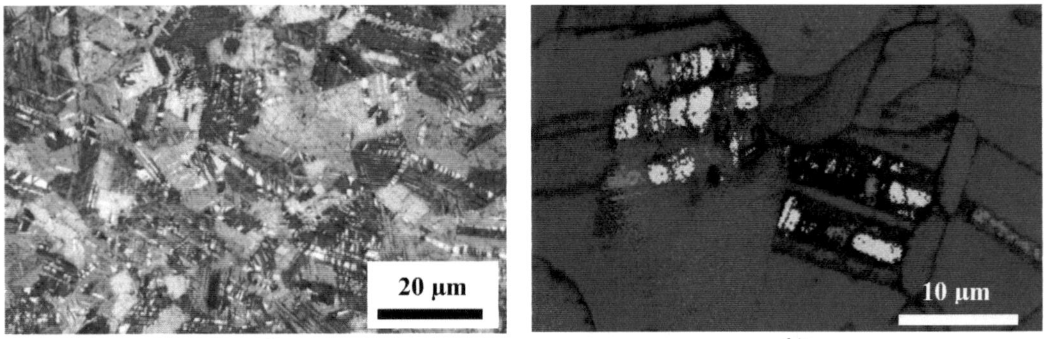

a) b)

Figure 1. Microstructure of heat T15-81 in initial state a) light microscope; b) EBSD analysis - identification of phases (α′ - yellow, ε – red; γ – blue).

The cold rolling affects the microstructure with deformation-induced transformation processes. Increasing strain caused by cold rolling induces twinning in austenite, twins further transforms to ε and ε transforms to α′. The microstructure of cold deformed specimens with 40% reduction is shown in figure 2. It is obvious, that the microstructure consists of the high volume fraction of martensite.

a) b)

Figure 2. Microstructures after cold rolling with 40% reduction
a) heat T15-81-40%; b) heat T15-82-40%.

According to the x-ray diffraction, ε martensite volume fraction in the initial state (before cold rolling) is 28 % for heat with lower aluminium content (T15-81) and 42 % for heat with higher aluminium content. BCC α′- martensite volume fraction in the initial state is similar for both heats: ca 6 – 8 %. Cold rolling leads to substantial increasing of α′ to 72 % and ε to 18 % for T15-81 and to 62 % (α′) and to 27 % (ε) for T15-82 respectively. Austenite content drops from 64 % to 10 % (for T15-81) and from 52 % to 12 % (for T15-82).

3.1.2. Annealing
Subsequent annealing after cold rolling leads to complete recovery of microstructure and restoration of austenite fraction. Following figure 3 shows the comparison of the microstructure of both heats after annealing at 750 °C for 1 hour and after annealing at 950 °C for 3 hours.

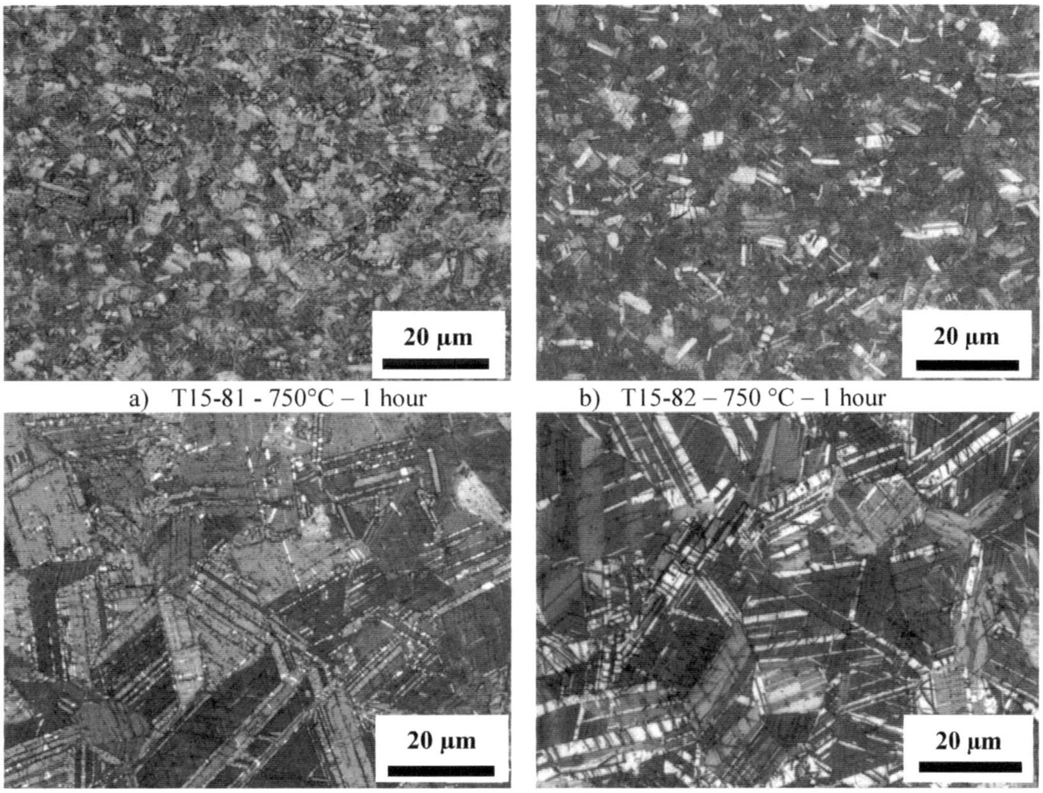

a) T15-81 - 750°C – 1 hour b) T15-82 – 750 °C – 1 hour
c) T15-81 - 950°C – 3 hour d) T15-82 - 950°C – 3 hour

Figure 3. Microstructures after annealing.

Annealing at a lower temperature for short time (1 h) leads to the microstructure with fine grains with an average size below 10 μm in comparison to rather coarse-grained microstructure after annealing at 950° C for a longer period (3 h). The x-ray diffraction results show the decrease of austenite content due to the longer annealing for both heats. All results from x-ray diffraction are summarised in table 2. Austenite decreases mainly on the expense of increasing α′ martensite in the case of heat T15-81 and the ε martensite for heat T15-82 respectively.

Table 2. Volume fraction of phases according to the X-ray diffraction

	750°C/1h		950°C/3h	
	T15-81	T15-82	T15-81	T15-82
α′	11.7	1.4	32.5	8.5
ε	26.2	33.8	29.6	48.4
γ	62.1	64.8	37.9	43.1

3.2. *Mechanical properties*

Evolution of mechanical properties during cold rolling are summarised in figure 4. Each column represents the yield and tensile strength after particular degree of reduction 0 %; 5 %; 10 %; 20 %; 30 %, 35 % and 40 %. The strengthening effect affects mainly the yield strength which increases from 258 MPa to 1472 MPa for heat T15-81 and from 304 MPa to 1278 MPa for heat T15-82.

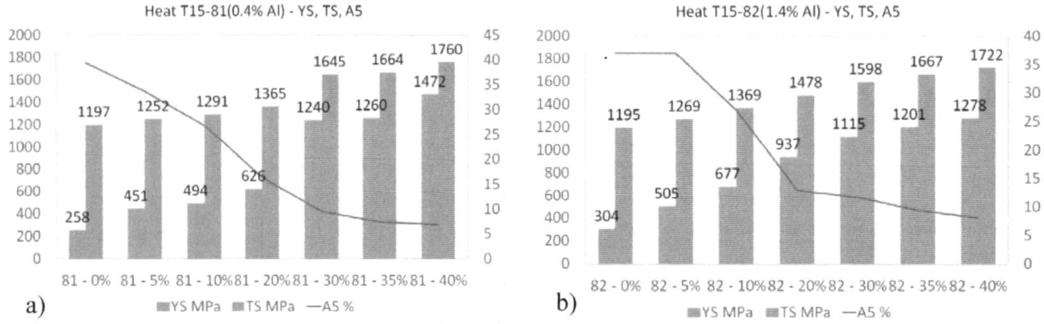

Figure 4. Mechanical properties of cold rolled sheets after reduction in the range 0 ÷ 40%: a) heat T15-81; b) heat T15-82.

3.3. *Welding*

In this experiment, laser welding was used to weld lap joints of two base metal compositions (heat T15-81 and T15-82) using laser power of 1650 Watts. In all of the welds, welding was done using a remote controlled solid state robotic laser. There were tried various metal combinations – both heats were combined, also various states – annealed and as cold rolled. The welding was done with and without a protective atmosphere. The overall look of weld line and cross section of the typical weld is shown in figure 5. The weld properties were tested by means of hardness measurement across the weld section and by means of minitensile test on samples which were taken from the top sheet, bottom sheet and the weld metal itself. The tensile strength of all samples made just from the weld metal varied from 880 to 1290 MPa with average tensile strength 1129 MPa. The average yield strength was 325 MPa. Lower values of tensile strength were achieved, when no shielding gas took place during the welding procedure.

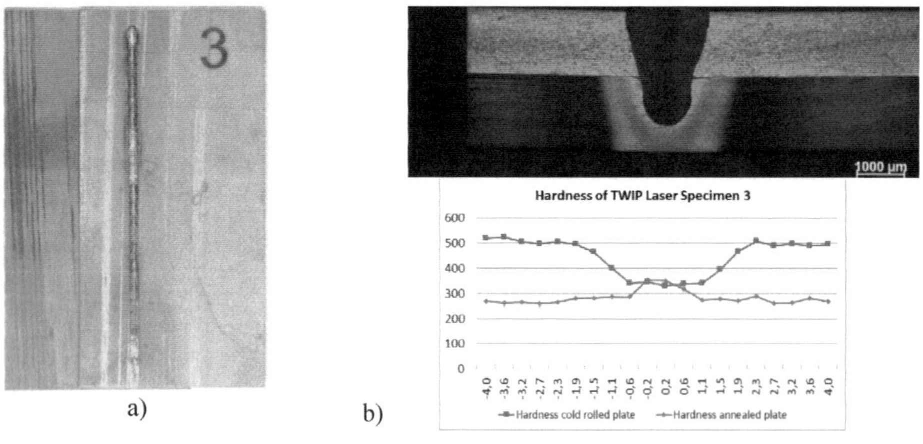

Figure 4. Laser welding of experimental sheets: a) configuration of weld seam, b) cross section of weld 3 with top annealed sheet + bottom cold rolled sheet and hardness values.

4 Conclusion

The strengthening behaviour and microstructure evolution of two heats of steel with different aluminium levels have been studied. The following phases have been identified in both heats: γ (FCC) and two crystallographic variants of martensite, ε (HCP) + α´ (BCC). The strengthening mechanism of both steels consists in increasing martensite content which is induced by the deformation during the cold rolling. Higher yield strength and tensile strength of heat T15-81 are related to the higher content of α´ martensite in this heat (72 %) in comparison to the heat T15-82 (62 %).

The combination of alloying elements in investigated manganese steel showed the higher content of ε martensite in steel with higher aluminium content. It is on the contrary to the results of Ryu et al. [9], where the positive effect of decreasing of ε martensite content by means of Al alloying was shown. Such different behaviour can be related to the dissimilar carbon content (Ryu et al. used 0.6 % of carbon) where carbon interplay (in some way) with aluminium. This phenomenon will be further investigated in the future research.

The tensile strength of both heats exceeds 1450 MPa in as cold rolled state, the elongation A5 was above 35 % for annealed state. Both heats proved good properties after laser welding.

Acknowledgements

These results were created under the project entitled Development of West-Bohemian Centre of Materials and Metallurgy No.: LO1412, financed by the Ministry of Education, Youth and Sports of the Czech Republic.

References

[1] Grassel O, Frommeyer G, Derder C, Hofmann H, *Journal of Physics*. IV 7 (1997) 383–388.

[2] Kang S, Jung, Y-S, Jun J-H, Lee Y-K, *Materials Science and Engineering* A 527 (2010) 745–751.

[3] Xiong Z P, Ren X P, Bao W P, Lia S X, Qu H T, *Materials Science and Engineering,* A 530 (2011) 426–431

[4] Hamada S A, Manufacturing, mechanical properties and corrosion behaviour of high Mn TWIP steels, *Acta Universitatis Ouluensis C 281* (2007).

[5] Niewielski G, Kuc D, Cebulski J, Lalik S, Microstructure and selected properties of Mn-Al duplex steels, *Archives of Materials Science and Engineering* 47/1, 2011, pp. 11-18.

[6] Frommeyerr G, Brux U, Microstructures and mechanical properties of high-strength FeMn-Al-C Light TRIPLEX Steels, *Steel Research International 77,* 2006, pp. 627-633.

[7] Saha D C, Cho Y and Park Y D. Metallographic and fracture characteristics of resistance spot welded TWIP steels. *Science and Technology of Welding and Joining*. 2013 VOL 18, NO 8.

[8] Asghari A,Zarei-Hanzaki A, Eskandari M, Temperature dependence of plastic deformation mechanisms in amodified transformation-twinning induced plasticity steel". *Materials Science and Engineering* A579(2013)150–156.

[9] Ryu J H, Kim S K, Lee CS, Suh D-W, Bhadeshia H K D H, Effect of aluminium on hydrogen-induced fracture behaviour in austenitic Fe–Mn–C steel. *Proceedings of the royal society A,* 469: 20120458 2013. http://dx.doi.org/10.1098/rspa.2012.0458

[10] Dzugan J, Konopik P., Rund M and Prochazka R *2015 ASME Pressure vessels and piping conf. - 2015* Determination of local tensile and fatigue properties with the use of sub-sized specimens, **Volume 1A: Codes and Standards** (New York: Amer. Soc. Mech. Eng.)

[11] Konopík P, Dzugan J and Prochazka R 2013 *METAL 2013 – 22nd Int. Conf. on Metallurgy and Mat. (Brno)* Determination of fracture toughness and tensile properties of structural steels by small punch test and micro-tensile test (Ostrava: TANGER Ltd.) pp 722 – 27

[12] Konopík P, Dzugan J and Prochazka R 2013 *Mat. Sci. and Tech. Conf. and Exh.* Evaluation of local mechanical properties of steel weld by miniature testing technique, pp 2404-11

[13] Konopík P and Džugan J 2012 *2nd Int. Conf. SSTT on Det. of Mech. Prop. of Mat. by Small Punch and other Miniature Testing Techn.* Determination of Tensile Properties of Low Carbon

Steel and Alloyed Steel 34CrNiMo6 by Small Punch Test and Micro-Tensile Test (Ostrava: Ocelot sro) pp 319-328.

[14] Podany P, Martinek P, Thermomechanical processing of micro-alloyed steel, *Materiali in Tehnologije*, 48 (6), 2014, pp. 855-859.

[15] Rund M et al 2015 *Procedia Engineering* Investigation of Sample-size Influence on Tensile Test Results at Different Strain Rates, Vol. 114 (Amsterdam: Elsevier science bv) pp 410-15

[16] Konopik P, Dzugan J and Rund M 2015 *METAL 2015 – 24th Int. Conf. on Metallurgy and Mat. (Brno)* Determination of fracture toughness in the upper shelf region using small sample test techniques (Ostrava: TANGER Ltd.) pp 710-15

IOP Conf. Series: Materials Science and Engineering **179** (2017) 012058 doi:10.1088/1757-899X/179/1/012058

Prediction of fracture in the shearing process using DEFORM and MARC software

I Poláková[1], M Urbánek[1], T Kubina[1], J Džugan[1]

[1]COMTES FHT a.s., Dobřany, 334 41, Czech Republic

ivana.polakova@comtesfht.cz

Abstract. The aim of this study is the evaluation of the parameters for the material model description in simulation of plasticity and fracture in shearing processes. The experimental measurements and the simulations are performed at room temperature for TWIP steel and 38MnVS6. The behaviour of materials is described using MARC/MENTAT and DEFORM software based on the finite element method. The material model is made up on the basis of tension test. The fracture criteria are based on the experimental failure during tension and torsion tests. The fracture models Normalized Cockcroft-Latham (NCL) and Oyane are used for subsequent simulation of shearing. The description of shearing comprises not only the initiation of a fracture but also the growth of the fracture until the material parts are separated. This issue is solved by means of two numerical methods, the softening and the element deletion.

1 Introduction

The finite element method (FEM) is widely used numerical methodology for the analysis of arbitrary metal forming processes. A numerical model for shearing process involves equations for accumulation of critical level of damage which lead to the initiation and the growth of a fracture in the workpiece. The prediction of a fracture in shearing process is analyzed in DEFORM and MARC software. The validation of accuracy and reliability of used methodology is verified through measurements.

The aim of this work is to describe the process of obtaining critical damage values for simulation of shearing process. Tensile and torsion test are carried out for obtaining the material model. Two damage models (Normalized Cockcroft & Latham and Oyane) for simulation of ductile fracture are used. The empirical hypothesis is that the ductile fracture occurs when the maximum damage value exceed the critical damage value [1]. The shearing simulation is subsequently calculated with the obtained critical damage values. The growth of the fracture in the simulation of shearing operation is ensured by means of two methods. The first method is the softening of material data and the second method is the element deletion. Damage softening is a method by which the flow stress of an element above this critical value will be reduced to a specified percentage. For element deletion, elements which exceed a defined critical damage value are deleted from the mesh in the simulation. The simulated and measured forces are afterwards compared for verification of used method.

The intention of the work is carry out the sufficiently accurate shearing simulation with the material and damage model from limited amount of experimental input data. This work appends plastic behavior of TWIP steel and 38MnVS6 and its damage critical values into the ductile fracture theory. It is expected that the introduced approach of obtaining data for shearing simulation can be useful for a wide range of application.

2 Material model

The experiments are performed on TWIP steel and 38MnVS6. The chemical compositions of tested materials are listed in **Table 1** and **Table 2**. The material of TWIP steel is provided as rolled sheet with the thickness 8 mm and the samples of 38MnVS6 are in the form of rod with diameter 10 mm.

Table 1. Chemical composition of TWIP steel.

C	Si	Mn	P	S	Cr	Mo	Ni	Cu	Al	As	B
0.116	1.58	15.09	<0.005	0.002	0.085	0.124	0.072	0.078	0.38	0.015	0.002
Pb	**Co**	**Nb**	**Ce**	**Sb**	**Sn**	**Ta**	**La**	**Ti**	**V**	**W**	**Zr**
0.090	0.008	0.022	<0.005	0.029	0.008	0.261	0.016	0.003	0.014	0.100	0.015

Table 2. Chemical composition of 38MnVS6

C	Si	Mn	P	S	Cr	Mo	Ni	Cu	Al	As
0.39	0.60	1.33	0.02	0.05	0.14	0.03	0.08	0.05	0.01	0.01

2.1 TWIP steel

The specimens for the tension and the torsion test are made from the tested material. In the case of TWIP steel, the part of the specimen subjected to tension has the length 60 mm, the width 12.5 mm and the thickness 8 mm. An extensometer with 25 mm gauge length is used to measure displacement change of the central part of the sample. This uniaxial tension test is used for obtaining load-displacement curves. The shape of true stress-strain curve up to the ultimate tensile strength is directly calculated from these measured curves. The tension test is simultaneously simulated and the true stress-strain curve is fitted by trial-and-error method until the load data from the experiment corresponded well with the simulation. The temperature of the tested material is 20°C and the rate of the sample holder is 0.05 mm/s. The same conditions set during the experiment are for the simulation. The measured load-displacement curves compared with the fitted data from simulation are presented in **Figure 1**. The measured length by extensometer elongates almost 9 mm before the fracture started on the sample.

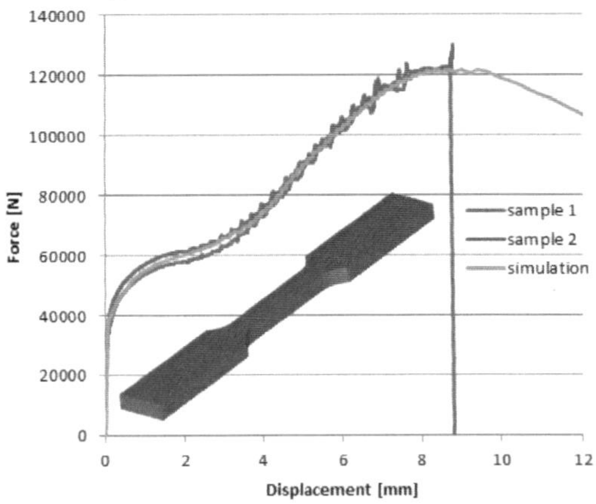

Figure 1. The comparison of measured and simulated tension test (TWIP steel).

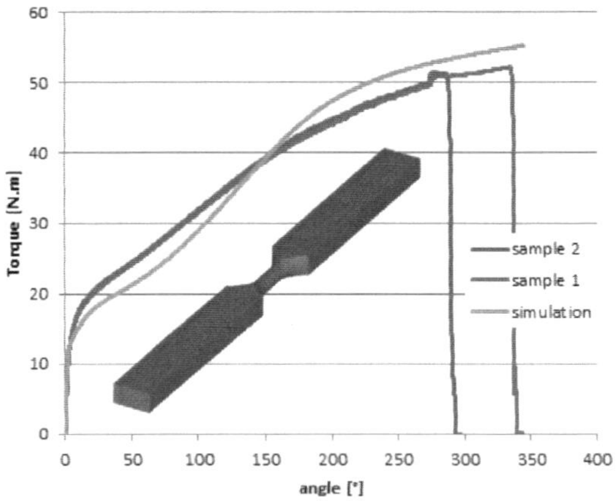

Figure 2. The comparison of measured and simulated torsion test (TWIP steel).

The sample for torsion test has the length 190 mm, the width 15 mm and the thickness 8 mm. The part of the sample subjected to torsion has the length 10 mm and the diameter 6 mm. The torsion test is carried out and simultaneously the simulation of the test is performed with the same process parameters. The rate of rotation is 2°/s. The material data model for the torsion test simulation is described by true stress-strain curves from the previous tension test. The comparison of measured and simulated torsion test is in **Figure 2**. The torsion angle of tested specimen is about 300° when the fracture appeares.

2.2 Material 38MnVC6

The smooth round bar has a diameter of 6 mm and a gauge length of 30 mm. The elongation record is obtained through a mechanical extensometer with the initial length 20 mm, see **Figure 3**. The extensometer provides the displacement vs. time curve while the machine provides the crosshead displacement and force-time records. Using the ZWICK 200 kN machine, the test is carried out at the speed of 0.033 mm/s. The extensometer measured length elongates 4.7 mm before the fracture.

A small-size specimen of 3 mm diameter and 10 mm gauge length is used for the torsion test. The specimen bent slightly during the test, which distorted the test data. Despite that, the torque data calculated from the piston angle appears to be in relatively good agreement with the simulation (**Figure 4**), showing that the deflection has no major effect. The torsion angle is ca 520° before the specimen fractures.

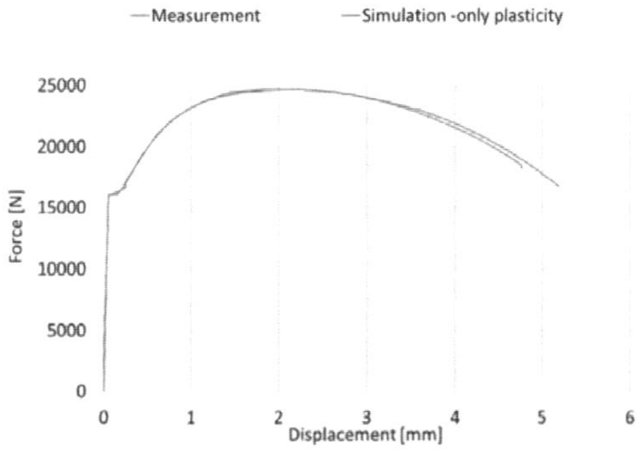

Figure 3. The comparison of measured and simulated tension test (38MnVS6 steel).

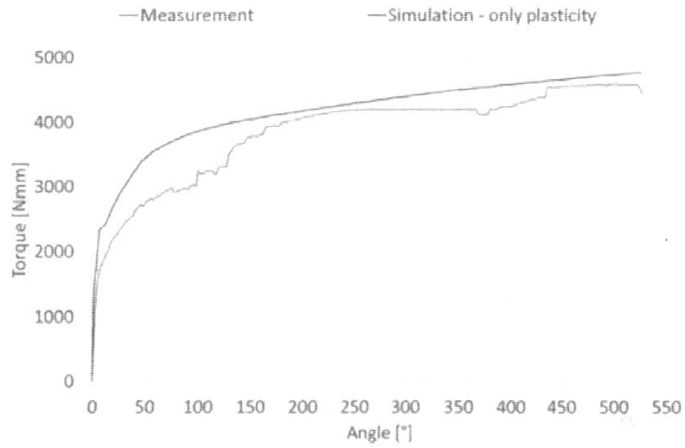

Figure 4. The comparison of measured and simulated torsion test (38MnVS6 steel).

3 Damage models

Many publications are available about the use of damage criteria [2]-[6]. The prediction of the fracture site and its propagation depends on the working conditions, i.e. deformation history, state of stress, temperature and strain rate sensitivity, etc. The fracture in the tested materials during the shearing process is studied by means of two damage criteria Normalized Cockcroft & Latham and Oyane. Damage is a cumulative parameter which evolves through the history material undergoes and its critical value needs to be determined experimentally through suitable workability tests, under defined and controlled conditions. This study examines a fracture criterion on the basis of tension and torsion tests. The quick drop of the load in the experiment shows the crack formation in the specimen. The simulation of performed tests can demonstrate an effect of the stress and strain growth during tests which lead to the fracture.

In the literature [7], it is described great effect of stress triaxility on fracture behavior of materials. The stress triaxility is defined as:

$$\eta = \frac{\sigma_m}{\sigma_{eq}} \tag{1}$$

where σ_m is hydrostatic stress and σ_{eq} is equivalent stress.

IOP Conf. Series: Materials Science and Engineering **179** (2017) 012058 doi:10.1088/1757-899X/179/1/012058

The damage criterion by Cockcroft & Latham (in normalized form) is defined:

$$D_C = \int_0^{\varepsilon_f} \frac{\sigma_1}{\sigma_{eq}} d\varepsilon \tag{2}$$

It takes accent on max. principal stress σ_1, it means that the principal stress cause the fracture in the specimen. The value ε_f represents the strain in the moment of fracture initiation.
The simulation of tensile and torsion tests were performed in the first iteration without damage fracture, where only cumulation of the damage factor for NCL criterion was calculated. The critical damage value is related to the moment of fracture in the simulation of torsion test which corresponds to the experimental fracture for each tested material. Consequently, the critical damage value was improved by calculation using modification presented in [7] of NCL equation (2).
Oyane damage criterion is based on stress triaxiality, which implicates the initiation of fracture:

$$D_C = \int_0^{\varepsilon_f} \left[1 + \frac{1}{a_0} \frac{\sigma_m}{\sigma_{eq}}\right] d\varepsilon \tag{3}$$

where a_0 is material coefficient to be determined experimentally.
The calculation of critical damage value for Oyane equation (3) needs to determine two parameters. The reason, why is chosen torsion test, is that the triaxiality of this test is approximately zero which enables simplify the equation: $D_C \approx \varepsilon_f$. This leads to the equality of critical damage value and the accumulated strain in the simulation of torsion test at the moment of fracture. Second parameter of (3) can be simply calculated by means of the fracture strain and triaxiality obtained from simulation of tension test.

4 Shear test

4.1 TWIP steel

Hydraulic shears is used for cutting of rolled sheets (**Figure 5**), with maximal hydraulic pressure 26(28) MPa, maximal cutting force 4.5MN and nominal vertical speed 5 mm/s (35 mm/s with accumulators). For measuring the shear force is used one analogue pressure transducer connected on incoming pipe to IBA system of pilot rolling mill. This configuration allows record the shear force during the test in dependence on time.

Several experimental cuts have been performed using above mentioned "shearing device". These cuts are numbered on a picture as 6 to 9 number, where cut 6 corresponds to the length of cut 210 mm, cut 8 to 185 mm and cuts 7 and 9 has equal length 63.5 mm.

Figure 5. Tested rolled sheet.

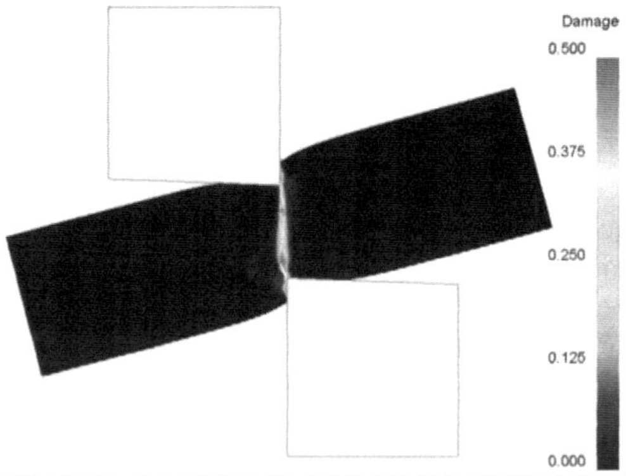

Figure 6. 2D numerical simulation of shearing.

The calculated critical damage values for both integral criteria NCL and Oyane are used in the subsequent simulation of fracture during shearing operation. The distribution of damage in the selected moment is seen in the **Figure 6**. Shearing operation is calculated using two methods: the softening of material data and the element deletion. **Figure 7** shows the comparison of the calculated force from numerical simulation and the measured force.

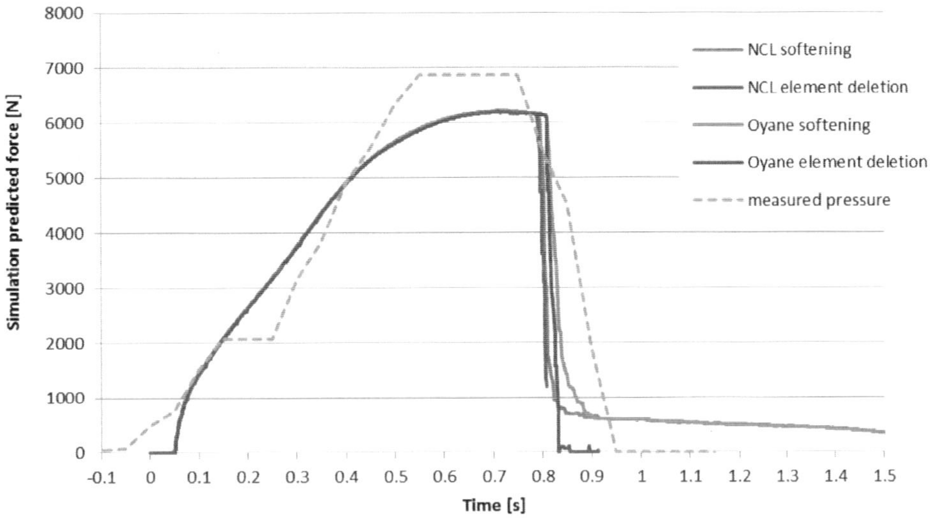

Figure 7. Measured and calculated force during shearing operation for selected methods.

4.2 Material 38MnVC6

The test is carried out at ambient temperature with the punch speed of 90 mm/s. The shearing die is placed on a plate load cell attached to the frame of an MTS 810 testing machine. The punch is connected to a moving piston and the test specimen of 10 mm diameter is placed into the fixture. The die clearance is 0.05 mm in accordance with the applicable standard. The model of the specimen comprises only one quarter of the actual piece (**Figure 8**) because its two symmetry planes. The shear plane has a much finer mesh than the other parts of the model. In the simulation model, the material is considered to be homogeneous, isotropic and elastoplastic. Its plasticity was characterized by the fitted flow stress curve. The parts of the fixture are considered rigid bodies for the calculation. The distribution of equivalent stress in the selected moment of simulation is seen in the **Figure 9**. **Figure 10** shows the comparison of the experimental forces and calculated force from numerical simulation.

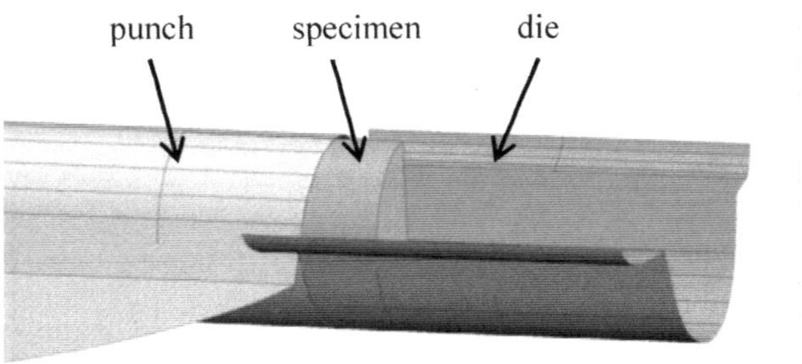

Figure 8. ¼ FEM model for shearing of rod.

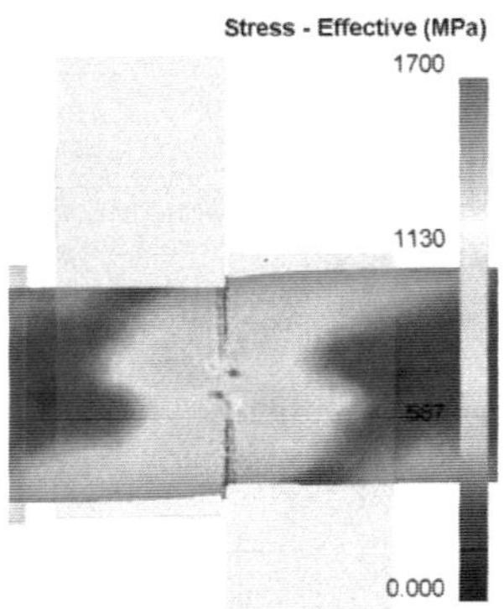

Figure 9. FEM simulation.

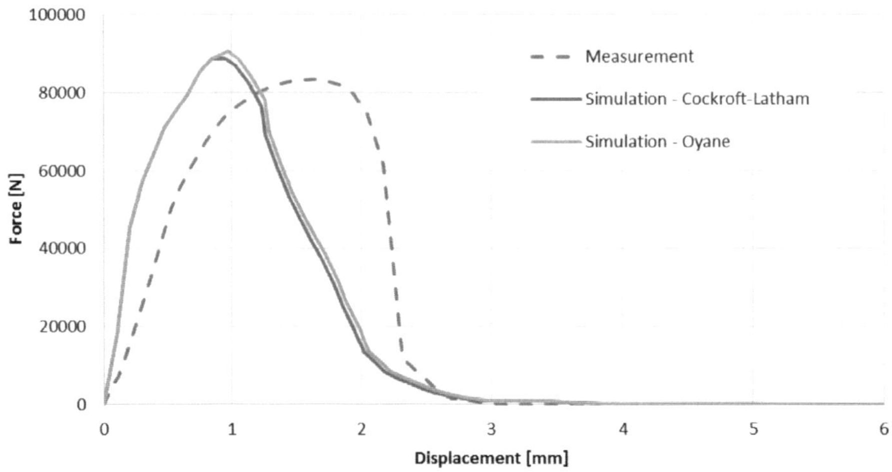

Figure 10. Comparison of measured and simulated results.

5 Conclusion

Fracture prediction plays an important role in the description of a shearing process. This study presents the procedure of obtaining critical damage values on the basis of only two experiments. Numerical simulations are performed with the use of uncoupled models, where the fracture is analyzed by combining plasticity and ductile damage. Material plasticity is characterized by flow stress curve obtained from a tensile test. The critical damage value for each material is identified from simulation of torsion test at the moment of fracture which corresponds to the failure during experiment.

The performed technological tests prove the usability of the chosen method of identifying the critical damage values. The measured and simulated data for shearing test were in a quite good agreement. This presented procedure can enable the prediction of forming limits for a wide range of application.

Acknowledgements
This paper was created by project Development of West-Bohemian Centre of Materials and Metallurgy No.: LO1412, financed by the MEYS of the Czech Republic.

References

[1] Yu S and Feng W 2011 Experimental research on ductile fracture criterion in metal forming *Frontiers of Mech. Eng.* **6(3)** pp 308-11.
[2] Graham S M, Zhang T, Gao X and Hayden M 2012 Development of a combined tension-torsion experiment for calibration of ductile fracture models under conditions of low triaxiality *Int. Journal Of Mech. Sci.* **54(1)** pp 172-181
[3] Pires F M A, Cesar de Sa J M A, Costa Sousa L and Natal Jorge R M 2003 Numerical modelling of ductile plastic damage in bulk metal forming *Int. Journal of Mech. Sci.* **45** pp 273-294
[4] Španiel M et al. 2014 Calibration of fracture locus in scope of uncoupled elastic–plastic-ductile fracture material models *Advances in Eng. Soft.* **72** pp. 95-108
[5] Ribadas H F and Rund M 2015 Evaluation of ductile fracture using micro-samples *Proc. Eng.* **114** pp 377-384
[6] Džugan J et al. 2015 Identification of ductile damage parameters for pressure vessel steel *Nucl. Eng. Des.* http://dx.doi.org/10.1016/j.nucengdes.2015.12.014
[7] Bai Y and Wierzbicki T 2015 A comparative study of three groups of ductile fracture loci in 3D space *Eng. Frac. Mech.* **135** pp 147-67

Evaluation of in-situ deformation experiments of TRIP steel

J Procházka, L Kučerová, M Bystrianský

RTI, UWB in Pilsen, Universitni 8, 30614 Pilsen, Czech Republic, EU

E-mail: jprochy@rti.zcu.cz, skal@rti.zcu.cz

Abstract: The paper reports on the behaviour of low alloyed TRIP (transformation induced plasticity) steel with Niobium during tensile test. The structures were analysed using in-situ tensile testing coupled with electron backscattering diffraction (EBSD) analysis carried out in scanning electron microscope (SEM). Steel specimens were of same chemical composition; however three different annealing temperatures, 800 °C, 850 °C and 950 °C, were applied to the material during the heat treatment. The treatment consisted of annealing for 20 minutes in the furnace; cooling in salt bath after the heating and holding at 425 °C for 20 minutes for all the samples. Untreated bar was used as reference material. Flat samples for deformation stage were cut out of the heat-treated bars. In situ documentation of microstructure and crystallography development were carried out during the deformation experiments. High deformation lead to significant degradation of EBSD signal.

1 Introduction

In-situ testing presents a new advanced method of material characterisation and testing. It can be performed in various kinds of microscopes to fit various purposes [1-4]. Scanning electron microscopes (SEM) provides suitable compromise between very high resolution but very small samples of transmission electron microscopy (TEM) and large observed area at low magnification of light microscopy. This is the reason why in-situ SEM testing is convenient for high strength steels analysis [3-5].

There are several grades of high strength steels with multiphase microstructures, whose mechanical properties depend strongly on volume fraction and morphology of individual phases and structural components. At least three grades of advanced high strength steels rely on carefully controlled amount of metastable retained austenite in the final microstructure [6-9]. To understand the behaviour of these steels under straining, it is important to establish how the phases react to external loading. Retained austenite in these steels tend to transform to martensite at room temperature during plastic deformation. The stability of retained austenite against this deformation induced transformation depend on the size, morphology and carbon content of each austenitic island, which makes the behaviour of these steels quite tricky to describe [10].

There are several challenges connected with preparation of large areas for electron backscattering diffraction (EBSD) analysis and testing of small samples. Deformation of ductile samples is not homogeneously distributed along the axe of the sample, the highest being local deformation around fracture area. To establish local deformation in the area observed during the test, measuring methodology has to be set up. Various sample shapes and geometries can be used for a particular deformation stage. Depending on the aim of the experiment, it might be practical to use either simple dog-bone shaped sample or some variations with more complicated geometry of the central part.

It should be also noted that every stage requires a minimal length of the sample, which is necessary to enable sample mounting into the stage.

2 Experimental details

2.1 Materials description

Low alloyed high strength TRIP (transformation induced plasticity) steel was used for this work. The steel had 0.2 % C and was further alloyed by 1.5 % Mn, 2 % Si and micro-alloyed by 0.06 % Nb. Manganese is typically added to TRIP steels to support retained austenite stabilization. Silicon is also very frequently found in TRIP steels as it increase the strength of the steel by solid solution strengthening and hinders cementite formation during the processing of the steel. The less carbides are formed the higher carbon content is available to diffuse into remaining austenite and thus increase its stability against martensitic transformation during final cooling of the steel. The original microstructure of the steel after casting, re-forging into the bars and annealing was ferritic-pearlitic.

The steel was further heat treated in the furnace by intercritical annealing at three temperatures of 800 °C, 850 °C and 900 °C for 20 minutes. All the samples were subsequently cooled to the temperature of 425 °C in a salt bath and 20 minutes long bainitic hold at this temperature took place in the second furnace. Resulting microstructures in all the samples consisted of the mixture of free ferrite, bainite, retained austenite and eventually M-A constituent (Figure 1). Around 10 % of retained austenite was found in the final microstructures by X-ray diffraction phase analysis.

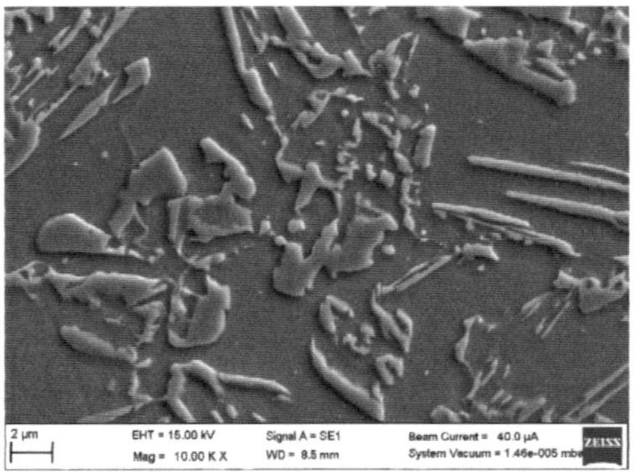

Figure 1. Multiphase TRIP microstructure, 800 °C annealing temperature.

Figure 2. Deformation stage.

2.2 Tensile testing

Deformation stage MTII/Fullam (Figure 2) placed in the SEM Zeiss EVO 25 equipped with energy dispesion X-ray spectrometer (EDX) and EBSD detectors (both from Oxford Analytical) was used for the experiment. This stage is designed for testing inside the SEM chamber with loading force up to 10 kN. Tensile tests described in this work were carried out at room temperature, however the stage enables heating of the samples up to 1200 °C during loading experiments. The stage was tilted 70 °towards EBSD detector during some of the measurement. When only SEM images of etched microstructure were taken, the samples were perpendicular to the electron beam.

Figure 3. Sample geometries used for in-situ tensile tests: plain (left) and curved (right).

The samples were cut using water jet in slices and then to the final shapes. The samples were customized to fit in the deformation stage. Two geometries were used, one for load-strain measurement and tensile strength evaluation, further indicated as 'plain' and the second one had a neck in the central part aimed to observe the fracture initiation mechanism, further indicated as 'curved' (Figure 3).
Prior the SEM analyses tensile tests were carried out to obtain mechanical properties using as-delivered plain samples. Elongation rate was kept $\dot{l} = 1\ \mathrm{mm/min}$. Prior to the in situ analyses, the samples were ground and polished using 1μm diamond paste. Samples dedicated to SEM analysis were etched in 3% Nital, while samples intended for EBSD analysis were electrolytically polished to obtain deformation-free surface.

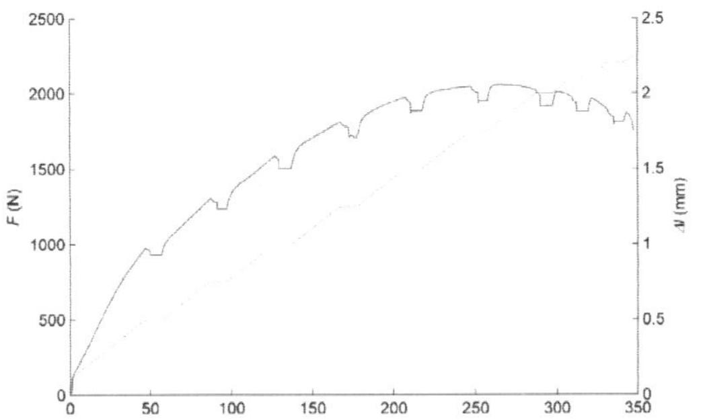

Figure 4. Time evolution of Load L and elongation Δl during tensile test with EBSD.

Figure 5. Emphasized points used for estimation of the strain induced in material.

The EBSD maps were obtained using tensile stage tilted to 70 ° and the tensile tests were interrupted several times to acquire desired data. The time evolutions of loading force F and elongation Δl are shown in Figure 4. During in-situ analyses the strain rate was set to lower value of $\dot{l} = 0.5\ \mathrm{mm/min}$. During the test several pauses had to be made to capture SEM images and perform EBSD measurements. In case of curved samples tests were paused at given elongations: $\Delta l = 0.5, 1.0, 1.25, 1.5, 1.75$ and 2.0 mm. In case of flat samples pauses were made at elongations $\Delta l = 0.5 - 4.0$ mm with step of 0.5 mm. In both cases when the ultimate strength of the specimen was reached, the tests were paused manually to observe the deformations and crack formation and propagation. The notches in the loading curves are caused by pausing the deflection therefore the stresses induced during deformation can relax. To prevent data log overflow, the planned pauses were set to 10 seconds each. Within this period, the sampling was paused manually allowing enough time (10-20 minutes) for EBSD maps and SEM images to be acquired. These manual pauses caused the sharp drops in the measured load in the scheduled pause period.
Stress-strain curves were evaluated from acquired data for plain samples using engineering stress $\sigma_e = \frac{F}{A_0}$ and strain $\varepsilon = \frac{\Delta l}{l_0}$, where A_0 and l_0 were calculated from sample dimensions, see Figure 3. Because of the shape of the curved sample the stress-strain curves could not be evaluated, unprocessed load-deflection curves are presented instead. The strain achieved in observed areas at each stop was estimated from position of three points chosen in every observed area (Figure 5).

3 Results

3.1 Effect of thermal processing on tensile strength

TRIP steel with niobium used in this work can possess very good combination of high ultimate tensile strength and excellent ductility after various heat or thermo-mechanical treatments. In the case of this work, tensile strength of three samples after heat treatments showed the strengths around 800 MPa with ductility approaching 40 % (Figure 6). Mechanical properties of the same steel in the original state before heat treatment are also shown in the graph by dark blue curve. It is apparent that original ferritic-pearlitic microstructure had lower strength and distinctively shorter area of homogeneous plastic deformation. The comparison of the curves support general explanation, that gradual transformation of retained austenite to martensite during plastic deformation postpones the onset of necking of the steel.

Once the continuous tensile test was performed, the stops for discontinuous testing were planed (Figure 7) and the tests were carried out. Each drop on the force – elongation curve corresponds to one pause in the test, during which either microstructure observation or EBSD analysis were performed.

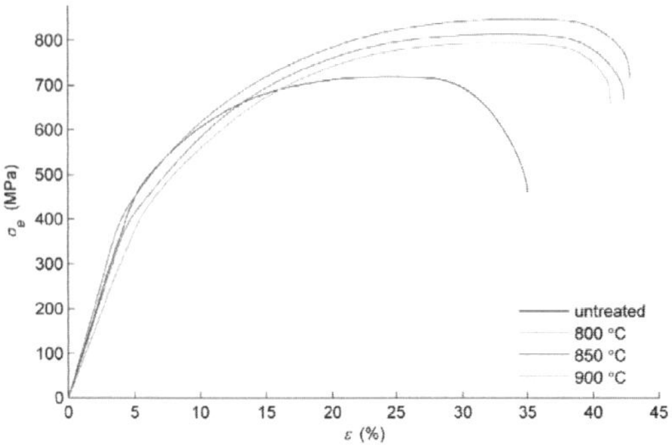

Figure 6. Engineering stress versus engineering strain.

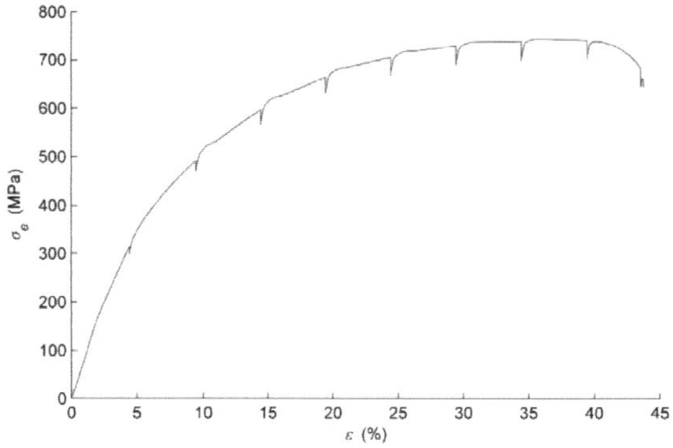

Figure 7. Stress-strain curve for EBSD experiment of the plain sample annealed at 900 °C.

3.2 Structure evolution prior to material failure

Microstructure development was observed during the pauses of tensile straining of the steel treated with 800 °C heating temperature. Representative stops were chosen and obtained microstructure images are shown in the Table 1. The images make it clear that cracks initiated at the highest load of 2 226 N, which corresponds to the ultimate strength of the steel. Fracture originated in this case from the free surface of the sample and the second crack just under the surfaces. Both of the origins used the presence of small hard grains in the microstructure and the cracks propagated along free ferrite – bainite interphase. Curved sample shape was used for this particular experiment to help to identify the most probable fracture initiation area.

The composite images in Figure 8 show the movement of individual phases and structural component during the test, for the sample heated to 900 °C (left) and 800 °C (right). It indicates that close to parallel bainitic laths are moving significantly less than individual smaller bulky items of the microstructure.

Table 1. Fracture initiation and crack propagation during final stages of the tensile test.

Δl [mm] *F* [N] ε [%]	Microstructure and fracture initiation	Δl [mm] *F* [N] ε [%]	Microstructure and fraction propagation
1.25 1 814 6.7	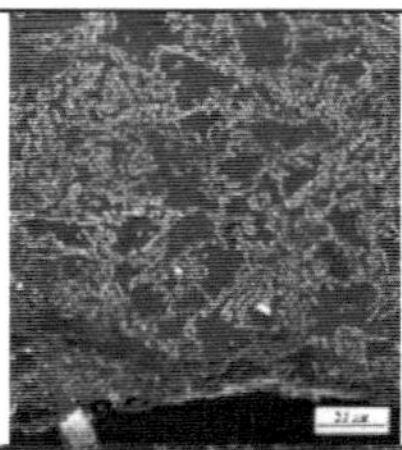	2.30 2 167 39.8	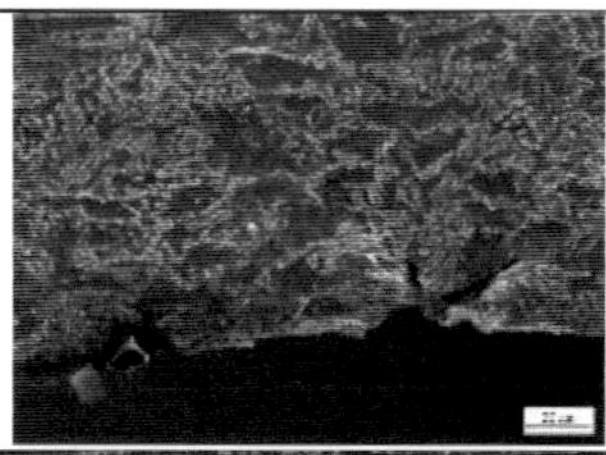
2.10 2 226 28.3	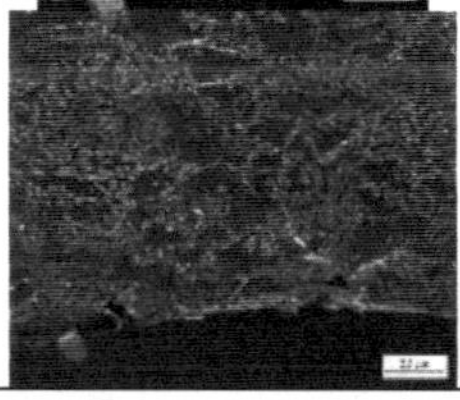	2.48 2 003 51.4	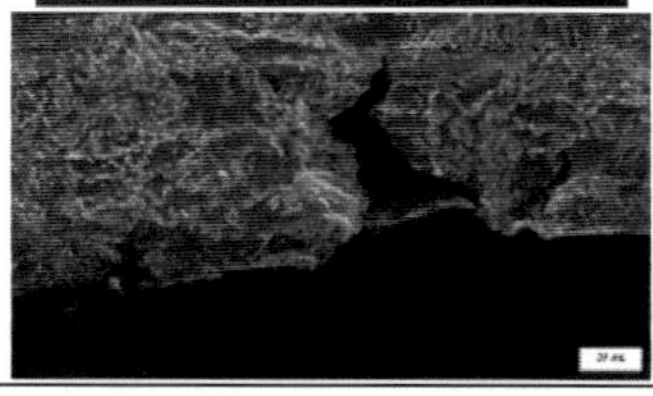

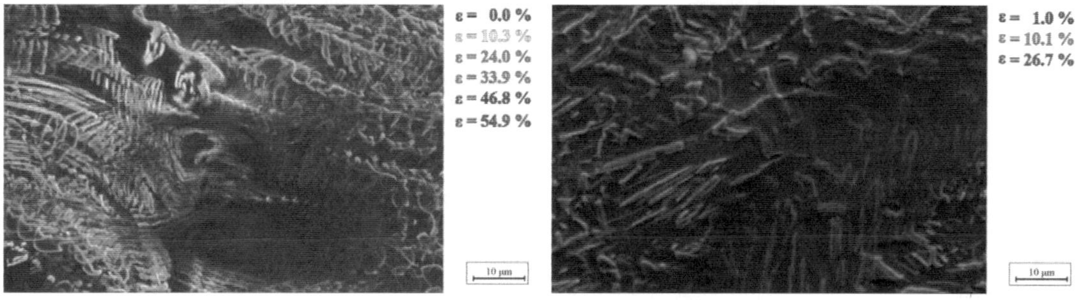

Figure 8. Composite figure of sample surface with various strain. The stress is applied in horizontal direction. Sample with 850 °C heating temperature (left) and 800 °C heating temperature (right).

The deformation occurs mainly in soft ferritic grains; harder areas of retained austenite or M-A constituent are dragged by the material flow. Small grains keep their shape and change only their position and orientation. Indicated deformation values were calculated from the gradual shifts of three chosen points during the straining (Figure 5).

Table 2. Band contrast, inverse pole figure and phase map derived from EBSD analysis of the microstructure development during tensile test of sample heated at 850 °C.

ε [%]	σ_e[MPa]	Band contrast	IPF-Z	Phase map
0	0			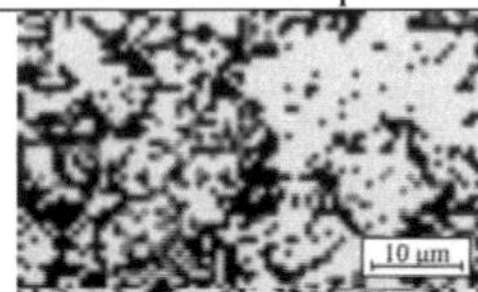
4.4	314	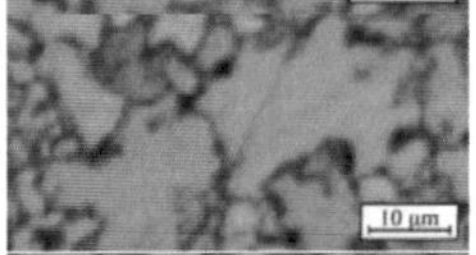		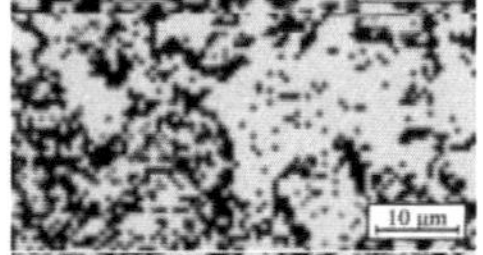
9.3	492			
14.4	597			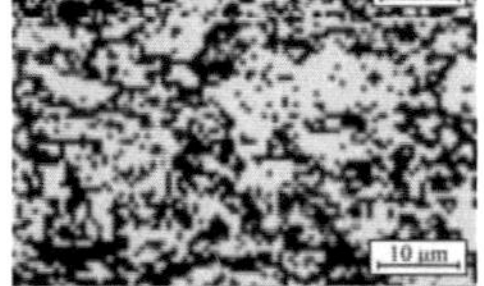
19.3	664			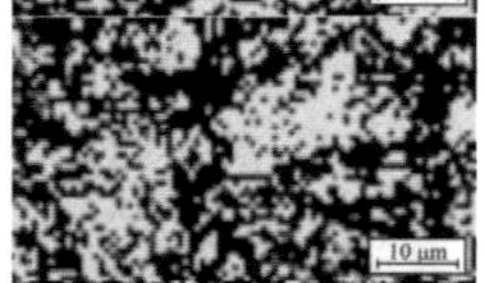
29.3	729		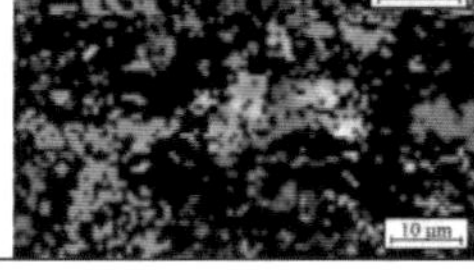	

Acquisition of a high quality EBSD pattern depends on several factors, one of the most important being perfectly planar and deformation free sample surface. As it turned out during the experiments with EBSD mapping, the smoothness of sample surface presents quite a significant issue at the later stages of the deformation. Documentation of EBSD maps during the nine stops of tensile test (Figure 7) is shown in Table 2.

Reasonable quality EBSD patterns were obtained up to about 10 % elongation of the sample. Further deformation creates visible surface relief which effectively disturbs EBSD mapping of the sample. As this value emerged as a critical one for several test of this steel, it seems to be a generally applicable value of elongation up to which reliable in-situ EBSD analysis can be performed on the multiphase steel based on ferrite – carbide free bainite microstructure. This is in good agreement with other works [11, 12], which also published EBSD maps obtained at smaller deformations up to about 10% only. The IPF-Z images (inverse pole figures in the direction perpendicular to sample surface) demonstrate that during the straining relatively large grains of free ferrite get fragmented into several sub-grains with various crystallographic orientations. The orientation misfit is originally small and increases gradually with increasing applied load. The sub-grains were further fragmented down at high loads.

4 Conclusion

In-situ tensile straining of three low alloyed TRIP steels with the same chemical composition, but various annealing temperatures of 800 – 950°C was performed. The microstructures consisted of ferrite, bainite and around 10% of retained austenite.

Two kinds of flat samples had been design and their surfaces prepared to high standards to enable EBSD acquisition. Classical dog-bone shaped flat samples were used for EBSD analysis and most of the microstructure observations. Microstructure development at the fracture and fracture initialization and propagation have been observed during the staining of curved sample Thinned middle area of the sample localized fracture into this particular area, making it easier to notice fracture initiation during the testing. Cracks initiated at sample surface and propagated along ferrite-bainite interphase.

It has been further found out, that bulky ferrite grains can split into several sub-grains with different lattice orientations during the straining. Lattice misfit between the sub-grains was gradually increasing with increasing load. EBSD analysis was possible up to 10% deformation, as the number of zero solutions increased significantly at higher deformations.

5 Acknowledgement

The present contribution has been prepared under project LO1502 'Development of the Regional Technological Institute' under the auspices of the National Sustainability Programme I of the Ministry of Education of the Czech Republic aimed to support research, experimental development and innovation.

6 References

[1] Chen Q Z, Chu W Y, Wang Y B, Hsiao C M 1995 *Acta Metall. Mater.* 43 4371-6.
[2] Yuan Z Z et al. 2006 *Mater. Charact.* 56, 79-83.
[3] Chena F F, Hwang J, Yu G P, Huang H H 1999 *Thin solid Films* 352 173-8.
[4] Guo E Y et al. 2013 *Mat.Sci.Eng. A* 580 159-68.
[5] Muránsky O, Šittner P, Zrník J, Oliver E C 2008 *ActaMater.* 56 3367–79.
[6] Aišman D et al. 2008 Proc Annals of DAAAM and Proceedings of the International DAAAM Symposium Ed. B. Katalinic DAAAM International Vienna 7-8.
[7] Kučerová L, Jirková H, Mašek B 2016 *Manufacturing Technology 16* (1), 145-9.
[8] Mašek, B et al. 2007 *Mechanical Behavior of Materials X*, 1 – 2, 345-346, 943-6.
[9] Jirková H et al 2015 *Mater Today* 2S, 627-30.
[10] Pereloma E V et al. 1999 *Mat.Sci.Eng. A* 273-275, 448-52.
[11] RYU J H et al. 2010 *Scripta Mater.* 63 297–9.
[12] LI W S et al. 2016 *Mat.Sci.Eng.* A 649 417-25.

Strain controlled cyclic tests on miniaturized specimens

R Procházka[1, 2] and J Džugan [1]

[1] COMTES FHT a.s., Průmyslová 995, 334 41 Dobřany, Czech Republic

[2] University of West Bohemia, Faculty of Mechanical Engineering, Department of Material Science and Technology, Univerzitní 8, Pilsen, Czech Republic

E-mail: radek.prochazka@comtesfht.cz

Abstract. The paper is dealing with strain controlled cyclic tests using a non-contact strain measurement based on digital image correlation techniques on proportional sizes of conventional specimens. The cyclic behaviour of 34CrNiMo6 high-strength steel was investigated on miniaturized round specimens with diameter of 2mm that were compared with specimens in accordance with ASTM E606 standards. The cycle asymmetry coefficient was R= -1. This application is intended to be used for life time assessment of in service components in future work which enables to carried out a group of mechanical tests from a limited amount of the experimental material. The attention was paid to confirm the suitability of the proposed size miniaturization geometry, testing set up and procedure. The test results obtained enabled to construct Manson-Coffin curves and assess fatigue parameters. The purpose of this study is to present differences between cyclic curves and cyclic parameters which have been evaluated based on conventional and miniaturized specimens.

1 Introduction

The current demands for cyclic properties of in service products are leading to use of a non-destructive techniques (NDT) or semi-destructive methods. The methods applied should not have any negative influence on its lifetime and can be used for a wide range of products. Base on NDT results, some relationships between mechanical properties and NDT properties can be found. That brings an advantage in some ways and disadvantages in others. The main issue is the accuracy in the mechanical properties evaluation which decreases with the increasing number of correlations and also there is a need for a large database of various materials. The problem becomes bigger in case of cyclic loading and lifetime prediction. In case of time dependent fatigue, there are many methods how to estimate the fatigue strength based on Charpy impact energy, hardness test or tensile properties. But there is no easy way to do that in case of low cycle fatigue test where cyclic parameters are needed.

To avoid using correlation relationships between two variables, the miniaturized testing technique was developed for the direct determination of mechanical properties. The idea was to use a size miniaturization geometry which is related to ASTM standard [1] and use the same loading conditions. That brings some simplification in evaluation processes and helps to extend the applicability of this method in case of acute shortage of experimental material.

Using latest technologies, the role of scale factor plays less important rule in evaluation processes until the ultimate tensile strength (UTS) point [2] where the grain size [3], surface condition, residual stress and strain rate effect were considered. Very good results obtained with the miniaturized tensile specimens were confirmed in in [4-8]. The weak influence of the scale factor is very promising in field of cycling loading where the normal stress does not exceed UTS, but on the other hand, the influence of probability of large specimens containing all-sized defects is still an open question.

To ensure that the surface conditions remain the same, in order to avoid surface scratches from the knife edges of an extensometer, a non-contact extensometer was employed for the strain controlled cyclic deformation. The Digital Correlation System (DIC) Mercury RT [9, 10] provides high measuring accuracy of measuring volumes from some square milimeters to square meters depending on the camera lens.

The aim of the study is to make the comparison between conventional and sub-sized specimens geometry for the future work dealing with the testing from in-service components using on-site sampling.

2 Materials and Methods

The experimental material was 34CrNiMo6 high-strength steel. Quasi-static tensile tests and strain controlled low cycle fatigue (LCF) tests were investigated here.

2.1 Material description

As the semi-product, the 34CrNiMo6 forged steel round bar with diameter of 95mm was used. The delivered condition of the steel was improved further through the quenching and tempering at 600°C to get optimum mechanical properties which are widely used in industry, see the tempering diagram in figure 1. After hardening a tempered fine-grained martensite was obtained. The chemical composition of the steel is summarized in table 1.

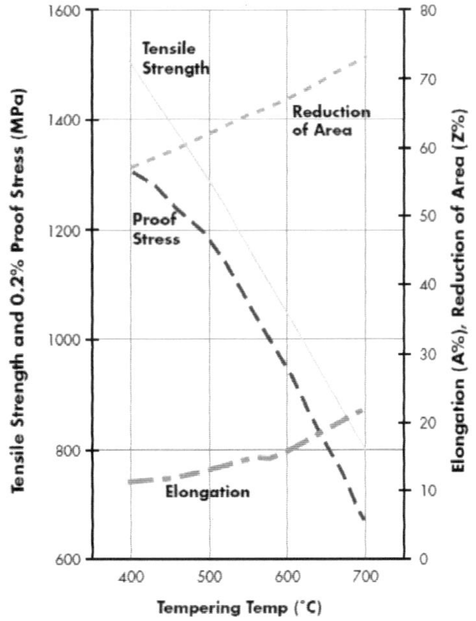

Figure 1. Tempering diagram [11].

Table 1. Chemical composition % of 34CrNiMo6 steel

C	Si	*Mn*	Ni	*P*	S	*Cr*	*Mo*
0.36	0.22	0.75	1.67	0.01	0.01	1.55	0.25

IOP Conf. Series: Materials Science and Engineering **179** (2017) 012060 doi:10.1088/1757-899X/179/1/012060

2.1 Methods

Firstly, the investigation of material homogeneity was carried out by means of tensile tests where the samples were extracted from the surface and middle section of the steel bar. Most crucial part was to obtain the Young modulus of elasticity for further evaluation of cyclic parameters. Secondly, the strain-controlled LCF testing was conducted on conventional and sub-sized specimens. Based on tensile and LCF results, Manson-Coffin curves were constructed and cyclic parameters were obtained. All tests were carried out at room temperature.

2.1.1 Tensile tests. Round test bars with the reduced gage section (8mm) were prepared in the longitudinal direction. Tests were conducted on electro-mechanical testing system Zwick Roell with a load capacity of 250kN. Samples were pulled at a constant strain rate of 0.0004s-1. The tests were performed according to ČSN EN ISO 6892-1 (close to ASTM E 8 – 01): Metallic materials – Tensile testing – Part 1: Method of test at room temperature. A longitudinal deformation of a loaded specimen was measured by means of mechanical extensometer with initial gauge length of 25mm. Dimensions of the samples were measured prior to testing and after the test for determination of stress and plastic behavior quantities determination (Young´s modulus - E, Proof stress - YS, Ultimate Tensile Strength - UTS, Elongation – A and Cross Section Reduction - CSR).

2.1.2 Strain-controlled LCF tests. Tests were conducted under variable strain controlled loading conditions with the cyclic asymmetry coefficient R= - 1 and the sinusoidal shape. The aim was to cover the total strain amplitudes from 0.35 to 2%. As the reference values, the conventional specimens 8mm in diameter was used in accordance with ASTM standards. The specimen geometry is illustrated in figure 2.

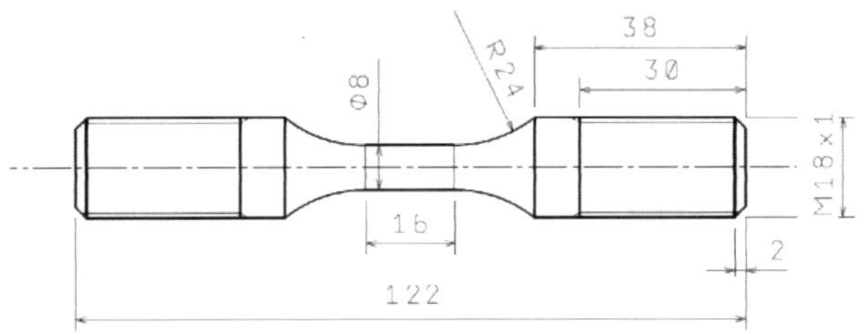

Figure 2. Conventional specimen geometry in accordance with ASTM standard E606, D8.

The sub-size specimen was designed to meet some recommendations from ASTM standard E606 as much as possible. The attention was also paid to design the reduced section which would fit in case of on-site sampling where the extracted sample has dimesions 4x20x40 mm. The draft of the sub-sized specimen geometry with 2mm diameter is depicted in figure 3. Because there is no acute shortage of experimental material in the current case, the larger geometry was used in experiment and its dimensions are depicted in figure 4.

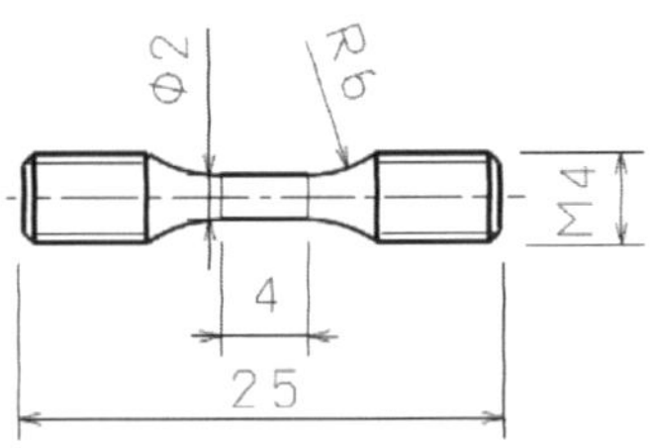

Figure 3. The draft of sub-size specimen geometry.

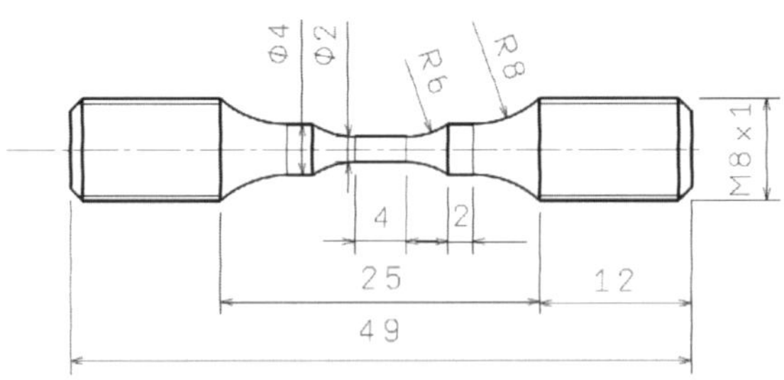

Figure 4. Sub-size specimen 2mm in diameter, D2.

In case of the sub-sized specimen, the initial gauge length of the optical extensometer was set to 3mm. The test setup as well as the detail of the specimen surfece are illustrated in figure5 and figure 6.

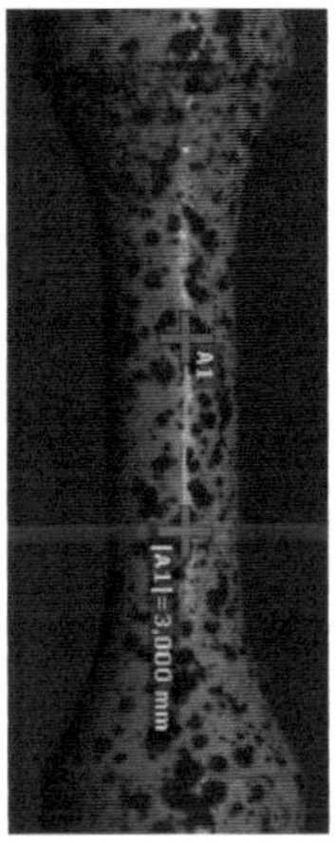

Figure 5. Initial gauge length of optical system.

Figure 6. The test setup of sub-size specimens includs servo-hydraulic testing machine and one CCD camera.

Further evaluation was done based on Basquin's law (1) where stress amplitude σ_a depends on number of cycles to failure N_f and fatigue strength exponent b and fatigue strength coefficient σ_f'.

$$\sigma_a = \sigma_f'(2N_f)^b \tag{1}$$

Mansson-Coffin relation in plastic deformation ε_{ap} is described by equation (2) where ε'_f is coefficient of ductility in fatigue and c is ductility exponent.

$$\varepsilon_{ap} = \varepsilon_f'(2N_f)^c \tag{2}$$

The total amplitude of the deforamtion ε_{at} can be expressed as

$$\varepsilon_{at} = \frac{\sigma'}{E} \cdot (2N_f)^b + \varepsilon' \cdot (2N_f)^c \tag{3}$$

where Young modulus E is needed. The cyclic deformation curve can be estimated based on equation (4) where K' is cyclic hardennig coefficient and n' is cyclic hardenning exponent.

$$\sigma_a = K' \cdot \varepsilon_{ap}{}^{n'} \tag{4}$$

The material cyclic properties were estimated in accordance with geometrical meaning of formula (1), (2) and (4). The parameters were determined based on power law regression analyses of variables in a fully-logarithmic scale.

3 Results and Discussion

Tensile test results show slight differences in tensile properties over the steel bar cross section area. The Young modulus that is required for further evaluation of low cycle fatigue tests was found to be 205GPa.

Table 2. Tensile test results of 34CrNiMo6 steel

	E GPa	YS MPa	*UTS MPa*	A %	*CSR %*
Surface	204.5	916.4	1040.3	18.6	64.2
Middle	205.0	892.9	1010.5	18.3	63.9

Relationships among variables have been found and the material cyclic properties of conventional and sub-size specimens were estimated based on formula (1), (2) and (3). For both specimen geometries the results are shown in table 3.

Table 3. Cyclic material properties of 34CrNiMo6 steel

	σ'f Mpa	b -	ε'f -	c -	K' MPa	n' -
D2	1175.6	-0.0683	0.42	-0.6481	1277,0	1.0409
D8	1049.3	-0.0491	0.46	-0.6368	1079.9	0.0716

The comparision of the cyclic bahaviour was done using cyclic stress-strain curves (CSSC) where the single regression through all points have been made in figure 7. As can be seen in figure 8, no significant differences can be observed.

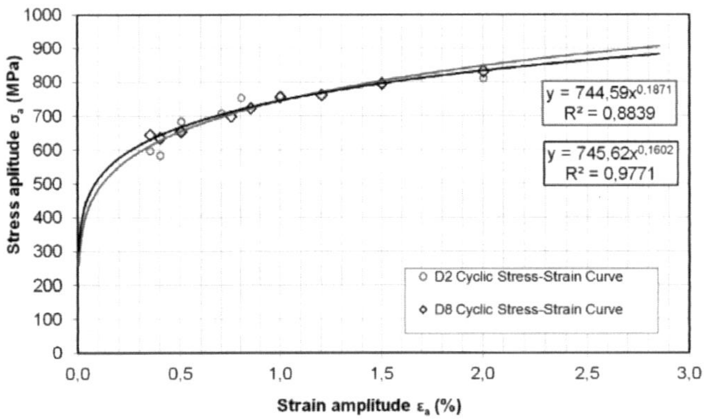

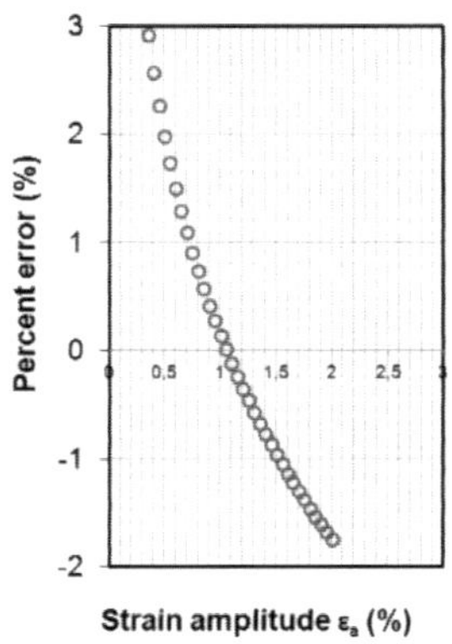

Figure 7. Cyclic stress-strain curves. **Figure 8.** Percent error of CSSC.

Using Manson-Coffin-Basquin equation (3), the fatigue life behaviour was estimated and the fully-logarithmic plot of the strain amplitude versus the number of cycles to failure was done in figure 9.

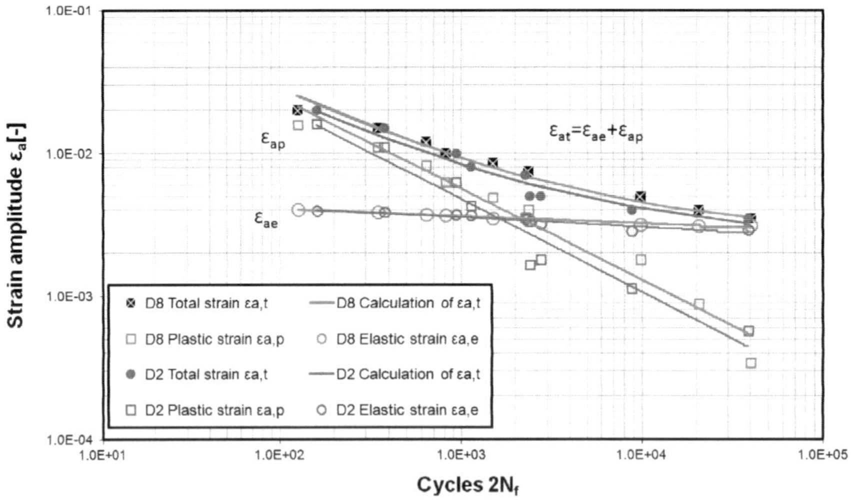

Figure 9. Manson-Coffin curves.

It should be noted that the differencies in cyclic behaviour, figure 8 and 9, may be caused by buckling at highest strain amplitudes and by surface condition effects which play an important role at high number of cycles to failure.

4 Conclusions

The possibility of using sub-size specimens to obtain cyclic material properties by LCF tests was presented in this paper. The results obtained with small sized specimens provide conservative results on the safe side in fatigue life behaviour estimation without any correlation. The results of cyclic material properties and CSSC obtained confirmed that the sub-size geometries are compareble to the conventional ones and representative results can be obtained even with small sized specimen geometries . In the future work, the material data obtained here will be used as an input for durability software analyses for finite element methods investigations.

Acknowledgements

This paper was created by project Development of West-Bohemian Centre of Materials and Metallurgy No.: LO1412, financed by the MEYS of the Czech Republic.

References

[1] American Society for Testing and Materials 2004 Standard Practice for Strain-Controlled Fatigue Testing, *E606-04e1 ASM International*

[2] Gussev M, Busby J, Field K, Sokolov M and Graz S 2015 Role of Scale Factor During Tensile Testing of Small Specimen *SPT1576*

[3] Kumar K, Madhusoodanan K and Rupani B 2006 Miniature Specimen Technique as an NDT Tool for Estimation of Service Life of Operating Presurre Equipment *International Conference &Exhibition on Pressure Vessel and Piping*

[4] Konopík P, Dzugan J and Prochazka R 2013 *Mat. Sci. and Tech. Conf. and Exh.* Evaluation of local mechanical properties of steel weld by miniature testing technique, pp 2404-11

[5] Dzugan J, Konopik P., Rund M and Prochazka R 2015 *ASME Pressure vessels and piping conf. - 2015* Determination of local tensile and fatigue properties with the use of sub-sized specimens, Volume 1A: Codes and Standards (New York: Amer. Soc. Mech. Eng.)

[6] Rund M et al 2015 *Procedia Engineering* Investigation of Sample-size Influence on Tensile Test Results at Different Strain Rates, Vol. 114 (Amsterdam: Elsevier science bv) pp 410-15

[7] Konopik P, Dzugan J and Rund M 2015 *METAL 2015 – 24th Int. Conf. on Metallurgy and Mat. (Brno)* Determination of fracture toughness in the upper shelf region using small sample test techniques (Ostrava: TANGER Ltd.) pp 710-15

[8] Konopík P and Džugan J 2012 *2nd Int. Conf. SSTT on Det. of Mech. Prop. of Mat. by Small Punch and other Miniature Testing Techn.* Determination of Tensile Properties of Low Carbon Steel and Alloyed Steel 34CrNiMo6 by Small Punch Test and Micro-Tensile Test (Ostrava: Ocelot sro) pp 319-328.

[9] Stamborska M, Lapin J, Bajana O and Losertova M 2015 Tensile Deformation Bahaviour of Ferritic-Pearlitic Steel Studied by Digital Image Cerrelation Method *Kovove materialy*

[10] Konopík P, Dzugan J and Prochazka R 2013 *METAL 2013 – 22nd Int. Conf. on Metallurgy and Mat. (Brno)* Determination of fracture toughness and tensile properties of structural steels by small punch test and micro-tensile test (Ostrava: TANGER Ltd.) pp 722 – 27

[11] Wakefield Metals 2014 Technical Handbook of Bar Products Tempering diagram Edition 1 p 44

Development of geometry of forming tools for extrusion of strip sheet by SPD process

S Rusz [1], M Salajka[1], J Džugan[2], O Hilšer[1], J Bořuta[1], M Pastrňák[1] and J Švec[1]

[1] VSB - Technical University of Ostrava, Faculty of Mechanical Engineering stanislav.rusz@vsb.cz Czech Republic
[2]COMTES FHT a. s., Průmyslová 995, Dobřany, 334 41

Corresponding author's E-mail: stanislav.rusz@vsb.cz

Abstract. On VSB -Technical University of Ostrava developed a method that uses the principle of severe plastic deformation to refine the structure and enhance mechanical properties of sheet metal strips. The greatest importance in practice represents an increase in yield strength and ultimate strength of sheet metal strips. The DRECE method (Dual Rolls Equal Channel Extrusion) is a newly developed method. Severe plastic deformation results in a high degree of the material deformation. The method can be used to produce metallic materials with a very fine grain structure. The paper analyses the effects of the values of angles of the newly developed forming tools on the achievement of mechanical properties in selected carbon steels by SPD process. The one type of steels (Ck55) was verified experimentally. Experiments were performed on the sheet metal strip with dimensions 58 (width) x 2 (thickness) x 1000 (length) mm with different inclination angle α.

1 Introduction

High deformation at comparatively low homological temperatures is an efficient method for manufacture of ultrafine grained (UFG) massive materials. News technologies, which use high deformation for obtaining of fine-grained structure, comprise namely the following authors [1].This research concerned the whole production of UFG materials, using Severe Plastic Deformation (SPD). These new technologies for production of semi-finished products with ultra-fine grained structure differ from conventional technologies. Even in scientific research, it is recognized that namely the controlled forming processes, including special processes, enabling control of technological parameters with regard to the structure and its further refinement, have currently the highest gradient of utilization efficiency of scientific research findings in practice [2, 3, 8]. In the metals like Cu and steel, the deformation by ECAR increases the strength and decreases elongation. Besides the DCAP (C2S2), CONFORM and ARB [3, 4] method for sheet metal forming, from the perspective of industrial practice new technologies are being intensively developed, to which belongs also the newly developed DRECE method (Dual Rolls Equal Channel Extrusion, see Fig. 1a, b), which were improved by the use of new forming tools with a new geometry (smaller angles in the zone of deformation, see Fig. 2). The issues of production of materials with UFG structure, especially in steels and alloys of non-ferrous metals in the form of sheet metal strip are analyzed in detail. Furthermore, a comparison is made of basic mechanical properties of the tested materials achieved after severe plastic deformation with the properties of these materials in the initial state. Analysis of the structure is performed on an optical microscope. The ECAR method is one of the similar methods of forming, approaching a continuous production process. Unlike the DRECE method it works with totally different inputs and also with hot forming, although it has a lower strengthening, but it lacks a dislocation strengthening as a source of grain refinement. From this perspective, this newly developed forming method (DRECE) is completely original.

2 Forming Equipment

The DRECE method is based on making use of the material's intensive plastic deformation, whereby this process is a combination of two known technologies, ECAP and CONFORM. As has already been mentioned earlier, the equipment is a prototype possessed by the Faculty of Mechanical Engineering of the VŠB – Technical University Ostrava. The equipment consists of a NORD gearbox and an electromotor with speed frequency converter which gives us the option to change the deformation rate even during the process, and thus allows us to flexibly react to the process progress. Other components include a plate clutch, a drive roller, two pressure rollers and a bottom and top forming tool. The pressure applied onto the front pressure roller is controlled by a pair of hydraulic cylinders; the pressure applied onto the rear pressure roller is controlled mechanically. This combination has so far proven itself when controlling the pressure in both non-ferrous alloys (on Al, Cu basis) as well as in sheet steel (DC01, Ck55). The machine's design of course allows hydraulic control to be applied also to the rear pressure roller. The entire DRECE equipment is shown in Fig. 1a, b. [5].

Figure 1. a) Forming device, b) Detail of the feeding device with induction heating.

In our experiments, we use samples of dimensions 58 mm x 2 mm – 1000 mm. However, used can be also sheet metal of smaller thickness, when the deformation zone range is adjusted by inserting pads placed underneath the pressure rollers. Since this is a prototype, in the initial stages of testing on this equipment, investigated were primarily optimal pressure forces of both pressure rollers, an effective interval between lubrications and also the impact of the process speed on the resultant structure of the material. The geometry of forming tools, in particular, the inclination angle α of the approach of the forming curve in the zone of deformation, is a very important factor influencing the efficiency of the SPD process (see Fig. 2) [5, 6].

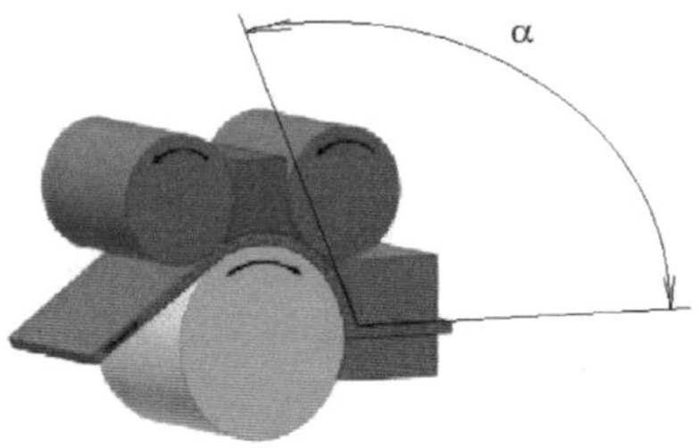

Figure 2. Scheme of deformation zone with different angle α.

2.1 Mathematical simulations

Preparation of geometric data for the formed material and equipment was performed in an engineering CAD program SolidWorks. Material data for the calculation were obtained by calculation based on the chemical composition of the material with the use of the program JMatPro. The program DEFORMTM, which analyses processes on the principle of finite elements, was used for computer modelling of the process itself and for subsequent numerical simulations. Simulation calculations predict for each moment of the process a distribution of temperature, strain, stress, and flow of the formed material. Moreover, it is possible to determine stress on instruments and evolution of the total forming forces.

2.2 Material data

Material data for the calculation were obtained on the basis of the chemical composition of both materials (for sheet - steel Ck55) using the program JMatPro. The program JMatPro creates a data file for the computer program DEFORM, which contains all the necessary material data, i.e. data on material flow stress, Young's modulus of elasticity, Poisson's ratio, thermal conductivity, heat capacity and thermal expansion. The simulation was conducted for an angle $\alpha = 108\,°$.

The basic equation for calculation of the strain intensity is in the form:

$$\varepsilon_{VM} = \frac{2\cdot n}{\sqrt{3}} \cdot \left[\cot\left(\frac{\phi}{2} + \frac{\psi}{2}\right) + \psi \cdot \frac{1}{\sin\left(\frac{\phi}{2} + \frac{\psi}{2}\right)} \right] \tag{1}$$

n - number of passes through a forming tool
$\phi = \alpha$ - forming tool angle in deformation zone
ψ - angle defining the curvature of the lower tool

a) **b)**

Figure 3. a) Scheme of method DRECE, b) Dethail of strip sheet.

2.3 Process data

All the rolls in the DRECE equipment are driven by a gearbox Nord. The speed of movement of the drive wheel is 1.72755 mm/s, which equals to the angular velocity of 0.01745 rad/s. Lubrication is performed with the use of lubricant Gleit HP 515, which is supplied continuously in small portions at the point of the deformation zone of friction between the sample and the tool. The value of the friction coefficient in the defect zone was 0.1. The rolls are roughened by shot blasting so that material is drawn by a friction force (friction coefficient has a value of 3).

2.4 Mathematical simulaton

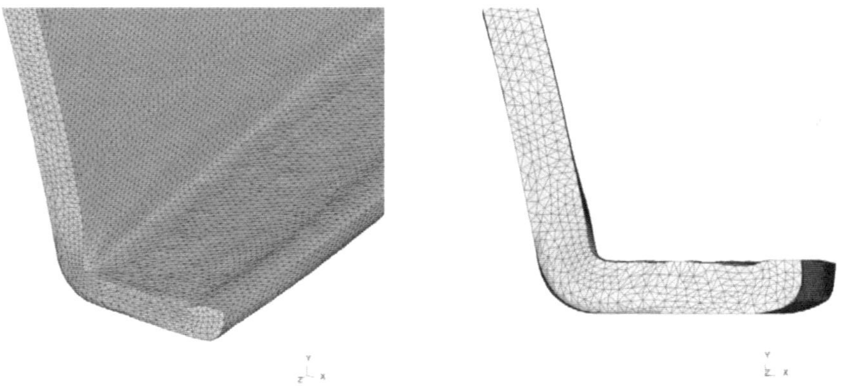

Figure 4. Simulation of forming process.

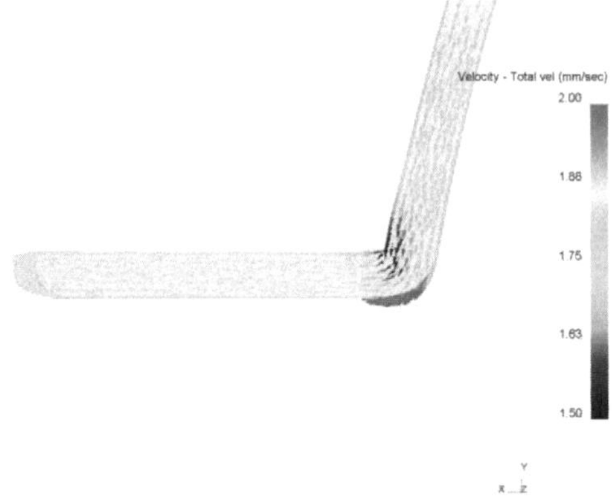

Figure 5. Material flow through forming tool.

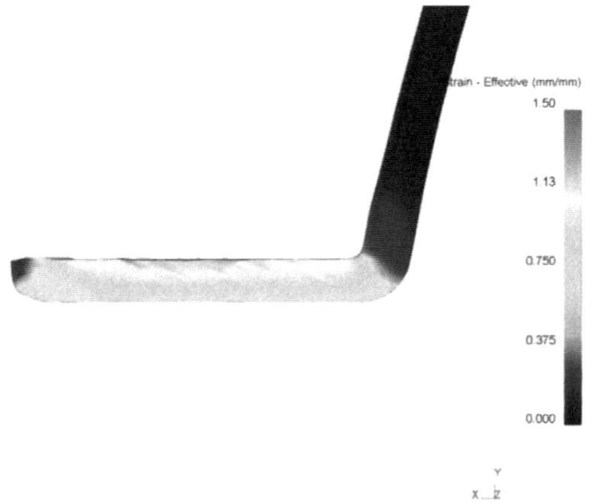

Figure 6. Effective plastic strain after 1st pass through forming tool.

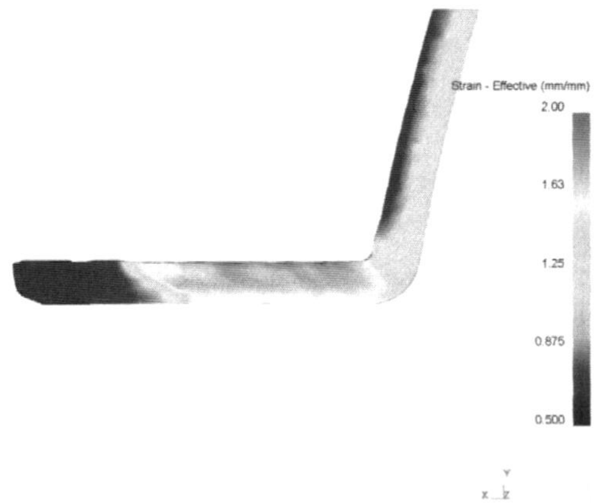

Figure 7. Effective plastic strain after 2nd pass through forming tool.

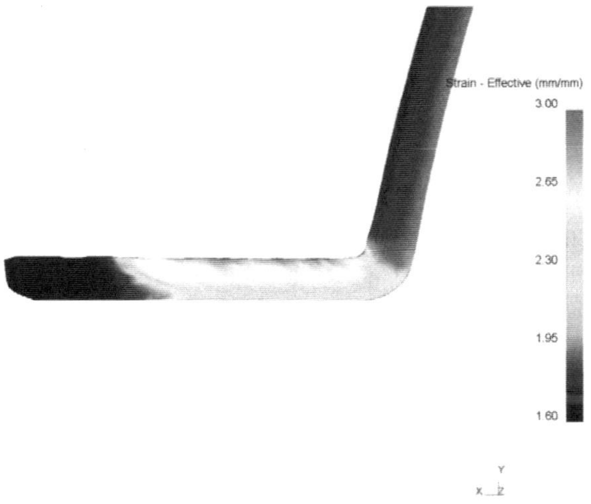

Figure 8. Effective plastic strain after 3th passes through forming tool

Table 1. Results of mathematical simulation on angle α = 118°

Eff. Plastic strain [-]	1st pass	2nd pass	3th pass
Middle of sample ssamplespecimens	1,35	1,92	2,93

2.5 Comments to results of mathematical simulation

The obtained magnitudes of effective plastic strain show an evident influence of the forming tool geometry on the strengthening of the formed material. The results are consistent with the theoretical knowledge when at the use of a lower forming angle of the tool a larger number of slip planes is formed and plastic deformation is more intense. The DRECE method is closing the gap between itself and the ECAP method. The obtained values of effective plastic strain should be regarded as indicative due to the accuracy of the entered boundary conditions.

3 Experimental conditions

The geometry of forming tools, in particular, the forming tool angle of the approach of the forming curve in the zone deformation, is a very important factor influencing the efficiency of the SPD process. This is currently newly developed forming equipment with 6 protected industrial designs of new geometries of the forming tools. Low carbon steel (Ck55) suitable for cold forming was verified experimentally. The chemical composition of the steel is given in Table 2 In the experiments the achieved mechanical properties and the structure at the forming tool angles 108 °, 113 ° and 118 ° were evaluated. It is known from the available literature data [2] that namely these angles determine the zone of plastic deformation. The effect of various angles of the forming tools influencing the basic mechanical properties, especially the yield stress $R_{p0.2}$, tensile strength R_m and ductility A_{80}, determined by tensile test, was tested also from this perspective.

Table 2. Chemical composition (wt %) of Ck55 steel.

Grade	C	Mn	Si	$P_{max.}$	$S_{max.}$	Al
Ck55	0,53	0,43	0,3	0,03	0,035	0,02

Attained yield strength for different angles forming tools depending on the number of passes are shown in Table 4 and Fig. 9

Table 4. Yield strength $R_{p0,2}$ on different tool angles.

Yield strength $R_{p0,2}$ [MPa]	α = **108°**	α = **113°**	α = **118°**
initial state	302	302	302
1×	435	401	400
2×	418	409	404
3×	431	397	435
4×		431	422

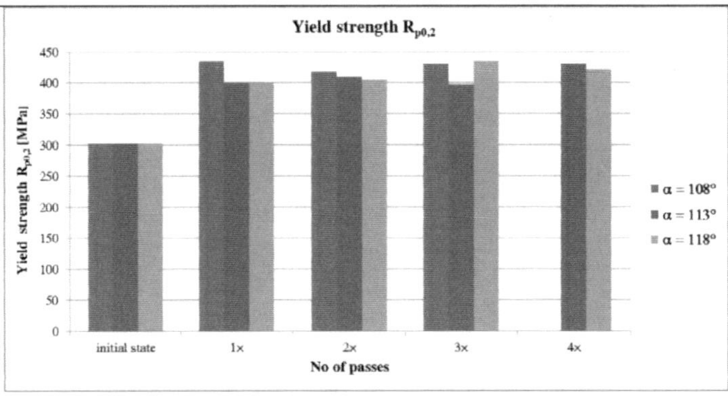

Figure 9. Yield strength depending on tool angles.

Table 5. Tensile strength R_m on different tool angles.

	α = **108°**	α = **113°**	α = **118°**
Initial state	431	431	431
1×	578	582	626
2×	591	591	652
3×	599	599	674
4×		603	683

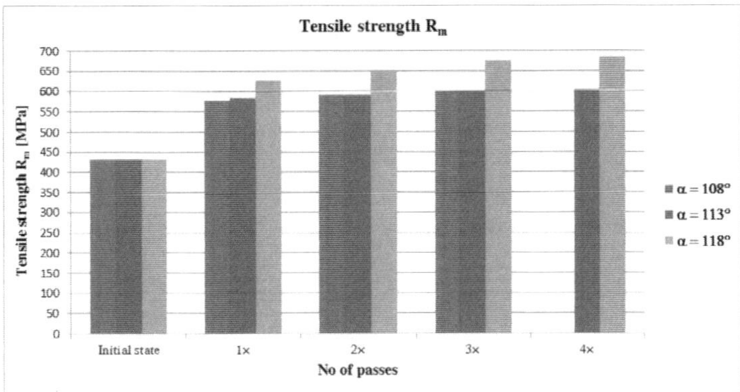

Figure 10. Tensile strength depending on tool angles.

Table 6. Ductility A on different tool angles.

	α = **108°**	α = **113°**	α = **118°**
Initial state	23,3	23,3	23,3
1×	23,3	17,3	10
2×	17,3	16,7	9
3×	15,3	16,7	9
4×		16	7,3

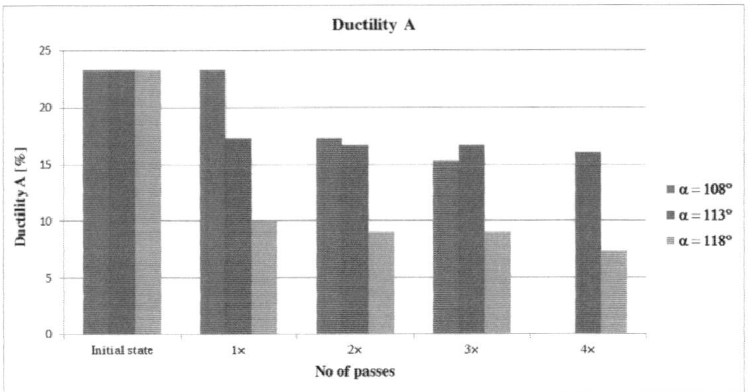

Figure 11. Ductility A depending on tool angles.

3.1 Comments on results

The obtained results show a significant influence of the forming tool geometry (angle α in the defect zone). At the angle of 108 ° a 45% increase of $Rp_{0.2}$ took place after the first pass through the forming tool in respect to the initial state. At the angles 113 ° and 118 ° this increase is 32%. After the second and third passes through the forming tool with a tool angle of 108 ° a slight decrease in the size of $R_{p0.2}$ takes place. At the angles of the forming tool of 113 ° and 118 ° a significant increase of $R_{p0.2}$ takes place till the 3rd and 4th passes, while the achieved values are identical with the value of $R_{p0.2}$ achieved at the tool angle of 108 ° after the first pass through the forming tool. This reflects unequivocally the efficiency of the forming process (SPD) - the influence of reduction of the angle of the forming tool from the perspective of achieving high material strengthening already after the first pass through the forming tool. At the angle of 108 ° it was not necessary to conduct an extrusion of the strip of the sheet in the 4th pass, since the value of Rp0.2 did not increase after the second or third passes in comparison compared with the value achieved after the first pass.

The tool geometry influence in a very similar way the magnitude of tensile strength R_m after individual passes through the forming tool. The biggest increase of R_m at all forming tool angles occurs after the first pass - by approx. 34% (angle 108 °), 35% (angle 113 °), or 45% (angle 118 °). Next passes do not have any substantial effect on the increase of the R_m at all forming tool angles. From the viewpoint of strengthening processes and application of knowledge into industrial practice, the most important is the increase in yield stress $Rp_{0,2}$ after the first pass through the forming tool. The biggest increase was achieved in accordance with presumptions at the angle α = 108 °. It is necessary to emphasise the high manufacturing accuracy of the forming tool (both top and bottom parts) and its holders. The tool design was for the purposes of protection of intellectual property protected by registration of industrial designs.

The achievement of an identical magnitude of ductility A after the first pass through the forming tool -23.3% at the angle α =108° is an important finding. At the angles of 113 ° and 118 ° the ductility A decreases by 21% or by 55%. After the 1st pass through the forming tool at the angle α = 108 ° it is possible to form successfully the semi-product (strip of the sheet) into the final shape of the machine part. This is the second important and significant finding for the application of the method DRECE in industrial practice. A drop of ductility after the first pass at the angles of 113 ° and 118 ° can be explained (interpret) by the ratio $R_{p0,2}/R_m$, which is unfavourable due to the ratio achieved at the angle of 108 °.

4 Metallographic analysis

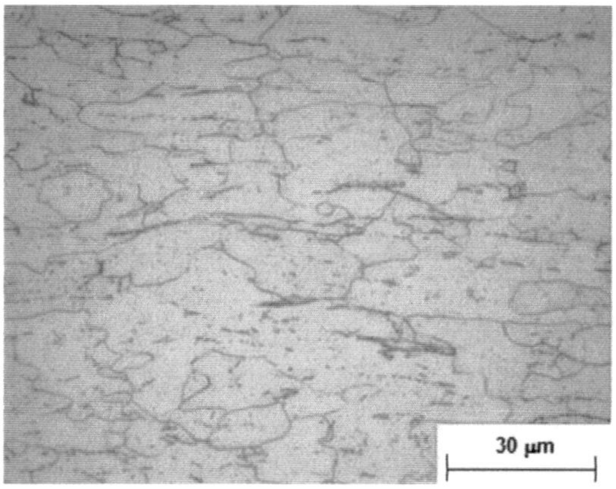

Figure 12. Metallographic analyze of steel Ck55, initial state (500×).

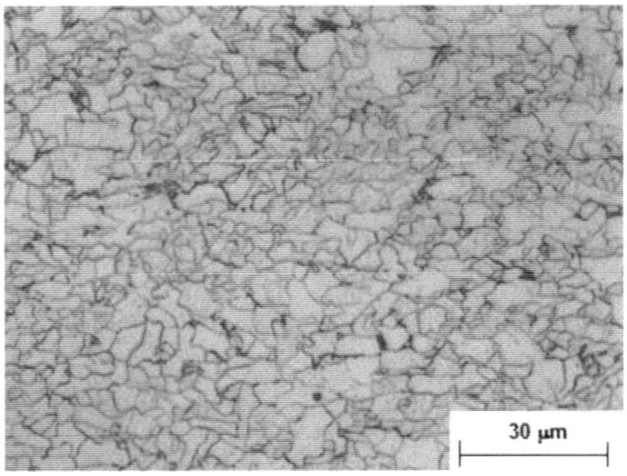

Figure 13. Metallographic analyze of steel Ck55, structure after 1. pass (500×).

4.1 Comments

For evaluation of structural characteristics the light microscope NEOPHOT 2 was used, with using of Nital etchant. Fig. 12 shows the structure of the analysed steel in the initial state and Fig. 13 shows the structure after the first pass through the forming tool for the angle α = 108 °. Average grain size is reduced from the original size of 30-40 μm to 5-8 μm. Refining of the structure corresponding to an increase in yield stress $R_{p0.2}$ achieved in experiments after the first pass was confirmed. In the next part of the experimental works an analysis of structure with the use of TEM and SAED method will be performed.

5 Conclusions

It can be postulated on the basis of the achieved results of mathematical simulation and experimental works that the proposed new geometry of the forming tool (forming tool angle α = 108°), it enables achievement of high degree of strengthening ($R_{p0.2}$) in the steel Ck55 already after the first pass through the forming tool. The achieved magnitude of strengthening in principle does not change substantially during the next passes. A smaller increase of strengthening occurs with the tools with angles α = 113 ° and 118 °. This finding is consistent with the physical model UFG materials according Ovidka [7]. To increase the hardening also contributes the interaction generated dislocations with precipitates that usually observes in these steels. The issue will be verified by SEM, TEM and SAED methods. The achieved results are very important from the perspective of the use of the forming method DRECE in industrial practice. This method makes it possible to substitute materials of higher quality (steels and alloys) by materials of lower quality, which after plastic deformation reach the parameters (mechanical and forming properties) of the higher quality materials.

Acknowledgements

This paper was created as a part of the project of Ministry of Education Pre-seed activity of the VSB-TUO II – Materials, No.CZ.1.05/3.1.00/14.0320 and within the project of the Ministry of Industry and Trade (MPO) No. 2A 1TP1/124 and project of Ministry of Education - Creation of an international team of scientists and participation in scientific networks in the sphere of nanotechnology and unconventional forming material No- CZ.1.07/2.3.00/20.0038

References

[1] Lee J C, Seok, H K, Han J H and Chung J Y Controlling the textures of the metal strips via the continuous confined strip shearing (C2S2) process. *Mater Res Bull* 2001; Vol. 36, pp. 997-1004.

[2] Saray O, Purcek G and Karaman I Principles of equal-channel angular sheet extrusion (ECASE): application to IF-steel sheets. *Adv Mater Sci* 2010, Vol. 25, pp. 42-51.

[3] Valiev R. Recent developments of severe plastic deformations techniques for processing bulk nanostructured material *Materials Science Forum* A 579 (2008) pp. 1–14

[4] Dzugan J, Procházka R and Konopík P Micro-Tensile Test Technique Development and Application to Mechanical Property Determination, Small Specimen Test Techniques, 6th Volume, STP 1576, Editors are Mikhail A. Sokolov and Enrico Lucon, pp. 12-29, DOI: 10.1520/STP1576-EB

[5] Rusz S, Cizek L, Michenka V, Dutkiewicz J, Salajka M, Hilser O, Tylsar S, Kedron J and Klos M New type of device for achievement of grain refinement in metal strip. *Advanced Materials Research.* Vol. 1127 (2015), Switzerland: Trans Tech Publications, s. 91 - 97. 10.4028/www.scientific.net/AMR.1127.91

[6] Rusz S, Klyszewski A, Salajka M, Hilser O, Cizek L and Klos M Possibilities of application methods DRECE in forming of non-ferrous metals, *Archives of Metallurgy and materials*, Vol. 60, 2015, ISSUE 4, DOI: 10.1515/amm-2015-0481

[7] Gutkin M Yu, Ovidko I A, Pande C S Theoretical model of plastic deformation processes in nanocrystalline material, Rev. *Adv. Mater. Sci.*, September 2001, Volume 2, pp. 80 – 102

[8] Donic T, Martikan M, Hadzima B New unique ECAP system with ultrasound and backpressure, 6th International Conference on Nanomaterials by Severe Plastic Deformation IOP Publishing IOP Conf. Series: *Materials Science and Engineering* 63 (2014) 012047, pp.1 - 16 doi:10.1088/1757-899X/63/1/012047

Influence of relative rolling reduction and thickness layers bimetallic plate at the non-uniformity of the strain after rolling process

D Rydz[1a], G Stradomski[1a], H Dyja[1a]

[1]Technical University of Czestochowa, Faculty of Production Engineering and Materials Technology Armii Krajowej 19, 42-200 Częstochowa, Poland, Institute of Plastic Forming Processes and Safety Engineering,

e-mail: rydz@wip.pcz.pl

Abstract. In the article were made numerical and laboratory tests of two-layers rolling process sheet composed of Al99,8 + M1E. Laboratory tests made with use of 150 mm diameter working rolls mill. To the modeling of the bimetallic plate rolling were taken the FEM Forge 2D software based on the theory of plasticity and MathCad program (in which to the rolling process modeling were used the mathematical model developed in the work [5] based on the theory of viscoelasticity). The aim of study was to determine the influence of layer thickness H_{T0}/H_{M0} and relative deformation ε on the uneven distribution of steel sheet deformation after rolling process. Calculations based on the theory of viscoelasticity allowed additionally take into account the impact of the delayed effects of the variation of viscoelastic deformation of layers of the bimetallic plate rolling process.

Keywords: bimetallic plate, rolling process, viscoelastic

1 Introduction

This paper will focus on the analysis of the second stage is a two-layer metal rolling process initially joined by explosive welding. Previous numerical studies were mainly based on calculations made using commercial software based on the theory of plasticity. In this paper we proposed to carry out modeling of two-layer metal rolling process based on plasticity theory and the theory of viscoelasticity. In the modeling process, bilayer sheet rolled taken to analyze the problem of uneven deformation of the layers in the double-layer plate according to the ratio of the layer thicknesses H_{M0}/H_{T0} (the ratio of the layer thickness of less resistance to deformation Al99.8 thickness of greater deformation resistance M1E) and the value of the relative ε deformation. In addition, the numerical researches based on the theory of viscoelasticity will determine the influence of the delayed effects on the variation of viscoelastic deformation of bimetallic plate layers during the rolling process. As a result, during rolling process of the two-layer sheet rolling the deformation of the individual layers are different, this cause the effect of bending on the exit of deformation gap [1, 5 ÷ 7]. The direction of bending depends on the location of the layer with higher deformation resistance (hard), the bent appear toward to this layers.

2 Description of the kinematic model of the rolling process of double layer viscoelastic bodies

In this paper it was assumed that the x and y axis directions are the principal directions of strains. Most of materials pairs forming a bimetallic sheet during the rolling process are nonuniform deformed, which causes bending of the rolled band after leaving the roll gap. In the present work for the analysis of the bimetallic sheet layers deformation during the rolling process is adopted the rheological Jeffreys model [3, 4]. It allows taking into account the dynamics of the rolling process for both homogeneous and composite sheet. In addition to the analysis of the bimetallic plate rolling process in this paper is used the Boltzmann's theory of bodies. Another assumption made in this work, is the adoption to the present analysis the incompressibility condition $\bar{\nu}(p)=\nu=0.5$. It results that the analysis assumes a flat state of stress and strain. Based on the assumptions were developed schematic kinematic model of two-layer metal rolling process, which is shown in Figure 1.

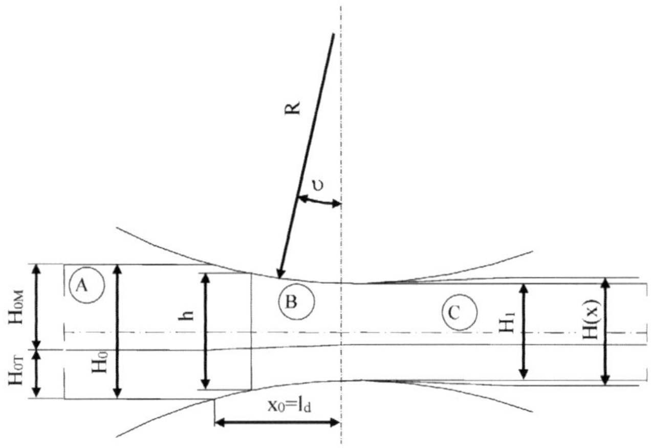

Figure1. Schematic viscoelastic deformation of bimetallic plate divided into zones of deformation A – zone without deformation before the surface of rolling gap entrance, B – zone of rolling gap and C – zone of delayed viscoelastic effects [5].

The rolling process of two-layer materials is a complex process, which is often accompanied by uneven deformation of the two layers of a bimetallic band. This is caused by different deformation resistance due to the different layers forming the bimetal [1, 2, 5 ÷ 7]. Soft layer deforms more and it can be written as:

$$\varepsilon_M > \varepsilon > \varepsilon_T \qquad (1)$$

where:

ε_T – Relative deformation of layers with higher plastic deformation resistance,

ε – Total relative bimetallic band deformation,

ε_M – Relative deformation of layers with minor plastic deformation resistance.

The relative deformation of each bimetal's sheet layer can be represented by the following formula [5]:

$$\left.\begin{aligned} \varepsilon_T &= \frac{H_{T0} - H_T}{H_{T0}} \\ \varepsilon_M &= \frac{H_{M0} - H_M}{H_{M0}} \end{aligned}\right\} \tag{2}$$

where:

$$\left.\begin{aligned} H_T &= 2\left(\frac{H_{T0}}{2} + R - \sqrt{R^2 - x^2}\right) \\ H_M &= 2\left(\frac{H_{M0}}{2} + R - \sqrt{R^2 - x^2}\right) \end{aligned}\right\} \tag{3}$$

$$\left.\begin{aligned} H_{T0} &= 2\left(\frac{H_T}{2} + R - \sqrt{R^2 - x_{T0}^{\ 2}}\right) \\ H_{M0} &= 2\left(\frac{H_M}{2} + R - \sqrt{R^2 - x_{M0}^{\ 2}}\right) \end{aligned}\right\} \tag{4}$$

After substituting the equations (4 and 5) to equation (3) and assuming that x ≤ R, the relationship (3) can be simplified to the form:

$$\left.\begin{aligned} \varepsilon_T &= \frac{x^2 - x_{T0}^2}{RH_T + x_{T0}^2} \\ \varepsilon_M &= \frac{x^2 - x_{M0}^2}{RH_M + x_{M0}^2} \end{aligned}\right\} \tag{5}$$

where:
R – roll radius,
x_{T0}, x_{M0} – the length of the rolling gap for the hard and soft layer,

because then it can written with some approximation that the thickness of the individual layers are:

$$\left.\begin{aligned} H_{T0} &\cong \left(H_T + \frac{x^2}{R}\right) \\ H_{M0} &\cong \left(H_M + \frac{x^2}{R}\right) \end{aligned}\right\} \tag{6}$$

The model takes into account the occurrence of late viscoelasticity effects, as a factor affecting directly the value of the deformation and indirectly on the curvature of bimetallic rolled sheet layers.

3 Modeling of the rolling process of bimetallic sheet in the program Forge 2D

The numerical investigations were made for three different ratios of the layer thicknesses H_{M0}/H_{T0} = 0.2, 0.6 and 1.0, and three values of the relative deformation ε = 5%, 10% and 15%. Numerical investigations were made for cold process in the ambient temperature. The comparison of results of laboratory tests and computer simulation show problem of uneven layers deformation in the bimetallic sheet rolling process. The effect of non-uniform deformation of sheet layers is the cause of bimetallic bending after leaving the roll gap. Strengthening curve parameters of tested materials were determined on the basis of chemical composition and set of from the Archenius equations [1]:

$$K(T,\bar{\varepsilon}) = K_0 \cdot (\bar{\varepsilon} + \bar{\varepsilon}_0)^n \cdot e^{-\beta T} \tag{7}$$

where:

$\bar{\varepsilon}_0$ – controlled deformation vector for next steps,
$\bar{\varepsilon}$ – total strain vector,
K_0, n, β - Strengthening curve parameters.

Frictions during the rolling process were modeled on the basis of the Tresca solutions and were determined from the equation [1]:

$$\tau = -m \frac{\sigma_0}{\sqrt{3}} \frac{\Delta V}{\Delta v} \tag{8}$$

where
τ – unity vector of friction forces,
σ_0 – base yield stress,
$\Delta V/\Delta v$ – parameter describing the slip of metal on the roll,
m – friction factor.

Numerical research made in the program 2D Forge were performed for the corresponding laboratory tests.

4 Material selection and the range of tests

The double-layer sheet rolling process was carried out on the duo laboratory rolling mill with work rolls diameter D = 150mm and Vw = velocity of 100 mm/s on samples after direct explosive welding connection method.

The experimental and numerical tests were made for bimetal sheet complex of Al99.8 and M1E layers.

Chemical compositions of the tested materials are shown in Table 1.

Table 1. The chemical composition M1E (PN-87/H-82120) i Al99,8 (PN-79/H-82160).

Material	Cu	Al	Zn	Sn	As	Sb	Si
M1E	99,9	-	max 0.003	max 0.002	max 0.002	max 0.002	-
Al99,8	max 0.02	99.8	max 0.05	-	-	-	max 0.13

Modeling and laboratory research were performed for three sets of samples with the following dimensions:

– I set: H_0 = 12 [mm], L_0 = 150 [mm] and H_{T0}/H_{M0} = 2/10,
– II set: H_0 = 8 [mm], L_0 = 150 [mm] and H_{T0}/H_{M0} = 3/5,
– III set: H_0 = 8 [mm], L_0 = 150 [mm] and H_{T0}/H_{M0} = 4/4,

where:

H_{M0} – the initial thickness of the layer (soft) with lower strain resistance,
H_{T0} – the initial thickness of the layer (hard) with higher strain resistance,
H_0, L_0– accordingly height and length of bimetallic plates.

5 The results and analysis

On the basis of laboratory and numerical studies, was determined the effect of layer thicknesses H_{T0}/H_{M0} and relative deformation ε on the value of the deformation of bimetallic plate forming layers fig. 2 and 3

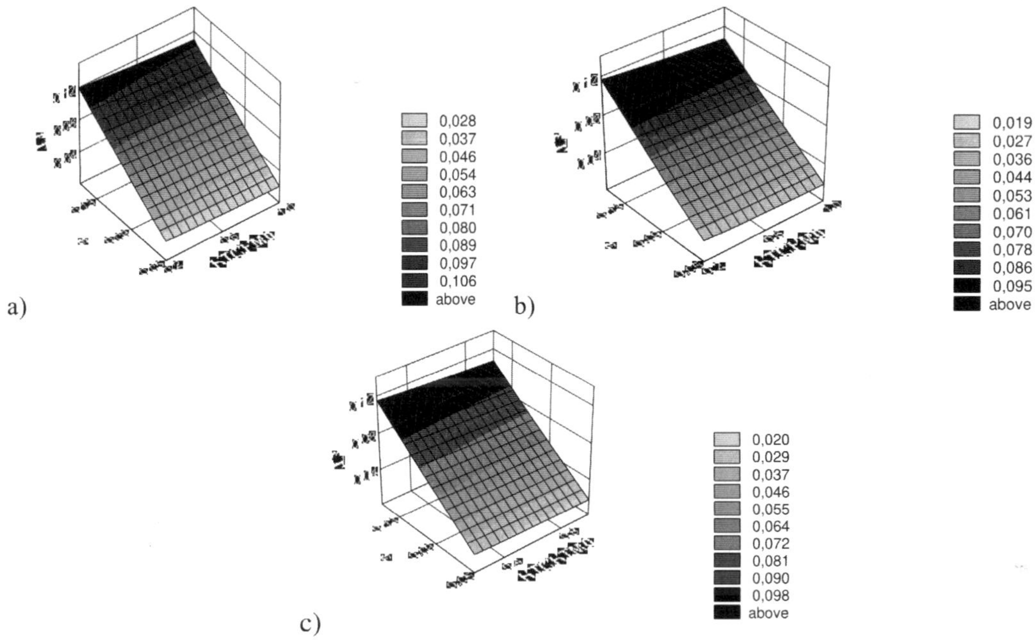

Figure 2. Influence of layer thicknesses and relative deformation on the value of hard layer deformation M1E a bimetal sheet a) according to the theory of plasticity, b) according to the theory of viscoelasticity c) the results of laboratory tests.

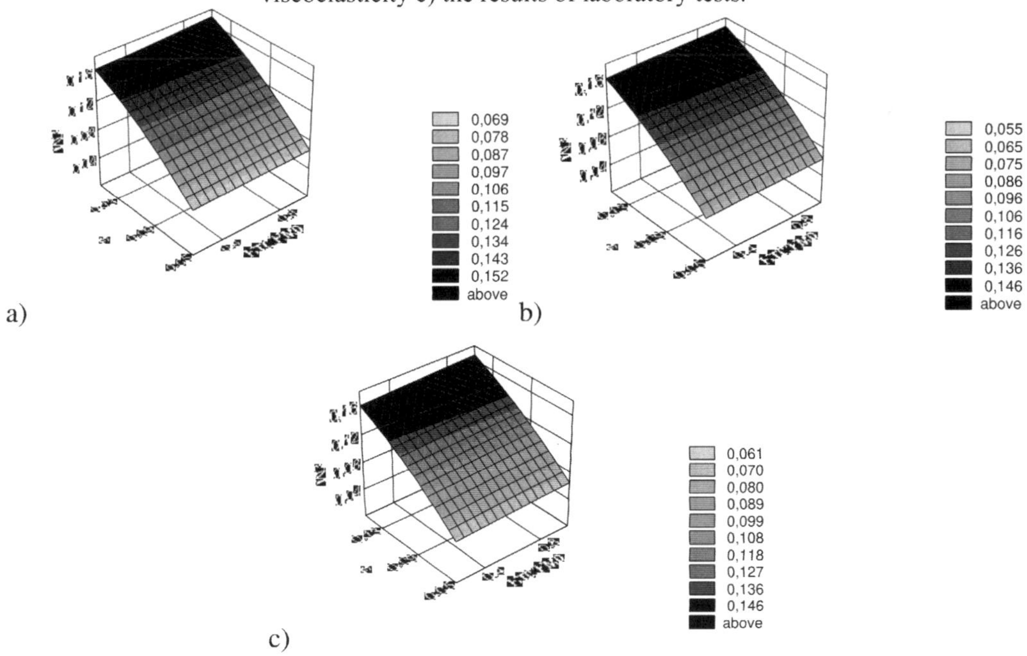

Figure 3. Influence of layer thicknesses and relative deformation on the value of soft layer deformation Al99,8 a bimetal sheet a) according to the theory of plasticity, b) according to the theory of viscoelasticity c) the results of laboratory tests.

On the basis the results of research presented on Figures 3 and 4, it can be concluded that both the solutions based on the theory of plasticity and viscoelasticity properly reflect the influence of layer thicknesses and relative deformation on the non-uniformity of bimetallic sheet layers deformation. Nevertheless, taking into account the delayed viscoelastic effects in the modeling of bimetallic plate rolling process allows to define more precisely the strain of hard and soft layers of M1E Al99.8.

Presented in Figures 3 and 4 results allow to conclude that the increase in layer thicknesses ratio H_{T0}/H_{M0} decreases the deformation value of a hard layer M1E, while increasing the strain value of the soft layer Al99, 8 Moreover, based on the numerical studies it can be concluded that the increase in values of relative deformation ε increases the uneven deformation of layers forming bimetallic plate.

Presented in the paper comparing the calculation results obtained using the model with the theory of plasticity and viscoelasticity theory and its experimental verification confirms the correctness of the results. It should be noted, however, that taking into account the effect of expansion of the material after leaving of the roll gap in the theory of viscoelasticity model makes the results of the calculations more exact and with less error.

6 Summary and Conclusions

The paper presents a theoretical and experimental analysis of double-layer sheet rolling process. For numerical studies based on the theory of plasticity was applied Forge 2D program, based on the finite element method. Numerical calculations based on the theory of viscoelasticity were performed in Mathcad program on the base of developed in the work [1] mathematical model. In this work, calculations were based on assumptions of the theory of small deformations and the adoption of the principle of superposition of Boltzmann, which is a heuristic law. Implementation of numerical calculations for bimetal sheet layers allowed to determine the value of deformation in the roll gap and delayed viscoelastic effects. Experimental verification of the results of calculations were performed on the basis of laboratory tests on laboratory duo mill with 150 mm work rolls diameter.

On the basis of numerical research results and experimental verification can be drawn the following conclusions:

- With the increase layer thicknesses ratio H_{T0}/H_{M0} of bimetallic plate and with an increase in the relative deformation ε uneven distribution of strain on the layer grows,
- Taking into account the delayed viscoelastic effects can increase the accuracy of the calculation. These effects cause affect on the value of the layers permanent deformation after rolling process of bimetallic plate.

References

[1] Dyja H., Mróz S., Rydz D.: „Technologia i modelowanie procesów walcowania wyrobów bimetalowych" Seria Metalurgia Nr 33, Prace Naukowe Wydziału Inżynierii Procesowej, Materiałowej i Fizyki Stosowanej, Częstochowa 2003,

[2] Dyja H., Pietrzyk M.: On the theory of the process of hot-rolling of bimetal plate and sheet, Journal of Mechanical Working Technology, Volume: 8 Issue: 4, p. 309-325, 1983

[3] Nowacki W.: Teoria pełzania, Warszawa, Arkady 1963

[4] Schiessel H., Metzler R., Blumen A., & Nonnenmacher T. F. Generalized viscoelastic models: Their fractional equations with solutions. Journal of Physics A: Mathematical and General, n.28, 1995,

[5] Rydz D.: Analiza procesu walcowania blach bimetalowych w oparciu o rozwiązania dla ciał lepkosprężystych, seria monografia nr 152, Politechnika Częstochowska, Wydawnictwo Politechniki Częstochowskiej, Częstochowa 2009,

[6] Skalmierski B., Rydz D.: Zastosowanie rozwiązań lepkosprężystych do opisania procesu walcowania blach, Seria: Metalurgia nr 39, V Międzynarodowa sesja naukowa: Nowe Technologie i Osiągnięcia w Metalurgii i Inżynierii Materiałowej, Częstochowa 2004,

[7] Skoblik R., Rydz D., Stradomski G.: Analysis of Asymmetrical Rolling Process of Multilayer Plates, Solid State Phenomena (JCR) Vol. 165 2010 p. 348-352 © Trans Tech Publications, Switzerland.

IOP Conf. Series: Materials Science and Engineering **179** (2017) 012063 doi:10.1088/1757-899X/179/1/012063

Influence of specimen dimensions on ductile-to-brittle transition temperature in Charpy impact test

S Rzepa[1], T Bucki[2], P Konopík[3], J Džugan[3], M Rund[3] and R Procházka[3]

[1]AGH University of Science and Technology, Kraków, Poland, EU
[2]Kielce University of Technology, Kielce, Poland, EU
[3]COMTES FHT a.s., Dobřany, Czech Republic, EU

E-mail: sylwrz94@gmail.com

Abstract. This paper discusses the correlation between specimen dimensions and transition temperature. Notch toughness properties of Standard Charpy-V specimens are compared to samples with lower width (7.5 mm, 5 mm, 2.5 mm) and sub-size Charpy specimens with cross section 3x4. In this study transition curves are correlated with lateral ductile part of fracture related ones for 5 considered geometries. Based on the results obtained, correlation procedure for transition temperature determination of full size specimens defined by fracture appearance of sub-sized specimens is proposed.

1 Introduction

Charpy test is one of the basic material tests used for material properties characterization. In order to avoid brittle failure under service conditions, transition temperature based on Charpy impact tests is determined for the materials exhibiting transition behavior. The size of the experimental material does not allow in all cases application of standard full size specimens (10x10 mm) and sub-size specimens have to be used. Although, some of sub-sized geometries are standardized, procedures for comparison of results obtained for various specimen geometries are not standardized due to ambiguous relations for different applications. Therefore, relations for specific areas have to be individually established.

The use of Charpy specimen geometry with different dimensions leads to development of correlation procedure of the results. A review of the literature shows the efforts to correlate results for standard Charpy with different width [1,2] and standard Charpy with sub-size Charpy [3-14]. Small specimen method is also used in another mechanical testing techniques [15,16, 17] to minimalize the size of tested material.

The aim of presented work was to study influence of Charpy specimen dimensions on ductile-to-brittle transition temperature. Standard Charpy-V specimens are compared to samples with reduced widths (7.5 mm, 5 mm, 2.5 mm) and sub-size Charpy specimens with cross section 3x4 mm. The paper proposes the correlation procedure for transition temperature determination of full size specimens based on fracture appearance of sub-sized specimens.

2 Experiment description

The fracture appearance transition temperatures (FATT) were investigated for Chromium steel P-AK1TD (10,5% Cr, 1.5-1.8% Ni, 1.6-2% W, 0.35-0.5% Mo). Figure 1 shows the microstructure of the steel investigated.

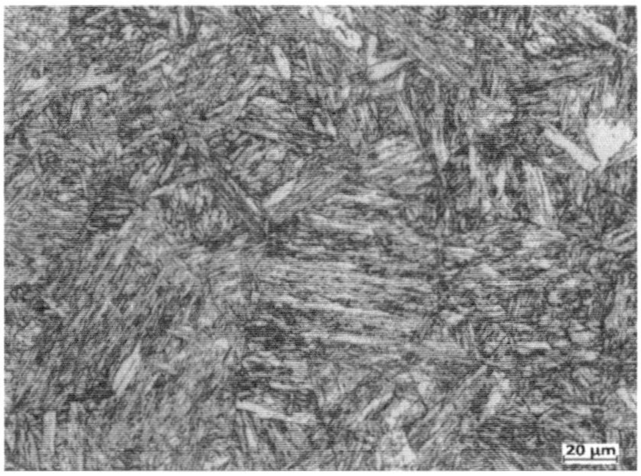

Figure 1. Microstructure of the P-AK1TD steel.

The tests were performed by using standard Charpy-V and sub-size Charpy specimens. Standardized Charpy-V specimens with various widths (10 mm, 7.5 mm, 5 mm, 2.5 mm) [18] and sub-size Charpy specimens with cross-section 3x4 mm [18], shown on the figure 2, were tested. Impact tests were conducted using a 300 J pendulum with striking edge radius 2 mm for standard specimens and 15 J pendulum with striking edge radius 2 mm for sub-size Charpy.

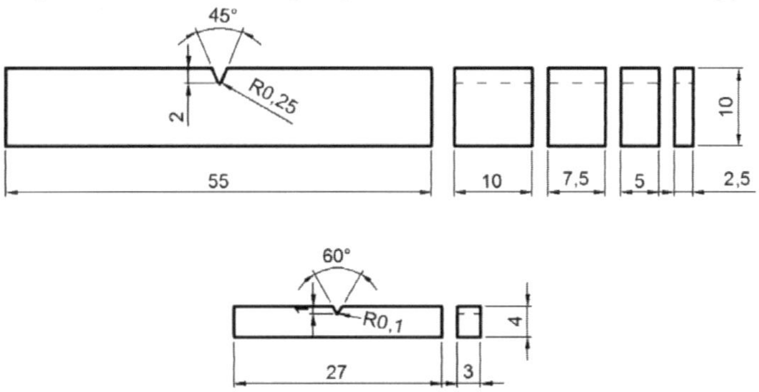

Figure 2. Dimensions of used specimens [18,19].

FATT (FATT 50) were determined for each batch by testing of 18 test pieces. Tests were carried out at about 10 testing temperatures for each batch. Fracture appearance and lateral ductile part thickness, marked ***x*** on the figure 3a were determined by microscopic measurements. The ***x***-value was obtained by dividing ductile part area at lateral edges of fracture by its length, as shown in figure 3b.

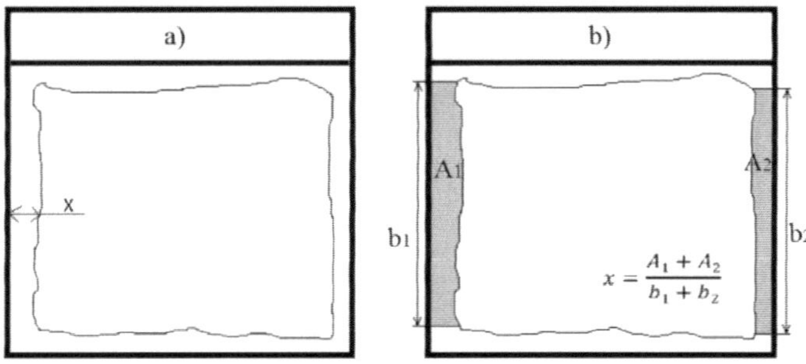

Figure 3. Fracture outline (a) lateral ductile part, (b) method of x thickness calculation.

3 Results and discussion

Figure 4 presents the fracture faces of specimens tested at various temperatures. Results suggest that lateral ductile part thickness - ***x*** is similar for all Standard Charpy specimens at the same temperature.

Figure 4. Fractures of tested specimens.

Table 1 and figure 5 present the relationship between temperature and ***x***-thickness for standard Charpy and for sub-size Charpy. These dependences can be approximated by fourth-degree terms shown in the figure 5.

Table 1. Measurements of *x*-thickness for standard Charpy and sub-size Charpy.

Temperature (°C)	x in standard Charpy (mm)	x in sub-size Charpy (mm)
-120	0.04; 0.06	0.08
-110	0.07	0.07
-100	0.09; 0.11	0.07
-90	0.13	0.18; 0.11
-80	0.15; 0.16	0.16
-70	0.18; 0.20	0.16; 0.30
-50	0.28	0.41; 0.55
-40	0.35; 0.36	0.64
-20	0.64; 0.82	-
0	1.53; 1.74	-

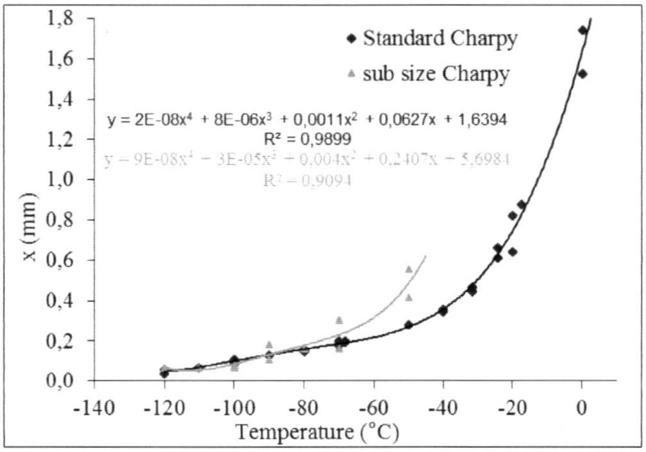

Figure 5. Relationship between temperature and ***x***-thickness for standard Charpy and sub-size Charpy

FATT and ***x***-thickness at corresponding transition temperature for each batch of specimens were measured and compared. Results are shown in table 2. Figure 6 presents dependence of FATT on ***x***-thickness for standard Charpy. It was found, that ***x***-thickness at FATT for standard Charpy is linearly proportional to sample width. In other words, fracture appearance transition in this material occurs, when ***x***-thickness is about 8.5 % of specimen width.

Table 2. Measurements of FATT and ***x***-thickness in FATT.

Specimen dimensions (mm)	FATT (°C)	x (mm)	x/W
10x10	-17.3	0.88	0.088
10x7.5	-24.4	0.66	0.088
10x7.5	-24.4	0.61	0.081
10x5	-31.6	0.46	0.092
10x5	-31.6	0.45	0.090
10x2.5	-68.3	0.20	0.080
10x2.5	-68.3	0.19	0.078
		Average: 0.085 (8.5 %)	
4x3	-71.5	0.21	0.070

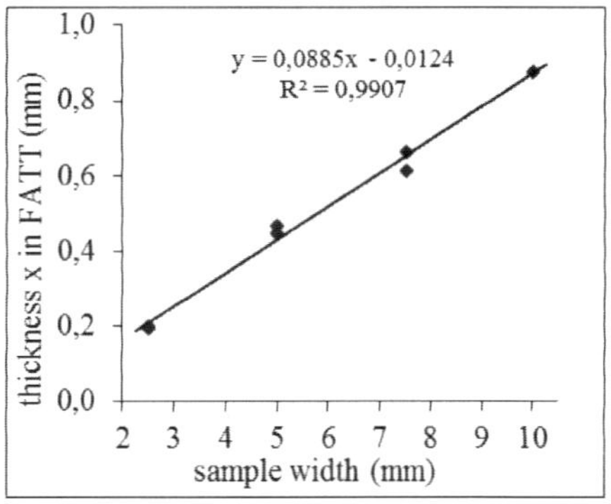

Figure 6. The dependence of FATT on ***x***-thickness for standard Charpy.

The assumption, that x/W ratio in FATT is constant for each standard Charpy gives a possibility to approximate FATT based on x-measurements of specimens with different widths. FATT depends on specimens' width, which can be calculated from x-thickness using simple equation (1). Figure 7 shows discussed dependence calculated from x-thickness measurements (figure 5). The correlations between standard Charpy width and transition temperature can be find in literature [1,2]. Figure 7 confirms, that presented correlation procedure gives similar results to correlations from the literature. Different trends for standard and sub-size Charpy seem to suggest influence of other parameters. Literature [20] confirms effect of notch depth, notch root radius and specimen height on transition temperature appearance, what can be observable at presented chart. The advantage of proposed correlation procedure, comparing to constant correlation curves is a possibility to find the relationship between FATT and specimens width for non - standard Charpy specimens and more precise results for non - standard steel.

$$W = \frac{x}{average\left(\frac{x}{W}\right)} \quad (1)$$

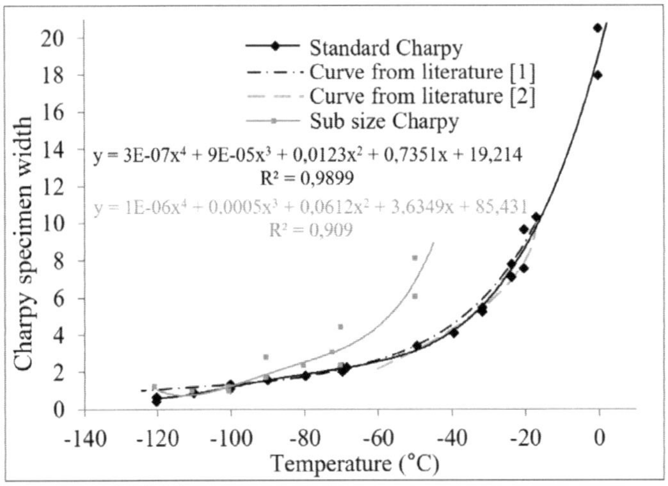

Figure 7. Correlations between Charpy width and FATT.

4 Conclusions

The correlation procedure for transition temperature determination of full size specimens based on fracture appearance of specimens with different width is proposed here. The correlation procedure assumes that the thickness of ductile part located on the edges of fracture in fracture appearance transition temperature is proportional to sample width. Using proposed procedure, there is possibility to approximate FATT temperature for each width of standard Charpy. Correlation for sub-size Charpy comparing to standard Charpy suggests the influence of specimens width, height, notch depth and notch root radius on transition temperature.

The advantage of proposed correlation procedure, comparing to constant correlation curves from the literature is a possibility to find the relationship between FATT and specimens width for non-standard Charpy specimens and more precise results for non - standard steel.

References

[1] Towers O L 1986 Testing of sub-size Charpy specimens: Part 1 - the influence of thickness on the ductile/brittle transition *Metal Construction* **18(3)** pp 171-176

[2] Wallin K 1994 Methodology for selecting Charpy toughness criteria for thin high strength steels: Part 1 - determining the fracture toughness *Jernkontorets Forskning, Report from Working Group* **4013(89)** p 28

[3] Konopik P, Dzugan J and Rund M 2015 Determination of fracture toughness in the upper shelf region using small sample test techniques *METAL 2015* **6(5)** pp 710-715

[4] Sokolov M A and Alexander D J 1997 An Improved Correlation Procedure for Subsize and Full-Size Charpy Impact Specimen Data *NUREG* **6379**

[5] Schubert L E, Kumar A S, Rosinski S T and Hamilton M L 1995 Effect of Specimen Size on the Impact Properties of Neutron Irradiated A533B Steel *J. Nucl. Mat.* **225** pp 231-237

[6] Konopik P, Dzugan J and Prochazka R 2013 Evaluation of local mechanical properties of steel weld by miniature testing technique *Materials Science & Technology* **27(31)** pp 2404-2411

[7] Corwin W R and Houghland A M 1986 Effect of Specimen Size and Material Condition on the Charpy Impact Properties of 9Cr1MoVNb Steel *ASTM, STP* **888** pp 325-338

[8] Louden B S, Kumar A S, Garner F A, Hamilton M L and Hu W L 1988 The Influence of Specimen Size on Charpy Impact Testing *J. Nucl. Mat.* **155(157)** pp 662-667

[9] Lucon E, Chaouadi R, Fabry A, Puzzolante J-L and Van Walle E 1999 Characterizing Material Properties by the Use of Full-Size and Sub-Size Charpy Tests: An Overview of Different Correlation Procedures *ASTM, STP* **1380**

[10] Sainte Catherine C, Hourdequin N, Galon P and Forget P 2002 Finite Element Simulations of Charpy-V and Sub-Size Charpy Tests for a Low Alloy RPV Ferritic Steel *Small Specimen Test Techniques* **1418** pp 107-136

[11] Schill R, Forget P and Sainte Catherine C 2002 Correlation between Charpy-V and Sub-Size Charpy Tests Results for an Un-Irradiated Low Alloy RPV Ferritic Steel *ECF 13* **1418** pp 54-68

[12] Lucon, E, Puzzolante J, Roebben G, and Lamberty A2005 Impact Characterization of Sub-Size Charpy V-Notch Specimens Prepared from Full-Size Certified Reference Charpy V-Notch Test Pieces *Journal of ASTM International*, **2(7)** pp 1-9

[13] Curll C H 1959 Subsize Charpy correlation with standard Charpy Watertown Arsenal Laboratories, *Technical Report* **112(95)**

[14] Sainte Catherine C, Poussard C, Vodinh J, Schill R, Hourdequin N, Galon P, Forget P 2002 Finite Element Simulations and Empirical Correlation For Charpy-V and Subsize Charpy Tests on an Unirradiated Low-Alloy RPV Ferritic Steel, *Small Specimen Test Techniques* **4** p 1418

[15] Džugan J, Procházka R and Konopík P 2014 Micro-Tensile Test Technique Development and Application to Mechanical Property Determination *Small Specimen Test Techniques* **6** pp 12-29

[16] Rund M, Procházka R, Konopík P, Džugan J and Folgar H 2015 Investigation of Sample-size Influence on Tensile Test Results at Different Strain Rates *Procedia Engineering* **114** pp 410-415

[17] Dzugan J, Konopik P., Rund M and Prochazka R 2015 *ASME Pressure vessels and piping conf. - 2015* Determination of local tensile and fatigue properties with the use of sub-sized specimens, **Volume 1A: Codes and Standards** (New York: Amer. Soc. Mech. Eng.)

[18] ASTM E23-94b 1995 *Annual Book of ASTM Standards* **03(01)** pp 137-156

[19] ESIS TC 5 2000 Proposed Standard Method for Instrumented Impact Testing of Sub-Size Charpy-V Notch Specimens of Steel *European Structural Integrity Society, Technical Committee n5 on Dynamic Testing Standards* **9**

[20] Sokolov M A, Nanstad R K 1996 On impact testing of subsize Charpy V - notch type specimens *Effect of Radiation on Materials* **17** pp 384-414

Effect of mechanical treatment on intergranular corrosion of 6064 alloy bars

P Sláma[1], J Nacházel[1]

[1]COMTES FHT a.s, Průmyslová 995, Dobřany, Czech republic

E-mail: peter.slama@comtesfht.cz

Abstract. Aluminium Al-Mg-Si-type alloys (6xxx-series) exhibit good mechanical properties, formability, weldability and good corrosion resistance in various environments. They often find use in automotive industry and other applications. Some alloys, however, particularly those with higher copper levels, show increased susceptibility to intergranular corrosion. Intergranular corrosion (IGC) is typically related to the formation of microgalvanic cells between cathodic, more noble phases and depleted (precipitate-free) zones along grain boundaries. It is encountered mainly in AlMgSi alloys containing Cu, where it is thought to be related to the formation Q-phase precipitates ($Al_4Mg_8Si_7Cu_2$) along grain boundaries. The present paper describes the effects of mechanical working (extrusion, drawing and straightening) and artificial aging on intergranular corrosion in rods of the 6064 alloy. The resistance to intergranular corrosion was mapped using corrosion tests according to EN ISO 11846, method B. Corrosion tests showed dependence of corrosion type on mechanical processing of the material. Intergranular, pitting and transgranular corrosion was observed. Artificial ageing influenced mainly the depth of the corrosion.

1 Introduction

Al-Mg-Si-type alloys (6xxx-series alloys) exhibit good mechanical properties, formability, weldability and good corrosion resistance in various environments. They frequently find use in automotive, aviation and other applications [1, 2]. Some of these materials are alloyed with copper to improve strength. In these alloys, particularly higher-copper alloys, increased susceptibility to intergranular corrosion (IGC) can be observed, most notably in unaged condition and less often in T6 temper condition. The effects of Cu as well as the opportunities for enhancing resistance to intergranular corrosion have received considerable attention in a number of studies [3-10]. Intergranular corrosion (IGC) is typically related to formation of microgalvanic cells between the cathodic more noble phases and the depleted (precipitate-free) zones along grain boundaries. It is encountered mainly in AlMgSi alloys which contain Cu, where it is thought to be linked to the formation of cathodic Q-phase ($Al_4Mg_8Si_7Cu_2$) along grain boundaries. The occurrence of phases along grain boundaries was observed using scanning transmission electron microscopy (STEM).

The impact of Cu addition and heat treatment on IGC was described in several papers [3-6]. The alloys contained 0.5-0.6 % Mg, 0.6-0.8 % Si, 0.2 % Fe, 0.2 % Mn and Cu at 0.02 through 0.7 wt. %. The occurrence of IGC was monitored in 2.5×78 mm^2 extruded flat bars. The effects of the cooling rate from the extrusion temperature were studied [3], as were the effects of artificial ageing [4, 5]. Corrosion tests were carried out according to EN ISO 11846. Corrosion was only monitored on the surface of the extruded parts. In an alloy with the Cu level of 0.02 %, no IGC was found. In an alloy with 0.2 % Cu, IGC occurred in dependence on the artificial ageing time, and changed into pitting corrosion. These findings suggest that between the occurrence of IGC and pitting corrosion, there is a region, in which no IGC occurs.

Hence, overageing (by increasing either temperature or time) permits transition from the region with IGC to a region with suppressed IGC.

The AA6056 material for aviation industry is used in overaged condition. It is supplied in the T78 state with enhanced resistance to IGC. According to the sources [7, 8], the T78 temper is achieved by two-stage ageing: 175 °C/6 h+210 °C/5 h.

In this research, the effect of thermomechanical treatment (extrusion, drawing, straightening and ageing) on intergranular corrosion in bars from EN AW-6064A (AlMg1SiBi) machinable alloy was studied. AlMgSi-type machinable alloys are used in automotive industry. Their improved machinability is imparted by alloying with Pb (6012 alloy) or with Bi+Pb (6262 and 6064 alloys). These alloys have higher alloy levels and contain more phases than the alloys studied in [3-10]. These phases include Bi and Pb cathodic particles. The resistance to intergranular corrosion was mapped using corrosion tests according to EN ISO 11846. Corrosion attack was observed on both the surface of the rods, and on the cross-section.

2 Experimental

The bars of 17 mm diameter were made by an industrial hot extrusion process using a multiple-hole die. The process temperature was 540–546 °C. Right after extrusion, the bars were water wave-cooled (T1 temper). The quenched bars were then drawn to the final diameter of 15 mm at 22 % reduction and straightened in Schumag straightening machine (T2 temper). The final operation was artificial ageing to T8 temper. Bars in conditions corresponding to each process step were gathered for testing. The chemical composition of the EN AW-6064A bars is shown in Table 1, the samples are listed in Table 2.

Table 1. Chemical composition of the alloy 6064A (wt.%).

Si	Fe	Cu	Mn	Mg	Cr	Pb	Bi
0.60	0.23	0.27	0.04	1.03	0.05	0.28	0.49

Table 2. Sample description.

Sample	Diameter	Temper	Description of thermomechanical processing
HA1	17 mm	T1	Extrusion, quenching
HB2	15 mm	T2	Extrusion, quenching, drawing
HF	15 mm	T2S	Extrusion, quenching, drawing, straightening

The bars were used in artificial ageing trials: single-stage and two-stage ageing from 160 °C/4h for underaged condition, to 180 °C/4h + 220 °C/4h for overaged condition.

The progress of ageing was monitored by HV5 hardness measurement using DURASCAN 50 hardness tester. Tests of resistance to intergranular corrosion were conducted in accordance with the EN ISO 11846 standard, method B [12]. For these tests, specimens of 2 cm length were made from the bars. Their cut surfaces were ground with P-1200 grinding papers. The original surface of the bar was not altered. Before testing, the specimens were degreased in acetone. In accordance with the standard requirements, they were etched with 5 % NaOH solution at 55 °C for 2 minutes. After a water rinse, they were placed in concentrated nitric acid for cleaning. The test itself involved submerging in a test solution for 24 hours at room temperature. The solution was 30 g NaCl/litre solution + 10 ml concentrated hydrochloric acid.

Metallographic sections were prepared on longitudinal cross sections through the specimens. The corrosion attacks on the bar surface as well as on the transverse cut surface were examined. The maximum corrosion depth was determined and documented using optical microscopy. The surfaces of specimens after corrosion testing were examined in JEOL JSM 6380 scanning electron microscope.

3 Results

3.1 Initial Microstructures

The microstructure of T8 temper bars upon drawing, straightening and ageing is shown in Fig. 1a. A micrograph of the phases is in Fig. 1b. The microstructure is fully recrystallized. The grains in the surface layer are relatively fine, with a size of 70 μm. In the centre, the grains are coarser, on the order of several hundred μm. Different grain sizes in the surface and in the interior are a typical occurrence in extruded bars from Al alloys. Typically, the surface layer contains coarse grains and the interior remains unrecrystallized [1, 2].

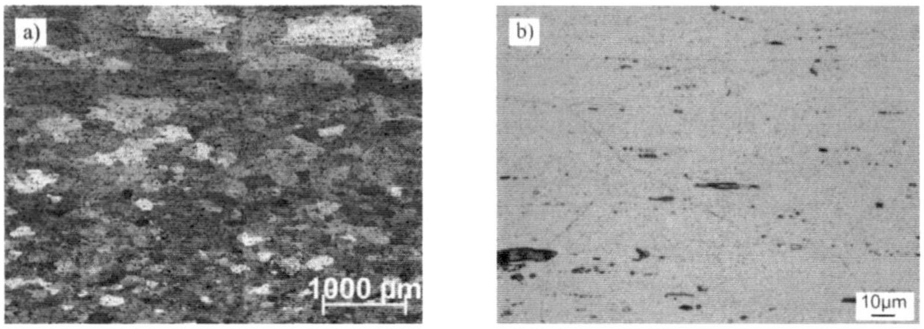

Figure 1. Micrographs of grains and phases in HC samples upon drawing and ageing: a) electrolytically etched with Barker's reagent, polarised light, b) etched with Dix-Keller's reagent.

The phases in the microstructure are banded and aligned in the extrusion/drawing direction. Large elongated particles consist of Bi or Bi+Pb. The small ones are alpha-$Al_{15}(Fe,Mn,Cu,Cr)_3Si_2$ particles. Other small particles are Mg_2Si particles. The Bi, Pb and alpha-$Al_{15}(Fe,Mn,Cu,Cr)_3Si_2$ particles are more noble, cathodic. The Mg_2Si particles are anodic. With cathodic particles, the matrix of the aluminium solid solution is etched away preferentially when placed in a corrosion environment. With anodic particles, they are the particles which are attacked. The microstructure may also contain cathodic Q-phase particles ($Al_4Mg_8Si_7Cu_2$). Figure 2b also shows minute particles along grain boundaries. EDS analysis revealed that they contain higher amounts of copper which suggests that they are Q-phase particles.

3.2 Corrosion tests of materials in initial condition

The initial condition evaluation was carried out on HF samples supplied in T2 (non-aged) condition. The surface corrosion is shown in Fig. 2.

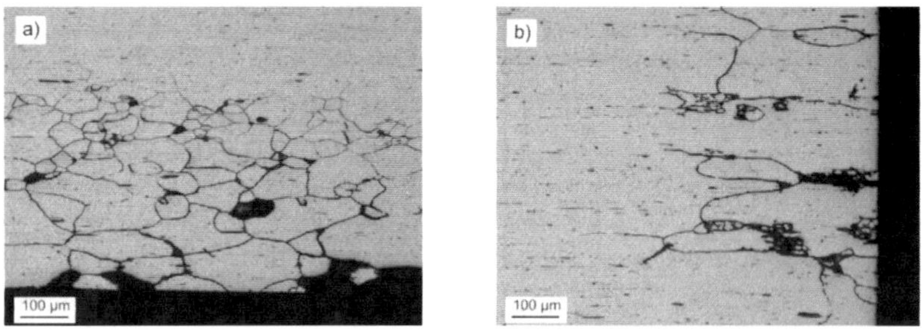

Figure 2. Corrosion attack in as-received bars HF, non-aged, a) surface, b) cross – section.

The surface of the non-aged HF sample shows extensive intergranular corrosion (IGC) with a depth of more than 420 μm. On the cross-section through the HF specimen, IGC with a depth of approx. 500 μm was found as well.

3.3 Corrosion tests after aging of specimens of extruded bars HA1

Surface corrosion of selected specimens in variously aged conditions is illustrated in Figure 3. The corrosion of the transverse cut surface is shown in Figure 4. In specimens in underaged condition, the most extensive surface corrosion was found, involving continuous IGC with a maximum depth of more than 300 µm. In the peak-aged condition, the depth of attack decreased and IGC ceased to be continuous. In overaged condition, only sporadic pitting corrosion can be observed with a depth of about 120 µm.

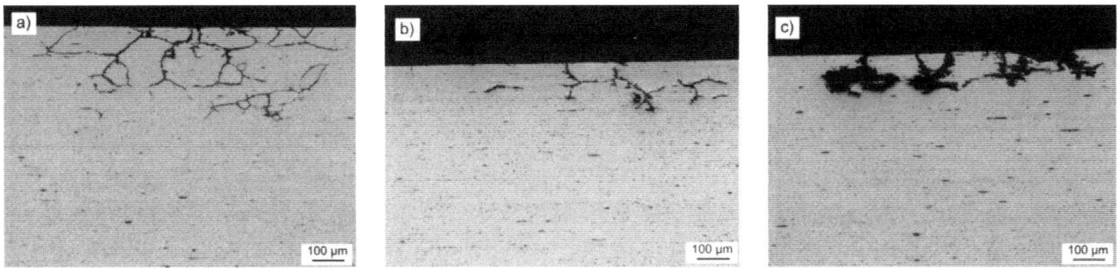

Figure 3. Corrosion on the HA1 bar surface after ageing: a) 180 °C/8h one-stage peak-aged, b) 160 °C/4h + 205 °C/4h two-stage peak-aged, c) 160 °C/4h + 220 °C/4h overaged.

The same type of corrosion was found on the transverse cut surface. However, the corrosion depth was larger there: more than 400 µm. The only exception was the overaged sample where the depth was less than 100 µm.

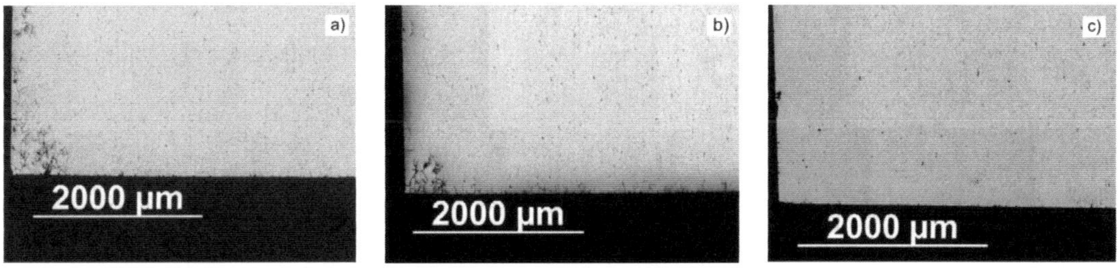

Figure 4. Corrosion on the HA1 bar cross-section after ageing: a) 180 °C/8h one-stage peak-aged, b) 160 °C/4h + 205 °C/4h two-stage peak-aged, c) 160 °C/4h + 220 °C/4h overaged.

3.4 Corrosion tests after aging of specimens of drawn bars HB2

Surface corrosion of selected specimens in variously aged conditions is illustrated in Figure 5. The corrosion of the transverse cut surface is shown in Figure 6.

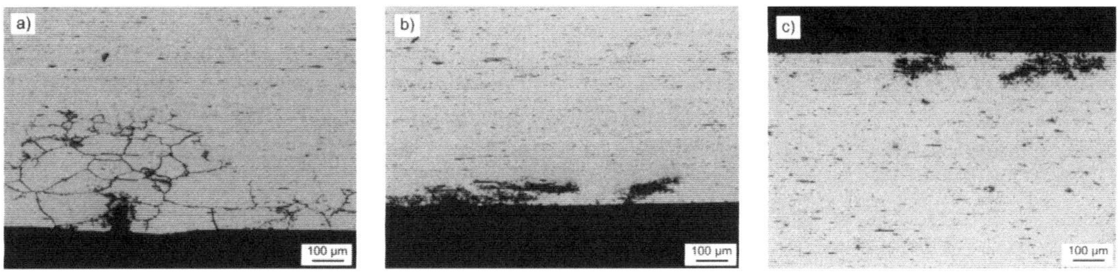

Figure 5. Corrosion on the HB2 bar surface after ageing: a) 180 °C/4h underaged, b) 180 °C/4h + 205 °C/4h two-stage peak-aged, c) 160 °C/4h + 220 °C/4h overaged.

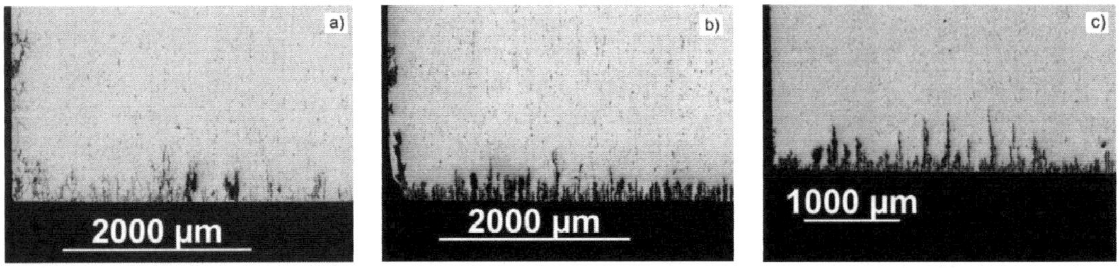

Figure 6. Corrosion on the HB2 bar cross-sectiom after ageing: a) 180 °C/4h underaged, b) 180 °C/4h + 205 °C/4h two-stage peak-aged, c) 160 °C/4h + 220 °C/4h overaged.

In HB2 drawn bars, IGC was found only in the underaged condition. This intergranular corrosion is not continuous. In the peak-aged and overaged conditions, the bar's surface only exhibits pitting corrosion with a depth of about 100 µm.
Besides that, corrosion spreads parallel to and beneath the surface, along the bands of coarse cathodic phases. Authors of [9] describe this type of corrosion as ELA (Exfoliation-Like Attack). On the transverse cut surface, corrosion is of the pitting type as well. It is much deeper and, again, more frequent in the near-surface areas. The depth of attack exceeds 700 µm (Figure 6).

3.5 Corrosion tests after aging of specimens of drawn and straightened bars HF

Surface corrosion of selected specimens in variously aged conditions is illustrated in Figure 7. The corrosion of the transverse cut surface is shown in Figure 8.

Figure 7. Corrosion on the HF bar surface after ageing: a) 180 °C/4h underaged, b) 180 °C/4h + 205 °C/4h two-stage peak-aged, c) 160 °C/4h + 220 °C/4h overaged.

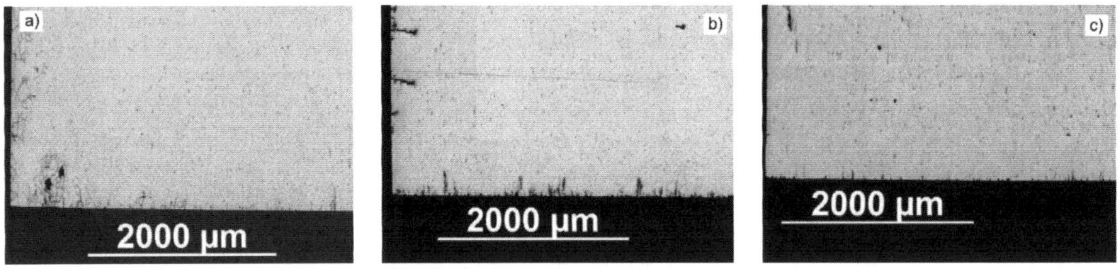

Figure 8. Corrosion on the HB2 bar cross-sectiom after ageing: a) 180 °C/4h underaged, b) 180 °C/4h + 205 °C/4h two-stage peak-aged, c) 160 °C/4h + 220 °C/4h overaged.

In specimens in underaged condition, there is deep IGC on the bar's surface, as well as on the transverse cut surface. In the peak-aged and overaged conditions, the bar surface only exhibits pitting corrosion that spreads perpendicularly to the surface to a depth of more than 300 µm. It is transgranular corrosion, as it penetrates the grains.

On transverse cut surfaces, the least extensive corrosion was found in the peak-aged condition (Fig. 8a). In the slightly-overaged condition, the corrosion is extensive and deep (Fig. 8b). In increasingly overaged specimens, the number and depth of corrosion attack locations decrease (Fig. 8c).

4 Discussion

The main mechanism of IGC is reported to be the formation of microgalvanic cells between cathodic more noble phases and depleted (precipitate-free) zones along grain boundaries. In this case, the key cathodic phase is the Q-phase ($Al_4Mg_8Si_7Cu_2$) which precipitates along grain boundaries. As a result, the grain boundary areas become depleted of Cu and other elements. In addition, a thin Cu film forms along grain boundaries and plays the key role in IGC growth and propagation [3, 4, 5, 6]. The entire precipitation process is thermally activated and depends on the diffusion of alloying elements. Its rate is described by Arrhenius equation. With increasing ageing temperature and time, the Q-phase precipitates coarsen and the volume fraction of the Cu film along grain boundaries decreases. Consequently, the susceptibility to IGC is reduced and the material typically exhibits only pitting corrosion.

EN AW-6064 alloy contains a number of other primary cathodic phases (Bi, Pb, alpha-$Al_{15}(Fe,Mn,Cu,Cr)_3Si_2$). Their arrangement in bands with short distances between phases helps the pitting corrosion to propagate larger depths, most notably beneath the transverse cut surface. In some cases, there were great differences between the corrosion attack on the bar's surface and on the transverse cut surface.

In the extruded bars (HA1), it was found that with increasing overageing the large-depth IGC changes into shallower pitting corrosion, which is in agreement with findings presented in [3-6]. In the overaged condition, the corrosion penetrations on the transverse cut surface were smaller. Sporadic pitting corrosion with a depth of about 100 μm was found.

In drawn bars (HB2), the transition from IGC to shallower pitting corrosion was observed as well. Unlike the specimens from bars which had not been drawn, all specimens in this group showed very deep corrosion (more than 700 μm) on their transverse cut surfaces (Figure 6).

In the drawn and straightened bars (HF), another type of corrosion was observed. In underaged bars, IGC was found on both the bar surface and the transverse cut surface. With ongoing ageing, IGC changes into pitting corrosion which – on the bar surface – propagates perpendicularly to the surface and by transgranular mechanism to a larger depth than pitting corrosion in drawn bars (Figure 7). This corrosion type corresponds to transgranular stress corrosion cracking (SCC) [13]. The difference can be attributed to the variation between internal stresses induced by drawing and straightening. Drawing typically induces tensile stress. Straightening, however, involves alternating bending loads and tensile and compressive stresses which lead to non-uniform residual stress that promotes corrosion propagation perpendicularly to the surface and to a larger depth. The transverse cut surface, unlike HB2 specimens, shows – in some cases – shallow sporadic pitting corrosion (Figure 8a, c).

5 Conclusion

Extruded and drawn bars from EN AW-6064A alloy were used for exploring the impact of thermomechanical treatment on intergranular corrosion (IGC). The effects of forming (drawing and straightening) and artificial ageing were mapped, along with the type of corrosion and corrosion depth on bar surface and its transverse cross-section. The corrosion tests were carried out in accordance with EN ISO 11486 – method B.

The results of the corrosion tests show that thermomechanical treatment affects both the type and depth of corrosion.

The bar surface exhibited three types of corrosion:

- IGC in underaged specimens: typically extensive corrosion with a depth of more than 300 μm.

- Pitting corrosion in more aged and overaged extruded/drawn bars, where the corrosion depth was approximately 100 μm.
- Transgranular pitting corrosion in more aged and overaged bars which had undergone final straightening. Here, the corrosion depth was larger and exceeded 300 μm.

With more intensive ageing and overageing (temperature, time), IGC changed into pitting corrosion in extruded/drawn bars. There was an adverse impact of the post-drawing straightening operation on the resistance to surface corrosion in the bars, evidenced by deep transgranular pitting corrosion. In most cases, the transverse cross-sections exhibited very deep pitting corrosion with depths up to 800 μm which followed the bands of coarse cathodic phases. Exceptions were found in severely overaged bars (extruded or extruded and straightened) which showed sporadic pitting corrosion with depths of approximately 100 μm.

Acknowledgements
This paper was created by project Development of West-Bohemian Centre of Materials and Metallurgy No.: LO1412, financed by the MEYS of the Czech Republic.

References

[1] Altenpohl D G 1998 Aluminum: Technology, Applications, and Environment: A Profile of a Modern Metal (Warrendale: Minerals, Metals, and Materials Society).
[2] Hatch J E 1984 Aluminium - Properties and Physical Metallurgy (Ohio: ASM) p 50 .
[3] Svenningsen G, Lein J E, Bjorgum A, Nordlien J H, Yu Y D and Nisancioglu K 2006 Effect of low copper content and heat treatment on intergranular corrosion of model AlMgSi alloys *Corrosion science* **48(1)** pp 226–242
[4] Svenningsen G, Larsen M H, Nordlien J H and Nisancioglu K 2006 Effect of high temperature heat treatment on intergranular corrosion of AlMgSi(Cu) model alloy *Corrosion science* **48(1)** pp 258–272.
[5] Svenningsen G, Larsen M H, Walmsley J C, Nordlien J H and Nisancioglu K 2006 Effect of artificial aging on intergranular corrosion of extruded AlMgSi alloy with small Cu content *Corrosion science* **48(6)** pp 1528–1543.
[6] Larsen M H, Walmsley J C, Lunder O, Mathiesen R H and Nisancioglu K 2008 Intergranular Corrosion of Copper-Containing AA6xxx AlMgSi Aluminum *J. Electrochem. Soc.* **155(11)** pp C550-C556.
[7] Guillaumin V and Mankowski G 2000 Influence of Overaging Treatment on Localized Corrosion of Al 6056 *Corrosion* **56** pp 12-23.
[8] Gallais C, Denquin A, Brechet and Lapasset G 2008 Precipitation microstructures in an AA6056 aluminium alloy after friction stir welding: Characterisation and modelling *Mater. Sci. Eng.* **A496** pp 77–89.
[9] Eckermann F, Suter T, Uggowitzer P J Afseth A and Schmutz P 2008 Investigation of the exfoliation-like attack mechanism in relation to Al–Mg–Si alloy microstructure *Corrosion Science* **50 (7)** pp 2085–2093
[10] Wang Z, Li H, Miao F, Sun W, Fang B, Song R and Zheng Z 2014 Improving the intergranular corrosion resistance of Al–Mg–Si–Cu alloys without strength loss by a two-step aging treatment *Mater. Sci. Eng.* **A590** pp 267-273.
[11] Halap A, Popović M, Radetić T,Vaščić V and Romhanji E 2014 Influence of the thermo-mechanical treatment on the exfoliation and pitting corrosion of an AA5083-type alloy *Materiali in tehnologije / Materials and technology* **48** pp 479–483
[12] EN ISO 11846:1995. Corrosion of metals and alloys. Determination of resistance to intergranular corrosion of solution heat- treatable aluminium alloys.
[13] ASM Handbook Vol. 13. Corrosion 1987 (Ohio: ASM)

IOP Conf. Series: Materials Science and Engineering **179** (2017) 012065 doi:10.1088/1757-899X/179/1/012065

Mechanical testing of anisotropy in ODS steel tubes

Z Špirit[1,2], M Chocholoušek[1], M Šíma[1]

[1]Centrum vyzkumu Rez s.r.o., Morseova 1245/6, Plzen, Czech Republic, EU

[2]Department of Material Science and Technology, University of West Bohemia, Plzen, Czech Republic, EU

E-mail: zbynek.spirit@cvrez.cz

Abstract. Anisotropy of fuel claddings made from Fe-9Cr and Fe-14Cr ODS (oxide dispersion-strengthened) steel for Generation IV reactors with coolants based on liquid metals was examined. Basic mechanical tests were performed at temperatures of 30 °C, 500 °C and 625 °C on ODS steel thin-walled tubes. Tensile tests were performed in axial and tangential directions in a vacuum to avoid oxidation of the small specimens. A scanning electron microscope with a small in-situ tensile testing device was used to conduct the tests. The result of this paper describes the basic mechanical properties of thin-walled tubes as ultimate tensile strength, yield strength, tensile modulus, etc. depending on the direction of loading.

1 Introduction

New types of Generation IV reactors are safer and more environmentally friendly than conventional reactors because they burn fuel more efficiently. These types include LFR (Lead-Cooled Fast Reactor System) reactors, which use heavy liquid metals (HLM) and their alloys (Pb, Pb-Bi, Pb-Li) as a coolant. Liquid metals contribute to safety through their internal properties, but they have a negative effect on construction materials because of the corrosive environment which they create. [1]

One of the candidate materials for cladding fuel rods are ODS steels. ODS steels are strengthened by oxidic dispersions with elements Al, Ti, Y – there is a synergistic effect between the oxidic phase and the solid solution which improves the mechanical and physical characteristics of the steel. These steels have a longer life and much higher resistance to cyclic degradation, especially at high temperatures, than alloys without dispersed oxidic phase formation. [1,2]

ODS steels are products of powder metallurgy where the sintering powder itself has a significant impact on the resulting mechanical, corrosive and microstructural properties of the material. Due to the complicated manufacturing methods of ODS steels, it is important to check the required properties.

This paper deals with the investigation of the anisotropy of the mechanical properties of thin-walled tubes of Fe-9Cr and Fe-14Cr ODS steels which have been made for the fuel cladding for Generation IV reactors. Due to the limited amount of material in European laboratories and the small size of real components it is necessary to perform the material testing on miniature specimens. With regard to the anticipated operating temperatures, the tests must be done at temperatures up to 625 °C. Oxidation at high temperatures on miniature specimens of ODS steels changes their material characteristics, and therefore it is best to perform tests without air. [3]

2 Material

Two thin-walled tubes with diameters of approximately 10.7 mm and a wall thickness of approximately 0.6 mm were used for the experiments. The material of the first tube, with the designation K49 Fe-9Cr is powdered ferritic-martensitic steel with oxidic phases based on Y_2O_3. The second material, designated K48-T2-M3 Fe-14Cr is also ferritic-martensitic steel with a higher proportion of chromium containing Y_2O_3 oxides. Both materials show good resistance to creep at high temperatures. The chemical composition of the major alloying elements of the two materials are shown in Table 1.

Table 1. Chemical composition of ODS steel.

Designation (material)	Fe [wt.%]	Cr [wt.%]	W [wt.%]	Ti [wt.%]	Y [wt.%]
K49 (Fe-9 Cr)	88.1	9	2	0.51	0.3
K48-T2-M3 (Fe-14 Cr)	84.4	14	1	0.26	0.3

3 Experiment

Miniature specimens for tensile testing in the axial and tangential direction of the ODS tube were made at Centrum Vyzkumu Rez (CVR) (Figure 1). The sample for tensile testing in the tangential direction is ring-shaped with a narrowing on the two opposite sides. The samples were dimensionally designed for Microtest 2000-EC, which is a part of the scanning electron microscope (SEM). The maximum allowable load of the testing device is 2 kN. The total initial cross-section area was designed to be approximately 1 mm^2 with regard to the maximum load capacity and to the expected ultimate strength of ODS steels. Considering the ODS tube thickness, 2 mm was selected as the width of the axial sample, and a width of 1 mm was chosen on each side in the tangential direction.

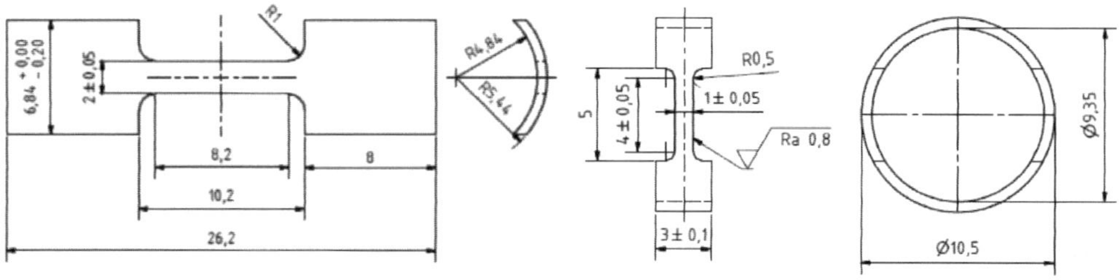

Figure 1. Miniature specimen for tensile testing in the axial direction (left) and in the tangential direction - ring test (right).

Tests were conducted in an SEM chamber at temperatures of 30 °C, 500 °C and 625 °C in a vacuum of $3 \cdot 10^{-2}$ Pa to avoid oxidation on the sample surface. The deformation speed of the crosshead was 0.4 mm/min (0.001 s^{-1}). Only the ultimate tensile strength (UTS) was measured on the samples in the tangential direction.

For samples in the axial direction, UTS and yield strength (YS), tensile modulus (E_{mod}), plastic elongation at break (A_{6mm}) and plastic elongation at UTS (A_g) were measured. To make this evaluation feasible, the sample was primarily designed with respect to the relationships between the initial cross-section and the initial measured length according to standard EN ISO 6892. The initial gauge length (L_0) was marked on the sample by using indentation imprints HV0.1. Measurement of the indentation distance was used instead of an extensometer. The measurement was made using SEM image analysis at selected load levels (see Figure 2).

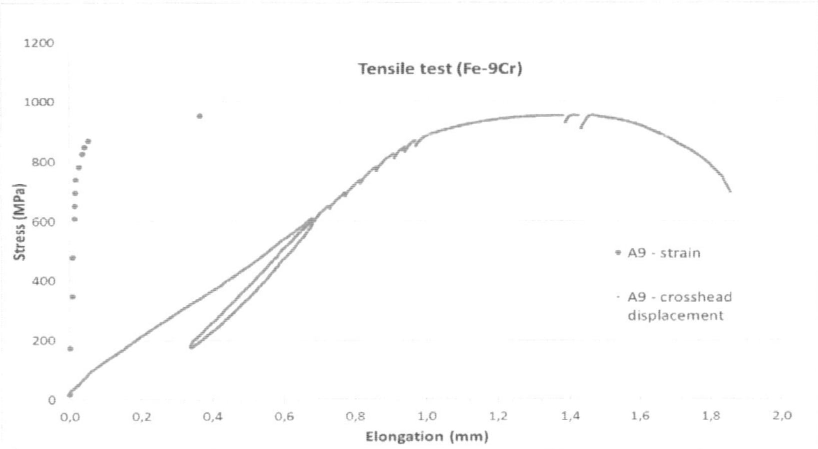

Figure 2. Comparison between elongations measured on L_0 (by SEM) and crosshead displacement.

4 Results

The resulting mechanical properties of the tensile test series performed on the ODS steel samples at temperatures of 30 °C, 500 °C and 625 °C are summarized in Table 2.

A_g and A_{6mm} elongations are measured only in the axial direction. A_{6mm} can be reliably evaluated only on samples where the break was in the centre of L_0. However, the samples often cracked beyond the middle third of L_0, therefore it was not possible to evaluate elongation at the break for all specimens and, particularly, there is no A_{6mm} evaluation at high temperatures. A_g is rounded to the nearest whole percent, and its character is purely informative, because its evaluation requires stopping the tension test at maximum load, but due to the flat load-displacement waveforms at UTS, such stopping is approximate, and so A_g-values are rounded to correspond to the accuracy of the measurement. Deformation measurement by SEM in the tangential direction on the ring shaped specimen was not possible, because the shape of the sample is complicated for both marking and mark observation. For the Fe-14Cr samples in the tangential direction at 500°C a more significant UTS decrease was observed than that of Fe-9Cr, and this corresponds to the appearance of the fracture surfaces (Figure 3).

Table 2. Mechanical properties of ODS steels at temperatures 30 °C, 500 °C and 650 °C.

Designation	Direction	Temperature (°C)	E_{mod} (GPa)	YS (MPa)	UTS (MPa)	HV0.1	A_g (%)	A_{6mm} (%)
Fe-9Cr	axial	30	226	829	956	364	8	14.8
	axial	500	118	499	637	-	3	-
	axial	625	78	316	455	-	4	-
	tangential	30	-	-	933	-	-	-
	tangential	500	-	-	617	-	-	-
	tangential	625	-	-	424	-	-	-
Fe-14Cr	axial	30	224	963	1049	373	7	16.8
	axial	500	152	652	737	-	2	-
	axial	625	99	442	558	-	1	-
	tangential	30	-	-	996	-	-	-
	tangential	500	-	-	623	-	-	-
	tangential	625	-	-	494	-	-	-

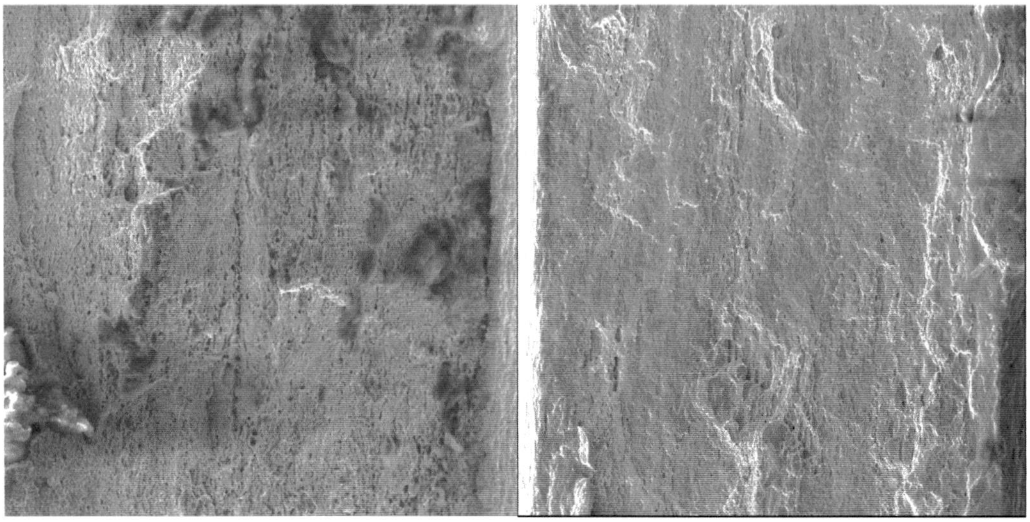

Figure 3. Fracture surface after testing sample A10 (Fe-9Cr - left) and sample B13 (Fe-14Cr - right) at 500 °C.

5 Conclusions

Mechanical tests on fuel cladding tubes made from Fe-9Cr and Fe-14Cr ODS steels show anisotropy between the axial and tangential directions (Table 2). The differences are higher for Fe-14Cr steel than that of Fe-9Cr steel. Moreover, a significant reduction of ductility occurs in Fe-14Cr steel at elevated temperatures. This reduction is most significant in the tangential direction at 500 °C.

Acknowledgements

The presented work was financially supported by the Ministry of Education, Youth and Sport Czech Republic Project LQ1603 (Research for SUSEN). This work has been realized within the SUSEN Project (established in the framework of the European Regional Development Fund (ERDF) in project CZ.1.05/2.1.00/03.0108).

References

[1] Katzer M *Generation IV Nuclear reactors*. [online]. Brno university of technology, Faculty of Mechanical Engineering, 2011 [cit. 2015-12-04].

[2] Nuclear energy agency organisation for economic co-operation and development http://www.oecd-nea.org/science/pubs/2015/7268-lead-bismuth-2015.pdf (*accessed 7. 10. 2015*).

[3] MATISSE WP4 Summary of first 18 months, [online] http://www.fp7-matisse.eu, 2014 [cit. 2016-4-06].

Computer simulation of combination extrusion of ENAW1050A aluminum

P Thomas

Department of Applied Computer Science and Armament Engineering, Faculty of Mechatronics and Machine Design, Kielce University of Technology, Al.1000-lecia P.P.7, 25-314 Kielce, Poland, EU

thomas@tu.kielce.pl

Abstract. Computer simulation of the combination extrusion process for ENAW-1050A aluminum alloy is presented. The tests were carried out for three values of relative strain in forward direction ε_1: 0.77, 0.69 and 0.59. For each value of relative strain ε_1, three different values of strain in backward direction, ε_2, were taken: 0.41, 0.52, 0.64. The effect of the relative strain degree on the development and values of the punch force was determined. It was demonstrated that the punch force increases with the increasing degree of relative strain in both forward and backward directions.

1 Introduction

In the recent years, considerable attention has been paid to aluminum and its alloys, which find ever more applications in many sectors. Their good mechanical properties and low specific mass are deciding factors for industrial users [1,2,3,4]. Science and technology advances of the last decade have contributed to the use of aluminum and its alloys in automotive subassemblies and components [5]. Current investigations are focused on creating aluminum-based materials of expressed qualities and on new forming technologies. In parallel with a growth in worldwide consumption of aluminum, there has been an increase in available computing power, contributing to a dynamic development of software. Numerous commercial software packages, including those implementing finite element modeling, have been developed for capacitive simulation. The most widely used packages in metal forming design and simulation include ABAQUS, DEFORM, FORGE 2&3, MARC Autoforge, MSC Superforge, and Qform [6,7,8,9].

2 Methodology

The objective of this study was to determine the effect of relative strain, ε_1 in the forward direction and ε_2 in the backward direction, on the values and development of the punch forces. The computer simulation consisted of free forward V_1 and backward V_2 flows of the material placed in the die as a result of the punch pressure on the material with the axial force F, according to diagram in Figure 1. The model for the QFORM 2D simulation was defined using the following stages:

- starting piece geometry and tooling shapes,
- flow curve for material,
- friction factor,
- number of the rectangular Lagrange mesh.

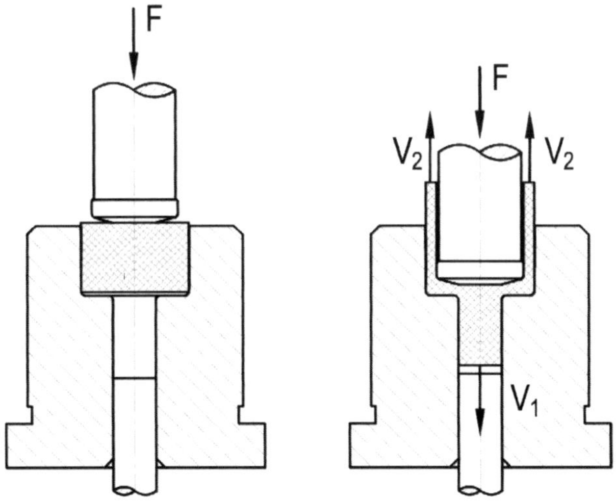

Figure 1. Schematic view of combination extrusion: a – initial stage, b – final stage; F – punch force, V_1 – forward flow of the material, V_2 – backward flow of the material.

The calculations ended when the thickness of the extrusion bottom was g=2 mm. This value, preset prior to the simulation, represents the ultimate distance, that is, the distance between the punch and the die insert in the axial direction. The geometrical model of the stock material was created in AutoCAD software and then imported into QFORM 2D. The material was defined by a mesh of 200 elements. The die inserts and the punch were taken to be rigid, non-deformable elements. Physical and mechanical properties of the material were defined as a model of a homogeneous material which experiences elastic-plastic deformation in accordance with the constitutive equation, $\sigma_p=152\varphi^{0.25}$. Material constants C and n were determined experimentally for aluminum ENAW-1050A used in the tests. Contact conditions between the deformed material and the tools were defined through the value of the friction factor, μ=0.28 (for pieces without lubrication). The friction factor was determined experimentally by the ring upsetting method. Langrange meshes were marked in the cross-sections of the pieces (22 horizontally and 11 vertically), as shown in Figure 2.

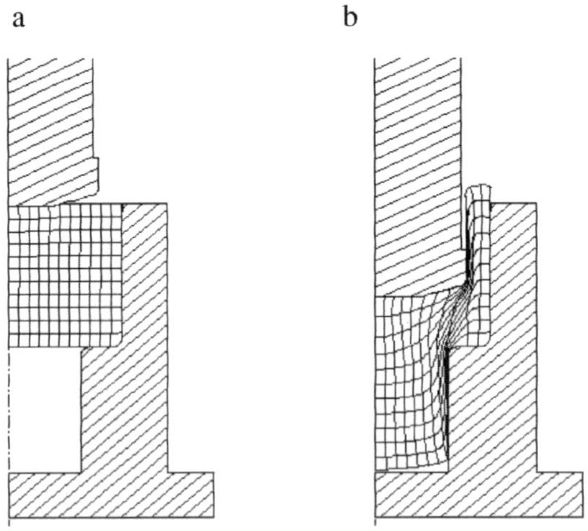

Figure 2. Lagrange mesh: a – initial condition, b – view of the deforming mesh.

The size of the mesh allows determining the areas of deformation in various zones of the extruded product. Since QFORM 2D program divides the extrusion process into 30 steps, the mesh deformations can be evaluated at any extrusion stage. Due to the axisymmetric deformation, the author defined only the right-hand part of the material and the tools used. This contributed greatly to the reduction of the computing cycle time. The computer simulation determined the punch forces for various values of relative strains, ε_1 (forward direction) and ε_2 (reverse direction).

3 Simulation of combination extrusion forces for various values of relative strain in the forward and backward directions

The simulation of the extrusion process was conducted with flat die inserts and flat-conical punches (15°). The stock material was assumed to have the following dimensions $d_0 \times h_0 = 24.95 \times 16$ mm. The end products obtained from the process, their shapes and dimensions are shown in Figure 3.

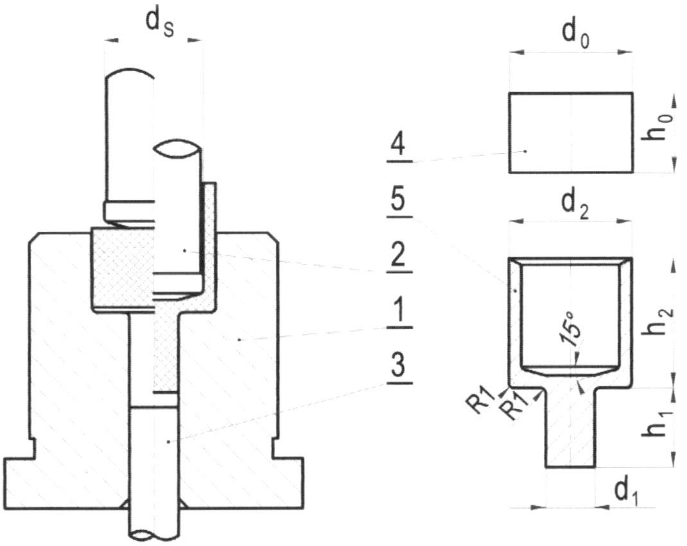

Figure 3. The scheme of combined extrusion of aluminum stampings using conical punch: 1 – die; 2 – flat-conical punch; 3 – ejector; 4 – billet; 5 – die stamping.

The end products had the following dimensions:

- the upper part of the product extruded in the backward direction: external diameter d_2=25 mm=const, internal diameters d_s = 16, 18, 20 mm,
- diameters of the bottom part of the product extruded in the forward direction: d_1 = 12, 14, 16 mm.

With the dimensions adopted for the stock material and the end products, the following relative strains were obtained

- backward extrusion, respectively:

$$\varepsilon_2=d_s^2/d_0^2=0.41;\ 0.52;\ 0.64 \tag{1}$$

- forward extrusion:

$$\varepsilon_1=(d_0^2-d_1^2)/d_0^2=0.77;\ 0.69;\ 0.59 \tag{2}$$

When the backward relative strain ε_2 increased, the wall thickness of the tubular section of the extrusion decreased; when the value of ε_1 was reduced, the diameter of the cylindrical section of the forward-extruded part increased. The maximum height of the cylindrical section extruded in the forward direction was assumed to be h_l = 14 mm. This value restricted further displacement of the material in the forward direction. The results of the computer simulation included the plots of the punch forces as a function of punch displacement during the combined process. Figure 4 shows the change in the punch force for different relative strain values in the backward direction ε_2=0.41; 0.52; 0.64, at constant strain value in the forward direction ε_l=0.77=const (d_l = 12 mm). The plots of forces (Figure 5 and 6) vary only in the value the relative strain ε_l: ε_l=0.69 (d_l=14 mm, Figure 5) and ε_l=0.59 (d_l=16 mm, Figure 6), respectively. It can be seen that the increase in the backward relative strain ε_2 contributes to the increase in contact forces primarily due to material hardening. The contact forces increase also with the increased forward extrusion relative strain ε_l.This is the result of the material hardening and the increase in the area of the flat part of the die insert, hindering the forward flow of the material.

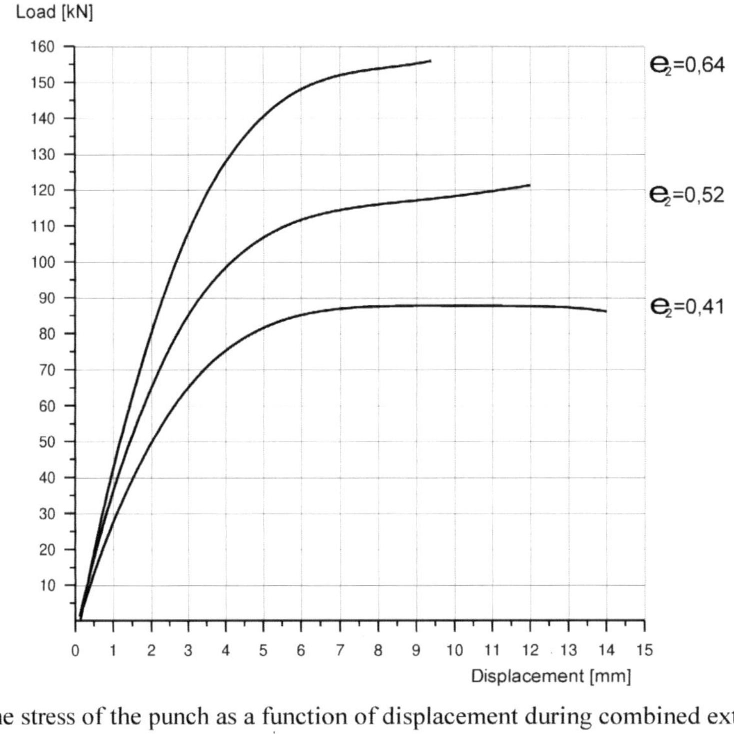

Figure 4. The stress of the punch as a function of displacement during combined extrusion for different degrees of backward strain ε_2 and constant value of forward extrusion ε_l = 0.77 (d_l=12mm).

The increase in backward extrusion relative strain ε_2 enhances the forward flow of the material. Earlier arrival of the material to the extruder ends the extrusion process for the displacement smaller than 14 mm, as shown in figures 4,5,6 for higher values of ε_2. In the initial stage of the combination extrusion process, the punch displacement causes a rapid increase in the contact force, followed by its stabilization. For the highest relative strain values ε_2 used, the pressure rises slightly, and for lower ε_2, the forces vary independently of the $\varepsilon_l$value.

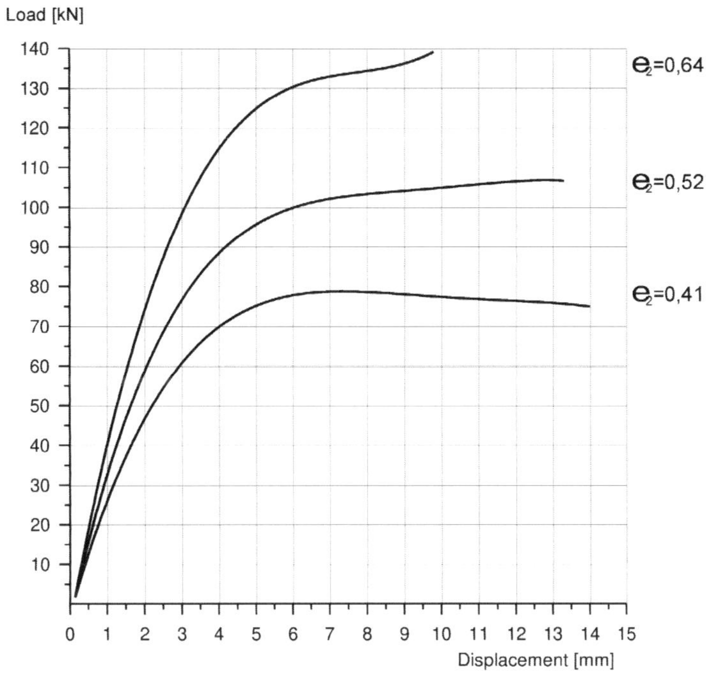

Figure 5. The stress of the punch as a function of displacement during combined extrusion for different degrees of backward strain ε_2 and constant value of forward extrusion ε_1 = 0.69 (d_1=14mm).

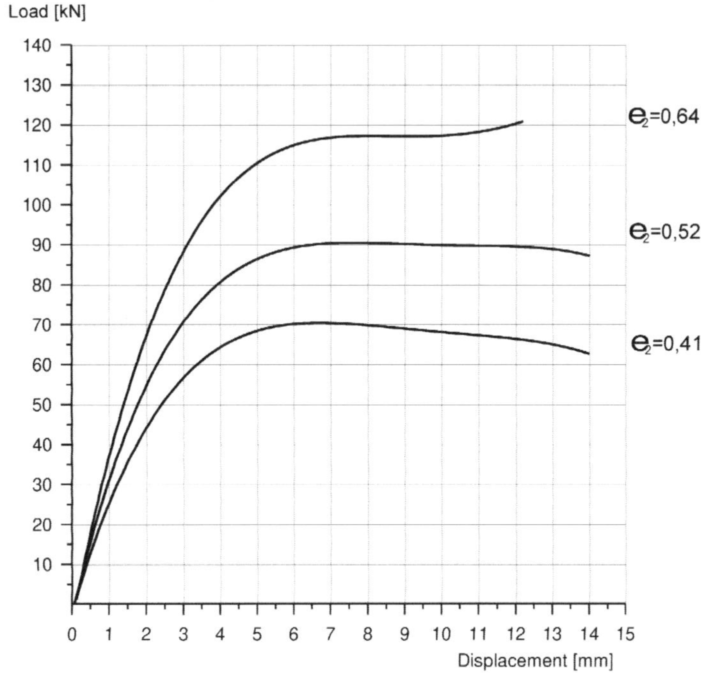

Figure 6. The stress of the punch as a function of displacement during combined extrusion for different degrees of backward strain ε_2 and constant value of forward extrusion ε_1 = 0.59 (d_1=16mm).

4 Conclusions

The following conclusions can be drawn from the simulation of combination extrusion of aluminum:

- an increase in the value of relative strain ε_2 in the backward direction contributes to material hardening, increasing the contact force,
- increased area of the flat die insert, which increases relative strains $\varepsilon 1$ in the forward direction, also contributes to the increase in contact force,
- for maximum relative strains ε_2 in the backward direction, material flow if more intense in the forward direction.

It has to be emphasized that the results presented in this paper show the advantages of the combination extrusion process over the technologies of forward and backward extrusions. Forming the tubular-cylindrical parts using combination extrusion eliminates the requirement of using two operations, in which first, forward extrusion is used to manufacture cylindrical part and then, in the second operation, tubular part is made using backward extrusion. Combination extrusion reduces the cost and time of the forming process. The reduction in the value of forming forces contributes to better durability of the tooling used.

References

[1] Dom F J 1994 Meltspun aluminium successful in racing pistons, Aluminium vol 70 no 9/10 pp 575-578.
[2] Brandt D 1992 Konstruieren mit Aluminium Schmiedestűcken, Umformtechnik vol 26 no 3
[3] Gabathuler J P 1997 Formed Components for the Car Industry, Aluminium vol 73 no 5
[4] Lenczowski B Pfannenműller T and Koch U 1997 Neue Aluminiumlegierungen fűr die Luftfahrt, Aluminium vol 73 no 5
[5] Neues Aluminiumrohr Speziell fűr KFZ-Konstruktionen 2004 Aluminium t 80 no 6
[6] Kocańda A 1999 Directions in metal forming, Mechanical review no. 2
[7] Kocańda A 2000 The selected directions of the development in metal forming processes, Mechanical review no 3 pp 6-11
[8] Kuzman K Pfeifer E Bay N and Hunding J 1996 Control of material flow in a combined backward can - forward rod extrusion, Journal of Materials Processing Technology vol 60 issue 1-4 june 15
[9] Pietrzyk M 2003 Numerical methods in the plastic working of metals, Publishing company AGH Kraków

Distribution of porosity and macrosegregation in slab steel ingot

M Tkadleckova[1], P Jonsta[2], Z Carbol[2], M Susovsky[2], K Michalek[1], L Socha [1], J Sviželová [1] and J Zwyrtek [1]

[1]VSB- Technical University of Ostrava, Faculty of Metallurgy and Materials Engineering, 17. Listopadu 15/2172, 708 33 Ostrava-Poruba, Czech Republic, EU
[2]VITKOVICE HEAVY MACHINERY a.s., Ruska 2887/101, Vitkovice, 703 00 Ostrava, Czech Republic, EU

E-mail: marketa.tkadleckova@vsb.cz

Abstract. The paper presents a new knowledge and experiences with verification and optimization of production technology of heavy slab ingot weighing 40 t from tool steel using the results of numerical modelling and of operational experiments at a steel plant in the company VÍTKOVICE HEAVY MACHINERY a.s. The final porosity, macrosegregation and the risk of cracks were predicted. Based on the results, the slab ingot can be used instead of the conventional heavy steel ingot. Also, the ratio, the chamfer, and the external shape of the wall of the new design of the slab ingot was improved, which enabled to reduce production costs while the internal quality of steel ingots was still maintained at very high level.

1 Introduction

The company VÍTKOVICE HEAVY MACHINERY a.s. (VHM) focused in recent years on an expansion of production of forgings with higher added value, which definitely includes also the forgings from tool steels finding their application mainly in general engineering, at the construction of machinery and equipment. This concerns e.g. thick steel plates, pulleys, disks, etc. Production of these demanding forgings requires knowledge and experience that can only be obtained through own research and development. However, with regard to the price of material and the weight of the forgings such research is very expensive. For forgings weighing from approx. 15 to approx. 60 tons it is necessary to use high-quality ingots weighing from 20 to 100 tons. The know-how for the production of heavy forgings requires knowledge and experience particularly in the field of metallurgy. It is necessary to observe the required chemical composition with respect to very strict requirements relating to the content of oxygen, sulphur, and other elements. Especially for larger forgings, it is required to obtain homogeneity and isotropy of properties, i.e. the minimum extent of segregations and very low content of non-metallic inclusions. Octagonal polygonal ingots are used most often for the forgings. In the case of tool steels, the ingots with a larger chamfer of the ingot wall with a relatively low ratio of the height of the ingot to its mean diameter (H / D) are recently preferred [1-4]. These ingot parameters ensure better conditions for charging of the molten steel from the ingot hot top to the ingot body during solidification. The result is a very small centerline porosity of the ingot. However, the larger diameter of ingots causes a greater central macrosegregation, and thus in possible non-homogeneity of mechanical properties of the forged product. For certain types of forgings (especially of plates) these macrosegregations can be limited by using slab ingots, which are distinguished by a characteristic A / B aspect ratio. It is expected that due to the relatively small width of the slab ingot, and thus its faster solidification in comparison to a polygonal ingot of the same mass, the occurrence of macrosegregations in the slab ingot will be smaller than in the polygonal ingot.

One of the ways, how to verify and optimize the production steps from the casting to the forming process, is the use of methods of numerical modelling [5-11]. The numerical modelling allows us to follow the character of the flow of the steel during the filling of the mould, to predict the temperature field during the casting and solidification, or to predict the final volume defects, which may occur in the final solid structure, such as porosity, macrosegregation, and cracks or hot tears.

Due to the huge plant effect from the results of the numerical modelling, in this study, the numerical modelling was also used for verification and optimization of production technology of heavy slab ingot weighing 40 t made from tool steel. The main aim of numerical modelling in ProCAST software realized under the conditions of the Department of Metallurgy and Foundry (DMF) and Regional Materials Science and Technology Centre (RMSTC) at the VSB-TU Ostrava was the optimization of the production of heavy steel ingots produced in VHM, especially focused on minimization of porosity and macrosegregations.

The paper summarizes also the research and development of the new design of the mould for slab steel ingot production. In the paper, the operational experiment and the results from plant experiment are also outlined. Based on the results of evaluation of complete series of numerical simulations, and based on their operational verification, it made it possible to reduce production costs while the internal quality of steel ingots was still maintained at very high level.

2 Operational experiment

The company VÍTKOVICE HEAVY MACHINERY a.s. is a traditional producer of large machine components. The typical products of this company are crankshafts, propeller and connecting shafts, rotor shafts for wind power plants, forged parts for containers of pressurizers, steam generators, heat exchangers and collectors for both conventional and nuclear power engineering. For these products, it is necessary to cast ingots weighing up to 200 tons. Steel plant of the VHM is equipped with EAF, LF, VD and VOD facilities. Ingots from 1.7 up to 200 tons are bottom cast. Typical steel grades for these products are structural carbon-manganese, low alloyed, middle alloyed and tool steels. The EAF capacity is 70 tons so the larger ingots are cumulated from two or three heats.

For the purposes of research, development, verification and optimization of production, an experimental slab ingot weighing 40 t was cast from tool steel at the VHM. The heat chemical composition of cast grade steel is documented in **Table 1.** The steel was produced and cast according to the VHM standard technological procedures valid for this steel grade. After stripping of the ingot from the mould, the ingot was cooled down by controlled cooling in a furnace for reduction of the stresses. Subsequently, the ingot was cut into two halves along the vertical axis. After flame cutting, the affected layer of the centre surface of the cut ingot was milled off for subsequent analysis. For an evaluation of the distribution of porosity, a penetration test was performed (see **Figure 1**). However, the test did not reveal the expected large porosity. It only highlighted the crack on the ingot, which was formed by milling. Similarly, due to the low content of sulphur in steel the Baumann sulphur print did not show the expected zones with its increased concentrations.

Table 1. The heat chemical composition of steel grade of the experimental ingot (wt. %).

C	Mn	Si	P	S	Cu	Ni	Cr	Mo	V	Al	Nb	N
0.37	1.45	0.26	0.011	0.001	0.16	1.03	1.96	0.20	0.007	0.018	0.005	0.0059

Figure 1. Section cut of the experimental slab ingot during the penetration test.

Due to the number of required chemical analyses, selection of required elements for determination of their concentration levels it was necessary to pay an increased attention to the choice of an optimal method of analyses. The method used in a standard manner for metallurgical analyses - analysis on laboratory optical emission spectrometers, which would ensure the efficiency and accuracy – was not applicable, since it required a sample in the solid state. Taking of the solid samples in the required quantity was technically not feasible. The partial solution consisted in the use of mobile spectrometers, either X-ray or optical emission, for determination of at least some elements. Its obvious advantage is the fact that the analyses can be performed directly on the ingot repeatedly and without taking of the samples.

For all the elements and individual methods, the possible achieved precision was determined experimentally by measurements of the control sample under conditions of repeatability (in the shortest time of repetition). The exception was the determination of phosphorus. The measured values were evaluated using robust procedures of the statistical software QC.Expert. The values presented in **Table 2**, such as "accuracy of the method", were calculated as half of the difference between the upper and lower quartiles (the interval, which contains 50 % of the measured results, the limits of the box in the box plots). Identical statistics was also used for processing the uncertainty of the results of individual measurements directly on the ingot.

Table 2. Average concentration of the elements and the accuracy of individual measurements (wt. %).

Element	Method	Average concentration of the element in the control sample	Method accuracy
C	Analyzer LECO	0.175	± 0.0015
C	OES – Spectrometry	0.175	± 0.01
Mn	Chemically	1.15	± 0.003
Mn	OES – Spectrometry	1.15	± 0.01
Si	OES – Spectrometry	0.26	± 0.01
P	Chemically	0.0087	± 0.0006
S	Analyzer LECO	0.0007	± 0.0001
Cu	OES – Spectrometry	0.11	± 0.01
Ni	OES - Spectrometry	0.18	± 0.01

3 Numerical modelling

In order to analyze the character of the predicted final internal structure of steel ingot, or the range of the volume defects, such as porosity and macrosegregation depending on the shape of the ingot, a comparison of the results of numerical modelling of internal structure of the conventional polygonal ingot with the slab ingot and with the new design of the slab ingot of the similar weight/steel grade/conditions of the casting was done. The numerical results were also confronted with the results of the plant experiment. Generally, numerical solution of each task was divided into three stages: 1. Pre-processing: it included geometry modelling and the process of generation of the computational mesh, and definition of calculation. 2. Processing: it involved computation in the solver. 3. Post-processing: it focused on the evaluation of the results. The conditions of numerical model settings were based on real conditions of the experimentally cast 40 tons steel ingot produced in VHM a.s. The details of the model settings were published in [12].

3.1 Solidification time

Although the temperature field at the end of filling was very similar for all three types of ingots, different times of solidification were obtained. The conventional polygonal ingot solidified in approx. 13 hours, while the total solidification time of the slab ingot was approx. 7.5 hours, or approx. 8.5 hours for a new design of the slab ingot, as it is evident from the **Figure 2**.

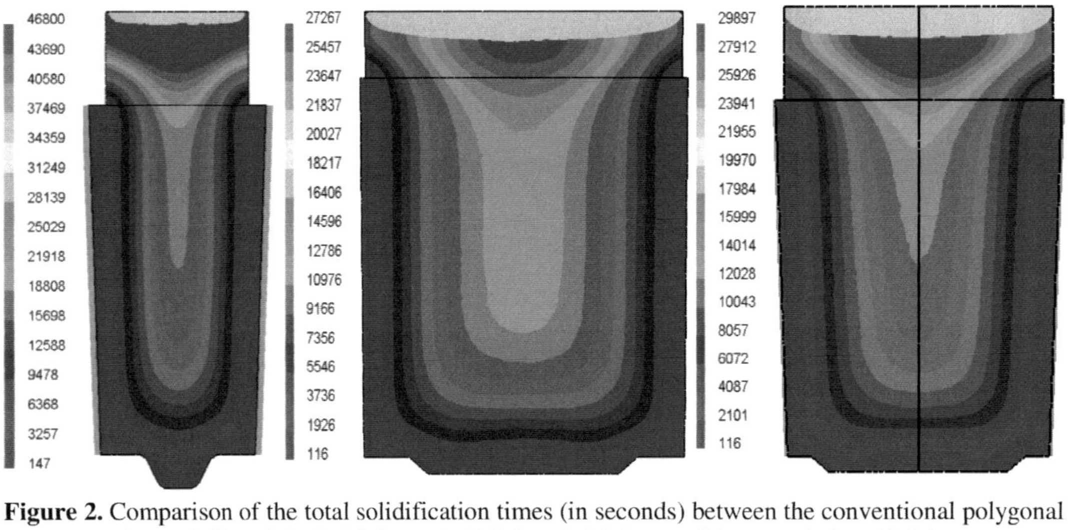

Figure 2. Comparison of the total solidification times (in seconds) between the conventional polygonal ingot (left), original slab ingot (middle) and slab ingot of the new design (right).

3.2 Character of final macrosegregation vs. porosity

Due to the shorter total solidification time, the final macrosegregation of all the elements in the slab ingots was smaller than in the conventional polygonal ingot. An example of a comparison of macrosegregation of phosphorus predicted by numerical modelling is shown in **Figure 3**. From **Figure 3** it is also evident that the macrosegregation of phosphorus in the slab ingot of the new design is minor in comparison with the original shape of the slab ingot. If we compare the content of the phosphorus detected from a real sample of the original slab steel ingot (see **Table 1**) with the numerical results, the content of the phosphorus corresponds to the border section of the slab ingot body. Small difference between the results can be caused by the following limitations of numerical modelling: no solid movement, no grain sedimentation, fully equiaxed dendrites, no columnar dendrites.

The final porosity in the original slab ingot was detected in higher volume range in comparison with the polygonal conventional ingot, as it is evident from **Figure 4**. On the other hand, the porosity in slab ingot of the new design brought the expected efficiency and it led to a lower volume of this volume defect. The differences/contradictions relate with the directional solidification from the bottom to the hot top of the ingot bodies and with the ability of the hot tops to fulfil the liquid metal in the last stages of the solidification into the ingot body.

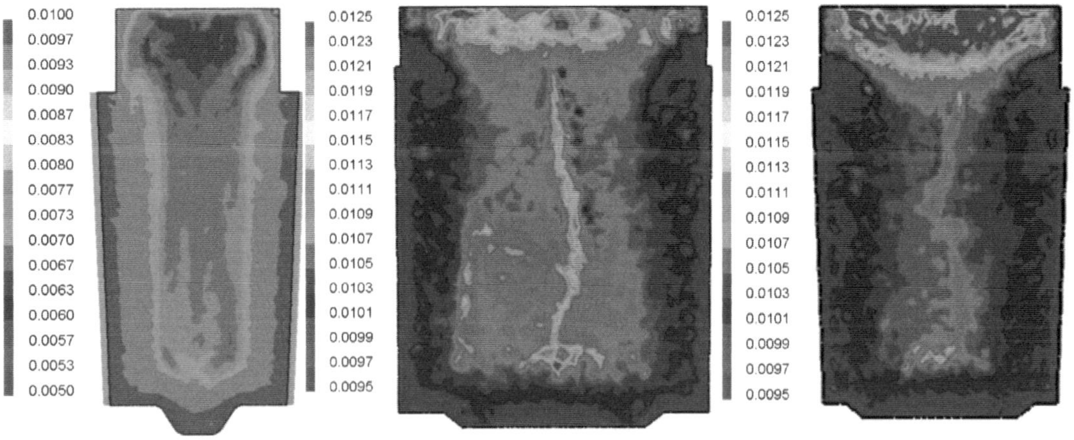

Figure 3. Comparison of the distribution maps of macrosegregation of phosphorus (in wt. %) between the conventional polygonal ingot (left), original slab ingot (middle) and slab ingot of the new design (right) The defined content of phosphorus in the simulation was for the conventional ingot 0.004 wt.%, for the slab ingots it was 0.01 wt. %.

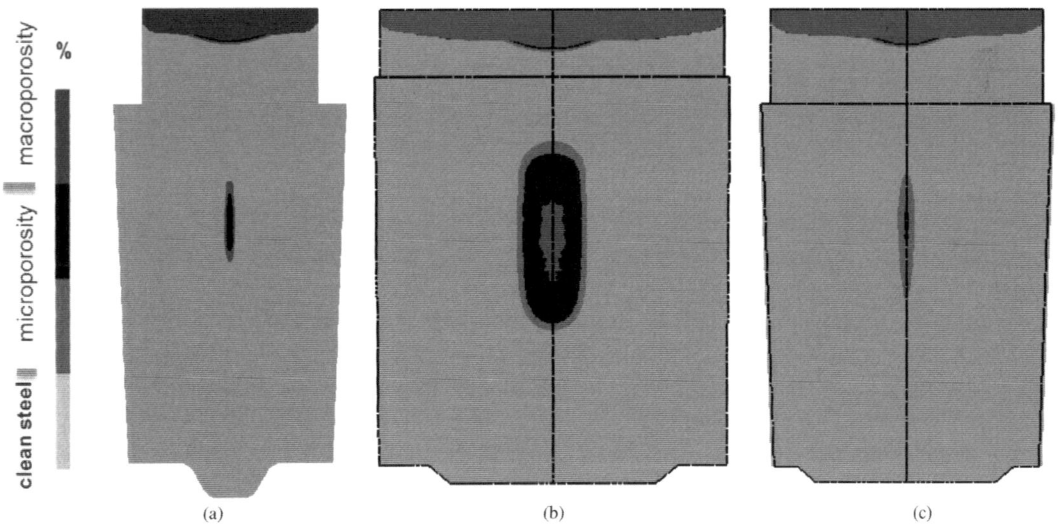

Figure 4. Comparison of the final porosity in (a) conventional polygonal ingot (b) original slab ingot (c) new design of slab ingot.

4 Conclusions

The paper was devoted to verification of production of the slab ingot from tool steel using the plant experiment and numerical modelling with a finite element method. The main reason of verification of casting and solidification of the 40-ton steel slab ingot was the possibility of replacement of the conventional heavy steel ingot actually used for the production of the special forgings by the slab ingot. The attention was also concentrated on rediction of the final porosity and macrosegregation in dependence on the new design of the mould for slab ingot. From the research, it was found that:

- the range of macrosegregation in slab ingots was lower than in the case of conventional polygonal heavy forging ingot of the same weight that were produced from the same steel grade;
- on the other hand, in the central axis of the ingot body of the original slab ingot, a large volume of microporosity was predicted in comparison with the conventional polygonal ingot. Nevertheless, the huge minimization of the porosity was achieved by the new design of the shape of the slab ingot. This microporosity can be eliminated by the following forging.

Acknowledgement

This paper was created with the financial support of the TA CR under the project No. TA04010035, in the Project No. LO1203 "Regional Materials Science and Technology Centre - Feasibility Programme" funded by Ministry of Education, Youth and Sports of the Czech Republic, and in the support of the projects of "Student Grant Competition" numbers SP2016/89 and SP2016/103.

References

[1] Tashiro K., Watanabe S., Kitagawa I. and Tamura I. 1983 Influence of mould design on the solidification and soundness of heavy forging ingots *ISIJ Inter.* **23** 312-321

[2] Kermanpus A., Eskandari M., Purmohamad H., Soltani M.A. and Shateri B. 2010 Influence of mould design on the solidification of heavy forging ingots of low alloy steels by numerical simulation *Materials and Design* **31:3** 1096-1104

[3] Baoguang S., Xiuhong K. and Dianzhong L. 2010 A novel technique for reducing macrosegregation in heavy steel ingots. *Jour. of Mater. Process. Technol.* **210:4** 703-711

[4] Kearney M., Crabbe M. and Talamantes-Silva J. 2007 Development and manufacture of large plate mill rolls *Ironmak. & Steelmak.* **34:5** 380-383.

[5] Song W., Zhang J. M., Liu Y., Wang S. X. and Wang B. 2015 Numerical simulation of solidification structure of 6.5 wt-%Si steel ingot slab *Ironmak. & Steelmak.* **42:9** 656-663

[6] Marx K., Rödl S., Schramhauser S. and Seemann M. 2014 Optimization of the filling and solidification of large ingots *La Metallurgia Italiana* **11-12** 11-19

[7] Bogdan O. 2010 Numerical Analysis of Casting Technology and A-segregation Prediction in AISI 4340 Forgings Products *Industrial Soft* (Montreal, Canada) 1-12 http://castingsnet.com/AISI4340-casting-report.pdf

[8] Lan P. and Zhang J. Q. 2014 Numerical analysis of macrosegregation and shrinkage porosity in large steel ingot *Ironmak. & Steelmak.* **41:8** 598-606

[9] Ludwig A., Wu M. and Kharicha A. 2015 On macrosegregation *Metallur. and Mater. Trans.* A **46A** 4854-4867

[10] Li W., Shen H., Zhang X. and Liu B. 2014 Modeling of species transport and macrosegregation in heavy steel ingots *Metall. and Mater. Trans. B* **45B** 464-471

[11] Meng Q., Wnag F., Li Ch., Li M., Zhang J. and Cui G. 2014 Numerical simulation of macrosegregation in water-cooled heavy flat ingot during solidification *JOM* **66:7** 1166-1174

[12] Tkadlečková M., Michalek K., Gryc K., Socha L. a Machovčák 2016 P. Prediction of qualitative parameters of slab steel ingot using numerical modelling *Metalurgija* **55:3** 395-398

IOP Conf. Series: Materials Science and Engineering **179** (2017) 012068 doi:10.1088/1757-899X/179/1/012068

Effect of silicon content on microstructure of low-alloy Q&P-Processed steels

I Vorel, Š Jeníček, J Káňa, K Ibrahim, V Kotěšovec, B Mašek

University of West Bohemia, Regional Technological Institute, Univerzitní 22, Plzeň, Czech Republic

E-mail: frost@rti.zcu.cz

Abstract. Today's requirements for the design of functional parts and components demand low weight and, at the same time, high strength. There are several heat treatment methods which can satisfy such requirements. These include TRIP heat treatment, long-time low-temperature austempering, and Q&P processing. It is the Q&P processing which delivers excellent results in terms of mechanical properties and light weight. It relies on stabilising retained austenite through partitioning of carbon between martensite and austenite. The carbon-enriched austenite then becomes a ductile constituent in the otherwise brittle martensitic matrix. A precondition for successful Q&P processing consists in sufficient silicon content in the steel, which precludes precipitation of undesirable cementite. Cementite would otherwise form as a result of enrichment of retained austenite with carbon. To ascertain the usefulness of higher silicon level in steel for Q&P processing, one can compare Q&P processes in steels with various levels of this element

1 Introduction

Silicon plays an important role in the heat treatment of modern high-strength steels. It is because it affects the precipitation of carbides caused by the state of disequilibrium in steels that had been rapidly undercooled to the region of martensitic or bainitic transformation. The cause of carbide formation is therefore the thermodynamic disequilibrium and the tendency of the material to relieve it. In this context, silicon retards precipitation because the solubility of silicon in cementite is negligible. Ultimately, this should mean that carbon, which would under normal conditions take part in the precipitation of cementite and other carbides, could be able – in steels with higher silicon level – diffuse to the surrounding austenite.

One of the treatment routes that rely on the above-described principle is Q&P processing. The Q&P process is based on the concept of carbon partitioning between martensite and retained austenite so that the latter becomes stabilized [1]. Retained austenite then exists as the ductile constituent in steel, and contributes to considerable ductility of the material on the whole. The Q&P process involves quenching to a temperature below the M_s to produce a certain fraction of martensite. The amount of martensite can be controlled by adjusting the quenching temperature below the M_s. The next step is partitioning, where the steel is reheated to a temperature above the quenching temperature but still lower than the martensite start temperature M_s. During this step, carbon should diffuse between martensite and austenite, and eventually stabilize the austenite [2,3,4]. The necessary condition for this mechanism is that carbon must not form carbides during the partitioning step because it would deplete the surrounding material of carbon. In such case, austenite would not become stabilized by carbon, and the ductility of the Q&P-processed steel would diminish. This necessary condition can be met by adding more than 1.5 wt. % silicon. Unless the steel contains a sufficient amount of silicon, the effectiveness of the Q&P process can be expected to be affected by this fact [5,6].

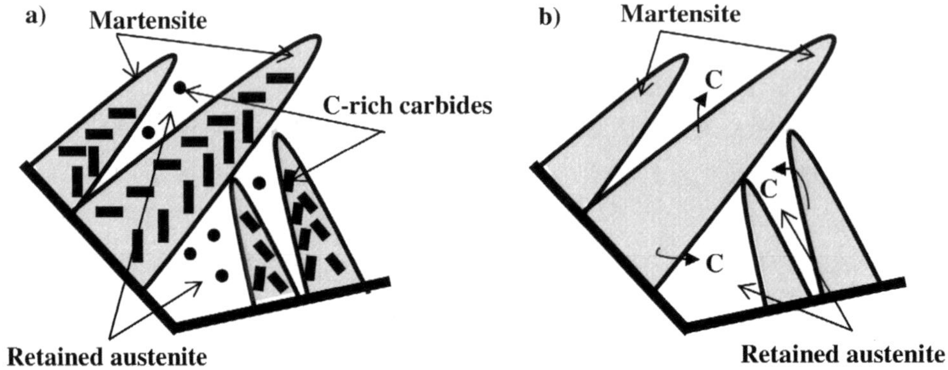

Figure 1. Effect of silicon content on the carbon partitioning mechanism during Q&P processing
a) low-silicon steel, b) high-silicon steel.

2 Experimental programme

The objective of this experiment was to compare the effects of silicon content on the microstructures of low-alloy steels with 0.43 % carbon, 1.33 % chromium, and 0.56–0.6 % manganese upon identical treatments involving the Q&P process (Table 1).

Table 1. Chemical composition of experimental steel.

Steel designation	C	Si	Mn	Cr	Mo	Nb	P	S	M_s [°C]	M_f [°C]
42SiCr	0.42	2.03	0.56	1.33	0.16	0.03	0.005	0.003	298	178
42CrMo	0.42	Max 0.4	0.6	1.2	0.15	-	0,03	0,04	305	184

Heat treatment routes for Q&P processing were proposed as shown in Fig. 2. The temperatures were chosen so as to obtain approximately 50 % martensite in the microstructure upon quenching and before partitioning. The partitioning temperature was selected to be below and above the M_s temperature that had been calculated by means of the empirical equations used by the JMatPro program.

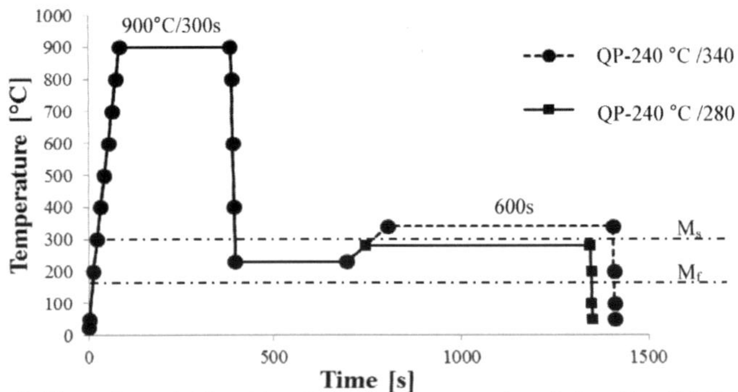

Figure 2. Experimental heat treatment sequences involving the Q&P process.

The partitioning time was 10 minutes. The experimental specimens were threaded round bars with a diameter of 8 mm. Prior to this experiment, they had been annealed at 1050 °C for 20 minutes.

Heating and cooling was carried out in a thermomechanical simulator with induction-resistance heating (Fig. 3).

Figure 3. Specimen clamped in the jaws of thermomechanical simulator.

3 Results and discussion

Microstructures of the specimens were examined using scanning electron microscopy. They were etched with 4 % picral, and 10 % $Na_2S_2O_5$ solution. Phase composition was examined using image analysis. For better resolution capability, the micrographs were processed by image transformation.

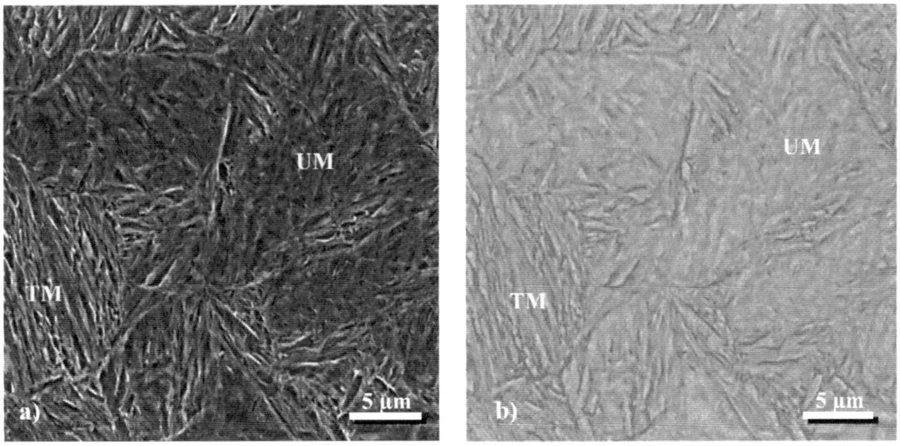

Figure 4. Micrograph of the 42CrMoS4 steel upon Q&P processing – $Q_{240\,°C}$-$P_{280\,°C/600\,s}$
a) Mixture of tempered lath martensite (TM), and untempered lath martensite (UM) – 5000× – SEM, b) phase contrast micrograph.

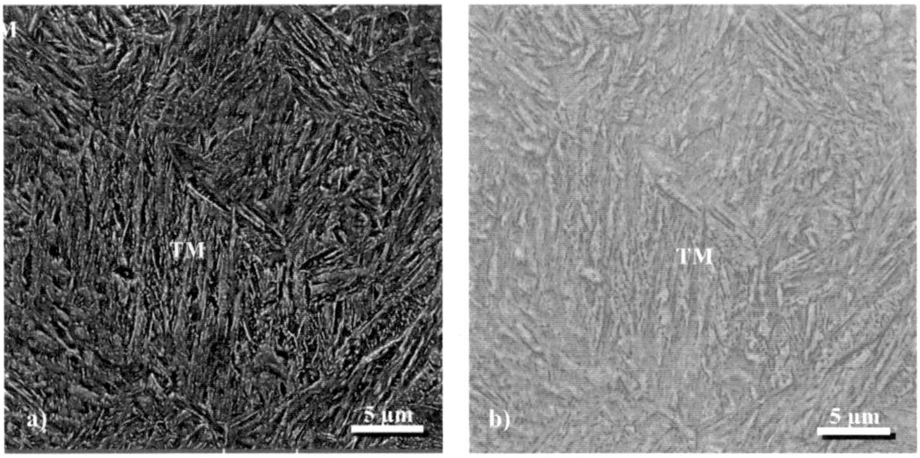

Figure 5. Micrograph of the 42CrMoS4 steel upon Q&P processing – $Q_{240\,°C}$-$P_{340\,°C/600\,s}$ a) Mixture of tempered lath martensite (TM) and bainite – 5000× – SEM, b) phase contrast micrograph.

Tempered martensite, untempered martensite, and bainite were identified in the specimens (Figs. 4, 5, 6, and 7) by comparing scanning electron micrographs from the specimens, and micrographs of martensitic structures from literature. In specimen 42CrMoS4-QP240/280, lath morphology was found in both tempered and untempered martensite (Fig. 4). Close examination of the microstructure revealed the presence of fine carbide precipitates. Specimen 42CrMoS4-QP240/340 contained laths of tempered martensite and, most probably, lower bainite as well. The latter, however, was impossible to distinguish from the former due to their close resemblance (Fig. 5). The comparison of specimens 42CrMoS4-QP240/280 and 42CrMoS4-QP240/340 led to an important conclusion that all austenite had transformed during the partitioning stage – most probably to a bainite-type structure due to the low silicon content in the material.

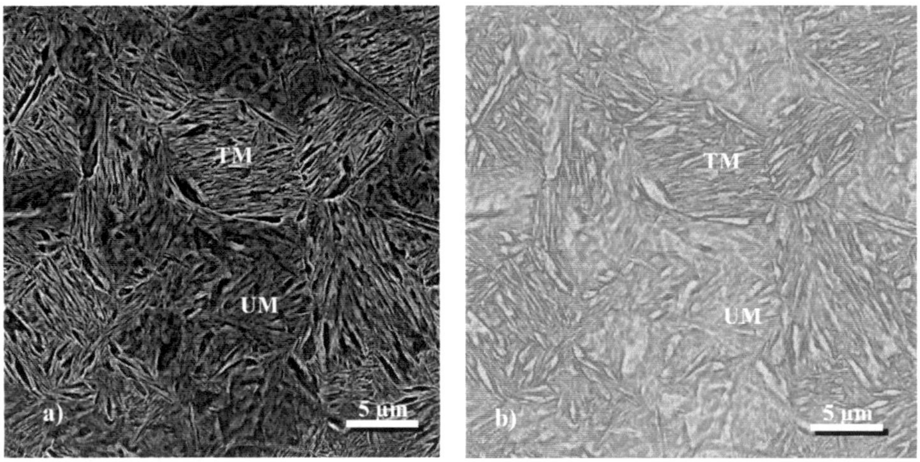

Figure 6. Micrograph of the 42SiCr steel upon Q&P processing-$Q_{240\,°C}$-$P_{280\,°C/600\,s}$ a) Mixture of tempered lath martensite (TM), and untempered lath martensite (UM) – 5000× – SEM, b) phase contrast micrograph.

Specimen 42SiCr-QP240/280 contained tempered and untempered lath martensite (Fig. 6). Unlike in specimen 42CrMoS4-QP240/340, the increased partitioning temperature in specimen 42SiCr-QP340 did not cause austenite to decompose into a bainite-type structure. Here, austenite remained present until the last step of the experimental sequence, and eventually transformed to martensite during quenching (Fig. 7). Close examination of the microstructures of specimens 42SiCr-QP240/280 and 42SiCr-QP240/340 did not prove that carbide precipitates were present, unlike in the 42CrMoS4 steel – which can be explained by the above-discussed higher level of silicon in 42SiCr.

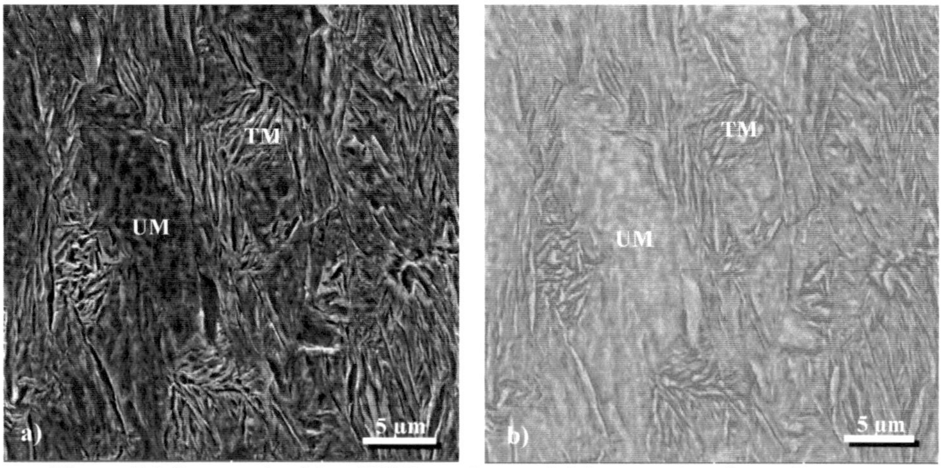

Figure 7. Micrograph of the 42SiCr steel upon Q&P processing-$Q_{240\,^{\circ}C}$-$P_{340\,^{\circ}C/600\,s}$ a) Mixture of tempered lath martensite (TM), and untempered lath martensite (UM) – 5000× – SEM, b) phase contrast micrograph.

4 Conclusion

The effect of silicon level on microstructural evolution during Q&P processing was examined in low-alloy steels that contained 0.42 % carbon, and very similar levels of manganese and chromium. The steels were quenched to the temperature that should provide approximately 50 % martensite in the resulting microstructure. The subsequent partitioning step took place below and above the M_s. In those specimens of the 42CrMoS4 and 42SiCr steels that were subjected to partitioning below the M_s, the resulting microstructure contained a mixture of tempered and untempered lath martensite. Fine carbide precipitates were found in the 42CrMoS4 steel which had low silicon level. Partitioning above the M_s led to differences in the microstructural evolution in the experimental steels. The microstructure of 42CrMoS4 probably comprised a mixture of tempered martensite and lower bainite formed by the decomposition of untransformed austenite during partitioning. On the other hand, the microstructure of 42SiCr – which had a high silicon level – consisted of tempered and untempered martensite.

Acknowledgements
This paper includes results achieved under the project LO1502 Development of Regional Technological Institute. The project is subsidised by the Ministry of Education of the Czech Republic from specific resources of the state budget for research and development.

References

[1] Kucerova L 2011 Comparison of Microstructures and Properties Obtained After Different Heat Treatment Strategies of High Strength Low Alloyed Steel *Journal of Iron and Steel Research International* vol. 18, pp. 427-431

[2] Speer J et al. 2003 Carbon Partitioning into Austenite after Martensite Transformation *Acta Materialia*, vol. 51, pp. 2611-2622
[3] Speer J et al. 2005 The "Quenching and Partitioning" Process: Background and Recent Progress *Material Reasearch* vol. 8 (4), pp. 417-423
[4] Tsuchiyma T et al. 2012 Quenching and partitioning treatment of a low-carbon martensitic stainless steel *Materials Science and Engineering* vol. 532, pp. 585-592
[5] Jirkova H, Kucerova L, Masek B 2012 Effect of Quenching and Partitioning Temperatures in the Q-P Process on the Properties of AHSS with Various Amounts of Manganese and Silicon *Materials Science Forum*, vols. 706-709, pp. 2734-2739
[6] Masek B et al. 2011 Improvement of Mechanical Properties of automotive Components Using Hot Stamping with Integrated Q-P Process *Journal of Iron and Steel Research International*, vol. 18 (1-2), pp. 730-734.

IOP Conf. Series: Materials Science and Engineering **179** (2017) 012069 doi:10.1088/1757-899X/179/1/012069

Anticorrosion nitrided layers on unalloyed and alloyed steels

P Wach, J Michalski, K Burdyński and A Ciski

Institute of Precision Mechanics, Duchnicka 3 St., 01-796 Warsaw, Poland

E-mail: wach@imp.edu.pl

Abstract. In the paper, nitrided layers on unalloyed and alloyed steels and their corrosion properties are presented. Nitrided layers in the controlled gas nitriding process on C10 and 42CrMo4 steels were formed. Two types of nitrided layers are presented: with nitride iron layers above and below 15 μm. Nitrided layer with nitride layer above 15 μm has good corrosion resistance, but after nitriding of machine parts were subsequently oxidised and impregnated. In the second type of nitrided layer, the surface layers of iron nitrides had a thickness of 3.0 to 11.0 μm. Nitrided layers with a surface layer of iron nitrides with the γ' (Fe_4N) structure were formed on unalloyed steel and investigated. The so-formed layers were subject to basic metallographic, X-ray diffraction and corrosion resistance studies carried out by electrochemical methods and in a neutral salt spray chamber. It was found that the layers consisting only of γ' phase had a good corrosion resistance. Necessary requirements for achieving an enhanced resistance comprise their complete tightness and thickness not lower than 9.0 μm. Thinner layers had good electrochemical properties but did not exhibit corrosion resistance in the salt spray chamber.

1 Introduction

Corrosion resistance is exhibited by nitrided layers with a surface layer of iron nitrides formed both on structural carbon steels and alloyed steels, and – to a lesser degree – on cast iron, because a continuous surface layer of iron nitrides is difficult to obtain on them. Nitrided machine parts undergo, most of all, pitting corrosion. In the process of destruction, the base corrodes, therefore the corrosion resistance depends on the tightness of the surface layer of iron nitrides [1]. The corrosion of elements with a nitrided layer made of an unalloyed steel, may be divided into three stages [2]:

- I initiation of corrosion in the pores of the iron nitrides' layer,
- II development of pitting corrosion through the pores in the nitrides' layer,
- III development of corrosion under the nitrides' layer.

The development of corrosion in the nitrided layer on carbon steels is caused by micropores and the phase composition of the iron nitrides' layer, contributing into formation of local galvanic cells [1,3].

In the industry, anticorrosive nitrided layers with a surface layer of iron nitrides with the $\varepsilon+(\varepsilon+\gamma'_{prec.})$ structure are widely used. In order for the layers to achieve a good corrosion resistance, they are subject to additional treatment, such as grinding, polishing [4,5], steam oxidation [6] and impregnation [7].

Grinding and polishing remove the porous layer. Oxidation leads to formation of a tight oxide layer on the surface. The impregnant penetrates the porous layer and efficiently seals the surface layers of iron nitrides, thus decreasing its susceptibility to the pitting corrosion. Additionally, by steam oxidation of the layers before their impregnation, its corrosion resistance is increased [8].

However, these operations significantly increase the cost of the process. Thus, studies aimed at production of compact layers of γ' iron nitrides or γ' iron nitrides with a limited content of the ε phase, which would not require any additional treatment and would have a good corrosion resistance, are purposeful. The paper presents the resistance of corrosion layers of the nitrided layers with a surface layer of iron nitrides with the $\varepsilon+(\varepsilon+\gamma'_{prec.})$ structure and a monophase surface layer of γ' type iron nitrides formed on C10 and 42CrMo4 steels.

2 Methodology and research material

Samples nitriding was performed using the Nitreg regulated gas nitriding method in the Nx609 industrial soaking furnace with the retort dimensions of ϕ600x900mm featuring computer process control, as manufactured by Nitrex Metal Inc. The process parameters are shown in Table 1. 1010 normalized steel (AISI standard) and 42CrMo4 toughened steels were used in the tests.

Samples of 42CrMo4 steels with thickness of nitrides layer about 20 μm after nitriding were subjected to oxidation and impregnation (process Nx620). Samples of nitrided 1010 steel with nitrides layer 3-12 μm were without impregnation and oxidation.

Table 1: Nitriding process parameters.

Process No.	Temperature [°C]	Heating atmosphere	Process temperature atmosphere	Nitriding potential value at the process temperature	Time [h]
Nx620	570	NH_3	NH_3	2,50	4
Nx300	490	NH_3-NH_{3diss}	NH_3-NH_{3diss}	1,98	4
Nx428	490	NH_3-NH_{3diss}	NH_3-NH_{3diss}	2,08	12
Nx301	490	NH_{3diss}	NH_3-NH_{3diss}	1,98	4
Nx369	530	NH_3-NH_{3diss}	NH_3-NH_{3diss}	1,24	4
Nx625	530	NH_3-NH_{3diss}	NH_3-NH_{3diss}	1,48	12
Nx354	560	NH_3-NH_{3diss}	NH_3-NH_{3diss}	0,82	4
Nx355	560	NH_3-NH_{3diss}	NH_3-NH_{3diss}	0,82	4
Nx418	560	NH_3-NH_{3diss}	NH_3-NH_{3diss}	0,92	8
Nx319	580	NH_3-NH_{3diss}	NH_3-$NH3_{diss}$	0,50	4
Nx322	580	NH_3-NH_{3diss}	NH_3-NH_{3diss}	0,80	4
Nx422	580	NH_3-NH_{3diss}	NH_3-NH_{3diss}	0,57	8

The metallographic studies of the nitrided layers were carried out on microsections perpendicular to the nitrided layer surface. On the microsections, the thickness of the surface layer of iron nitrides was determined and the structures of the layer and the base were observed. The structure of the nitrided layers was revealed using Nital (4% HNO_3 solution in C_2H_5OH). The structural observations were carried out using Neophot-2 and Nikon ECLIPSE LV150 optical microscopes. The thickness of the surface layers of iron nitrides was determined as an arithmetic mean of at least 6 measurements. The measurement error was estimated by calculating standard deviation.

Studies of the surface topography were carried out using a DSM 942 scanning microscope from Zeiss, LEO.

Studies of the phase composition were carried out using an XRD 3003T-T X-ray diffractometer from Seifert and a Dron 2.0 diffractometer. During the phase analysis, CoKα radiation with a wavelength of λ = 1.79021 Å and a penetration distance of ~20 μm was used.

Potentiodynamic electrochemical measurements of the corrosion resistance were carried out using an electrochemical microcell with a working surface of 0.066 cm^2. Electrochemical tests were conducted using a three-electrode cell with the sample as the working electrode, a platinum electrode and saturated calomel electrode (SCE) as reference electrode. Open circuit potential (E_{ocp}) and polarisation curve from the potential of approx. 500 mV more cathodic than the E_{ocp}, towards the anode, with a potential change rate of 1 mV/s, were recorded. The polarisation curve was recorded up to approx. +2000 mV. A 3% aqueous solution of NaCl, imitating seawater and simulating the conditions of salt spray chambers to a significant degree, was used as the electrolyte. During the investigations, the corrosion current density i_{kor} and the passivation current density i_{pas} were determined, and the degree of corrosion protection β was calculated:

$$\beta = \left(1 - \frac{i_{kor}^{azot}}{i_{kor}^{o}}\right) \cdot 100 \ \ [\%] \qquad (1)$$

where:

i_{kor}^{azot} - the corrosion current density of a nitrided material,

i_{kor}^{o} - the corrosion current density of a non-nitrided material.

For a preliminary evaluation of the improvement of electrochemical properties of the nitrided layers, the following parameters and their corresponding values were selected:

- corrosion currents $i_{kor} < 0.75$ μA/cm^2,
- degree of protection β >80%,
- passive currents $i_p < 100$μA/cm^2,
- open circuit potential $E_{ocp} > -100$ mV.

The studies of the corrosion resistance to the neutral salt spray (NSS) effect were carried out in an accredited IMP laboratory according to the PB/2-50B/LB-3 procedure: „Resistance to the neutral salt spray (NSS). Metals, metallic and other inorganic coatings." According to this procedure, accelerated studies in a neutral salt spray of the corrosion resistance are carried out for: metals and their alloys, metal coatings (anodic and cathodic), non-metallic coatings, inorganic coatings applied to a metal surface (conversion, oxide anodic). Visual inspection of the samples (three pieces from each process) were carried out in equal time intervals every 24 h or to the occurrence of defined corrosive changes. The appearance of the samples was evaluated according to the PN-EN ISO10289:2002 standard "Methods For Corrosion Testing Of Metallic And Other Inorganic Coatings On Metallic Substrates – Rating Of Test Specimens And Manufactured Articles Subjected To Corrosion Tests".

3 Results of experiments

3.1 Metallographic and X-ray diffraction studies

Samples of 42CrMo4 steel nitrided in the Nx620 process had thick subsurface layer of iron nitrides. Their thickness was about 20 μm (Figure 1). X-ray analysis confirmed that at the sample surface there is sub-surface nitrided layer consisting of the ε and γ' phases (Figure 2).

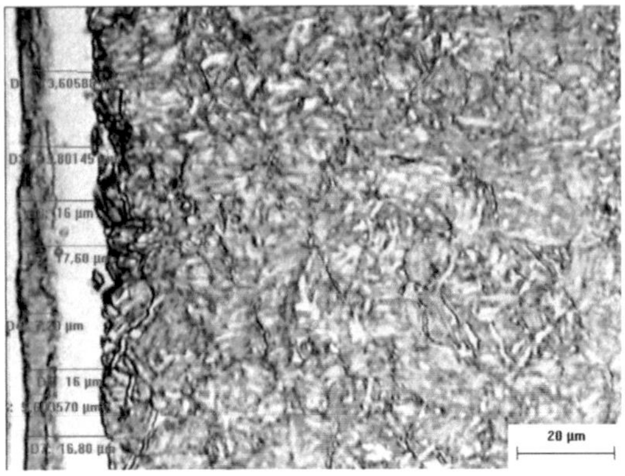

Figure 1. Microstructure of nitrided layer on 42CrMo4 steel.

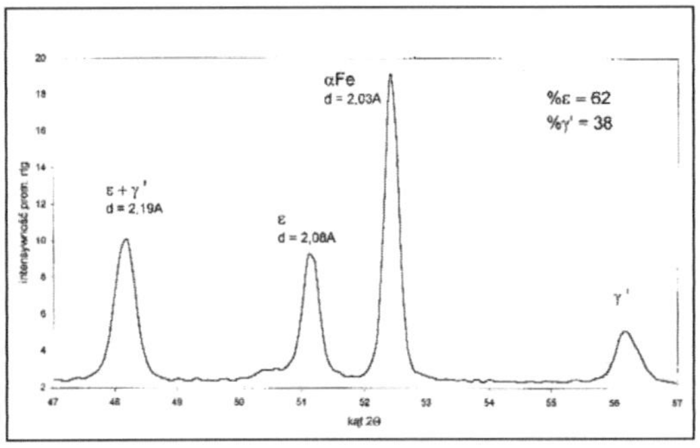

Figure 2. X-ray diffraction scanes of nitrided layer on 42CrMo4 steel.

In case of 1010 steel, after the processes of the controlled gas nitriding, a thickness of the surface layers of iron nitrides from 3.5 to 11.8 µm was obtained. The thicknesses of the surface layers of iron nitrides are shown in figure 3.

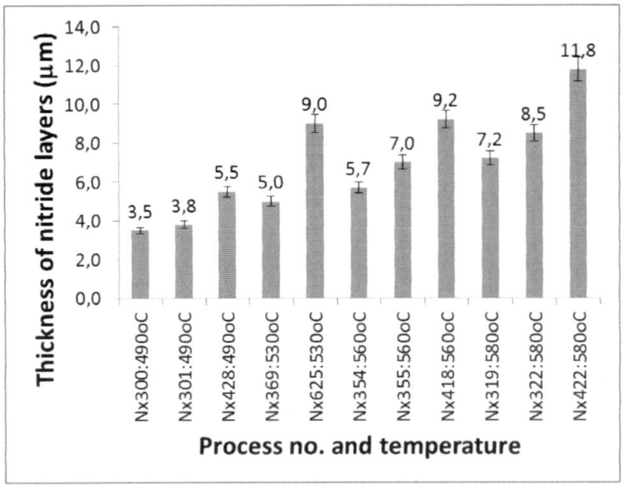

Figure 3. Thicknesses of the surface layer of iron nitrides obtained after the processes.

All nitrided layers had a surface layer of iron nitrides of γ' type, confirmed by the X-ray diffraction studies. In figure 4, exemplary structures of the nitrided layers from representative Nx301, Nx625 and Nx418 processes are shown.

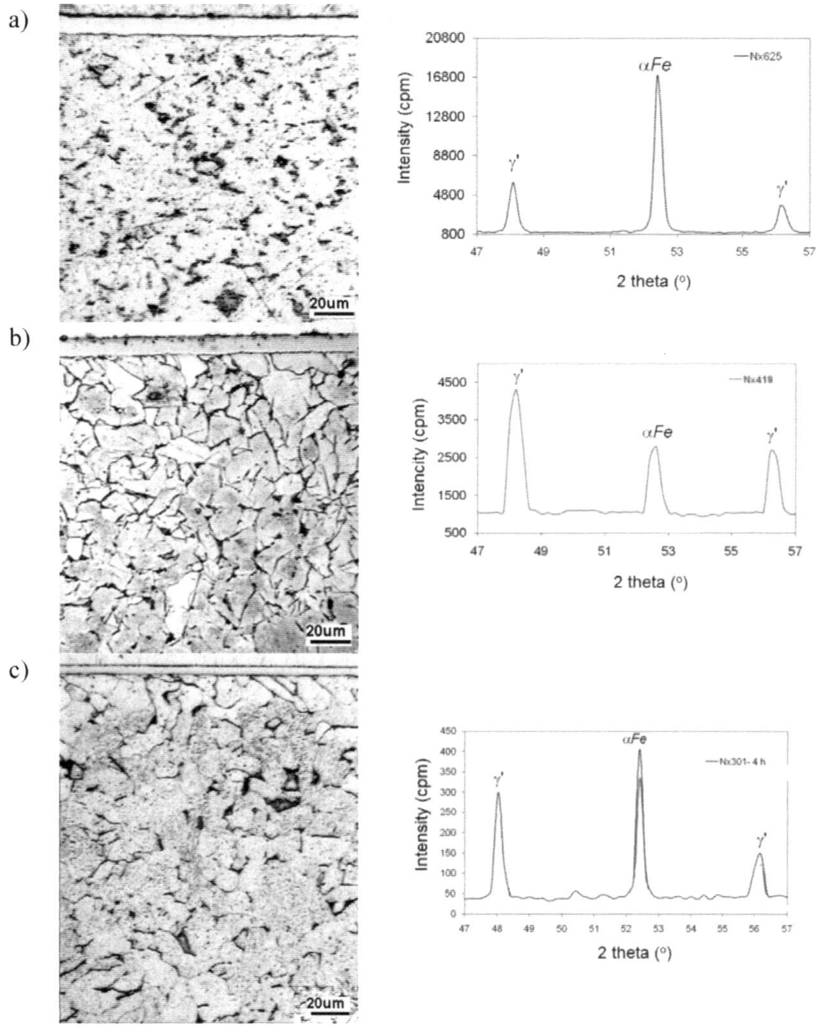

Figure 4. Structures and diffractograms of phase composition of the surface layers of iron nitrides on the 1010 steel after the processes: (a) Nx625, (b) Nx418, and (c) Nx301.

The layers formed below 560°C did not have a porous zone. On the other hand, those formed at a higher temperature had a porous zone from 0.5 to 2.0 μm.

3.2 Corrosion resistance studies

In figure 5, the electrochemical parameters and corrosion current (Figure 5a), passivation current (Figure 5b), open circuit potential (Figure 5c) and the degree of corrosion protection (Figure 5d) obtained in a 3% NaCl solution, characterising the corrosion resistance of the nitrided steel with iron nitrides' layers with a thickness of 3.5÷5.7 μm, are shown.

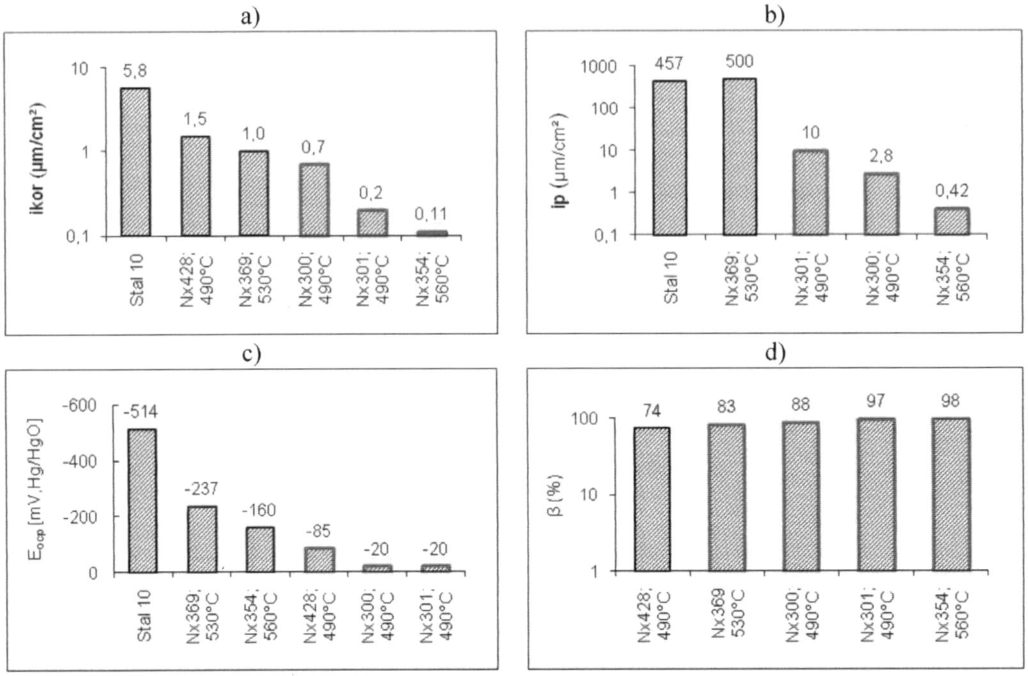

Figure 5. Basic electrochemical parameters and corrosion current (a), passivation current (b), open circuit potential (c), degree of corrosion protection (d), obtained in a 3% NaCl solution for the nitrided 1010 steel with iron nitrides' layers with a thickness of 3.5÷5.7 µm.

All four criteria of electrochemical resistance were met by the iron nitrides' layers with a thickness of 3.5÷3.8 µm obtained on the 1010 steel, nitrided in the Nx300 and Nx301 processes. The iron nitrides' layer with a thickness of 5.7 µm obtained in the Nx354 process exceeded the limit value of the open circuit potential E_{ocp}. On the other hand, it was characterised by the lowest corrosion current i_{kor} and passivation current i_p, and the highest degree of protection β.

In figure 6, the electrochemical parameters and corrosion current (Figure 6a), passivation current (Figure 6b), open circuit potential (Figure 6c) and the degree of corrosion protection (Figure 6d) obtained in a 3% NaCl solution, characterising the corrosion resistance of the nitrided 10 steel with iron nitrides' layers with a thickness of 7.0÷11.8 µm, are shown.

All four criteria of electrochemical resistance were met by the iron nitrides' layers with a thickness of 8.5 µm obtained on the 1010 steel, nitrided in the Nx322 process. The iron nitrides' layer with a thickness of 9.0 µm obtained in the Nx418 process exceeded the limit value of the open circuit potential E_{ocp}. On the other hand, it was characterised by the lowest corrosion current i_{kor} and passivation current ip, as well as the highest degree of protection β, while the iron nitrides' layer with a thickness of 7.0 µm, obtained in the Nx355 process, also exceeded the limit value of the corrosion current i_{kor}. Moreover, it was characterised by the highest open circuit potential E_{ocp}, a relatively low passivation current i_p, as well as the highest degree of protection β.

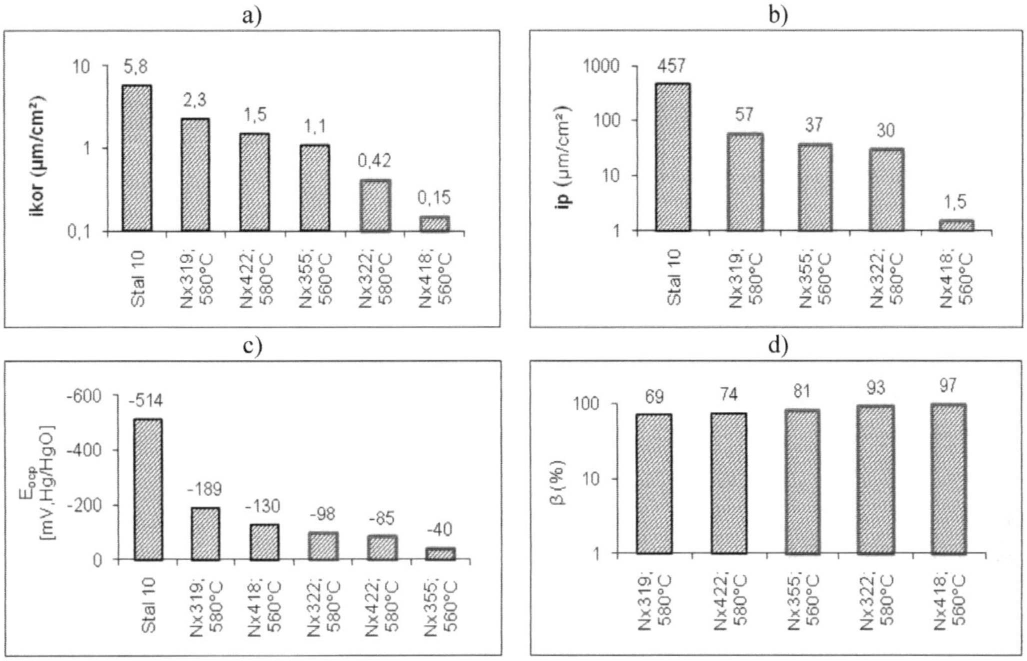

Figure 6. Basic electrochemical parameters and corrosion current (a), passivation current (b), open circuit potential (c), degree of corrosion protection (d), obtained in a 3% NaCl solution for the nitrided 1010 steel with iron nitrides' layers with a thickness of 7.0÷11.8 μm.

As it results from the investigations, in the group of layers with thicknesses of 3.5÷5.7 μm, the assumed criteria of the corrosion resistance were met by the layers with a thickness of 3.5 and 5.8 μm formed at a temperature of 490°C. In the group of layers with thicknesses of 7.0÷11.8 μm, the assumed criteria of the corrosion resistance were met by only one layer with a thickness of 8.5 μm formed at a temperature of 580°C. Thus, there is no distinct dependence between the thickness of an iron nitrides' layer and the process temperature, and the layer's electrochemical parameters.

Nitrided layers formed on the 1010 steels at temperatures of 490, 530, and 560°C (selection after a preliminary electrochemical assessment) were subject to studies in the salt spray chamber. The Nx625 process is a process at a temperature of 530°C elongated to 12 h. The results of these studies are reported in table 2.

A better corrosion resistance on 1010 steel was obtained for the layer formed in the Nx625 process than for that of Nx418 process, in spite of the same thickness of iron nitrides (9.0 μm). The difference in the behaviours of the layers resulted from the presence of a porous zone (1.0 μm) on the surface of the nitrides' layer formed in the Nx418 process. As one can see in figure 7, the layer obtained in this process at a temperature of 560°C had more local discontinuities and porosities (Figure 7a) than the payer from the Nx625 process carried out at a temperature of 530°C (Figure 7b).

Table 2: Results of the salt spray chamber studies of the nitrided layers on 1010 and 42CrMo4 steel.

Process No./Temp./Time	Appearance of the samples after exposure for:			
	24 h	**48 h**	**72 h**	**96 h**
3.5÷6.0 µm layers of iron nitrides 1010 steel				
Nx301/490°C/4h	1 corrosion spot	Corrosion on approx. 45% of the surface	Corrosion on approx. 45% of the surface	Corrosion on approx. 60% of the surface
Nx428/490°C/4h	No change	Corrosion on approx. 50% of the surface	Corrosion on approx. 60% of the surface	Corrosion on approx. 70% of the surface
7.0÷9.0 µm layers of iron nitrides 1010 steel				
Nx418/560°C/8h	No change	Corrosion on approx. 20% of the surface	Corrosion on approx. 40% of the surface	Corrosion on approx. 70% of the surface
Nx625/530°C/12h	No change	No change	No change	Corrosion on approx. 50% of the surface
20 µm layers of iron nitrides 42CrMo4 steel				
Nx620/570°C/4h	No change	No change	No change	No change Corrosion on approx. 10% of the surface after 240h

In case of 42CrMo steel with a thick surface layer of the iron nitrides (20 µm), after the oxidation and impregnation the high corrosion resistance (above 240 h in salt spray chamber) was achieved.

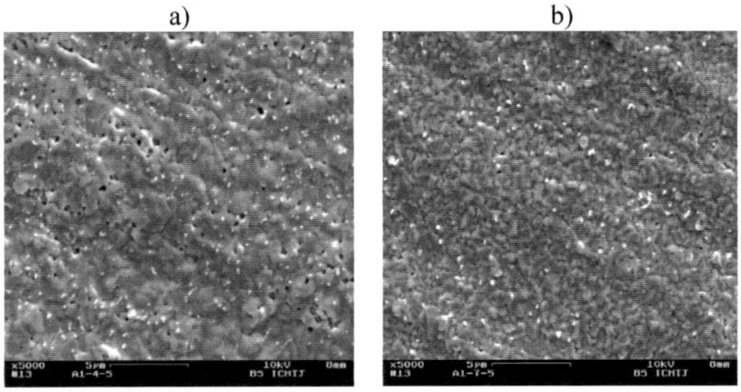

Figure 7. Appearance of the surfaces of the nitrided layers formed in Nx418 (a) and Nx625 (b) processes.

4 Conclusions

- Forming of thick subsurface layers of iron nitrides by nitriding with subsequent processes of oxidation and impregnation allows to achieve a very high corrosion resistance (over 240 h in salt spray chamber).
- The stage of heating in the nitriding process has a great influence on the formation of iron nitrides layer, especially on their final phase composition and morphology. Conducting of heating to a fixed process temperature with the adjusting of the nitriding potential guarantees formation from the beginning of tight subsurface layer of iron nitrides with a defined phase composition.

- Monophase surface iron nitrides' layers with the γ' structure, formed on unalloyed steels have a good corrosion resistance (to 72 h in salt spray test).
- A necessary requirement for a good resistance of nitrided unalloyed steel consists in a formation of the tight surface layers of iron nitrides.
- Unalloyed steels with a tight layer of iron nitrides with a thickness of 9.0 μm have a higher corrosion resistance (confirmed by the studies in the salt spray chamber) than that of the thinner layers.
- It is more favourable to generate nitrided layers with a good corrosion resistance at temperatures below 530°C than at higher temperatures. It results from a higher propensity to form porous zones at higher temperatures of the nitriding process.

Acknowledgements
The research and its dissemination is an ongoing program design TANGO NitroCor "Innovative nitrided layers of new generation with increased corrosion resistance produced on machine elements", funded by the National Centre for Research and Development in the years 2015-2018

References
[1] Praca zbiorowa pracowników IMP, Procesy obróbki cieplno-chemicznej, tom 2, Podręcznik dla uczestników Studium Korespondencyjno-Słuchowego, Warsaw, 1987
[2] H.-J. Spies: 6 - Corrosion behaviour of nitrided, nitrocarburised and carburised steels, Thermochemical Surface Engineering of Steels, Edited by: Eric J. Mittemeijer and Marcel A. J. Somers, 2015, p. 267-309 ISBN: 978-0-85709-592-3
[3] M.L. Doche, V. Meynie, H. Mazille, C. Deramaix, P. Jacquot: "Improvement of the corrosion resistance of low-pressure nitrided and post-oxidized steels by a polymer impregnation final treatment", Surface and Coatings Technology 154, 2002, p. 113-123,
[4] Bo Lin He, Ying Xia Yu, Jian Ping Shi: "The Effect of QPQ Salt-Bath Nitriding on Microstructure and Wear Resistance of 3Cr2W8V Steel", Materials Science Forum, vol. 628 - 629, 2009, p. 715-720
[5] Wei Cai, Fanna Meng, Xinyan Gao, Jing Hu: "Effect of QPQ nitriding time on wear and corrosion behavior of 45 carbon steel", Applied Surface Science, vol. 261, 2012, p. 411-414
[6] Yang Li, Liang Wang, Dandan Zhang, Lie Shen: "Improvement of corrosion resistance of nitrided low alloy steel by plasma post-oxidation", Applied Surface Science, vol. 256, 2010, p. 4149-4152
[7] J. Iwanow, J. Senatorski, J. Tacikowski: "Impregnated nitride layers – the new modification of layers resistant against the wear and corrosion damges", Proceedings of the 9th International Seminar Nitriding Technology Theory&Practice, Warsaw, 2003, p. 87 – 102 ISBN 83-916386-2-6
[8] Jiqiang Wu, Han Liu, Xuemei Ye, Yating Chai, Jing Hu: "Enhancement of corrosion resistance for plasma nitrided AISI 4140 steel by plain air plasma post-oxidizing", Journal of Alloys and Compounds, vol. 632, 2015, p. 397-401

IOP Conf. Series: Materials Science and Engineering **179** (2017) 012070 doi:10.1088/1757-899X/179/1/012070

The influence of heat treatment on properties of cold rolled alloyed steel and nickel superalloys sheets used in aircraft industry

K Zaba[1, a], I Dul[2] and S Puchlerska[1]

[1]AGH University of Science and Technology, Krakow, Poland
[2]Pratt&Whitney Rzeszów S.A, Rzeszow, Poland

[a]krzyzaba@agh.edu.pl

Abstract. Superalloys based on nickel and selected steels are widely used in the aerospace industry, because of their excellent mechanical properties, heat resistance and creep resistance. Metal sheets of these materials are plastically deformed and applied, inter alia, to critical components of aircraft engines. Due to their chemical composition these materials are hardly deformable. There are various methods to improve the formability of these materials, including plastic deformation at an elevated or high temperature, or a suitable heat treatment before forming process. The paper presents results of the metal sheets testing after heat treatment. For the research, sheets of two types of nickel superalloys type Inconel and of three types of steel were chosen. The materials were subjected to multivariate heat treatment at different temperature range and time. After this step, mechanical properties were examined according to the metal sheet rolling direction. The results were compared and the optimal type of pre-trial softening heat treatment for each of the materials was determined.

1 Introduction

Heat treatment is the main factor constituting alloys microstructure, and consequently, their end properties. In the case of nickel superalloys and high alloyed steels, where applications require particularly high properties of the product, each stage of product forming should be monitored, and the result should be controlled in details. Heat treatment in a technological line of producing critical parts of jet engines is an example of special processes, that is, they cannot be controlled directly. The results are known after the process. Heat treatment processes give materials initial and interoperational technological characteristics, indicative of their ability to process and final properties defining the usefulness of the product. About their result decide all phases of the process – heating, holding time and cooling.

Nickel superalloys and high alloyed steels are materials used in the aerospace industry because of their properties, which include very good resistance to high-temperature oxidation and good mechanical properties at elevated temperature. Demanding components for the aerospace industry are produced from sheets made of these alloys formed in the processes of metal forming [1].

Due to the complex chemical composition, sheet made of nickel superalloys and of high alloyed steel have a large capacity for strain hardening. During the plastic deformation at a temperature lower than the recrystallization temperature of the material, there is the phenomenon of strain hardening. Not only the mechanical properties are changing but also magnetic, electric or corrosion resistance. Authors [2] investigated effects of solution and intermediate heat treatments on the properties of

Inconel 718. Material was subjected to solution heat treated in the range 917-1038°C and subsequently given intermediate and/or aging heat treatments. The studies on the alloyed steel formability of at high temperatures were carried out by [3-4]. The authors proposed a simulation method of metals deformation on the Gleeble® 3800 simulator. The aim of the study was to recreate the course of changes in temperature and strain, which is subject to the material during high temperature metal forming.

Materials for the plastic forming may be heat treated either before forming them to obtain the required technological properties, during shaping to renew the technological properties, and after plastic forming processes, in particular in order to achieve the desired operating characteristics.

In the case of nickel superalloys, short-term heat treatment was predominantly used prior to final aging step with a view to obtaining the phase γ' of a fine dispersion. One of the main objectives of the heat treatment of nickel superalloys for plastic processing is to obtain adequate dispersion of carbides and their types. It is generally a multi-stage treatment carried out in the temperature range 1040-1100°C, depending on the alloy [5-7].

Heat treatment of stainless steel is usually performed based on the type of steel and the reasons for carrying out the procedure. Heat treatment such as annealing, quenching and tempering, increased ductility and corrosion resistance of the material. The properties can be modified during the manufacture of the alloy, or during heat treatment after the manufacturing process. Heat treatment of stainless steel is carried out under controlled conditions to avoid carburizing, decarburization and coating surfaces with scale.

The essence of the research presented in the publication was to seek opportunities to improve the technological properties of the material being processed, and above all, to improve the plastic properties of batch sheets. The search for methods to improve the properties of deformed material is fully justified by the results of preliminary tests of experimental industrial process of rotary forming, which showed that the loss of plastic properties is a serious limitation of the efficiency and effectiveness of the process.

2 Methodology

For research samples from sheets made of the following materials were chosen:

1. AMS 5599
2. AMS 5596
3. AMS 5510
4. AMS 5604
5. AMS 5536

The shape, size and direction of sampling for testing the mechanical properties are shown in figure 2.1.

In the first stage of the study, the mechanical properties of metal sheets in the initial conditions were identified. For this purpose, a static tension test on machine Zwick/Roell Z050 was conducted. The machine was operated and controlled by program testXpert® 2, which was also used to collect data from the tensile test. Rockwell hardness measurements were also performed.

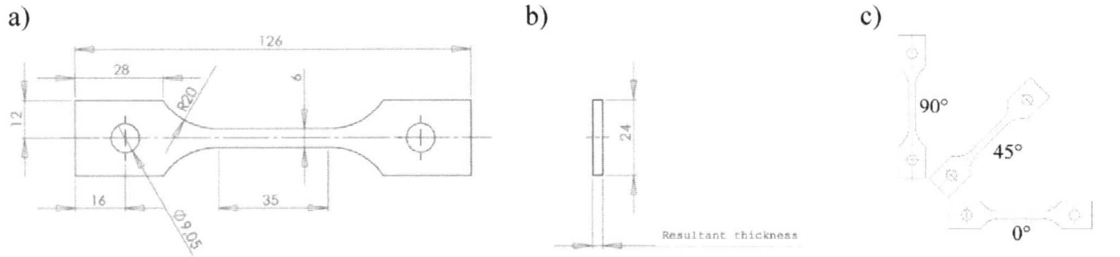

Figure 2.1. Shape and dimensions of the samples used for testing
a) the dimensions of the sample, b) the thickness of the sample, c) sampling direction.

Afterwards, a study on the impact of heat treatment on the mechanical properties of the metal sheets were made. Studies were realized in 2 bars vacuum furnaces used for aviation production with cylindrical molybdenum or graphite chamber and horizontal loading. Furnaces had gas mixer and a heat exchanger located within the housing. The dimensions of the working chamber was 900x780x1800 mm. The operating range of the furnace was 500-1180°C. The required temperature uniformity is defined by ranges:

1. (500÷720) +/-8°C
2. (721÷1180) +/-14°C

Figure 2.2 shows the patterns of heating and cooling for the heat treatment.

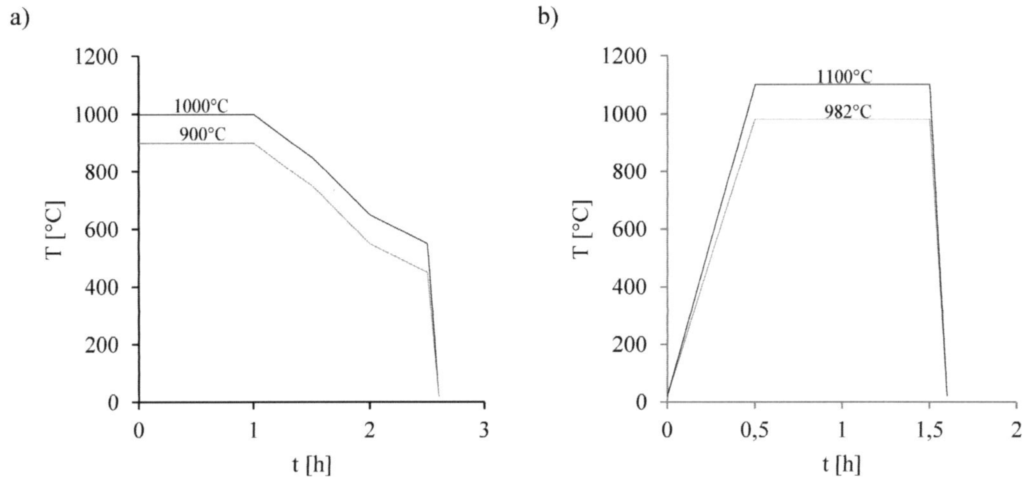

Figure 2.2. Patterns for heating and cooling in the thermal treatment
a) annealing b) solutioning.

Summary plan of the heat treatment are shown in table 1.1.

Table 1.1. Plan of the heat treatment (HT = heat treatment).

Material	HT[a] 1	HT[a] 2	HT[a] 3	HT[a] 4	HT[a] 5	HT[a] 6
AMS 5599	annealing 900°C	annealing 1000°C	vacuum 10^{-4} Tr 982°C+Ar 60min RCW[b]	vacuum 10^{-4} Tr 1100°C+Ar 60min RCW[b]	vacuum 10^{-4} Tr 982°C+Ar 60min RCO[c]	vacuum 10^{-4} Tr 1100°C+Ar 60min RCO[c]
AMS 5596	annealing 900°C	annealing 1000°C	vacuum 10^{-4} Tr 982°C+Ar 60min RCW[b]	vacuum 10^{-4} Tr 1100°C+Ar 60min RCW[b]	vacuum 10^{-4} Tr 982°C+Ar 60min RCO[c]	vacuum 10^{-4} Tr 1100°C+Ar 60min RCO[c]
AMS 5510	annealing 900°C	annealing 1000°C	vacuum 10^{-4} Tr 982°C+Ar 60min RCW[b]	vacuum 10^{-4} Tr 1100°C+Ar 60min RCW[b]	vacuum 10^{-4} Tr 982°C+Ar 60min RCO[c]	vacuum 10^{-4} Tr 1100°C+Ar 60min RCO[c]
AMS 5604	annealing 900°C	annealing 1000°C	vacuum 10^{-4} Tr 982°C+Ar 60min RCW[b]	vacuum 10^{-4} Tr 1100°C+Ar 60min RCW[b]	vacuum 10^{-4} Tr 982°C+Ar 60min RCO[c]	vacuum 10^{-4} Tr 1100°C+Ar 60min RCO[c]
AMS 5536	annealing 900°C	annealing 1000°C	vacuum 10^{-4} Tr 982°C+Ar 60min RCW[b]	vacuum 10^{-4} Tr 1100°C+Ar 60min RCW[b]	vacuum 10^{-4} Tr 982°C+Ar 60min RCO[c]	vacuum 10^{-4} Tr 1100°C+Ar 60min RCO[c]

[a] heat treatment
[b] rapid cooling water
[c] rapid cooling oil

Conditions for the heat treatment varied by temperature and cooling medium. After the heat treatment, a static tensile testing was carried out again, and Rockwell hardness was tested.
The results are presented in the form of tables and figures.

3 Results and discussion

Table 3.1. shows the chemical composition of the alloy to be tested.

Table 3.1. The chemical composition of alloys intended for research.

Material	Element [%]											
	Ni	Cr	Fe	Mo	Nb+Ta	C	Mg	Si	Al	Ti	Co	remaining
AMS 5599	58	23	5	10	4,14	0,1	0,5	0,5	0,4	0,4		b*
AMS 5596	55	21	b*	3,3	5,5	0,08	0,35	0,35	0,7		1,0	b*
AMS 5510	12	19	b*			0,08				0,7		b*
AMS 5604	5,0	17,5	b*		0,45	0,07	1,0	1,0				b*
AMS 5536	50	21	18	9								b*

* balance

Figure 3.1 presents a summary of the stress-strain curves for the samples in initial conditions and after heat treatment. Due to slight differences between the tensile results of samples at a different direction relative to the rolling direction, results were averaged.

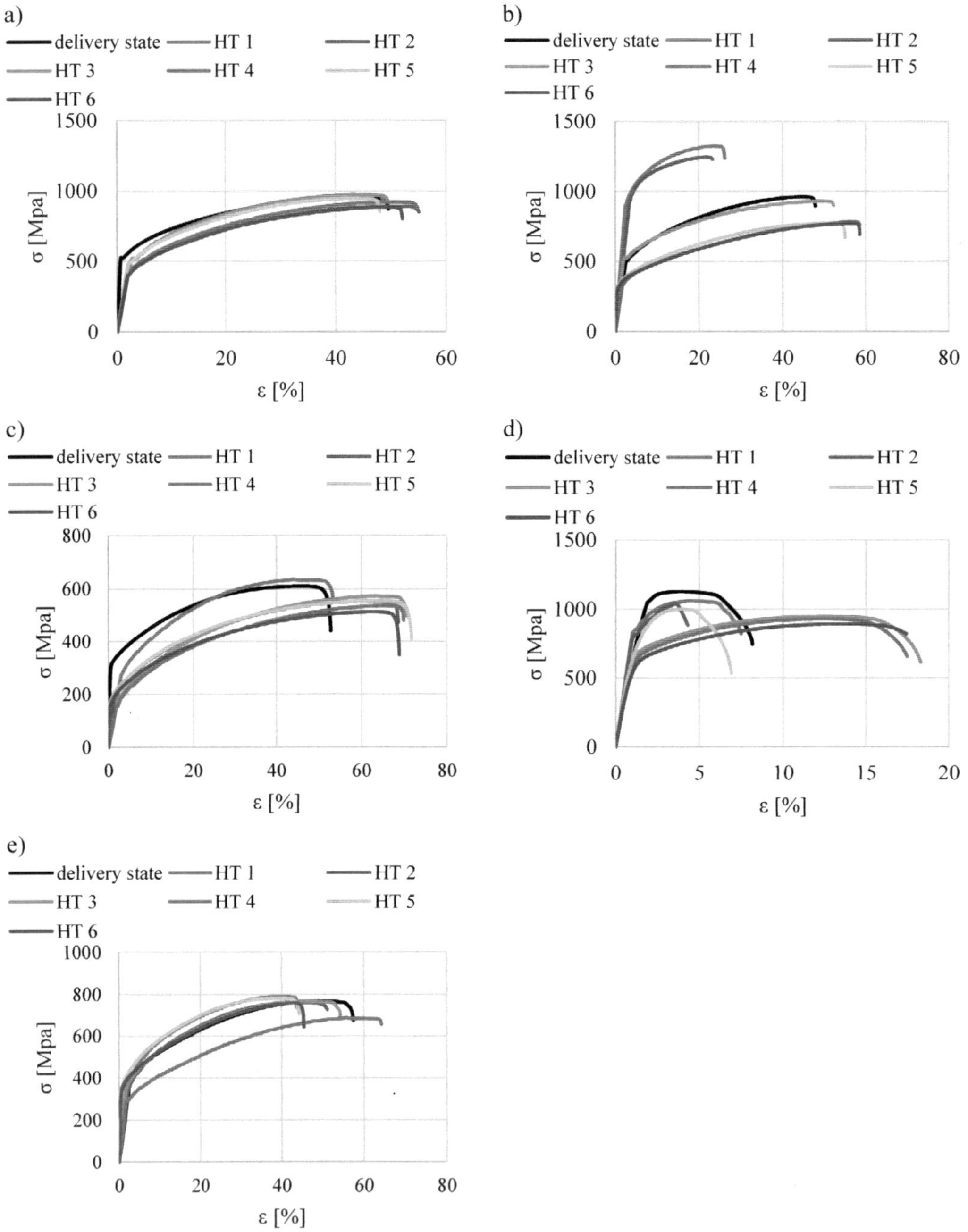

Figure 3.1. Summary diagrams of the stress-strain curves for the samples in initial conditions and after heat treatment
a) AMS 5599, b) AMS 5596, c) AMS 5510, d) AMS 5604, e) AMS 5536.

Analyzing the above results that the mechanical properties, it can be concluded that the use of the softening heat treatment bring the expected results, but not in every case the heat treatment significantly improved technological properties of the tested materials. The best results were obtained for alloys of AMS 5596, AMS 5510 and AMS 5536. Table 3.2 summarizes the optimum softening heat treatment and mechanical properties of the alloys in initial conditions and after heat treatment.

Table 3. 2. Summary of the mechanical properties of the samples in initial conditions and after heat treatment.

Material	AMS 5599		AMS 5596		AMS 5510		AMS 5604		AMS 5536	
Selected heat treatment	HT 2		HT 6		HT 6		HT 6		HT 4	
	YTS [Mpa]	E [%]	YTS [Mpa]	E [%]	YTS [Mpa]	E [%]	YTS [Mpa]	E [%]	YTS [Mpa]	E [%]
Mechanical properties in initial conditions	946	50	932	51	610	52	1147	5	767	55
Mechanical properties after heat treatment	493	49	300	38,4	152	68	510	16	273	63

Based on the results of Rockwell hardness it can be concluded, as in the case of the tensile tests results, that the application of the softening heat treatment bring the expected results, but not in every case the heat treatment significantly improved technological properties of the tested materials. The best results were obtained for the AMS 5596 and AMS 5510. Table 3.3 summarizes the optimum softening heat treatment and Rockwell hardness of the alloys in initial conditions and after heat treatment.

Table 3.3. Summary of the Rockwell hardness of the samples in initial conditions and after heat treatment.

Material	AMS 5599	AMS 5596	AMS 5510	AMS 5604	AMS 5536
Selected heat treatment	HT 6	HT 6	HT 6	HT 3	HT 6
	HRA	HRA	HRA	HRA	HRA
Hardness in initial conditions	60,3	58,5	46,7	64,0	48,1
Hardness after heat treatment	50,8	38,4	16,3	64	40,7

4 Summary

The results indicate the possibility of controlling the technological properties by prior heat treatment of metal batch. Type of heat treatment may be chosen based on the basic characteristics of alloys describing the evolution of the properties as a function of heat treatment parameters, namely temperature, cooling rate, and the sequence of treatments. Parameters are subject to optimization based on measurements of mechanical properties. The study is taking into account constraints resulting from the operational requirements and technical and technological possibilities.

Acknowledgment

The work carried out within the framework of agreement No. 11.11.180.004/2016.

References

[1] Zaba K Nowosielski M Puchlerska S Kwiatkowski M Kita P Głodzik M Korfanty K Pociecha D and Pieja T 2015 Investigation of the mechanical properties and microstructure of nickel superalloys processed in shear forming *Archives of Metallurgy and Materials* 60 (4) pp 2627-2644

[2] Muller J F and Donachie M J 1975 The effects of solution and intermediate heat treatments on the notch-rupture behavior of Inconel 718 *M.J. Metall and Mat Trans A* 6: 2221.

[3] Hojny M and Glowacki M 2011 Modelling of Strain-Stress Relationship for Carbon Steel Deformed at Temperature Exceeding Hot Rolling Range *J. Eng. Mater. Technol.* 133 (2) pp. 021008-1 – 021008-7

[4] Hojny M and Glowacki M 2008 Computer modelling of deformation of steel samples with mushy zone *Steel Research International* 79 (11) pp 868-874

[5] Dul I 2009 Application and processing of nickel alloys in the aircraft industry *Welding Technology Review* 81 (7-8) pp 67-73

[6] Sims C T 1984 A history of superalloy metallurgy for superalloy metallurgists *General Electric Company Schenectady New York*

[7] Pirowski Z Uhl W Wodnicki J Gwizdz A and Jaskowiec K 2011 Effect of heat treatment on structure of the Inconel 740 type alloy *The Transactions of the Foundry Research Institute* 51 (2)

Institute of Physics
Dirac House, Temple Back
Bristol BS1 6BE UK

ISSN: 1757-8981
ISBN 978-1-5108-3717-1

9 781510 837171

CURRAN ASSOCIATES INC.
proceedings.com